DATE DUE

Lecture Notes in Physics

Edited by J. Ehlers, München, K. Hepp, Zürich
R. Kippenhahn, München, H. A. Weidenmüller, Heidelberg
and J. Zittartz, Köln

100

Einstein Symposion Berlin

aus Anlaß der 100. Wiederkehr
seines Geburtstages
25. bis 30. März 1979

Edited by
H. Nelkowski, A. Hermann, H. Poser,
R. Schrader, and R. Seiler

Springer-Verlag
Berlin Heidelberg New York 1979

Editors

H. Nelkowski
Institut für Festkörperphysik
Technische Universität Berlin
Straße des 17. Juni 135
1000 Berlin 12

H. Poser
Institut für Philosophie, Wissenschaftstheorie,
Wissenschafts- und Technikgeschichte
Technische Universität Berlin
Straße der 17. Juni 135
1000 Berlin 12

A. Hermann
Historisches Institut, Abteilung für Geschichte
der Naturwissenschaften und Technik
Universität Stuttgart
Friedrichstraße 10/IV
7000 Stuttgart 1

R. Schrader
R. Seiler
Institut für Theorie der Elementarteilchen
Freie Universität Berlin
Arnimallee 3
1000 Berlin 33

Veranstalter des Einstein Symposions Berlin:

Der Senator für Wissenschaft und Forschung
Freie Universität Berlin
Technische Universität Berlin
Berlin (West)

ISBN 3-540-09718-X Springer-Verlag Berlin Heidelberg New York
ISBN 0-387-09718-X Springer-Verlag New York Heidelberg Berlin

Library of Congress Cataloging in Publication Data
Einstein Symposion, Berlin, 1979.
Einstein Symposion, Berlin.
(Lecture notes in physics; 100)
English or German.
Bibliography: p.
Includes index.
1. Physics--Congresses. 2. Einstein, Albert, 1879–1955--Anniversaries, etc.
I. Nelkowski, H., 1921- II. Series.
QC1.E39 1979   530.1   79-25206
ISBN 0-387-09718-X

Printing and binding: Beltz Offsetdruck, Hemsbach/Bergstr.
2153/3140-543210

VORWORT

Mit dem Einstein Symposion Berlin wurde der hundertsten Wiederkehr des
Geburtstages von Albert Einstein gedacht.

In den fast 20 Jahren, die er in Berlin verbrachte, hat Einstein seine
größten wissenschaftlichen Triumphe erlebt und seine bittersten menschlichen Erfahrungen gemacht. Seine Ideen haben unsere Vorstellungen von
Zeit und Raum in revolutionierender Weise geändert. Ihre Wirkung reicht
weit über die Lösung der physikalischen Probleme hinaus, die wir Einstein
verdanken. Das Symposion hatte sich deshalb das Ziel gesetzt, aus physikalischer, wissenschaftstheoretischer und historischer Sicht aufzuzeigen,
welche Prägung zentrale Themen durch sein Werk und Wirken erfuhren und
welchen Einfluß diese auf unser heutiges Denken haben.

Der vorliegende Band enthält die Beiträge des Symposions einschließlich
der Manuskripte von M. Bunge und A. Polikarov, die an der Teilnahme verhindert waren.

Die Autoren, die Herausgeber und der Verlag danken dem Einstein-Estate
für die Genehmigung, aus Werken des von ihm verwalteten Einstein-Nachlasses zitieren zu dürfen.

                                        Die Herausgeber

INHALT

ANSPRACHE DES SENATORS FÜR WISSENSCHAFT UND FORSCHUNG

Dr. Peter Glotz, Berlin

Albert Einstein hat in den hinter uns liegenden Wochen viel Ehre an
sehr vielen Orten erfahren. Und es sind ihm auch Ehrungen widerfahren,
gegen die er Einspruch erhoben hätte. Alle, die ihn aus einer weltanschaulichen Ecke für sich vereinnahmen wollen, irren: Er war gewiß enga-
giert für Frieden und Menschlichkeit. Aber seine wissenschaftliche,
politische, menschliche Existenz ist keine Legitimation für politische
Ideologien - es sei denn für die Grundbedingung der Forschung: für To-
leranz.

Wir haben aus der nicht mehr aufzählbaren Fülle von Büchern, Zeitungs-
und Zeitschriftenaufsätzen, von Rundfunk- und Fernsehsendungen ein deut-
liches Einstein-Bild herausfiltern können. Berlin hat in diesem Jahr
zum 100. Geburtstag Einsteins mit einer Ausstellung der großen Vier der
Naturwissenschaften - Liese Meitner, Max von Laue, Otto Hahn und Albert
Einstein gedacht - Weggefährten in Leben und Wissenschaft.

Ich will zu Beginn unseres Symposiums, das ja zwar auch Ehrung, aber
doch mehr *wissenschaftliche* Aufarbeitung von Einstein ist, mich nicht
mehr zu jenem Einstein äußern, der die Schreiber, die Redner, die Be-
richterstatter und die Ideologen so anregt: Der Gelehrte mit dem wirren
Krauskopf, unkonventionell gekleidet, der die Schule und beinahe den
zweiten Bildungsweg nicht schafft und so nebenher im Berner Patentamt
die Theorien der Physik auf den Kopf stellt. Einsteins Biographie und
sein Habitus widerspricht bürgerlichen Vorstellungen ganz und gar. Man
erkannte das Genie an, zum Vorbild taugte er nicht. Er hat unser Jahrhundert und ihre Physik geprägt - pathetisch gesagt. Die eigentümliche
Anziehungskraft dieser Forscherfigur setzt sich facettenhaft zusammen
aus überragender wissenschaftlicher Leistung, aus politischem Engagement
und Emotionen bis hin zu Irrtümern und der Offenheit, mit der Einstein
sich dem Epitheton "Genie" beharrlich entzog und sich der Welt und der
Fachwelt als private Person präsentierte. Einstein ist mit dieser Haltung populär geblieben. Warum sonst verkaufen sich die Posters mit Einstein heute noch so gut? Was also noch zu Einstein sagen? Ich wil die
Frage nach dem Einfluß der gesellschaftlichen Bedingungen auf die Leistung eines Wissenschaftlers stellen. Wir erinnern uns: Die Ruhe und die
Kommunikationschance der Berliner Jahre, die Ruhe, aber auch die gewisse
Isolation für das Genie im Institut für advanced studies in Princeton.
Dies waren für einen Wissenschaftler optimale Rahmenbedingungen. Ohne

Zweifel hat Einsteins unglaubliche Kreativität den Hauptanstoß für seine
Leistungen gegeben. Es stellt sich aber die Frage, insbesondere an die
Wissenschaftspolitiker heute, welche Rolle bei dieser überragenden For-
scherleistung die "scientific community" spielte? Wir wissen heute, daß
ein Gegensatz zwischen "einsamen Genie" und "wissenschaftlicher Gesellschaft" eine falsche Gegenüberstellung wäre. Wir wissen, daß Einstein
die Berliner Zeit als beglückend empfand durch die permanenten Gespräche,
den regen Gedankenaustausch, durch das Zusammensein von bedeutenden, an-
regenden Wissenschaftlern. Was lernen wir daraus für die Forschungslandschaft in Berlin? Müssen wir nicht vermehrt institutionelle Rahmen schaffen für die Entfaltung von Talenten unter diesem Kommunikatonsaspekt?
Es gibt viele Vorwürfe gegen schlechte Arbeitsbedingungen der Forschung
heute. Der wissenschaftliche Nachwuchs, so lautet einer dieser Vorwürfe,
könne sich unter den gegebenen Bedingungen nicht befriedigend entwickeln.
Ich warne in diesem Zusammenhang vor Nostalgie, ich warne davor, von
Einsteins Zeit schnurgerade Parameter zum Heute zu ziehen. Vieles in der
Wissenschafts- und Forschungslandschaft hat sich verändert. Die Universitäten haben einen Funktionswandel erfahren, der sich auf Studien- und
Arbeitsbedingungen auswirkt. Wir bilden heute - anders als in Einsteins
Zeiten, anders als in den 60er Jahren, sehr viel mehr junge Menschen
akademisch aus - nicht zu Genies, zu kleinen Einsteins oder Savignys,
sondern damit sie mit einem wissenschaftlich fundierten Rüstzeug für Be-
rufe lernen. Ich höre den Einwand: Massengesellschaft. Ich antworte:
Industriegesellschaft. Ich höre: Gleichmacherei. Ich antworte: Chancen-
gerechtigkeit. Mit der sozialen Öffnung der Hochschulen haben sich die
Politiker den Vorwurf eingehandelt, das Erbe Einsteins zu gefährden.
Aber wie paßt eine solche Berufung auf anscheinend ideale Bedingungen
zusammen mit der Biographie dieses Gelehrten, der

- sagte: "Die idealen Arbeitsbedingungen als Wissenschaftler hat man
  als Leuchtturmwächter - einen bezahlten Beruf, aber genug Zeit...",

- der die Höchstleistung bei einem full-time-job am Berner Patentamt
  erbrachte,

- der es ungeheuer schwer hatte, Dozent zu werden, weil er die normalen
  Qualifikationen nicht vorweisen konnte?

Er schuf sie sich in der Freizeit, mit Forschung. Sage mir keiner, daß
heute die Politiker die äußeren Voraussetzungen für die Gewinnung eines
Nobel-Preises nicht etwas erleichtert hätten! Gerade der Lebenslauf
Albert Einsteins eignet sich nicht zur Stützung larmoyanter Vorwürfe
gegen die Wissenschaftspolitik. Ich komme darauf zurück. Sicher: wir
müssen dafür sorgen, daß Genialität und Kreativität eines Gelehrten

nicht stranguliert werden durch Routineverpflichtungen des akademischen
Unterrichts oder durch Überbelastung mit akademischer Verwaltungsarbeit.
Die Gefahr ist gewiß heute groß. Ich sehe auf der anderen Seite aber auch
die Gefahr, wollen wir ehrlich sein, daß *jeder* Genialität und Kreativität für sich reklamiert und alle entsprechenden Freiräume fordert und
die Routine von sich wegschiebt.

Berlin ist ein gutes Beispiel dafür wie sehr wir versuchen, der Wissenschaft die notwendigen Freiräume dafür zu erhalten. Dies hat in Berlin
Tradition; und damit ich nicht in Verdacht komme, mich selbst zu loben,
will ich ein historisches Beispiel wählen, um zu zeigen, wie sehr sich
die Dinge gleichen. Vor 90 Jahren führte Heinrich Hertz, der gerade in
Karlsruhe seine elektronischen Wellen entdeckt hatte, Berufungsverhand-
lungen mit dem legendären Friedrich Theodor Althoff. Danach schrieb
Hertz in sein Tagebuch:

> *Noch eins muß ich erwähnen: Die Liberalität, mit welcher man mir*
> *in Bezug auf die Verpflichtung, Vorlesungen zu halten, entgegen-*
> *gekommen ist. Als ein Kollege der Fakultät wünschte, ich solle*
> *doch auch dies und jenes lesen, sagte Althoff: 'Nein, wir legen*
> *Herrn Professor Hertz nur die Verpflichtung auf, die Experimental-*
> *physik zu lesen und das Laboratorium zu leiten, er soll freie Zeit*
> *für seine Arbeiten behalten...' Und zu mir allein auf seinem Zim-*
> *mer...: 'Überladen Sie sich nicht mit Kollegien, es ist zwar ge-*
> *setzlich, daß jeder Professor ein Publikum lesen muß, aber selbst*
> *darin sieht man durch die Finger.' Und ein andermal sagte er:*
> *'Machen Sie, daß im Laboratorium etwas geleistet wird, wie Sie es*
> *bisher getan haben, dann ist es uns im Ministerium ganz gleich, ob*
> *es heißt, es stehe leer oder nicht.' Ich sah aus all diesem, wie*
> *viel Wert man auf die wissenschaftliche Seite meiner Tätigkeit*
> *legte.'*

Ich unterstreiche alles, was die Liberalität in diesem Zitat betrifft.
Ich wünsche uns den Althoff'schen Mut; und die Institution der demokratischen Öffentlichkeit, die Souveränität, leere Laboratorien nicht immer
und *nur* als "Skandale" zu skandieren - damit die Politiker den Althoff'-
schen Mut nicht mit ihrer politischen Existenz bezahlen müssen! Die Universitäten sind heute nicht mehr abgeschlossene, vom Treiben der Gesellschaft abgeschirmte säkularisierte "Klöster", in denen der Wissenschaftler in Einsamkeit denkt und forscht. Heute haben wir es mit Ausbildungs-
zentren und Großforschungseinrichtungen zu tun, deren Leistungen für die
Konkurrenzfähigkeit unserer Volkswirtschaft entscheidend sind. Ich will
keine Mißverständnisse erzeugen: Die Hochschulen haben auch weiter eine
kritische Funktion. Sie sind nicht *nur* Dienstleistungsbetriebe für Ausbildung und Forschung. Wo wir Nischen erhalten können, in denen im Dil-
theyschen Stil gearbeitet werden kann, da sollten und wollen wir es tun.
Nur: Die Erinnerung an Einstein sollte nicht zur Legitimation für Nost-
algische Träume der Wissenschaftspolitik benutzt werden.

Ein letzter Aspekt: Ich wünsche mir von dieser Arbeitstagung im Ergebnis, daß die Physik sich als attraktive, zukunftsweisende Wissenschaft der Gegenwart erweist - gerade am Forschungsplatz Berlin, wo wir uns bemühen, gute Arbeitsbedingungen für die Physik zu schaffen. Unsere Anstrengungen gelten sowohl der anwendungsorientierten Forschung als auch der Grundlagenforschung; und sie zielen genauso darauf, die Voraussetzungen für gute Berufschancen für junge Physiker in der angesiedelten Industrie unserer Stadt zu schaffen. Sicher waren Einstein und seine Zeitgenossen bewegt von der Fortschrittsbegeisterung einer Zeit, die vom Naturwissenschaftler ein Sesam-öffne-dich für eine bessere Zukunft erwartete. Nicht von ungefähr kam es nach 1900 aus dem Zeitgefühl heraus zu einem emotionalen Engagement für den technischen Fortschritt bei den Besten in der Naturwissenschaft. In der Kunst etwa spiegelte der Futurismus eine Fortschrittsgläubigkeit, die wir heute nicht mehr verstehen. Daß diese Versionen einer technischen Revolution später den Namen Krieg und Atombombe tragen würden - dies lag den Gedankengängen damals fern. Einstein hat seine Ahnungen um die Gefährdung des Menschen durch die Technik politisch nicht mehr umgesetzt. Heute sind Technik - Euphorie und Fortschrittsglaube dahin. Das Pendel ist ausgeschlagen, an die Stelle der Begeisterung ist Ernüchterung und Abkehr bis zum Kulturpessimismus getreten. Ein Hauch von Entmenschlichung scheint den technischen Innovationen anzuhaften. Fortschritt und technische Neuerungen werden auf die Folgen befragt - zu Recht. Technische Neuentwicklungen, die Naturwissenschaft und damit auch die Physik laufen Gefahr, gleichgesetzt zu werden mit Zivilisationskrankheit - zu Unrecht.

Angesichts dieses gesellschaftlichen Phänomens müssen Wissenschaft und Gesellschaft gemeinsam dahin wirken, daß nicht rückwärtsgewandte Ideologien sich durchsetzen. Die Wissenschaft muß ihren Platz als *Aufklärerin* im Bewußtsein der Menschen erhalten, sie muß ihr Tun reflektieren und der Gesellschaft, mithin der Politik Lösungen anbieten, die dem Menschen und seinen Bedürfnissen gerecht werden und sozial vertretbar sind - das heißt, sie muß alles tun, daß *Angst nicht Motor des Rückschritts* wird. Wir würden dies nicht überleben. Der notwendige Dialog von Wissenschaft und Politik ist mir umso wichtiger, weil wir in Berlin hierin richtungweisend sein können. Wir haben die geistigen und institutionellen Kapazitäten für ein solches Zusammenwirken. Die Berliner Wissenschaftspolitik ist darauf gerichtet, mit dem Ernst-Reuter-Zentrum auf hoher Ebene eine solche wissenschaftliche Gesprächsmöglichkeit zu schaffen, die sich mit dem Forschungsort Princeton sehr wohl vergleichen läßt und sich orientiert an der fruchtbaren Gesprächssituation des Berlin von damals, als Einstein, Meitner, von Laue und Hahn hier wirkten: Unser Center für

advanced studies wird die Rechtsform einer privaten Stiftung haben.
Grundstück, Gebäude sowie 8 hochdotierte Stipendien sind bereitgestellt.
Hier sollen Gelehrte von internationalem Rang reflektieren und relati-
vieren können. In Einbringung ihrer fachlichen Spezialisierungen, aber
interdisziplinär und weit über das eigene wissenschaftliche Arbeitsfeld
hinaus. Wir erwarten uns davon für Wissenschaft und Forschung wesentliche Impulse und neue Perspektiven für die einzelnen Fachgebiete der dort
kommunizierenden Gelehrten und darüber hinaus und über Berlin hinaus
wissenschaftliche Fingerzeige in die Zukunft. Wir sind sicher, daß wir
uns mit dem Zentrum eng an komplexe internationale Wissenschaftsentwick-
lungen zu halten vermögen. Eine wissenschaftliche Elite soll ein bis
drei Jahre ohne die Alltagsbelastungen des normalen Wissenschaftsbetrie-
bes schöpferisch tätig sein können. Die Berliner Wissenschaftslandschaft
wird davon nicht unberührt bleiben, dies garantiert die enge Verknüpfung
mit den Berliner Universitäten und mit der vorhandenen Forschungslandschaft. Wir treten damit den Beweis an: Soziale Öffnung der Hochschulen
und Bildungsstätte und Eliteförderung widersprechen sich *nicht*.

Hier in Berlin, im ungeliebten Preußen, hatte Einstein ideale Arbeitsbedingungen gefunden. Wir bemühen uns, an diese Tradition wieder anzu-
knüpfen. Wir begreifen diese Gründung daher auch als Verpflichtung zur
Wiedergutmachung mit dem Ziel, die Lücken zu füllen, die die Zäsur von
1933 bis 1945 in die deutsche Wissenschaftswelt gerissen hat. Wir knüpfen bewußt an die Vorzeit zur Barbarei an. Es trifft viele andere wie
mich ein Verdikt Einsteins gegenüber dem Deutschland, aus dem er ver-
trieben wurde. Er attestierte unserem Land seine Verachtung mit dem
Satz, es sei unmöglich, aus diesen Kerlen Demokraten zu machen. Auch
nach dem 2. Weltkrieg gab er uns keine Chance. Ich hoffe, wir können beweisen, daß dieses Urteil zu den Irrtümer des großen Einstein gehört.
Die letzten 30 Jahren geben uns jedenfalls die Legitimation zu sagen:
Wir haben es versucht.

BEGRÜSSUNG DURCH DEN PRÄSIDENTEN DER FREIEN UNIVERSITÄT BERLIN

Professor Dr. Eberhard Lämmert

Herr Senator, sehr verehrte Gäste aus dem In- und Ausland, meine Damen
und Herren!
A l b e r t   E i n s t e i n   zu Ehren veranstalten der Senat von Berlin und die Berliner Universitäten ein mehrtägiges Symposion. Es soll zur
100-jährigen Wiederkehr seines Geburtstages sein Leben im Zusammenhang unserer Geschichte beleuchten und die Auswirkungen seines Nachdenkens auf
unser Jahrhundert umreißen. Die Gemeinsamkeit, mit der Staat und Universitäten sich zu dieser Veranstaltung in Berlin zusammenfinden, hat ihren
historischen Sinn. Das goldene Zeitalter der Physik, das mit dem Entwurf
der Relativitätstheorie und der Entdeckung der Quantenmechanik begann,
fand zu Beginn des Jahrhunderts in Berlin einen Mittelpunkt, als die Preußische Kultusverwaltung, die Friedrich-Wilhelm-Universität und die Preußische Akademie der Wissenschaften in einem ideenreichen Zusammenwirken
hier eine höchst ergiebige Konzentration naturwissenschaftlicher Forschung
ermöglichten, die selbst den Weltkrieg überdauerte und erst im nationalsozialistischen Gewaltregime ihr Ende fand. Wir beginnen diese Veranstaltung in einem Gebäude, in dem erst seit wenigen Monaten die Bücherschätze,
die der Preußische Staat seinen Bürgern zur Verfügung stellte, der Öffentlichkeit wieder zugänglich sind. Und wir danken dem Präsidenten der Stiftung Preußischer Kulturbesitz, Professor   K n o p p , und dem Hausherrn
dieser Staatsbibliothek, Generaldirektor Professor   V e s p e r , ganz
besonders dafür, daß sie uns die Gelegenheit geben, für diese Gemeinsamkeit, die einst für die Entfaltung des wissenschaftlichen Lebens in Berlin
ein so ungewöhnlich tragfähiges Fundament abgab, heute ein neues Zeichen
zu setzen. Von der Großzügigkeit dieser Bibliothek werden Sie sich in jeder Richtung überzeugen können, denn Sie alle sind anschließend an diese
Eröffnung zu einem Empfang in der Halle dieser Bibliothek herzlich gebeten.

Albert Einstein konnte von einer so ausgezeichneten Arbeitsstätte wie der
Eidgenössischen Technischen Hochschule Zürich nach Berlin nur gewonnen
werden, weil man ihm hier versprechen konnte, daß die Universität, die
Akademie und die gerade begründeten Forschungsinstitute der Kaiser-Wilhelm-
Gesellschaft von den zehn Leuten, die auf der ganzen Welt die Relativi-
tätstheorie wirklich verstünden, acht schon in Berlin versammelt hatten.

M a x   P l a n c k   war mit   E m i l   F i s c h e r , mit dem Theologen   A d o l f   v.   H a r n a c k   und den weitsichtigen preußischen
Kulturpolitikern   A l t h o f f   und   S c h m i d t - O t t   der

Stifter dieser Generationsgemeinschaft bedeutender Physiker, für die
Einstein Kristallisationskern und Solitär zugleich werden sollte: unter
seinen Altersgenossen sehr verschiedenen Temperaments M a x v o n
L a u e , L i s e M e i t n e r , O t t o H a h n , E r w i n
S c h r ö d i n g e r der Mann, der am wenigsten in einen regulären
Wissenschaftsbetrieb sich einfügen mochte und den doch am ehesten eine
Ahnung streifte, daß ein neuer Denkschritt im Bereich der Physik Folgen
für eine neue Bestimmung der menschlichen Erkenntnismöglichkeiten überhaupt haben würde. So war es nur folgerichtig, daß seine Theorie weit über
seine Disziplin und sogar über die Wissenschaft hinaus Anhänger und leidenschaftliche Gegner fand und ins intellektuelle und künstlerische Leben
seiner Zeit entschieden ihre Spuren zog.

Ein Physiker ist strenger als andere Wissenschaftler daran gebunden, seine Erkenntnis in überprüfbarer Formelsprache festzulegen. Ein Physiker
hat aber auch vor anderen Wissenschaftlern die große Chance, für seine
Theorie eindeutige experimentelle Bestätigungen oder Widerlegungen finden
zu können.

Mitten im Kriege hat die London Royal Astronomical Society es auf sich
genommen, diese Überprüfungen vorzunehmen, und englische Astronomen haben
genau 60 Jahre vor diesem unserem Symposion in Westafrika und in Brasi-
lien diese Bestätigung geliefert. Mitten im Ersten Weltkrieg hat das No-
bel-Preis-Komitee Max von Laue, bald nach dem Kriege Einstein und im härtesten Kriegsjahr 1944 Otto von Hahn ausgezeichnet: Deutlicher ist kaum
zu belegen, daß Wissenschaft nicht ortsgebunden arbeitet und daß sie politischen Interessen und Machtverhältnissen nicht nachgeordnet ist. Diese
internationale Gemeinsamkeit ernsthafter Erkenntnissuche soll sich in diesem Symposion einmal mehr bewähren, und darum danken wir an dieser Stelle
schon den Gästen aus Übersee und aus vielen europäischen Ländern, daß sie
mit ihren Beiträgen bereit sind, dieser Veranstaltung den Charakter zu
geben, der einer wissenschaftlichen Erörterung von Einsteins Leben und
Werk gebührt.

Denn gerade die Internationalität, auf der jede scientific community bestehen muß, ist die Gewähr dafür, daß Wissenschaftler in der Lage sind,
auch über politische Fronten hinweg Fürsprecher und Helfer für viele einzelne und ganze Völker zu werden, wenn deren Menschenrechte schutzlos geworden sind. Berlin ist ein Ort, an dem dies als Mahnung aus der Vergangenheit besonders gründlich herüberragt, aber auch als Auftrag für Gegenwart und Zukunft deutlicher als anderswo zu erfahren ist.

Deshalb begrüße ich unter den Gästen des heutigen Tages stellvertretend
und ausdrücklich den Vorsteher der Jüdischen Gemeinde zu Berlin, Herrn

G a l i n s k i , und danke ihm dafür, daß er mit seinem Erscheinen ein
Zeichen der Versöhnung setzt, zu dem Einstein nach den unsäglichen Verbrechen der Deutschen am jüdischen Volke noch nicht bereit war.

Einstein hat sein wissenschaftliches Ansehen unermüdlich dazu benutzt,
der politischen Erniedrigung des Menschen ebenso in den Weg zu treten wie
seiner Beschränktheit durch eingefahrene Denkgewohnheiten. So trat er für
Zionisten, aber auch für Sozialisten, für Pazifisten und unter Anfechtung
selbst für die kriegführenden Mächte ein, um die Welt vor der Erpressung
durch Hitler zu bewahren. Er tat es auf eine Weise, die ihn aus dem Verdacht hebt, begrenzten Interessen zu dienen. Er tat es mit demselben Gewissen, das ihn als Physiker - nach den Erschütterungen, die er in das
Denkgebäude seines Zeitalters gebracht hatte - ernstlich und zeitlebens
danach drängte, neue Erkenntnis bis zu den Schönheiten einer einfachen
Harmonielehre voranzutreiben.

Wissenschaft wirft, indem sie neue Lösungen anbietet, auch neue Widersprüche
auf. Einsteins Versuche, zu einer einheitlichen Feldtheorie zu gelangen,
markieren gerade in ihrer Nichterfüllung seinen brennenden Wunsch, zu
einem vollkommenen harmonischen Rechenschaftsbericht des Seins zu gelangen.

Wir werden also eines Wissenschaftlers ansichtig, der Unruhe in die Welt
brachte mit dem Ziel, ihr dauerhafteren Frieden zu sichern. Auch sein
Weltstaatgedanke gehört deshalb deutlich in den Zusammenhang wissenschaftlicher Verantwortung für die Folgen eines Denkvorstoßes aus jahrhunderte-
langer Sicherheit in noch nicht beherrschte neue Dimensionen.

Dieses Symposion wird, wenn es nicht nur resümieren, sondern Erkenntnis
weiter vortreiben will, diese Verantwortung mit zu tragen und - so steht
zu hoffen - mit zu formulieren haben. Daran ist den Veranstaltern dieses
Symposions, in ihrer wissenschaftlichen und ihrer politischen Zuständigkeit, gemeinsam gelegen.

Ohne konkrete Achtung für nahestehende Menschen bleibt aber auch die Ver-
sicherung wissenschaftlicher Verantwortung für die Menschheit abstrakt
und allzu leicht gesagt. Darum gilt unser Gruß und Dank zuletzt denen,
die Einsteins Leben umsorgt und die in persönlichem Austausch mit ihm
den Ertrag seiner wissenschaftlichen Arbeit gesichert und ausgebreitet
haben. Wiederum stellvertretend für die Schüler und Mitarbeiter Einsteins,
die unter uns sind, möchte ich dazu einen Abwesenden grüßen, dem Alters-
unbilden die Reise nach Berlin verwehrten, obwohl er gern und gewiß zu
unserem Gewinn unter uns wäre, und der durch mich Ihnen seine guten Wünsche sagen läßt: W a l t e r G e r l a c h . Er hat nicht nur Einsteins Arbeiten von 1908 bis 1930 als genauer Beobachter und Experimenta-
tor ausfalten helfen, er hat auch vor nunmehr 30 Jahren eine erste Brücke

zwischen Princeton und Einsteins Geburtsstadt zu schlagen versucht, und Einstein hat es ihm gedankt.

Einsteins Sache war die Versöhnung mit den Deutschen nicht. Den freundschaftlichen Zusammenhalt mit den wenigen Kollegen, die er aufrecht fand, hat auch der nationalsozialistische Furor nicht zerstört. Wissenschaft mit Charakter zu betreiben, war seine Sache und sein Appell. So, nur so, hat sie die Chance, Erkenntnis in menschlichen Gewinn umzusetzen. Dies geschieht nicht von selbst, darum wünsche ich es mit besonderem Nachdruck diesem Symposion.

EINSTEIN'S THEORY OF GRAVITATION

Jürgen Ehlers*,Max-Planck-Institut für Physik und Astrophysik, München

1. Preface

Among his many important contributions to Physics Albert Einstein considered his theory of spacetime and gravitation as his most significant
achievement. In a lecture delivered in 1976, P.A.M. Dirac called this
theory "probably the greatest scientific discovery that was ever made".

I shall attempt to review the main assumptions on which this theory is
based, describe some of its principal consequences, and consider a few
of its unsolved problems which are presently topics of research. I shall
restrict myself essentially to Einstein's classical field theory of
spacetime and gravity, considered here as a theory to be applied to
macroscopic objects and processes. Some historical remarks will be
inserted, but the purpose is to describe Einstein's theory systematically
as it is presently understood, not to review its origin and develop -
ment [1].

2. Basic Assumptions About Spacetime And Local, Nongravitational Physics

The basic concepts and assumptions of Einstein's general theory of relativity (GR) can be described in two steps. The first one consists of
specifying those mathematical structures of spacetime which are to
represent the *spatio - temporal metric* and the *gravitational - inertial
field*, as well as the rules used in generalizing *non - gravitational* , *local
laws of physics* from their gravity-free, special-relativistic form to a
general-relativistic one. In the second step the metric, or gravitational
field, is related to matter through the Einstein-Hilbert *field - equation* .

Let us consider the first step. As in Newtonian physics and in the special theory of relativity (SR), *events* are represented as points of a
real, four-dimensional, connected, smooth Hausdorff-manifold M. Following
Euclid's "definition" of a point one can circumscribe the intended meaning of "event" as "a process without parts" [2] .

Particles and *light rays* are represented as curves in M("world lines").
Similarly, extended *bodies* are represented as cylindrical "world tubes"

---

* I dedicate this paper to my daughter, Kathrin Ehlers.

(see Fig. 1)

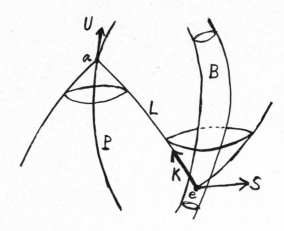

Fig. 1. Spacetime diagram representing a particle P, a body B and a light
ray L emitted at the event e∈B and absorbed at a∈P.

Coordinates x^a are used to set up, at least locally, a one-to-one
correspondence between events and points (x^a) of R^4. In general, several
coordinate systems or "charts" are needed to cover M. Coordinate systems
may be subjected to arbitrary smooth, invertible transformations [3]
$(x^a) \to (x^{a'})$.

The spacetime manifold M is assumed to be equipped with a *Lorentzian
metric* [4]

$$g = g_{ab} dx^a \otimes dx^b, \tag{1}$$

i.e. a symmetric, 2-covariant tensor field with signature (+ - - -)
which defines on each tangent space M_p of M an *inner product*

$$(X,Y) \to X \cdot Y = g(X,Y) = g_{ab} X^a Y^b. \tag{2}$$

Eq. (2) determines at each event the *null cone* and the classification of
vectors into *timelike*, *spacelike* and *null vectors*. This classification naturally extends to curves and other subsets of M. Thus in Fig. 1 the
tangent vector K of L at e is null $(L^2=0)$, the tangent U of P at a is
timelike $(U^2>0)$, and S at e is spacelike $(S^2<0)$. B is a "timelike
cylinder", *i.e.* the union of the members of a 3-parameter family of
timelike curves. Half of the null cones at e and a, respectively, are
indicated.

The field of null cones, *i.e.* the conformal structure of (M,g), determines the *causal structure* (see section 6); the local distinction between
space and time; *angles* and *relative velocities* between particles passing

through (nearly) the same event; as in special relativity theory. The
metric itself determines, via arc lengths, proper *times* along particle
world lines (clocks) and proper *lengths* along spacelike curves representing, *e.g.*, instantaneous configurations of measuring tapes (see
Fig. 2).

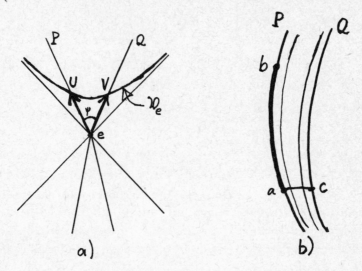

a)   b)

Fig. 2. a) Two colliding particles have final 4-velocities, or world
directions, U,V at e; U,V are elements of the (Lobatschewskian) *velocity
space* $\mathcal{V}_e$. Their relative *speed* v is related to the pseudo-angle (Lob.-
distance) ψ by cos hψ= U·V/$(U^2V^2)^{1/2}$, v= tghψ. b) The *time* between a and
b on P is $\int_b^a |g|^{1/2}$, the *length* of the string with ends P, Q "at a" is
$\int_a^c |g|^{1/2}$.

Before proceeding with the description of GR it is useful to recall a
few statements from SR. In SR, the spacetime manifold M is taken to be
diffeomorphic to R^4, and it is assumed that there exist global coordinate
systems $(x^a) = (x^o,x^\alpha) = (t,\vec{x})$ such that

$$g_{\mu\nu} = \eta_{\mu\nu} = \text{diag}(1,-1,-1,-1) \tag{3}$$

With respect to such *inertial coordinates*, a free particle moves according
to the *law of inertia*,

$$\frac{dx^a}{ds} = u^a, \; \frac{du^a}{ds} = 0, \; u_a u^a = 1. \tag{4}$$

Maxwell's equations take the form

$$\partial_{[a}F_{bc]} = 0, \qquad \partial_b F^{ab} = J^a \tag{5}$$

so that

$$\partial_a J^a = 0. \tag{6}$$

They imply that a *light ray* in vacuo, defined by means of a WKB asymptotic approximation to (5), is given by a null straight line, *i.e.* it obeys the first two equations (4) and $u_a u^a = 0$, which justifies the name *light cone* instead of null cone. The *stress-energy-momentum tensor* T of any kind of matter or field obeys the relations

$$T^{[ab]} = 0, \quad \partial_b T^{ab} = 0. \tag{7}$$

Dirac's equation for the state of a free, spin $\frac{1}{2}$ particle with mass m reads, in 2-component spinor notation,

$$\partial_{A\dot{A}}\, \psi^A = \frac{m}{\sqrt{2}}\, \chi_{\dot{A}}\,, \qquad \partial^{A\dot{A}}\, \chi_{\dot{A}} = -\frac{m}{\sqrt{2}}\, \psi^A\,. \tag{8}$$

Here, $\partial_{A\dot{A}} = \sigma^a{}_{A\dot{A}}\, \partial_a$ denotes the spinor derivative operator containing the components $\sigma^a{}_{A\dot{A}}$ of the isomorphism $\sigma : S \otimes \bar{S} \to V^C$ which maps the spinor space $S \otimes S = \{\phi^{A\dot{A}}\}$ onto the complexified space of 4-vectors, $S = \{\psi^A\}$ and $\bar{S} = \{\chi^{\dot{A}}\}$ being the "complex conjugate" spinor spaces associated with the two fundamental representations of SL_2 (C), the 2-fold covering group of the restricted Lorentz group $O^{\uparrow}_{+}(1,3)$.

These examples illustrate that, according to SR, physical laws take a standard, Poincaré-invariant form if referred to inertial coordinates. That is, the group of isometries of (R^4, η), the *Poincaré group* , besides describing the symmetry of spacetime itself, is also the symmetry group of the laws of physics, in accordance with Einstein's *special principle of relativity* [5] of 1905.

In SR the metric η is an absolute, constant, non-dynamical entity, in contrast to the dynamical (state-) variables exemplified by u^a, F_{ab}, T^{ab}, ψ^A, $\chi_{\dot{A}}$.η is specified once and for all, irrespective of the physical processes to be described, not capable of any changes governed by field equations.

The fundamental heuristic idea from which general relativity originated, conceived by Einstein in 1907 [5] and elaborated by him until 1915 [6], is that *inertia* and *gravity* are but different names for the *same universal property of matter*. This idea is suggested and supported by the composition independence of the world lines of freely falling test particles already noticed by S. Stevin ($\sim$1600) and G. Galilei ($\sim$1630), and more generally by the universal proportionality of inertial and passive gravitational masses of test bodies which was established with ever increasing precision by I. Newton (1686, 10^{-3}), F.W. Bessel (1832, 10^{-5}), R.v. Eötvösz *et al.* (1922, 10^{-8}), R.H. Dicke *et al.* (1964, $3 \cdot 10^{-11}$), V. Braginski *et al.* (1971, 10^{-12}), and also by the terrestrial redshift experiments by R.V. Pound and G.A. Rebka (1960), R.V. Pound and J.L. Snider (1965, 10^{-2}) and the rocket experiment by R.F.C. Vessot and M.V. Levine (1976, $2 \cdot 10^{-4}$). (For details see, *e.g.*, [7].)

These experimental facts can be taken to show that the description of inertia as an absolute, integrable affine structure (straight world lines, parallelity) of spacetime or, equivalently, as a class of global inertial frames of reference, and the completely different description of gravity as one force among others, is fundamentally inadequate – as illustrated so lucidly by Einstein's elevator experiment –; and that a more satisfactory theory should account for inertia and gravity in terms of a single, indecomposable structure. The kind of this structure is indicated rather directly by the phenomena: Free fall experiments exhibit that spacetime is endowed with a *path-structure*, *i.e.* some kind of field which determines, for each event e and each world direction (velocity) at e, a unique path (="unparametrized curve") through e in that direction. Assuming, as the simplest idealization of the empirical facts, the path structure to be "infinitesimally isotropic" or, equivalently, "infinitesimally scale-independent", it follows ([8], [9]) that the paths are the ("deparametrized") geodesics of a class $\tilde{D}=\{D\}$ of projectively equivalent [10], symmetric linear connections on M. One is thus led to a *geodesic path structure* as a candidate for the sought-for mathematical representation of the inertial-gravitational field. (This is geometrically very plausible anyway, since a manifold with a geodesic path structure is the most natural differential-geometrical analogue of a projective space.)

On the other hand the validity of SR in "macroscopically infinitesimal" regions of spacetime suggests that M has a metric [6] g, as discussed above. These two basic structures, a Lorentzian metric and a geodesic path-structure, which arise from independent physical considerations, are *compatible* - in the sense that the light cone of an (arbitrary) event e bounds the set of possible free-fall paths through e - exactly if the free-fall paths coincide with the geodesics determined by the metric g [11] . Therefore, if one wishes to represent the combined inertial-gravitational field in the manner just outlined, then, in order to account for the facts that 1) any real gravitational field is inhomogeneous and causes freely falling particles to be relatively accelerated, and 2) the gravitational field depends on the distribution of matter in spacetime, one has to take the *metric* g postulated in eq. (1) to be a *dynamical field* the associated curvature of which represents the gravitational field gradient (tidal field). Thus one arrives at *Einstein's description of gravity* .

In accordance with the preceding considerations the world lines of freely falling test particles are characterized by the variational principle

$$\delta \int ds = 0 \tag{9}$$

or, equivalently, by the Lagrangean

$$L(x^a, \dot{x}^b) = \frac{1}{2} g_{ab}(x) \dot{x}^a \dot{x}^b \tag{10}$$

or the Hamiltonian

$$H(x^a, p_b) = \frac{1}{2} g^{ab}(x) p_a p_b \ . \tag{11}$$

These equations show that the tensor field g, besides determining times, angles and lengths, also plays the part of a *gravitational-inertial potential*. The Euler-Lagrange equations of eq. (10),

$$\ddot{x}^a + \Gamma^a_{bc} \dot{x}^b \dot{x}^c = 0, \tag{12}$$

contain the Christoffel symbols Γ^a_{bc} formed from the g_{ab} and their first derivatives. They are components of the Riemannian connection [3] associated with g. Because of eq. (12) one may call this connection the gravitational-inertial *field strength*, and the Γ^a_{bc} its components relative to (x^a). This is an example of a physically important object which is neither a tensor nor a spinor. It is non-local in the sense that it does not have a coordinate-independent "value" at a spacetime point, in accordance with Einstein's elevator argument showing that the gravitational field can be "transformed away" at an event, like inertial forces in Newtonian theory.

The connection determines not only the geodesics, but also the *parallel transport* of vectors and tensors along curves, and hence the *covariant derivatives* of tensor fields.

(12) implies that the relative acceleration of two nearby, freely falling test particles with separation vector ξ^a is given by the equation of *geodesic deviation*

$$\ddot{\xi}^a = R^a{}_{bcd}\dot{x}^b\dot{x}^c\xi^d \tag{13}$$

in which the Riemann *curvature tensor* $R^a{}_{bcd}$ of g_{ab} is formed from the $\Gamma^a{}_{bc}$ and their first derivatives. In contrast to electrodynamics, in gravidynamics not only the *tidal field* (=field gradient = $R^a{}_{bcd}$), but also the potential g_{ab} has measurable, pointwise, tensorial values, whereas the field strength ($\Gamma^a{}_{bc}$) does not. In fact, given an event e, there always exist coordinate-systems such that, at e, eq. (3) and $\Gamma^a{}_{bc} = 0$ hold; such coordinates are said to be *inertial at e*.

Eqs. (4) and (12) exemplify how local physical laws can be generalized from SR to GR: One retains their form at the origin e of an inertial (at e!) coordinate system or, equivalently one replaces, with respect to arbitrary coordinates, the flat metric η and the covariant deriva-tives with respect to its connection by g and the associated covariant derivatives, to be denoted by D_a. Applying this to (5) and (7) gives, respectively,

$$D_{[a}F_{bc]} = 0, \quad D_b F^{ab} = J^a \tag{14}$$

and

$$T^{[ab]} = 0, \quad D_b T^{ab} = 0. \tag{15}$$

(14) implies

$$D_a J^a = 0 \tag{16}$$

and expresses charge conservation, as in SR. That is, the implication (5) $\Rightarrow$ (6) and the property of (6) of being equivalent to a conservation law are preserved under the "general-relativization" $(\eta,\partial) \rightarrow (g,D)$. Sim-ilarly, the implication that light rays are null geodesics is preserved; thus also in GR one may speak about light cones, and use (10) or (11) combined with the constraint L = 0 or H = 0 as a basis for geometrical optics. The second of eqs. (15), however, although it is the formal analog of (7), does not express the conservation of the material energy-momentum the density of which is measured by T^{ab}; it is not equivalent to any covariant, integral conservation law.

The "minimal coupling" rule $(\eta,\partial) \to (g,D)$ is similarly ambiguous as the simple Schrödinger quantization rule $(q,p) \to (q\cdot,-i\hbar\partial\cdot)$, for a formally analogous reason: In contrast to the ∂_a, the D_a do not commute. This non-commutativity is an essential property of gravity (and similar gauge fields), just as the non-commutativity of quantum coordinates and momenta is characteristic for the micro-dynamics of matter. In both cases the correspondence rule is a useful aid in generalizing theories to a new domain (only), but not a "law".

We shall continue the discussion of local, non-gravitational physical laws in a curved spacetime in the next two sections.

As an afterthought, I should like to mention two alternative ways of motivating physically the fundamental assumption of GR that the metric, the existence of which is postulated because of the approximate, local validity of SR, has a non-vanishing curvature determined by the distribution of matter: 1) One uses the terrestrial redshift measurements to argue that those frames of reference relative to which Maxwell's equations have their standard, SR-form, coincide with the local inertial frames attached to freely falling particles, and one infers from this and the inhomogeneity of the Earth's gravitational field that local inertial frames at different events are relatively accelerated, which implies curvature. 2) One accepts Maxwell's equations in vacuo or that light rays are null geodesics determined by g_{ab}, and one infers from the observed light deflection by the Sun that the null cones are distorted near matter; this implies conformal curvature, and consequently also metric curvature (and excludes, *e.g.* Nordstrøm's theory). (Several additional reasonings can be found in the literature.)

3. Frames, Orientations, Bundles and Spinor Fields.

For several purposes it is useful, for some even necessary, to use *frames* on spacetime. A frame (tetrad, Vierbein) $E = (E_a)$ is a set of four linearly independent vectors E_a at some event, and a frame field on some open set U of M is a (smooth) function assigning to each event p of U a frame at p. Just as coordinate systems assign numbers to events, so frame fields serve to assign numerical components to vectors, tensors, connections etc. Coordinate systems and frame fields can be chosen independently; it is not necessary to use the coordinate frame $(\frac{\partial}{\partial x^a})$ once coordinates x^a have been chosen.

The general linear group $GL_4(R)$ acts freely on the manifold LM of all frames of M; under this action a frame E at p runs through the "fibre" consisting of all frames at p. Let $\pi: LM \to M$ be the map sending the frame

E at p into p . The structure given by $(LM,M,\pi,GL_4(R))$ exemplifies a
principal fibre bundle with bundle space LM, base space M, projection π
and structure group $GL_4(R)$; see Fig. 3. It is called the *bundle of
linear frames* of M.

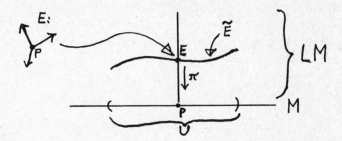

Fig. 3. The bundle of linear frames with a frame field $\tilde{E}$ on U.

Clearly, a Lorentz metric g on M is characterized by the subset $L^{(g)}M$
of its *orthonormal frames* in LM. $L^{(g)}M$ together with the Lorentz group
O(1,3) gives rise to the *orthonormal frame bundle* of g over M. Conversely,
any reduction of the bundle of linear frames to an O(1,3)-bundle defines
a Lorentz metric.

If the null cones of a Lorentz manifold (M,g) can be partitioned into
"future" and "past" halves in a continuous manner, (M,g) is said to be
time-orientable; it then admits exactly two "opposite" time orientations.
In such a case the orthonormal frame bundle can be reduced further to
that of *ortho-chronous* orthonormal frames with structure group $O^{\uparrow}(1,3)$.
(Fig. 1 shows the past null half cone of a and the future null half cone
of e, *e.g.*) Analogously one can define *space-orientability* [13] and
spacetime-orientability, with bundle reductions belonging to the ortho-
chorous $(O_o(1,3))$ and the proper $(O_+(1,3))$ Lorentz groups [14] , respectively. A Lorentz manifold admits either none or exactly one, or all
three kinds of orientations. In the last case it is said to be *orientable*.

A time orientation is a prerequisite for the formulation of a spectral
condition for the 4-momenta of particles, a Clausius inequality (law of
non-negativity of entropy production density), and for defining the total
charge of a localized current distribution J^a (by an integral over a
future-oriented spacelike hypersurface). For these and other reasons
it is customary to assume *spacetime* to be a *time-oriented* Lorentz manifold.

Any representation of O(1,3) or $O^{\uparrow}(1,3)$ *etc.* gives rise to corresponding
fields, defined as functions on the respective bundle spaces with appropriate transformation properties (see, *e.g.*, [15]) .

Let (M,g) be an oriented spacetime and denote by $O_+^\uparrow M$ the bundle of its oriented orthonormal frames. Further, let h : SL_2 $(C) \to O_+^\uparrow (1,3)$ be a covering homomorphism for the restricted Lorentz group $O_+^\uparrow (1,3)$. Then a *spinor structure* on (M,g) is an $SL_2(C)$-bundle over M together with a smooth map k from its bundle space SM onto $O_+^\uparrow M$ such that, for all $A \epsilon SL_2(C)$, h(A)∘k= k∘A. The manifold SM of *spin frames* is then mapped by k onto that of oriented orthonormal frames, $O_+^\uparrow M$, in a two-to-one manner. See Fig. 4

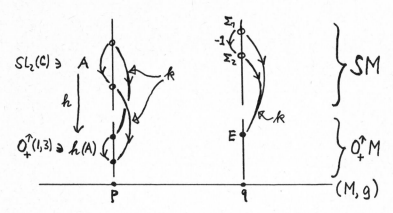

Fig. 4. A spinor structure on (M,g). The actions of A and h(A) are k-related, as illustrated by means of the fibres over p. The fibres over q show the two-fold covering of $O_+^\uparrow M$ by SM which parallels h.

A spinor structure exists, for a non-compact M, exactly if (M,g) admits a global, continuous orthonormal frame field [16] . The connection on (M,g) or, equivalently, on $O_+^\uparrow M$ induces a unique *spinor connection* on any spin frame bundle SM. One can define *spinor fields* via representations of $SL_2(C)$, and these can be differentiated covariantly. Thus one can, in particular, adapt Dirac's equation (8) to GR, replacing $\partial_{A\dot{A}}$ by the spinor covariant derivative $D_{A\dot{A}}$. One can motivate convincingly, following R. Penrose [17] and taking into account the experiments showing that for a Fermion a rotation by 2π is not the identity [18] , that physical *spacetime does admit a spinor structure*.

The point of these remarks is that, under mild topological restrictions on M, spinor fields are as natural objects associated with a curved spacetime as tensor fields. (This point of view, emphasized and applied in "classical GR" in particular by R. Penrose [17] , opposes the frequently taken attitude according to which tetrad fields, frames and spinors do not "fit naturally" into GR.)

The formulation of the "field kinematics" of GR in terms of principal bundles and their associated bundles allows one to consider GR as a *gauge theory*, and thus to compare its structure - which differs considerably from Yang-Mills-type gauge theories - with other such theories. (See, *e.g.*,[15].) (As far as I can see such gauge considerations have not led to a deeper understanding of GR as such, nor in particular of the role of the Poincaré group in GR. The physical role of *translations* in SR, as space-time displacements, is taken over in GR by *diffeomorphisms* of spacetime. I at least fail to see that the use of affine bundles with affine (in Cartan's sense) connections changes this fact, nor does it help me to appreciate it more deeply. The approximate local validity of SR appears to be due to small curvature, without the need for a modified translation or Poincaré group. Of course, these remarks are not intended to pass any judgement on theories other than Einstein's, with or without gauge.)

4. Classical Descriptions of Matter and Radiation

To describe planets, stars, galaxies and intergalactic matter, the main objects of interest of a macroscop gravitation theory, one needs a theory of bulk matter. Moreover, a formalism to describe electromagnetic as well as particle radiations is required, both since such radiations are important components of the universe and because they convey to us most of the information about distant matter.

Two descriptions of matter have been elaborated in GR rather extensively: *Continuum mechanics* and a (Maxwell-Boltzmann type) *kinetic theory*, both in sufficient generality so as to permit the inclusion of equilibrium and non-equilibrium *thermodynamics*. For systematic accounts of these theories, see *e.g.*,J. Ehlers [19] , [20] , R.K. Sachs and H. Wu[21], B. Carter and H. Quintana [22] , G. Maugin [23] , W. Israel and J.M. Stewart [24] and the papers quoted there. For thermodynamics of black holes, see P. Candelas and D.W. Sciama [25] .

Electromagnetic radiation can be described semi-phenomenologically either by means of kinetic theory as a photon gas, or as an ensemble of Maxwell fields, with compatible results.

In this section only a few remarks will be made.

A *perfect fluid* is described, as in SR, by a (proper) energy density ρ, a scalar pressure p and a mean 4-velocity U^a. The corresponding stress-energy-momentum tensor is then, with c = 1,

$$T^{ab} = (\rho+p)U^a U^b - pg^{ab}. \tag{17}$$

Eq. (15) in this case is equivalent to a local energy balance and the GR-generalization of Euler's equation.

The state of an assembly of particles is given in kinetic theory by a *distribution function* $f(x^a, p^b)$, a non-negative function on the tangent bundle TM of spacetime obeying some Boltzmann equation. The corresponding stress-energy-momentum tensor is

$$T^{ab}(x) = \int p^a p^b f(x,p) \pi, \qquad (18)$$

where π is a measure on the fibre (momentum space) over x.

If, in case (17), $|p| < \rho$ or, in (18), the support of f in each tangent space is contained in the future half of the null cone, then $T^a_{\ b}(x)$ maps the interior of that half cone into itself, which implies that for all local observers the energy density T^{oo} is positive, the magnitude of the density of linear momentum is less than T^{oo}, and the stresses are also dominated by T^{oo}. This is an example of an *energy condition* imposed frequently on $T^a_{\ b}$.

The basic variables of relativistic thermodynamics are a scalar, nonnegative *temperature* T and an *entropy current density* 4-vector S^a. The latter is assumed to satisfy the Clausius inequality

$$D_a S^a \geq 0. \qquad (19)$$

As indicated here and in section 2, most parts of classical physics have been adapted to GR, and the generalized theories have found applications in astrophysics, particularly in the theory of stellar structure and cosmology. It appears that all of classical physics, as far as it has been developed within SR (and this is a non-trivial restriction), can be incorporated coherently into GR. The problem of the relation of GR to quantum physics, even if gravity is considered as an external field, appears to be far less understood (see *e.g,* [26], [27]).

It is perhaps worth pointing out that as long as one works with *local* laws - differential equations - , one can usually derive without much difficulty exact GR-analogues of theorems of classical physics. However, when *integrals* are involved -*e.g.* , when one wants to define a total 4-momentum of an extended body or system, difficulties arise due to the path dependence of parallel transport. Only in some cases such as in Dixon's dynamics of extended bodies [29] these difficulties have been overcome satisfactorily, after considerable efforts. These difficulties show up, *e.g.*, in any attempt to derive a geodesic law for test bodies by performing a limiting process based on a treatment of extended bodies, which in turn is necessary since point particles are incompatible with

Einstein's field equation.

5. The Einstein-Hilbert Field Equation of Gravitation

To complete the review of the basic assumptions of GR begun in section
2 we have to consider how the "inertial-gravitational-metrical" field
g_{ab} is related to matter, or to other fields, according to Einstein
and Hilbert.

Before stating the field equation I should like to recall some mathematical theorems which can be used to motivate and to characterize that
equation.

In the sequel, $R_{ab} = R^c{}_{acb}$ denotes the contracted (Ricci) curvature
tensor, $R = g^{ab}R_{ab}$ denotes its trace, and $G^{ab} = R^{ab} - \frac{1}{2} g^{ab}R$ denotes
the Einstein curvature tensor which satisfies the contracted Bianchi
identity.

$$D_b G^{ab} = 0. \tag{20}$$

A. If a functional $g \rightarrow K$ assigns to each smooth Lorentzian metric field
g a tensor field (of arbitrary valence) K such that the components of
K at a point p depend smoothly on $g_{ab}, \partial_c g_{ab}, \ldots, \partial_{c_1 \ldots c_r} g_{ab}$ at p,
then K is obtainable from g, the curvature tensor, and the covariant
derivatives of the latter up to order r, *pointwise* by the elementary
operations of tensor *algebra*. (E.B. Christoffel, 1869 [30].)

Corrollaries. A_1. There are no such "differential concomitants" K of
g of (exact) order one. A_2. There are *no concomitants of order r ≥ 2*
which are *linear* in g. A_3. The general second-order, 2-contravariant
concomitant of g which is *quasilinear* in g is given by $AR^{ab}+Bg^{ab}+\Lambda g^{ab}$,
where A, B, Λ are constants.

B. The general second order, 2-contravariant concomitant K^{ab} of g_{ab}
which satisfies identically $D_b K^{ab} = 0$ is, for dimension 4, given by
$K^{ab}=AG^{ab} + \Lambda g^{ab}$ with $A \neq 0$. (Lovelock 1972 [31].)

C. $G^{ab} = -\frac{\delta}{\delta g_{ab}} (\sqrt{-g} R)$. ($\frac{\delta}{\delta_{..}}$ denotes the Euler-Lagrange, or variational, derivative.) (Hilbert 1915 [32].)

According to Einstein and Hilbert g_{ab} *is related to matter by*

$$G^{ab} = T^{ab}. \tag{21}$$

(Units and dimensions have been chosen such that c = 1 and Newton's
constant $G = (8\pi)^{-1}$ so that all quantities are powers of time.) This
equation was communicated to the Royal Society of Sciences in Göttingen
by Hilbert on November 20, 1915, and to the Royal Prussian Academy of

Science in Berlin by Einstein on November 25, 1915. (For the history
leading up to eq. (21) see W. Pauli [33] and J. Mehra [34].)

One simple *physical motivation of eq.* (21) runs as follows [19] :
Poisson's equation $\Delta U = \frac{1}{2}\rho$ says that the trace of the gravitational
tidal field tensor, which transforms the relative position vector of
two freely falling test particles into their relative acceleration,
equals minus one-half the mass density. For a local observer with 4-
velocity v^a in GR the corresponding trace is, by eq. (13), $-R_{ab}v^av^b$.
Two quantities in GR correspond, under Newtonian conditions, to the
mass density, viz. $T_{ab}v^av^b$ and $T = T^a{}_a$. So, to be safe, take a constant
λ and take as the GR-analogue of Poisson's equation,

$$R_{ab}v^av^b = \frac{1}{2}(\lambda T_{ab} + [1-\lambda] T g_{ab})v^av^b$$

or, since this should hold for all observers,

$$R_{ab} = \frac{1}{2}(\lambda T_{ab} + [1-\lambda]T g_{ab}).$$

This last equation is compatible with the physical requirement (15) and
the identity (20) only if $\lambda = 2$, since in general T is a non-constant
function. This, indeed, gives (21).

Of course, the argument given is not a "derivation" of (21) which is
an *axiom* of GR; but it does show that (21) is reasonable; the argument
can be inverted to show that (21) implies, in a certain sense, Poisson's
equation locally.

Independently of the preceding argument, the mathematical facts listed
above show: If one wants to have a field equation of the form $K^{ab} = T^{ab}$
where K^{ab} is a concomitant of g_{ab} of order ≥ 1, then (i) K^{ab} cannot be of
first order, (ii) K^{ab} cannot be linear in g_{ab}. If one requires, in addition, that K^{ab} be of second order and obey $D_b K^{ab} = O$ so that (15) is
incorporated into the field equation, then one necessarily obtains

$$G^{ab} + \Lambda g^{ab} = T^{ab}, \tag{22}$$

i.e. Einstein's equation modified by a *cosmological term* (only). The
symmetry of both sides of (22) *follows*, as does the *quasilinearity*
of the left-hand-side, in both of these motivational arguments. Moreover,
eq. (22) is the only second-order field equation of the form
$K^{ab}[g..] = T^{ab}$ which is derivable from an *action principle* in which the
gravitational Lagrangean density is a scalar differential concomitant
of g_{ab} of any finite differential order, provided T^{ab} can be obtained
as a variational derivative of some matter Lagrangean with respect
to g_{ab}, as in many cases it can.

Let us, then, accept (21). (For most purposes it does not matter whether
one takes (21) or (22) as long as Λ is sufficiently small. I do not wish
to discuss specifically the Λ-term here.) Then, of course, eqs. (15)
follow as universal laws of matter. Moreover, if $T^a_{\ b}$ satisfies the
energy condition of section 4, then (21) implies positive mean space
curvature and that, on average, test particles accelerate towards each
other, in the sense of a contraction; this expresses locally that gravity is attractive. (For negative $T^a_{\ b}$ it would be repulsive for all test
particles.)

The nonlinearity of (21) is of fundamental significance for the following
reason: Since the local law of motion (15) is implied by the field equation (21), *the field equation*, combined with equations of state, *determines the motion of the sources*; there is no room for additional "equations
of motion". Therefore, *the nonlinearity of the field equation* by itself
implies that bodies interact gravitationally, in fundamental contrast
to the manner in which interactions come about in the standard forms
of Newton's or Maxwell-Lorentz's theories. (In GR, the interaction between bodies is incorporated into the field equation similarly as charge
conservation is in electrodynamics.)

One can summarize the basic assumptions of Einstein's theory of spacetime
and gravity as follows [3] : A *spacetime model* is an equivalence class of
objects $(M,g,T,...)$ each consisting of a time-oriented, connected Lorentz
manifold (M,g) and a collection of tensor fields $T,...$ satisfying the
field equation (21) (or (22)), an energy condition (see section 4) and
a number of non-gravitational, local physical laws. $(M,g,T,...)$ and
$(M',g',T',...)$ are considered as equivalent if and only if there exists
a diffeomorphism Φ mapping M onto M' such that, with respect to the derivative of $\Phi,g,T,...$ are mapped into $g',T',...$ Physical interpretation rules
have been indicated in sections 2 and 4; numbers to be compared with
observations are to be represented in the theory as scalar invariants
(see, *e.g.*, [19], section 2.10). The symmetry of a spacetime model is
given by the (Lie-) group of isometries leaving $T,...$ invariant.

It emerges from this description that whereas the Lorentz groups
$0(1,3)$, $0^{\uparrow}(1,3)$ form part of the structure associated with GR spacetime
models, the Poincaré group does not (see, however, section 7); translation
invariance is destroyed by gravity, at least according to Einstein's
theory. It seems to me, for this and other reasons, that reformulations
of GR as a "Poincaré-invariant theory with a large, non-abelion gauge
group" , which are certainly useful as a preliminary tool to relate GR
formally to "the rest of physics", revoke the main conceptual advance

made by Einstein in taking the step from SR to GR, viz. that there is
no non-dynamical, flat metric, but that the spacetime metric is a
dynamical field just like other fields of physics; geometry should
not be separated from everything else as being an a priori, absolute
structure [7] . (I find it rather paradoxical that sometimes "relativists"
are accused of assigning a preferred role to geometry. With at least
as much justification one may say that only relativists treat geometry
like other physical structures, by not considering the metric as the only
absolute, universal *external* field which is not influenced by anything
else.)

6. Some Consequences And Problems of GR

The field equation (21) implies that rapidly oscillating, locally
plane gravitational waves, superimposed on a slowly varying background
metric, propagate in matter-free regions without dispersion along the
null-geodesics of the background [35] , and that the smallest angle such
that their polarization patterns are invariant under rotations about that
angle is π (see, *e.g.*, [36]). These facts can be taken to give a precise
meaning to the statement that the gravitational field in GR belongs to
mass zero and spin 2 without having to refer to a flat-space approximation. The linear vacuum ($G_{ab}=0$) perturbations off the trivial solution
$g_{ab}=\eta_{ab}$ are identical with the solutions of the Poincaré-invariant,
classical, free field theory of mass zero and spin 2.

Geodesically complete, empty spacetimes ($G_{ab}=0$) representing plane
gravitational waves have been found and discussed in detail [37].Cylin-
drically symmetric spacetime models representing material sources with
gravitational waves have also been constructed [38] . Gravitational wave
solutions without symmetry have also been determined [39] ; but no single
exact spacetime model representing a physically reasonable spatially
bounded source emitting gravitational waves has yet been found - a GR-
analogue of a Hertz oscillator is still missing. Formal approximate
solutions have been computed, but without a proof that they do indeed
approximate a (rigorous) solution. It has been possible to give a fairly precise meaning to the assertion that gravitational waves carry energy-momentum (for an outline of the arguments and references, see
ch. VIII of [36]), but some difficulties remain; I shall come back to
this problem in the next section.

The *initial value problem* for the vacuum field equation, the coupled
Einstein-Maxwell system, the Einstein-Euler-system (eq. (21) with source
(17) and an equation of state p = f (ρ)) and the Einstein-Boltzmann-

system (eq. (21) with source (18) and Boltzmann-eq. for f) has been discussed extensively (see [2], [40] for reviews).

Consider, as an example, the *Einstein-Euler system*

$$G^{ab} = (\rho+p(\rho))U^aU^b - p(\rho)g^{ab} \tag{23}$$

with $g_{ab}U^aU^b = 1$ and $|p(\rho)|<\rho, 0\leq p'(\rho)\leq 1$. An *initial data set* $(N,h_{\alpha\beta},k_{\alpha\beta},$
σ,t_α) for this system consists of a Riemannian 3-manifold $(N,h_{\alpha\beta})$ and
tensor fields $\sigma(\geq 0)$, t_α , $k_{\alpha\beta}(=k_{\beta\alpha})$ on N obeying the *constraint equations*

$$R(h) + (k^\alpha_\alpha)^2 - k^\alpha_\beta k^\beta_\alpha = 2\sigma , \tag{24}$$

$$\nabla_\beta(k_\alpha^{\ \beta}- k_\gamma^{\ \gamma}\delta_\alpha^{\ \beta}) = t_\alpha$$

and the energy inequality $\sigma^2 > h^{\alpha\beta} t_\alpha t_\beta$.

A *development* $(M,g_{ab},\rho, U^a;i)$ of an initial data set is an Einstein-Euler
spacetime (M,g_{ab},ρ,U^a) satisfying (23), together with a smooth embedding
i: N→S of N into M such that S is a *Cauchy surface* for that spacetime
which carries the correct initial data. (*I.e.* if N_a is the unit normal
of S in (M,g_{ab}), then the pull-back by i of the metric induced in S
by g_{ab} should equal - $h_{\alpha\beta}$, the pull-back of the second fundamental form
(extrinsic curvature) of S should equal $k_{\alpha\beta}$, and the pull-backs of
$T_{ab}N^aN^b$ and $T_{ab}N^b$ should equal, respectively, σ and t_α .) This is illustrated in Fig. 5.

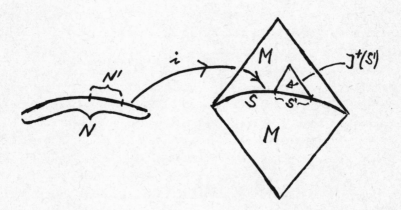

Fig.5. An initial data set and its maximal Cauchy development. i embeds
N and the data on N into M and data on i(N) = S.

The (elliptic) equations (24) which restrict the initial data corre-
spond to the constraints $\vec{\nabla}\cdot\vec{B} = 0$, $\vec{\nabla}\cdot\vec{E} =\rho$ of electrodynamics and contain
part of the Newtonian, Coulomb-like gravitational interaction. Under some

technical conditions (differentiability etc.), Y. Choquet-Bruhat and
R.P. Geroch [41] , and S.W. Hawking and G.F.R. Ellis [2] have established that *any initial data set determines a unique, maximal development*
(unique up to diffeomorphisms, as stated at the end of section 5) depending stably on the data. If N' is an open, relatively compact part of
N, then its image i(N') = S' in the Cauchy surface S determines its own
maximal development. The future part of this development consists of
those events for which all past-directed timelike curves without a past
endpoint hit S'. This statement expresses *GR-causality*.

The preceding description shows both the similarities and the difference
between initial value problems in SR and in GR. It shows in particular
how, at the classical level, precise statements about causality are
possible in spite of the fact that the conformal structure is itself
dynamical. (It does not seem to be clear how causality can be formulated
meaningfully in a quantum-version of GR - if such a version exists.)

7. Isolated Systems. The n-Body Problem in GR

One of the main problems of any theory of gravity is to describe isolated
systems of one or several gravitationally interacting, macroscopic bodies.
In fact, almost all quantitative empirical information about details of
the gravitational interaction so far has been obtained, and in the fore-
seeable future is expected to be obtainable, by observing and theoretically modeling systems like the solar system or the binary pulsar discovered by R.A. Hulse and J.H. Taylor in 1974 [42] .

Treatments of relativistic n-body systems by formal, conceptually as well
as mathematically questionable approximation methods, based on an Ansatz
like $g_{ab}=\eta_{ab} + h_{ab}$ with a "small" deviation h_{ab} of the physical metric
g_{ab} from an auxiliary flat background metric η_{ab} and a representation
of the bodies as point particles, singularities or extended fluid
balls, are as old as GR itself. It is usual to distinguish the weak-field,
slow motion approximation levels for bound systems by introducing a small
parameter $\varepsilon= \frac{v}{c}$ (where v is a speed typical for the relative motions within the system) and to denote as the *m-th post-Newtonian approximation* that
one in   which all terms in the equations of motion $D_b T^{ab}$ = 0 are taken
into account which are smaller than the Newtonian terms by factors of
order up to ε^{2m}.

The *first post-Newtonian approximation* (1PNA) for nearly spherical,non-
rotating bodies has been obtained by several methods which all agree
with the celebrated *Einstein-Infeld-Hoffmann equations* [43] . (These
results had been obtained earlier already. Later investigations have

clarified somewhat the assumptions involved in the derivations.) It forms
the theoretical basis for the discussion of the *standard tests* (redshift
of light from the Sun, light deflection and radar time delay by the solar
gravitational field, non-Newtonian perihelion advances) of GR in the
solar system, and more recently for "measuring" the mass of the pulsar
and its companion in the binary pulsar system PSR 1913+16 [44]. *GR agrees
with the observations*, as far as the standard tests are concerned, within
the accuracy of observation which has now reached 1% or even a little
better, provided the formal 1PNA is a valid approximation to at least
the same accuracy, which is generally assumed but has not been rigorously
established.

Recent calculations [45] - [48] have shown that if one assumes, within
the 1PNA, the bodies to be axially symmetrical and rotating, one obtains
spin-precession formulae related to "magnetic type" gravitational fields.
These results extend the pioneer-work by Lense and Thirring on "dragging
of inertial frame" - effects and might be amenable to observational or
experimental tests in the near future [44], [49] .

Until recently the 2PNA and the $2\frac{1}{2}$ PNA suffered from divergences. By
an - at least formal - improvement of the interation scheme [50] these
two levels of approximation have now been worked out [51] free of divergences. The $2\frac{1}{2}$ PNA is of particular interest since its eqs. of motion
contain the lowest-order *gravitational radiation reaction terms*; it had
been worked out for the first time in [52] . Opinions as to whether the
presently available approximations suffice to treat quantitatively such
radiation reaction effects, in particular whether they permit the compu-
tation of the rate-of-change of the orbital period of the binary pulsar -
which has been observed as $\dot{P}_b = (-2.7\pm0.4)\cdot10^{-12}$ [53] - with sufficient
accuracy, differ considerably. (I think they do not.) For discussions
of the theoretical objections, see [54], [55] .

A *rigorous treatment of isolated systems* within GR requires (a) a description of extended bodies, (b) a specification of boundary conditions
expressing the isolatedness of the system, (c) means to judge whether
there is incoming or outgoing gravitational radiation. The need for (a)
arises from the fact that point particles are not compatible with the
basic assumptions of GR as laid down in sections 2 and 5. A framework
for satisfying (a) is available (see [29],[55] and the papers cited
there). Proposals concerning (b), in the form of definitions of *asymptotic flatness* and consequences thereof, are also available; for a
recent review see A. Ashtekar and R. Hansen [56] and for critical comments
see [55]. Covariant criteria for radiation, (c), have also been given

(*e.g.* in [57]). However, in spite of all this we do not have a rigorous theory of isolated systems in GR since it is not known whether solutions of the field equation (21) exist which (d) contain several spatially compact bodies and (e) are asymptotically flat in any of the proposed senses. Also, those parts of the theory which have been carried out rigorously (theory of bodies, asymptotics, initial value problem) have not been connected with each other or with approximation methods. (Work in these directions is beginning now, however.) To do this requires a combination of differential geometry, functional analysis, non-linear partial differential equations, guided by physical intuition. This appears to be a challenging field of research where real observations can be confronted with richly structured theory. Combination of "analytical" with numerical work in this area has been started [58] and is likely to play a major part in the development of the subject.

A question which has attracted much attention is whether the total 4-momentum (Arnowitt-Deser-Misner, or ADM-4 momentum), defined in terms of asymptotically flat initial data (see section 6), is timelike if the source-data satisfy an energy-condition. Partial (affirmative) answers to this "positivity-of-energy problem" have been given by Y. Choquet-Bruhat and J. Marsden [59] , and by R. Schoen and S.T. Yau [60] , after a lot of work which paved the way. A similar, physically even more important question has been raised with respect to the Bondi-Sachs 4-momentum at future null infinity the change of which accounts for the 4-momentum carried out of the system by gravitational radiation. It has not been answered yet. From the point of view of physics it would seem to be particularly important to establish an *energy-momentum balance between radiation and sources* which uses covariantly defined concepts of GR only and goes beyond merely formal approximation calculations [8] . This problem, and related ones, have been discussed in [55] .

The *asymptotic structure of spacetime* is closely related to the theory of *black holes*. One may consider systems consisting of black holes rather than ordinary bodies. These questions will not be considered here, however, except for the remark that black holes are probably the simplest objects of relativistic mechanics, corresponding to mass points in Newtonian gravitational theory.

If a spacetime is asymptotically flat in a sufficiently strong sense, then its *asymptotic symmetry group* can be taken to be the Poincaré group, whereas according to weaker definitions the asymptotic symmetry group is infinite-dimensional. (For details see, *e.g.*, [56]) It is not yet known which type of asymptotic flatness corresponds to which class of physical systems. But at any rate *this* asymptotic role of the Poincaré group is not related to the empirically well-founded "macroscopically local" validity of SR. (A similar remark applies, in my opinion, to the Poincaré group as a holonomy group.)

8. Connections Between GR And Observations

Some connections between GR and observations and experiments have been mentioned in sections 2 and 7. Here some supplementary remarks will be made. More detailed surveys, references to the original papers and discussions of future prospects of testing relativistic properties of gravity can be found, *e.g.*, in the reviews by J.J. Shapiro and C.M. Will[7].

A number of very precise *null experiments* can be interpreted as providing evidence for the existence of one and only one second rank tensor field coupling directly and universally to matter, called the metric of spacetime; or for exactly one symmetric linear connection (or the path-structure determined by it) compatible with that metric; or for the universal form of non-gravitational, local physical laws in local inertial frames of that metric; or for the constancy of scalar parameters occuring in those laws. Such arguments, and the figures of merit assigned to the experiments, depend on the a priori theoretical framework adopted to analyze the experiments, and are not "purely empirical", of course. Consider the *Eötvös experiment*, *e.g.* If one uses Newtonian mechanics to set up the equilibrium conditions for the torsion balance which is attached to a non-inertial, rotating frame, assigning inertial masses m_i and independent passive gravitational masses m_g to the test bodies, one can state the result of the Braginski-Panov experiment as saying that the ratios $\frac{m_g}{m_i}$ for Aluminium and Platinum differ from each other by less than $2 \cdot 10^{-12}$. Using the weak-field, slow motion approximation of GR - essentially eqs. (15) combined with the analogue of eq. (17) for an elastic body - to analyze the experiment, amounts to essentially the same thing, except that now necessarily one obtains 1 instead of $\frac{m_g}{m_i}$. *I.e.*, to obtain a figure of merit one has to analyze the experiment by means of a theory which differs from the one to be tested in at least one-logically arbitrary-assumption which has to be stated. For lack of space I shall list below some important experimental results in the customary jargon and only indicate the underlying assumptions (which are stated in Will [7]). The *Hughes-Drever experiment* can be said to show that nucleonic masses are isotropic to within 10^{-23}, or that electromagnetic interactions of nucleons (with each other and external fields) in an external static spherically symmetric gravitational field are locally Lorentz invariant to high accuracy. The *terrestrial redshift measurements* by Pound, Rebka and Snider can be said to show that a small frame of reference in which Maxwell's equations are valid in their standard, SR-form, has the same acceleration relative to a terrestrial laboratory as a freely falling test particle to within 10^{-2}. Limits on the constancy of weak, electromagnetic and strong coupling constants have been inferred from fission yields of old radioactive substances.

Measurements testing the *Newtonian approximation to the metric* in weak-field, quasistationary systems via the second-order Doppler and/or gravitational time dilation effects include solar redshift measurements and the Vessot-Levine rocket experiment (accuracy $2 \cdot 10^{-4}$).

The light-deflection and radar-time-delay measurements are used to test some of the *linear contributions to the metric in the first post-Newtonian approximation*. Here the largest accuracy, of 0.1%, has been reached in time-delay observations (Shapiro 1979).

The value of the "anomalous" perihelion advance of Mercury is used to check the only *non-linear contribution to the metric* (in fact, to g_{oo}) *in the first, post-Newtonian approximation*. The GR-value agrees with observations to within 0.5%, provided the contribution due to the solar quadrupole moment is neglected. This last assumption is still a matter of controversy.

The value of the rate of change of the orbital period of the binary pulsar can be used to test equations purporting to describe the *gravitational radiation reaction* effect on the emitting material sources. The observed value (see section 7) is in agreement with a straightforward application of Einstein's "quadrupole formula" to within the observational error of about 20%. Whether this derivation is theoretically correct is a matter of controversy, as was mentioned before; but at any rate the measurements by J.H. Taylor, L.A. Fowler and P.M. McCulloch represent a breakthrough into the domain of observable gravitational radiation effects, sixty years after Einstein's theoretical prediction of the existence of gravitational waves.

In view of the preceding summary of some observational tests of GR I should like to end this review by quoting Albert Einstein [61] :

> *The scientific theorist is not to be envied. For Nature,*
> *or more precisely experiment, is an inexorable and not*
> *very friendly judge of his work. It never says "Yes" to*
> *a theory. In the most favourable cases it says "Maybe",*
> *and in the great majority of cases simply "No". If an*
> *experiment agrees with a theory it means for the latter*

"Maybe", and if it does not agree it means "No". Probably every theory will some day experience its "No" - most theories, soon after conception.

Einstein's theory of gravitation has not received its "No" yet.

Notes and References

1. For a concise, historically oriented account of the conceptual development leading to the special and general theories of relativity, see, *e.g.* the classic review [1] by H. Weyl and the very succinct sections 1 and 50 in W. Pauli's article [33].

2. This intuitive idea can be formalized by defining events as minimal Cauchy pre-filters of a lattice of processes [63].

3. For mathematical details see, *e.g.*, [2].

4. Historically, this terminology is inappropriate. "Metrics" of this kind seem to have been considered and applied in Physics first by H. Minkowski [3].

5. A mathematically precise formulation of this and other "principles", in the spirit of mathematical physics, has been proposed by A. Trautman [4].

6. One could also assume with Weyl [12], that instead of a metric only a conformal structure is given, so that the relative scales of times, lengths, masses etc. in different, infinitesimal regions are not fixed a priori by the choice of the geometry, but are to be determined from the dynamics. This intriguing idea has, however, not lead to useful physics so far, and will therefore not be pursued here.

7. In this context Einstein has remarked (see [62] , Autobiographisches, footnote on p. 76):"Bei der engeren Gruppe zu bleiben und gleichzeitig die kompliziertere Struktur der allgemeinen Relativitätstheorie zugrunde zu legen, bedeutet eine naive Inkonsequenz. Sünde bleibt Sünde, auch wenn sie von sonst respektablen Männern begangen wird."

8. A promising approach to this problem has recently been made by M. Streubel and R. Schattner [28] .

[1]   Weyl, H.: 50 Jahre Relativitätstheorie. Naturwissenschaften **38**, 73 (1950).

[2]   Hawking, S.W. and G.F.R. Ellis: The large scale structure of spacetime. Cambridge 1973.

[3]   Minkowski, H. Lecture at the Mathem. Society in Göttingen, 1907; Ann.d. Phys. **47**, 927 (1915). Lecture at the Assembly of Natural Scientists in Cologne, 1908; Phys. Ztsch. **10**, 104 (1909)

[4]   Trautman , A. p. 179 in Mehra, J., Ed.: The Physicist's Conception of Nature. Dordrecht 1973.

[5]   Einstein, A., Jahrb.f.Rad. u. El. **4**, 411 (1907).

[6]   Einstein, A., Ann. d. Phys. 49, 769 (1916).

[7]   C.M. Will, p. 1 in Bertotti, B., Ed.: Experimental Gravitation, New York 1974; also: p. 24 in Hawking, S.W. and W. Israel, Eds.: General Relativity, Cambridge, 1979; Shapiro, J.J., contribution to the Proceedings of the Einstein Centennial Symposium held in Princeton, to be published by Princeton University Press.

[8]   Ehlers, J. and E. Köhler, J. Math. Phys. 18, 2014 (1977).

[9]   Coleman, R.A. and H. Korte, Jet Bundles and Path Structures, to appear in J. Math. Phys.

[10] Ehlers, J., and A. Schild, Commun. Math. Phys. 32, 119 (1973).

[11] Ehlers, J., F.A.E. Pirani and A. Schild, p. 63 in O'Raifeartaigh,L., Ed.: General Relativity, Oxford 1972.

[12] Weyl, H., Sitz. Ber. Preuß. Akad. Wiss. Berlin 465, 1918.

[13] Geroch, R.P., 7 in Sachs, R.K., Ed.: General Relativity and Cosmology, New York, 1971.

[14] Streater, R.F. and A.S. Wightman: PCT, Spin And Statistics, And All That, New York, 1964.

[15] Trautman, A., contribution to the Einstein Centennary Volume to be published by Cambridge University Press.

[16] Geroch, R.P., J. Math. Phys. 9, 1739 (1968); 11, 343 (1970).

[17] Penrose, R., in DeWitt, C. and J.A. Wheeler, Eds.: Battelle Recontres in Mathematics and Physics: Seattle 1967, New York, 1968.

[18] Rauch, H., A. Zeilinger, G. Badurek, A. Wilfing, W. Bauspiess and U. Bonse, Phys.Lett. 54A, 425 (1975); Werner, S.A., R. Collela, A.W. Oberhauser and C.F. Eagin, Phys.Rev.Lett. 35, 1053 (1975); Klempt, E., Phys.Rev. D13, 3125 (1976).

[19] Ehlers, J., p. 1 in Israel, W., Ed.: Relativity, Astrophysics and Cosmology, Dordrecht, 1973.

[20] Ehlers, J., p. 213 in Shaviv, G. and J. Rosen, Eds.: General Relativity and Gravitation, New York 1975.

[21] Sachs, R.K. and H. Wu: General Relativity for Mathematicians, New York-Heidelberg-Berlin, 1977.

[22] Carter, B. and H. Ruintana, Proc. Roy. Soc. London A331,57 (1952).

[23] Maugin, G., Gen.Rel.Grav. 4, 241 (1973); 5, 13 (1974).

[24] Israel, W. and J.M. Stewart, Ann. of Physics 118, 341 (1979).

[25] Candelas, P. and D.W. Sciama, Gen.Rel.Grav. 9, 183 (1978).

[26] Isham, C., Annals of the New York Academy of Sciences 302, 114 (1977).

[27] Gibbons, G.W., p. 639 in Hawking, S.W. and W. Israel, Eds.:General Relativity, Cambridge, 1979.

[28] Streubel, M. and R. Schattner: Connection between asymptotic structure and material sources, preprint, Max-Planck-Institut für Physik und Astrophysik, München, Sept. 1979.

[29] Dixon, G., p. 156 in Ehlers, J., Ed.: Isolated Gravitating Systems in General Relativity, Amsterdam, 1979.

[30] Christoffel, E.B., Crelle's Journ. 70, 46 (1869).

[31] Lovelock, D., J. Math. Phys. 13, 874 (1972).

[32] Hilbert, D., Nachr. Ges. Wiss. Gött. p. 395, 1915.

[33] Pauli, W.: Relativitätstheorie in Enz. d. mathem.Wissensch.V/19, Leipzig 1921.

[34] Mehra, J., p. 92 in Mehra, J., Ed.: The Physicist's Conception of Nature, Dordrecht, 1973.

[35] Choquet-Bruhat, Y., Commun. Math. Phys. 12, 16 (1969).

[36] Misner, C.W., K.S. Thorne and J.A. Wheeler: Gravitation, San Francisco, 1973.

[37] Witten, L., Ed.: Gravitation: an introduction to current research, New York, 1962. (See ch. 2)

[38] Marder, L., Proc. Roy. Soc. A244; A246, 133 (1958).

[39] Robinson, J. and A. Trautman, Proc. Roy, Soc. A265, 463 (1962); Robin, J. and J.R. Robinson, Int. J. Theor. Phys. 2, 231 (1969).

[40] Fischer, A.E. and J.E. Marsden, in Hawking, S.W. and W. Israel, Eds.: General Relativity, Cambridge, 1979.

[41] Choquet-Bruhat, Y. and R.P. Geroch, Commun. Math. Phys.14, 329 (1969).

[42] Hulse, R.A. and J.H. Taylor, Ap. J. Lett. 195, L51 (1975).

[43] Einstein, A., L. Infeld and B. Hoffmann, Ann. of Math. 39, 65 (1938). - See also Kerr, R.P., Nuovo Cimento 16, 26 (1960).

[44] Taylor, J.H., L.A. Fowler and P.M. McCulloch, Nature 277, 437 (1979).

[45] Barker, B.M. and R.F. O'Connell, Phys. Rev. D12, 329 (1975).

[46] Börner, G., J. Ehlers and E. Rudolph, Astron. Astrophys. 44, 417 (1975).

[47] McCrea, J.D. and G. O'Brian, Gen. Rel. Grav. 9, 1101 (1978).

[48] Caporali, A. and N. Spyrou, to be published.

[49] Van Patten, R.A. and C.W.F. Everitt, Phys.Rev.Lett. 36, 629 (1976). See also later progress reports.

[50] Ehlers, J., in Proceedings of the Intern. School of Gen.-Rel. Effects in Physics and Astrophysics: Experiments and Theory (3rd Course). Max-Planck-Institut für Physik und Astrophysik, MPI-PAE/Astro 138 (1977).

[51] Kerlick, D., to appear in Gen. Rel. Grav.

[52] Chandrasekhar, S. and F.P. Esposito, Ap. J. 160, 153 (1970).

[53] Taylor, J.H., lecture presented at the Second Marcel Großmann Meeting on General Relativity and Gravitation, ICTP, July 1979.

[54] Ehlers, J., A. Rosenblum, J.N. Goldberg and P. Havas, Ap. J. 208, L77 (1976).

[55] Ehlers, J., contribution to the Proceedings of the Ninth Texas Symposium on Relativistic Astrophysics, to be published in the Annals of the New York Academy of Sciences.

[56] Ashtekar, A. and R.O. Hansen, J. Math. Phys. 19, 549 (1978).

[57] Sachs, R.K., Proc. Roy, Soc. A270, 103 (1962); Leipold, G. and M. Walker, Ann. Inst. Henri Poincaré 27, 61 (1977).

[58] Smarr, L., Annals of the New York Acad. of Sciences 302, 569 (1977).

[59] Choquet-Bruhat, Y. and J. Marsden, Commun. Math. Phys. 51, 283 (1976).

[60] Schoen, R. and S.T. Yau, Commun. Math. Phys. 65, 45 (1979).

[61] Quoted from Hoffmann, B. and H. Dukas: Albert Einstein, the human side, Princeton, 1979.

[62] Schilpp, P.A., Ed.: Albert Einstein, Philosopher - Scientist, p. 76, New York, 1949.

[63] Mayr, D., Dissertation, University of Munich, 1979; See also Yodzis, P., Proc. Roy. Irish Acad. A75, 37 (1975).

RECENT ADVANCES IN GLOBAL GENERAL RELATIVITY: A BRIEF SURVEY

Roger Penrose, Mathematical Institute, Oxford, U.K.

Theoretical research into Einstein's general relativity has made some
very significant advances over the past 20 years or so. During this
period, attention has been concentrated almost entireley on Einstein's
original 1915 theory rather than on the various later attempts to generalize it to some kind of unified field theory. This research has revealed a number of desirable and remarkable properties, both physical
and mathematical, that the original theory possesses. (It is hard to
see that many of these could survive if the theory were to be generalized, unless new insights, as deep as those that led Einstein to his
1915 theory, may eventually be invoked in making such a generalization.)

Many of these recent advances have been into the global rather than the
local structure of the theory. Early work into general relativity had,
to a large degree, been concerned with finding exact solutions to the
field equations - always a difficult task owing to the complicated non-
linear nature of these equations and to the fact that establishing
equivalence between two such solutions depends upon making general coordinate transformations. In the study of global questions, however, one
is not concerned with particular exact solutions but with general qualitative properties. The detailed solution of Einstein's equation becomes
of secondary importance and attention is concentrated, instead, on geometrical and topological properties of space-times.

One of the topics that has received a great deal of attention in recent
years is the study of black holes. Here, exact solutions, notably the
Schwarzschild solution and its generalization to that of Kerr, *do* also
play important roles, first as models indicative of the kind of behaviour
to be expected in more general situations, and second, remarkably, as the
final states into which *all* black holes finally settle down.

The familiar Schwarzschild solution, as extended to within the Schwarzschild radius $r = 2m$ by means of, say, Eddington-Finkelstein coordinates [1,2], provides the basic (spherically symmetrical) model of the
final collapse of a massive star to form a black hole (see Fig. 1). This
model was put forward by Oppenheimer and Snyder [3] in 1939, but it had
not been clear to what extent its two salient features, namely the horizon at $r = 2m$ and the curvature singularity at $r = 0$, were stable
properties under perturbations of the model or mere artefacts of the

very special assumption of exact spherical symmetry.

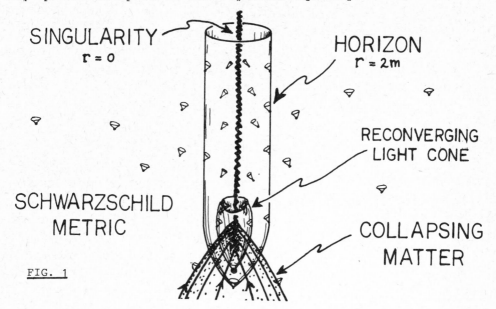

SINGULARITY
r = 0

HORIZON
r = 2m

RECONVERGING
LIGHT CONE

SCHWARZSCHILD
METRIC

COLLAPSING
MATTER

FIG. 1

In fact, with regard to the singularity it has been possible to show [4,5,6] using global techniques of differential topology, that the presence of a space-time singularity is, indeed, a stable property. For this one requires a suitable criterion signifying that, in some suitable sense, collapse past a point of no return has actually taken place. The existence of a "trapped surface" is one such criterion [4], and of a point whose future light cone eventually begins to reconverge in every direction [5], is another. These criteria can also be used in the reverse direction in time in order to establish that the big-bang singularity, also, is not merely a feature of the high symmetry that is assumed for the standard cosmological models, but is a rigorous consequence of Einstein's thoery [7,5,6] as applied to (seemingly) very convincing astrophysical data (such as the nature of the 3^0K background radiation). Such singularity theorems can also be used in other contexts and are very broadly [6] applicable. But they are somewhat negative in the sense that they give very little hint as to the detailed nature of the predicted singularities. In recent years there have been some improvements in this (and other) respects [8], over the original results.

The status of the existence of a horizon in asymmetric gravitational collaps is not so clear however. The question is whether the singularities that arise in collapse can ever be "naked", $i.e.$ visible to an observer at large distances. If there is some general principle preventing this (according to the laws of classical general relativity), at least in cases of "generic" collapse, then we say that *cosmic censorship* holds true [9], and the forming of horizons about singularities is a necessary consequence of the theory. But at present, despite some encouraging partial evidence to support it, cosmic censorship remains an unproved conjecture [8] probably the most important unsolved problem of classical general relativity theory.

The general argument that irretrievable gravitational collapse always results in a black hole rather than in a naked singularity does, strictly speaking, require the cosmic censorship assumption. However, most astrophysicists seem quite prepared to accept the former possibility as the "lesser of two evils" - and what theoretical evidence that there is seems, so far, to support them. Having accepted that black holes can exist in very general situations of gravitational collapse, it may be further presumed that, if left undisturbed on its own, a black hole will settle down into a stationary configuration. Granting this, one can then appeal to a succession of remarkable theorems [10-13] which, in effect, establish the precise form of the resulting space-time metric as that known as the *Kerr metric* [14,6], depending upon only two parameters a and m, where m refers to the mass and am, to the angular momentum ($a = 0$ giving the Schwarzschild metric). The gravitational field of the initial collapsing body would require many more parameters (infinitely many, in principle,) for its detailed description, but almost all this information gets washed out in the final black-hole state, the quadrupole and higher multipole moments being radiated away in gravitational radiation emitted as the configuration finally settles down to that of Kerr.

An unexpected bonus, from the computational point of view, is that the geodesic equations can be reduced simply to quadratures in the Kerr metric [15], owing to the presence of an extra constant of the motion for particle orbits (Carter's constant), and this helps immensely in the analysis of the metric. A further constant for null orbits enables the rotation of light polarization to be integrated explicitly [16]. Furthermore, the wave equation [15] and its generalization for all the various spins [17,18], and also the Dirac equation [19], all separate in the Kerr background, so that the solution of these equations becomes a comparatively straightforward matter. All this seems a remarkable gift from Einstein's 1915 theory!

When gravitational waves are emitted by an isolated system, some energy is lost by the system. Indeed, it seems to have been this demonstration, by Bondi [20] and Sachs [21] in around 1960, that gravitational waves carry positive mass-energy, that largely initiated the renewed interest into general relativity of more recent years. Since gravitational energy cannot be localized, the study of such questions requires some form of asymptotic analysis. A convenient way of handling asymptotic questions in general relativity is the introduction of a conformal boundary to space-time, yielding a smooth conformal manifold-with-boundary in most of the cases of interest [22]. For a radiating isolated system this boundary consists of two pieces I^- and I^+, each of which is a null hypersurface of topology $S^2 \times \mathbb{R}$ (see Fig. 2). The geometry and other data at I^- describe the incoming radiation fields and at I^+, the outgoing radiation fields. In this way, the Bondi-Sachs energy-momentum law, and certain generalizations such as to incorporate other massless fields interacting with gravity, may be simply derived [23].

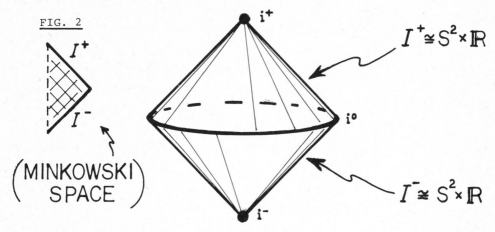

FIG. 2

I^+

I^-

$\left(\begin{array}{c}\text{MINKOWSKI} \\ \text{SPACE}\end{array}\right)$

i^+

$I^+ \cong S^2 \times \mathbb{R}$

i^0

$I^- \cong S^2 \times \mathbb{R}$

i^-

Important problems remain to be solved, however, such as establishing, in an unambiguous way, the relation between the Bondi-Sachs energy-momen-tum and the more familiar energy-momentum integral [24,25] defined at spacelike infinity (though significant advances have been made recently [26]), the establishment of a "positive energy theorem" that would ensure that the total Bondi-Sachs energy is always non-negative (which would considerably strengthen the results already established [27] for the more familiar spacelike "positive energy conjecture"), and obtaining a proper understanding of angular-momentum and mass-centre in general relativity. The difficulty that arises in connection with this last problem occurs because the rotation elements of the Poincaré group cannot be uniquely identified as asymptotic symmetries of a general asymptotically flat space-time. The problem does not arise for the translation elements, which is why a unique energy-momentum concept does arise. But the con-cept of an asymptotic rotation is (retarded) time-dependent, so there seems to be no clear way of identifying the different angular momentum concepts that are defined for different (retarded) times [28,29].

The origin of all these difficulties is the fact that the group of asymptotic symmetries of an asymptotically flat space-time, defined as the appropriate conformal symmetries of I^+ (or of I^-), is not the Poincaré group, but an infinite-parameter group known as the Bondi-Metzner-Sachs (BMS) group [20,28], in which the Poincaré rotation elements cannot be uniquely singled out. This prompted Newman to initiate another approach to asymptotic "symmetries" involving a complexification of the asymptotic space-time parameters [30,31]. Although this has not, as yet, led to a clarification of angular momentum it has had a number of other interesting consequences. It provides a remarkable concept known as H-space - which may be regarded as the complex "vacuum space-time" reconstructed solely from the self-dual part of the gravitational radiation field. This approach has some remarkable mathematical aspects to it, but its physical significance remains somewhat obscure. One of its important properties is that it provides a direct link with the theory of twistors [32] and points the way towards a general construction of non-linear fields in terms of deformations of twistor space (or of bundles over twistor space) [33,34].

Another instance of the value of introducing conformal boundaries to space-time is in the study of cosmological models. Here, the boundary can describe either a space-time singularity or infinity (Fig. 3). In the case of the big-bang singularity we obtain a conformal boundary which is a spacelike hypersurface. This provides a simple understanding of the *particle horizons* [35,36] that these models possess, whereby an observer in the early stages of the expansion can receive signals only from a limited, but steadily increasing, number of particles in the universe.

FIG. 3

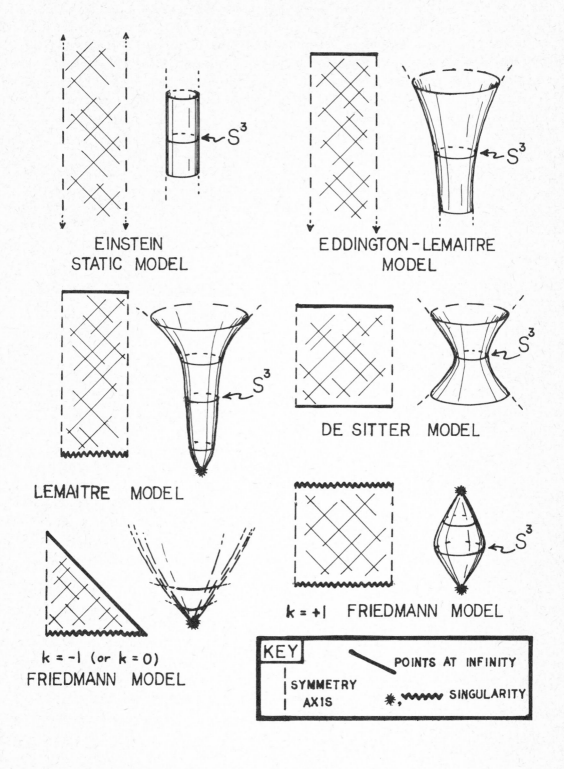

EINSTEIN
STATIC MODEL

EDDINGTON-LEMAITRE
MODEL

S^3

S^3

LEMAITRE MODEL

S^3

DE SITTER MODEL

S^3

$k = +1$ FRIEDMANN MODEL

S^3

$k = -1$ (or $k = 0$)
FRIEDMANN MODEL

KEY

SYMMETRY
AXIS

POINTS AT INFINITY

SINGULARITY

Several different methods [6,8] have been suggested for attaching boundaries to space-time. The one that I have been adopting here, the *causal
boundary* [37], assigns the same future end-point to two timelike or null
curves if and only if their (chronological) pasts coincide; and, accordingly, the same past end-point whenever their (chronological) futures
coincide. Another method that had gained widespread acceptance is the
b-boundary [38] (which provides only singular points and not points at
infinity) and its conformal generalization [39] (which provides also
points at infinity). This method has encountered certain difficulties
[40], however, when applied to cosmological models, and it is hard to
use in practice. The causal boundary has the advantage of simplicity
and the fact that it relates directly to physically important concepts
(*e.g.* horizons).

One of the implications of the cosmic censorships assumption is the
area theorem for black holes [41,42]. This asserts that the surface
area of the horizon is non-decreasing with time. Any disturbance to a
black hole will serve to increase its surface area. For a Kerr hole,
the formula for this area is

$$A = 8\pi m(m + (m^2 - a^2)^{1/2}) .$$

In the non-rotating case $a = 0$, so A is simply proportional to
m^2 . The mass of a non-rotating black hole thus grows whenever the
hole is disturbed. This need not be so in the rotating case, however,
and it is possible to extract some mass-energy from the hole at the
expense of slowing down its rotation. This can be done either by suitable
projecting particles into the hole [9,43] or by means of external fields
("superradiance") [44,45].

This irreversible behaviour of a black hole's horizon relates to the
second law of thermodynamics. By means of thought experiments [46] and
considerations of quantum field theory in a curved (black-hole) background [47] the Bekenstein-Hawking formula

$$S = \frac{1}{4} \frac{kAc^3}{\hbar G}$$

for the entropy of a black hole of surface area A has been derived
(k = Boltzmann's constant, $\hbar$ = Planck's constant/2π, G = Newton's
gravitational constant and c = velocity of light). This implies that
a black hole should also have a non-zero temperature (inversely proportional to the mass, for a non-rotating hole) though for a black hole

arising in stellar collapse, this temperature would be utterly insigni-
ficant ($\sim 10^{-8}$ °K for a 10 $M_\odot$ hole). In principle, in an ever-expanding
cosmological model, such a black hole would eventually radiate away all
its mass because of this non-zero temperature (Hawking radiation) though
this process would take over 10^{53} times the present age of the universe
for a hole arising from stellar collapse [47]. On the other hand if very
tiny holes (of mass $\sim 10^{15}$g and radius $\sim 10^{-13}$cm) were produced in a
highly chaotic big bang, then their final explosive disappearance could
have observational significance [48]. (It should be remarked that the
area theorem of classical general relativity is being violated here,
which is legitimate because the normal local positivity assumption for
energy is violated for *quantum* fields.)

For reasons connected with the second law of thermodynamics and the
nature of gravitational entropy, however, it is highly reasonable to
infer that the big bang singularity was *not* of such a chaotic nature
[42,49], so such small black holes would seem to be very improbable.
(This is also consistent with direct observation [50].) A suitably
strong version of the cosmic censorship assumption implies that singularities divide themselves into two quite distinct types: past and
future. (This is essentially equivalent to the assumption that the universe is globally hyperbolic [42,49], which incidentally implies that
the topology of the universe is unchanging with time [51].) The past-type
singularities would apparently have to be uniform and unchaotic in nature
while the future singularities would not be so constrained and would indeed be expected to be very chaotic and non-uniform.

No prior theoretical reason for such an asymmetrical behaviour of singularities has been put forward, however. It would seem that the answer
to this puzzle (which would include an answer to the question of the
origin of the second law of thermodynamics) must lie in the finding of
a correct theory that unifies general relativity with quantum mechanics.
Unfortunately, despite many heroic efforts, we seem as far away from such
a theory as ever [52].

References

[1] A. S. Eddington, Nature 113 (1924) 192

[2] D. Finkelstein, Phys. Rev. 110 (1958) 965

[3] J. R. Oppenheimer and H. Snyder, Phys. Rev. 56 (1939) 455

[4] R. Penrose, Phys. Rev. Lett. 14 (1965) 57

[5] S. W. Hawking and R. Penrose, Proc. Roy. Soc. A314 (1970) 529

[6]   S. W. Hawking and G. F. R. Ellis, The Large Scale Structure of Space-time (Camb. Univ. Press 1973)

[7]   S. W. Hawking and G. F. R. Ellis, Astrophysics J. $\underline{152}$ (1968) 25

[8]   F. J. Tipler, C. J. S. Clarke and G. F. R. Ellis, Singularities and Horizons - a Review Article (to appear in G R G Einstein Centenary Volume 1979).

[9]   R. Penrose, Revista del Nuovo Cimento, Serie I, $\underline{1}$ Numero Speciale (1969) 252

[10]   W. Israel, Phys. Rev. $\underline{164}$ (1967) 1776

[11]   B. Carter, Phys. Rev. Lett. $\underline{26}$ (1971) 331

[12]   S. W. Hawking, in Black Holes: Les Astres Occlus, Les Houches Summer School 1972 (Eds. C. M. DeWitt and B. S. DeWitt, Gordon & Breach, New York 1973)

[13]   D. C. Robinson, Phys. Rev. Lett. $\underline{34}$ (1975) 905

[14]   R. P. Kerr, Phys. Rev. Lett. $\underline{11}$ (1963) 237

[15]   B. Carter, Commun. Math. Phys. $\underline{10}$ (1968) 280

[16]   M. Walker and R. Penrose, Commun. Math. Phys. $\underline{18}$ (1970) 265

[17]   S. Teukolsky, Ann. New York Acad. Sci. $\underline{262}$ (1975) 275

[18]   S. Chandrasekhar, Proc. Roy. Soc. $\underline{A358}$ (1978) 421, 441

[19]   S. Chandrasekhar, Proc. Roy. Soc. $\underline{A349}$ (1976) 571

[20]   H. Bondi, M. G. J. van der Burg and A. W. K. Metzner, Proc. Roy. Soc. $\underline{A269}$ (1962) 21

[21]   R. K. Sachs, Proc. Roy. Soc. $\underline{A270}$ (1962) 103

[22]   R. Penrose, Proc. Roy. Soc. $\underline{A284}$ (1975) 159

[23]   R. Penrose, in: Relativity Theory and Astrophysics, $\underline{1}$ (ed. J. Ehlers, Amer. Math. Soc. 1967)

[24]   L. D. Landau and E. M. Liftschitz, The classical Theory of Fields (Pergamon Press, Oxford 1962)

[25]   R. Arnowitt, S. Deser and C. W. Misner, in: Gravitation: An Introduction to Current Research (Wiley, New York, 1962)

[26]   A. Ashtekar and R. O. Hansen, J. Math. Phys. $\underline{19}$ (1978) 1542

[27]   R. Shoen and S. T. Yau, Commun. Math. Phys. $\underline{65}$ (1979) 45

[28]   R. K. Sachs, Phys. Rev. $\underline{128}$ (1962) 2851

[29]   R. Penrose, in: Group Theory in Non-Linear Problems (Ed. A. O. Barut, Reidel, Dordrecht 1974)

[30]   E. T. Newman, Gen. Rel. and Grav. $\underline{7}$ (1976) 107

[31]   M. Ko, E. T. Newman and K. P. Tod, in: Asymptotic Structure of
       Space-time (Eds. F.P. Esposito and L. Witten, Plenum, New
       York 1976)

[32]   R. O. Hanson, E. T. Newman, R. Penrose and K. P. Tod, Proc. Roy.
       Soc. A363 (1978) 445

[33]   R. Penrose, Gen. Rel. and Grav. 7 (1976) 31

[34]   R. S. Ward, Phys. Lett 61A (1977) 81

[35]   W. Rindler, Mon. Not. Roy. Astr. Soc. 116 (1956) 6

[36]   R. Penrose, in:Battelle Recontres, 1967 Lectures in Mathematics
       and Physics (Eds. C. M. DeWitt and J. A. Wheeler, Benjamin
       New York 1968)

[37]   R. Geroch, E. H. Kronheimer and R. Penrose, Proc. Roy. Soc.
       A327 (1972) 545

[38]   B. Schmidt, Gen. Rel. and Grav. 1 (1971) 269

[39]   B. Schmidt, Commun. Math. Phys. 36 (1974) 73

[40]   B. Bosshardt, Commun. Math. Phys. 46 (1976) 263

[41]   S. W. Hawking, Commun. Math. Phys. 25 (1972) 152

[42]   R. Penrose in: Theoretical Principles in Astrophysics and Rela-
       tivity (Eds. N. R. Lebovitz, W. H. Reid and P. O. Vandervoort,
       Chicago Univ. Press, Chicago 1978)

[43]   R. Penrose and R. M. Floyd, Nature Physical Science 229 (1971)
       117

[44]   C. W. Misner, Phys. Rev. Lett. 28 (1972) 994

[45]   B. Ya. Zel'dovich, Zh. Eksp. Teor. Fiz. 62 (1972) 2076; Engl.
       Transl.: Soviet Phys.-JETP 35 (1972) 1085

[46]   J. D. Bekenstein, Phys. Rev. D7 (1973) 2333, D9 (1974) 3292

[47]   S. W. Hawking, Commun. Math. Phys. 43 (1975) 199.

[48]   M. J. Rees, Nature 266 (1977) 333

[49]   R. Penrose, in: General Relativity, An Einstein Centenary Survey
       (Eds. S. W. Hawking and W. Israel, Cambridge Univ. Press,
       Cambridge 1979)

[50]   B. J. Carr, Astrophys. J. 206 (1976) 8

[51]   R. Geroch, J. Math. Phys. 11 (1970) 437

[52]   C. J. Isham, R. Penrose and D. W. Sciama, Eds., Quantum Gravity,
       and Oxford Symposium (Clarendon Press, Oxford 1975)

SPEKULATIONEN ÜBER DIE MÖGLICHKEIT EINES NICHT EUKLIDISCHEN RAUMES VOR EINSTEIN

Imre Toth, Universität Regensburg

> *Que esas regiones bárbaras, donde la*
> *tierra es madre de monstruos, pudieran*
> *albergar en su seno una ciudad famosa,*
> *a todos nos pareció inconcebible.*
>
> J. L. Borges, *El Immortal*

1. Nicht-Euklidische Geometrie und Nicht-Euklidische Entwicklungs-Strukturen

Die Theorie der Kegelschnitte wurde von den antiken griechischen Geometern ohne jeden äußeren Anlaß aus innenmathematischen, rein ästhetischen Gründen entwickelt. Nach 2000 Jahren hat sie Kepler in fertigem Zustand vorgefunden und mit ihrer Hilfe die Bewegung der Planeten erklärt. Sowohl die Zeitgenossen als auch die späten Nachfolger Keplers waren von der Tatsache zutiefst beeindruckt, daß die Theoreme eines begrenzten geometrischen Bereichs nun eine völlig unvoraussehbare und erfolgreiche Anwendung in der Beschreibung der Struktur des Sonnensystems gefunden hatten. In seiner Gedenkrede auf Kepler sprach Einstein von der *Bewunderung und Ehrfurcht*, die er vor dieser *rätselhaften Harmonie der Natur*,[1] also vor der Übereinstimmung der Kegelschnittlehre mit den Gesetzen der Planetenbewegung empfindet.

Auch Einstein und Minkowski haben in der Mathematik ihrer Zeit Geometrien nicht-euklidischer Räume vorgefunden, und die Anwendung dieser mathematischen Theorien auf die Beschreibung des physikalischen Universums hat ein vielleicht noch größeres Erstaunen ausgelöst als die Kepler'schen Ergebnisse. *Wie ist es möglich, daß die Mathematik, die doch ein von aller Erfahrung unabhängiges Produkt des menschlichen Denkens ist, auf die Gegenstände der Wirklichkeit so vortrefflich paßt. Kann denn die menschliche Vernunft ohne Erfahrung durch bloßes Denken Eigenschaften der wirklichen Dinge ergründen?* - fragte Einstein in seiner Rede über *Geometrie und Erfahrung*.[2] Und was er einst über Kepler schrieb, betraf eigentlich sein eigenes Werk: *Es scheint, daß die menschliche Vernunft die Formen erst selbständig konstruieren muß, ehe wir sie in den Dingen nachweisen können.*[3]

Man könnte glauben, die beiden Fälle seien nichts als eindrucksvolle Illustrationen eines an sich normalen, sogar banalen wissenschaftshistorischen Phänomens: Ausgehend von einem partikulären Gebiet der

Wirklichkeit wird durch Abstraktion eine allgemeine Theorie gewonnen,
die potentiell die Möglichkeit unvoraussehbarer und unerwarteter Anwendungen auf völlig neuen Gebieten in sich enthält.

Dieser Eindruck ist nicht ganz falsch. Die Analogie ist jedoch begrenzt
und im Grunde genommen irrelevant, sogar irreführend. Wesentlich ist
nämlich der fundamentale Unterschied, der die beiden Fälle einander ent-
gegensetzt. Das Zusammentreffen einer Ebene mit einem Kegel ist ein
höchst realistisches Ereignis und es findet in derselben Welt statt, wo
auch die Planeten die Sonne umkreisen. Dagegen ist das Zusammentreffen
von präfabrizierten nicht-euklidischen Geometrien mit dem physikalischen
Universum auf einer schwarzen Tafel ein im vollständigen Sinne des Wortes surrealistisches Ereignis.[4]

Erstens beschreibt eine solche Geometrie nicht die Eigenschaften eines
einzigen geometrischen Körpers, sondern die vollständige Struktur eines
ontologisch abgeschlossenen Universums geometrischer Objekte. Jede Geometrie ist an sich eine prästabilierte Kosmologie. Jedoch ist das Universum ihrer Objekte mindestens vorläufig, ein *univers* des geometrischen
discours bloß, und so allein für das Denken und nicht für die Beobachtung da. Die nichteuklidischen Geometrien sind also keine Erkenntnisse,
die als abstrakte Endergebnisse kosmologischer Triangulationen im himmli-
schen Niltal[5] oder im irdischen Kidrontal[6] entstanden wären; (s. Figur)

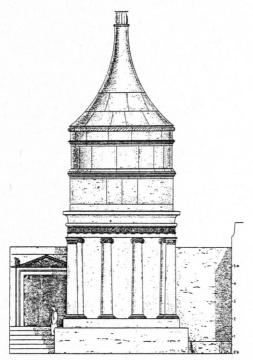

Pseudosphäre im Kidrontal:
Abshalom's Grab

(N. Avigad, The Ancient Monuments
in the Kidron Valley, Fig. 52)

diese deduktiven Satzsysteme sind zwar alle in einem euklidischen Ge-
dankenuniversum entstanden, jedoch war die Absicht, die ihren Aufbau
leitete, weder die nachahmende Abbildung des physikalischen Universums,
noch die Beschreibung möglicher Welten, sondern die detaillierte Darstellung eines (mindestens in bezug auf den euklidischen) völlig unmöglichen Raumes, eine Beschreibung des Nichtseienden - mit absoluter
Genauigkeit durchgeführt.[7]

Die Wissenschaft gilt im allgemeinen als eine seriöse Angelegenheit, da
es dem Wissenschaftler nicht erlaubt ist, sich bewußt mit nicht existierenden und noch weniger mit Gegenständen zu beschäftigen, die von vornherein als unmöglich gelten. Solche Exkursionen in das Imaginäre und das
Unmögliche sind nur einem Künstler erlaubt. Es ist jedoch eine historische Tatsache, daß in dem Jahrhundert zwischen 1733 und 1826 Saccheri,
Lambert, Schweikart, Taurinus und unabhängig von ihnen Thomas Reid,[8]
ein System entwickelt haben, dessen Sätze formale Negationen der strikt
euklidischen Theoreme sind, und so eine Welt beschreiben, die nur in der
Imagination ihrer Verfasser enthalten war. Saccheri glaubte noch, daß
dieses System einen inneren Widerspruch enthalten müßte. Lambert war von
seiner Inkonsistenz schon nicht mehr überzeugt, während Schweikart und
Taurinus bereits dessen Konsistenz zugegeben haben; letztere nahmen jedoch an, alle seine Sätze seien falsch. Um im folgenden die Ausdrucksweise zu vereinfachen, werde ich die Gesamtheit der den euklidischen
formal entgegengesetzten Sätze mit dem (wahrscheinlich von Gauss und
Bolyai stammenden) Ausdruck *antieuklidisches System*[9] bezeichnen, -
falls dem euklidischen System der Wert der Wahrheit und dem ihm entgegengesetzten der Wert der Falschheit zugeordnet wird. Vor der Begründung
der eigentlichen nichteuklidischen Geometrie haben nicht nur Saccheri,
Lambert, Taurinus, sondern auch Gauss, Bolyai und wahrscheinlich auch
Lobatchevsky ein antieuklidisches System entwickelt; mit der Zeit hatten
sie bereits seine Konsistenz erkannt, ihm jedoch den Wert der Falschheit zugeordnet. Das antieuklidische System wurde so im Bewußtsein seines fiktiven Charakters und seiner Falschheit, quasi als literarisches
Werk entwickelt.

Es ist in der Tat einleuchtend, daß die Negation des euklidischen Parallelenaxioms notwendig falsch sein muß, wenn das Parallelenpostulat
selbst von vornherein als wahr angenommen wird. Jedoch kam es darauf
an zu beweisen, daß die Wahrheit, die dem euklidischen Parallelenaxiom
zukommt, die Modalität der Notwendigkeit an sich absolut besitzt. Der
Terminus *notwendig* bezeichnet das Nicht-anders-sein-können; sein Gegensatz stellt die schlichte Unmöglichkeit dar. Der Modalität der Unmöglichkeit entspricht im Bereich der Satzsysteme die Inkonsistenz oder

Absurdität. Nur wenn das antieuklidische System inkonsistent ist, darf behauptet werden, daß das euklidische Postulat eine notwendige Wahrheit darstellt. In diesem, und nur in diesem Fall, kommt der Wert der Wahrheit ausschließlich und allein dem euklidischen und keinem anderen Satz zu.

Das Ziel, Unizität und Notwendigkeit der euklidischen Wahrheiten zu beweisen, kann sowohl von historischem als auch von epistemologischem Standpunkt aus als hinreichender Grund für die Entwicklung des antieuklidischen Systems gelten. Die erfolgreiche Durchführung eines solchen Inkonsistenzbeweises hätte uns die Möglichkeit gegeben, zwischen Geometrie als Literatur und Geometrie als Wissenschaft mit Hilfe eines objektiven Kriteriums zu unterscheiden. Das hätte bedeutet: im Bereich der Geometrie muß die Fiktion von alleine zusammenbrechen; die Falschheit einer geometrischen Aussage, auch wenn diese von vornherein nicht erkannt wurde, muß dadurch offenbar werden, daß sie zu zwei unvereinbaren Folgerungen führt. Das ist jedoch in der Literatur nicht der Fall. Der Text eines Romans kann in bezug auf die Wirklichkeit durchaus falsch sein, muß jedoch keinen Widerspruch enthalten.

Diese Hoffnung ist nicht in Erfüllung gegangen. Die Inkonsistenz des antieuklidischen Systems konnte und kann auch nicht bewiesen werden. Im Gegenteil: Es wurde bewiesen, daß die Inkonsistenz des antieuklidischen, also des falschen Systems unvermeidlich die Inkonsistenz des wahren, des euklidischen Systems nach sich ziehen würde. Die logische Solidarität von Fiktion und Wirklichkeit, die Verbundenheit auf Leben und Tod eines Universums lauter wahrer mit dem negierten Universum lauter falscher Sätze - dies ist ein erstaunliches, in der Literatur völlig unbekanntes Phänomen. Es wäre in der Tat ungewöhnlich, aus der Inkonsistenz einer frei erfundenen, apokryphen Biographie auf die Inkonsistenz der ihr entgegengesetzten wahren Biographie zu schließen. Wegen dieser Unmöglichkeit eines Inkonsistenzbeweises also konnte die Geometrie das antieuklidische System trotz seiner evidenten Falschheit nicht mit Hilfe einer Immunreaktion automatisch aus seinem Körper eliminieren. Damit hat sich jedoch die Motivation, die den Aufbau des antieuklidischen Systems ursprünglich rechtfertigte, als hinfällig erwiesen. Das antieuklidische System stand somit unzerstört da, ohne sein bloßes Dasein und noch weniger seine Zugehörigkeit zur Geometrie mit irgendeinem ethischen, wissenschaftlichen oder pragmatischen Grund rechtfertigen zu können. Hätte man dem antieuklidischen System den Status einer Geometrie zugeordnet, so hätte man damit stillschweigend eine immerhin *more geometrico* gestaltete literarische Tätigkeit für den weiteren Ausbau der Geometrie erlaubt. Wäre es so geschehen, wäre, im Grunde genommen, noch

alles in Ordnung geblieben, - so lange allein die natürliche Geometrie
als wirklich und wahr, die künstliche Geometrie der literarischen Fiktion als irreal und falsch erklärt worden wäre.

Mit der Begründung der eigentlichen nichteuklidischen Geometrie ist jedoch eine völlig unerwartete und auch völlig unvoraussehbare Situation
entstanden. Um die nichteuklidische Geometrie begründen zu können, mußte
den antieuklidischen Sätzen ohne jeden Beweis der Wert der Wahrheit zugeordnet werden. Das ist aber noch nicht alles. Die Zuordnung der Wahrheit ist zwar notwendig, jedoch nicht hinreichend, um die nichteuklidische Geometrie zu etablieren. Es hätte nämlich sein können, daß in diesem Fall die euklidische die falsche Geometrie geworden wäre. Mit der
Konstituierung der nichteuklidischen Geometrie wurde jedoch die euklidische Geometrie nicht als falsch eliminiert, sondern weiterhin als wahr
bezeichnet und zusammen mit dem entgegengesetzten System als gleichbe-
rechtigter Bestandteil der mathematischen Wissenschaft aufbewahrt. Mit
dem Terminus *nichteuklidisch* werde ich die den euklidischen formal entgegengesetzten Sätze dann und nur dann bezeichnen, wenn ihnen die Wahrheit simultan mit den ursprünglichen euklidischen Sätzen zugeordnet ist.
Die eigentliche nichteuklidische Geometrie ist als Ergebnis einer son-
derbaren historischen Entwicklung entstanden, die eine der euklidischen
Genesis entgegengesetzte, eine *nichteuklidische Entwicklungsstruktur*,
aufweist. Es ist, als ob ein Buch, dessen Verfasser als Literat durch
Negation der biologischen Eigenschaften reeller Tiere bewußt Fabeltiere
fingiert hätte, - auf einmal, von seinem Verfasser selbst, neben und
zusammen mit dem großen Brehm zu einem gültigen, echt wissenschaftlichen Lehrbuch der Zoologie erklärt worden wäre.[10]

Es stellt sich die Frage, was kann unter diesen Umständen nicht allein
die Entstehung, sondern darüber hinaus die Aufnahme der nicht-euklidischen Geometrie historisch, epistemologisch oder ethisch rechtfertigen?
Kann der sonderbare Wunsch, einem fingierten Raum Wirklichkeit, einem
im Bewußtsein der Falschheit ausgesprochenen Satz Wahrheit zuzuordnen,
als hinreichend betrachtet werden, die Konstituierung der nicht-euklidischen Geometrie historisch zu erklären und wissenschaftstheoretisch
zu motivieren?

2. <u>Das Parallelenproblem: Die historische Dynamik seiner Entwicklung
   und seine Rolle bei der Vorbereitung der nicht-euklidischen Geo-
   metrie</u>

Die nichteuklidische Geometrie ist in der geschichtlichen Entwicklung
als abschließender Akt der reiterierten und immer wieder gescheiterten
Lösungsversuche des sogenannten Parallelenproblems entstanden. Wie ich

im folgenden nachweisen möchte, ist durch das Parallelenproblem allein
die Konstituierung der nicht-euklidischen Geometrie allerdings nicht
erklärbar. Das Parallelenproblem betrifft die folgende metamathemati-
sche Frage: Ist der Fundamentalsatz, den Euklid als Axiom zu den Grundlagen seiner Geometrie gestellt hat, eine logische Folge der übrigen
geometrischen Axiome, oder aber ist er von diesen Axiomen logisch unabhängig? Verzichtet man auf das euklidische Axiom (und auch auf seine
Negation), so bildet die logische Konjunktion der übrig gebliebenen un-
beweisbaren Grundsätze das Axiomensystem der sogenannten *absoluten Geometrie Bolyais*. Das Parallelenproblem stellt also die Frage, ob der euklidische Satz ein Theorem der absoluten Geometrie Bolyais oder aber ein
unentscheidbarer Satz ist, der in der absoluten Geometrie weder als wahr
bewiesen noch als falsch widerlegt werden kann. In seiner klassischen
Form ist das Problem allerdings nicht als Alternative, sondern als eine
kategorische, jedoch falsche metamatematische Konjektur dargestellt:
Das euklidische Parallelenpostulat ist kein Axiom, sondern ein Theorem
der absoluten Geometrie Bolyais.

In der Literatur herrscht die Auffassung, das Parallelenproblem wäre
erst nach der Publikation der *Elemente* als kritische Reaktion auf das
euklidische Werk entstanden. Tatsache ist jedoch, daß das Problem bereits eine Generation vor Euklid entstanden ist,[11] und seine *Elemente*
selbst enthalten die erste, im übrigen tadellose Antwort auf die in dem
Problem enthaltene Frage.

Im 4. Jahrhundert vor Christus, im Zeitalter von Plato, Aristoteles und
Eudoxos war der Axiomencharakter des späteren euklidischen Postulats
noch nicht erkannt. Die Gesamtheit der Grundsätze, die von den griechischen Geometern explizit oder stillschweigend als unbewiesene und unab-
weisbare Axiome akzeptiert wurden, war im Wesen mit dem Axiomensystem
der absoluten Geometrie Bolyais identisch. Damit wurde stillschweigend
zugegeben, die strikt euklidischen Sätze seien Theoreme der absoluten
Geometrie. (Mit dem Terminus *strikt euklidisch* bezeichne ich eine Aussage, die entweder identisch mit dem euklidischen Parallelensatz oder
ohne diesen Satz nicht beweisbar ist). Gelten jedoch nur die Axiome der
absoluten Geometrie als unbewiesen und unbeweisbar, so muß der Beweisvorgang mindestens eines strikt euklidischen Satzes einen logischen
Zirkel enthalten. Wahrscheinlich in dem Mathematikerkreis um Eudoxos
wurde zum ersten Mal entdeckt, daß der Beweisvorgang des ersten strikt
euklidischen Satzes der *Elemente*, Elem. I 29, bereits mit einem solchen
logischen Kurzschluß behaftet ist. Dieser Satz drückt eine fundamentale
Eigenschaft der euklidischen Parallelen aus: Sind zwei komplanare Geraden (in beiden Richtungen)nicht-inzident, so bilden alle gemeinsamen

Schneidenden des Geradenpaares mit den Geraden Innenwinkel, deren Summe
auf einer Seite der Schneidenden mit zwei rechten Winkeln gleich ist.
Die Aufgabe, für diesen Satz einen absolut geometrischen Beweis zu finden, hat sich schon damals als selbständiges Problem konstituiert, seit
dem Ende des 19. Jahrhunderts wurde sie mit dem Terminus *Parallelenproblem*
bezeichnet.

Eine bedeutende Anzahl von Stellen aus dem *corpus aristotelicum*[12] gibt
uns Anlaß, anzunehmen, daß die Geometer um Eudoxos schon den indirekten
Weg versucht haben, um das Problem zu lösen: Ist das Theorem Elem. I 29
aus den akzeptierten Axiomen direkt nicht ableitbar, so ist vielleicht
seine Negation in demselben Rahmen widerlegbar. Sie haben es also unternommen, die Folgerungen einer anti-euklidischen Hypothese zu entwickeln,
in der Hoffnung, diese mit der Aufweisung einer Inkonsistenz zerschlagen
zu können. Als Ausgangspunkt der indirekten Beweisversuche wurde außerhalb des Satzes Elem. I 29 seine unmittelbare Folge, der Fundamentalsatz Elem. I 32.2 gewählt: Die Winkelsumme beträgt in jedem Dreieck zwei
rechte Winkel.

Den voreuklidischen Geometern ist es tatsächlich gelungen, die eine
Hälfte dieser allgemeinen anti-euklidischen Hypothese ad absurdum zu
führen: Beträgt die Winkelsumme in einem Dreieck mehr als zwei rechte
Winkel, so sind zwei nichtinzidente und komplanare Geraden inzident.[13]
Die Versuche, die andere Hälfte der anti-euklidischen Hypothese zu widerlegen, *die Winkelsumme des Dreiecks ist kleiner als zwei rechte*, sind
sicherlich gescheitert. Merkwürdig ist jedoch, daß die griechischen
Geometer niemals der Illusion zum Opfer gefallen sind, es wäre ihnen
gelungen, die Inkonsistenz dieser antieuklidischen Hypothese nachzuweisen.

Der Mißerfolg führte sie in irgendeiner Weise zur folgenden positiven
und richtigen metamathematischen Vermutung: Aus den gegebenen, also
aus den Axiomen der absoluten Geometrie, ist die Wahrheit des euklidischen Fundamentalsatzes Elem. I 32.2 nicht beweisbar, und so ist auch
die logische Alternative: "euklidischer oder anti-euklidischer Satz",
nicht entscheidbar.

Als einen ersten Ausdruck dieser tatsächlichen Unentschiedenheit zitiere
ich eine Stelle aus den *Zweiten Analytiken* (II 2; 90a13-34) von Aristoteles: *Das Wesen des Dreiecks besteht in seiner Winkelsumme, und diese
Winkelsumme kann einen Wert haben, der entweder zwei rechte Winkel beträgt oder nicht; sie kann also entweder gleich oder größer oder aber
kleiner als zwei rechte Winkel sein.* Einige Seiten weiter (anal.post.
II 8, 93a32-35) ist die Unentschiedenheit der Alternative bereits in

Form eines ungelösten Problems ausgedrückt: *Welcher der beiden sich widersprechenden Sätze ist wahr, der Satz, der die Gleichheit, oder aber der Satz, der die Ungleichheit der Dreieckswinkelsumme mit zwei rechten Winkeln behauptet?* Aristoteles läßt die Frage offen und später in der *Nicomachischen Ethik* (VI 5, 1140b13-15) behauptet er, daß das *Urteil*, das diese Alternative zu entscheiden berufen ist, von menschlichen Leidenschaften *nicht korrumpiert* werden kann.

Mit keinem Wort deutet Aristoteles - weder an den angeführten noch an irgendeiner anderen Stelle - darauf hin, er würde den euklidischen Fall zu Ungunsten der anderen als die einzige in Betracht kommende Wahrheit bevorzugen. Seine Zurückhaltung ist auffallend, und kann mit zufälligen Versäumnissen nicht erklärt werden. Aristoteles führt häufig falsche oder absurde Sätze an und versäumt niemals auf ihre Falschheit explizit hinzuweisen, auch dann nicht, wenn diese eine notorische Falschheit oder Absurdität ausdrücken. Erwähnt er zum Beispiel, sei es nur *en passant*, die Hypothese der Kommensurabilität der Quadratdiagonale, beeilt er sich, unverzüglich ihre Absurdität in Erinnerung zu bringen.

Endlich erwähne ich das Problem 30.7 (956a15-26), das von der folgenden Frage eingeleitet wird: *Weswegen freuen wir uns nicht, wenn wir die Gleichheit der Winkelsumme des Dreiecks mit zwei rechten Winkeln betrachten oder wenn wir diese Gleichheit mit Hoffnung erwarten?* Die Antwort des Verfassers ist unerwartet und überraschend: *Da wir dieselbe Lust empfinden würden, wenn die Winkelsumme des Dreiecks nicht gleich mit zwei rechten Winkeln wäre* (sondern z.B. mit drei oder sogar mehr). - Der angeführte Text drückt nicht nur die explizite Anerkennung der Unentschiedenheit dieser Alternative aus, sondern enthält darüber hinaus die erste historische Aussage, in der die Möglichkeit der Wahrheit eines anti-euklidischen Satzes als eine durchaus annehmbare, von vornherein nicht ausschließbare Alternative anerkannt wird. Bis Gauss, Bolyai und Lobatchevsky bleibt die Haltung von Aristoteles einmalig, einzig und unübertroffen.[14]

Zu den eben angeführten Stellen möchte ich noch die folgende Bemerkung hinzufügen: 54 mal zitiert Aristoteles den klassischen euklidischen Satz über die Gleichheit der Dreieckswinkelsumme mit zwei rechten Winkeln. Er führt ihn jedoch nur deswegen an, um damit den Begriff der Allgemeingültigkeit (oder, zweimal, den der Ewigkeit) und niemals den Begriff der absoluten Notwendigkeit, des Nichtandersseinkönnens zu illustrieren. Dagegen ist das andere Lieblingsbeispiel seines Repertoires, die Inkommensurabilität der Quadratdiagonale (47 mal zitiert) unveränderlich als Prototyp der absoluten Notwendigkeit und die entgegengesetzte

Hypothese als Prototyp der absoluten Unmöglichkeit angeführt. Besonders
auffallend ist jedoch die Tatsache, daß die den euklidischen Theoremen
elem. I 29 und elem. I 32.2 entgegengesetzten antieuklidischen Sätze
niemals von einem ablehnenden logischen Satzprädikat (falsch, absurd,
unmöglich) begleitet sind. Diese anti-euklidischen Sätze sind die einzigen, die von Aristoteles ohne ein begleitendes logisches Satzprädikat
eingeführt werden. Ohne es separat und explizite zu behaupten, führt er
diese Sätze als in der Tat unentschiedene Aussagen an.

Das Parallelenproblem wurde bekanntlich zum ersten Mal von Euklid gelöst. Er hat einem fundamentalen Satz, der sich mit Hilfe der absoluten
geometrischen Axiome weder beweisen noch widerlegen ließ, ohne jeden
Beweis den Wert der Wahrheit zugeordnet und diesen Satz als ein neues
Axiom den vorhandenen absolut geometrischen Axiomen hinzugefügt. Die
Notwendigkeit dieses Aktes stellte sich für ihn als ein nicht beweis-
barer Imperativ in Form einer Forderung dar: Es wird gefordert, diesem
Fundamentalsatz den Wert der Wahrheit ohne Beweis zuzuordnen, (*princi-
pium petere*), da die bekannten Theoreme der Geometrie, (i.e. der euklidische Geometrie), sonst ohne Zirkel nicht zu beweisen sind. Auffallend
ist auch, daß das Parallelenaxiom in seiner originellen, euklidischen
Formulierung[15] nur einer der beiden anti-euklidischen Hypothesen formal entgegengesetzt ist. Es handelt sich um die sogenannte Hypothese des
spitzen Winkels, um eine Aussage also, die in Form eines unbewiesenen,
wahren Satzes dem Parallelenaxiom der hyperbolischen Geometrie äquivalent ist.[16] Die formale Negation des euklidischen Parallelenpostulats
führt so nicht zu zwei, sondern zu einer einzigen, nämlich zur hyperbolischen Geometrie. In diesem Sinne ist es gerechtfertigt, allein die
hyperbolische Geometrie mit dem Terminus nichteuklidische Geometrie zu
bezeichnen.

Aus dem Kommentarwerk von Proklos wissen wir, daß diese freie Entscheidung von Euklid nach der Erscheinung seiner *Elemente pro* und *contra*
heftig diskutiert wurde.[17] Es gab sicher Mathematiker, die die Zuordnung der Wahrheit zu dem euklidischen Fundamentalsatz aus Gründen
der intuitiven Evidenz bejaht haben.[18] Es schien ihnen evident, daß
zwei komplanare und zueinander konvergierende Geraden sich irgendwo
schneiden müßten, falls sie endlos verlängert würden. Es gab jedoch
auch Mathematiker, die das Gegenteil, die Unbeweisbarkeit der Inzidenz,
sogar die effektive Nicht-inzidenz von zueinander konvergierenden komplanaren Geraden mit Hilfe eines Zenon'schen Gedankenganges provokativ
zu beweisen glaubten.[19] Andere wieder fanden, daß der euklidische
Satz eines Beweises deswegen bedarf, weil er nicht mit den Charakteri-
stiken eines Axioms ausgezeichnet ist. Diese Meinung hat sich in den

folgenden Jahrhunderten etabliert und durchgesetzt. - (Ich möchte darauf hinweisen, daß im historischen Sprachgebrauch auftauchende Ausdrücke *beweisen, streng beweisen* u.ä. alle mit dem präzisen technischen Ausdruck: *als Theoreme der absoluten Geometrie Bolyais beweisen* gleichwertig sind und von guten Mathematikern auch immer so verstanden wurden).

Im Gegensatz zu den voreuklidischen waren die nacheuklidischen Geometer alle von dem Erfolg ihrer Lösungsversuche überzeugt. Wahrscheinlich unter dem überlieferten Einfluß der voreuklidischen Geometer hat man hie und da wieder versucht, das Problem mit Hilfe eines indirekten Beweisverfahrens zu lösen.

Die Kette der so elaborierten antieuklidischen Sätze brach jedoch sehr früh ab, da die Verfasser der Versuche immer in irrtümlicher Weise einen Widerspruch entdeckt zu haben glaubten. Erst mit Saccheri, 1733, fängt man an, das antieuklidische System schrittweise und planmäßig aufzubauen, - immer in der Absicht und Hoffnung, das so erzeugte geometrische Opferlamm werde ohne Anwendung äußerer Gewalt, von alleine zu Grunde gehen, um so zu beweisen, die euklidische sei die einzig mögliche wirkliche und wahre Welt.

Der hohe wissenschaftliche Wert, der dem Parallelenproblem zugeordnet wurde, wurde in der zweiten Hälfte des 18. Jahrhunderts mit der stereotypen Behauptung begründet, dieses Problem habe alle großen Mathematiker der vergangenen zwei Jahrtausende intensiv beschäftigt; da deren Lösungsversuche alle gescheitert sind, ist der Ertrag des jeweiligen Verfassers, dem die Lösung gelang, selbstverständlich um so wertvoller. Durch fast endlose Wiederholungen hat diese Behauptung für die späteren Historiker die Merkmale einer belegten historischen Tatsache gewonnen, die geeignet war, das Interesse für das Parallelenproblem am Ende des 18. und am Anfang des 19. Jahrhunderts zu rechtfertigen. Die nähere Untersuchung der einschlägigen Literatur zeigt jedoch, daß diese Behauptung falsch ist. Erst ab etwa 1760 zeigt sich ein ansteigendes Interesse für das Parallelenproblem, und auch zu dieser Zeit nur in dem damals besonders vom mathematischen Standpunkt aus noch zurückgebliebenen Deutschland.[20] Während der vorausgehenden Jahrhunderte war das Parallelenproblem ein ausschließliches Anliegen der Kommentatoren Euklids. Damit gehörte es seit dem 16. Jahrhundert zu dem von Mathematikern besonders verachteten Gebiet der scholastischen Pedanterien. Wenn einige bedeutende Mathematiker, v.a. im Bereich der islamischen Kultur,[21] sich doch mit diesem Problem beschäftigt haben, so nur gelegentlich, und auch hier meistens als Übersetzer oder Komentatoren Euklids.

Die bedeutendsten Mathematiker des 16., 17. und 18. Jahrhunderts haben

vom Parallelenproblem, falls śie seine Existenz überhaupt zur Kenntnis
nahmen, nur mit Verachtung geredet. Arnauld[22] qualifizierte in seiner
besonders einflußreichen Geometrie (1667, 1683) das Parallelenproblem
als Zeitverschwendung und nutzloses Kopfzerbrechen ab. Ähnliche Äußerungen finden wir in der Geschichte der Mathematik Montucla's,[23] in den
Schriften von Malebranche,[24] d'Alembert,[25] Lacroix[26] und anderen.
Sie sahen darin ein rein scholastisches Problem, metaphysische Schika-
nen, eine Angelegenheit mittelmäßiger, unbegabter Verfasser von elemen-
tarmathematischen Lehrbüchern. Das Problem war für sie mathematisch
irrelevant, trivial und steril. Ausgeschlossen, daß die Mathematik etwas
gewönne, wenn eine von niemandem je bezweifelte Wahrheit mit einem ebenso tadellosen wie überflüssigen Gedankengang bewiesen würde.[27]

Wie kann unter diesen Umständen erklärt werden, daß in Deutschland in
der zweiten Hälfte des 18. Jahrhunderts das Interesse für das Parallelenproblem auf einmal erwacht und gegen die Verachtung der maßgebenden Fach-
leute ständig zunimmt und sich endlich durchsetzt?

3. Göttliche Allmacht und geometrische Struktur des Universums

Der Ursprung dieser eigenartigen Entwicklung befindet sich höchstwahr-
scheinlich in dem 1196 publizierten *Führer des Unschlüssigen* von Moses
Maimonides, - Naturwissenschaftler, Arzt und Philosoph, der seine Hei-
mat, Spanien, wegen Judenverfolgungen verlassen mußte und in ägyptischer
Emigration gestorben ist.[28] Gott ist zwar allmächtig und absolut frei,
behauptet Maimonides (I 73), trotzdem konnte und kann auch er nichts
erschaffen, was gegen die Gesetze der aristotelischen Logik oder der
euklidischen Geometrie verstoßen würde. So konnte Gott zum Beispiel
kein Quadrat schaffen, in dem die Diagonale gleich mit einer der Seiten
wäre.[29]

Ein solches Quadrat zeichnet die hyperbolische Geometrie aus: Es ist
eigentlich ein entartetes Quadrat; seine vier Seiten sind unendliche
Geraden und keine der Seiten ist mit einer anderen inzident. Die be-
nachbarten Seiten sind darüber hinaus untereinander strikt-parallel;
(ihr Parallelismus ist jedoch nicht transitiv) sie schließen jedoch
einen ebenen Bereich ein, dessen Inhalt endlich ist - die oberste Grenze aller Quadratinhalte. Zweifelsohne wußten bereits die alten Geometer
(was immerhin erst von Saccheri als ein Theorem der absoluten Geometrie
bewiesen wurde), daß ihr geometrisches *Wesen* eine induktive Eigenschaft
der geometrischen Figuren darstellt: Ist ein einziges Dreieck oder Vier-
eck euklidisch bzw. hyperbolisch oder elliptisch, so sind es alle.[30]
Das geometrische Wesen ist vererbbar, oder in einer kosmoligischen Terminologie ausgedrückt: Der Raum ist isotrop, homogen und seine Krümmung

konstant. Jede Geometrie ist implizite eine Kosmometrie. Das von Maimonides zitierte Quadrat ist so stellvertretend für das ganze Universum
da. Gott kann kein nichteuklidisches Quadrat und infolgedessen auch
kein nichteuklidisch strukturiertes Universum schaffen. Maimonides macht
jedoch den *Unschlüssigen* darauf aufmerksam, daß diese Tatsache den absoluten Charakter der göttlichen Freiheit und Allmacht keineswegs beschränkt.

Das Beispiel des nichteuklidischen oder besser gesagt antieuklidischen
Quadrates ist im *Führer des Unschlüssigen* nicht durch Parthenogenese
erschienen. Eine merkwürdige Stelle in *de caelo* (I 12,281b 5-7) von
Aristoteles deutet auf eine vermutlich bereits vorhandene Erkenntnis
griechischer Geometer hin, nach der die Kommensurabilität der Quadratdiagonale eine notwendige Folge der antieuklidischen Hypothese ist, die
die Ungleichheit der Winkelsumme des Dreiecks mit zwei rechten Winkeln
behauptet.[31] In ihren Kommentaren *ad locum* haben Alexandros und Simpli-
cius zwei konkrete Beispiele von solchen antieuklidischen Quadraten angeführt. Das eine ist das entartete Quadrat der elliptischen, das andere
das der hyperbolischen Ebene. In dem ersten ist die Quadratdiagonale
das Doppelte der Seite, in dem zweiten ist sie mit der Seite gleich.
Beide Figuren stellen das maximale Grenzquadrat der jeweiligen nichteuklidischen Ebene dar. Da Maimonides ein ausgezeichneter Kenner von
Aristoteles und Alexandros war, und da auch andere Beispiele aus dem
de caelo in demselben Werk zitiert sind, ist es plausibel anzunehmen,
daß er das Beispiel des antieuklidischen Quadrates aus Aristoteles direkt oder aus Alexandros übernommen hat. Damit wäre sein Ursprung erklärt.

Das wichtigste ist jedoch nicht das Beispiel selbst, sondern der Kontext, in den es eingebettet ist. Was hat Maimonides veranlaßt, die Möglichkeit oder Unmöglichkeit eines nichteuklidisch strukturierten Universums geometrischer Objekte mit der Freiheit des Allmächtigen Gottes
in Verbindung zu bringen? Ein Teil der Antwort befindet sich in der
Eudemischen Ethik (II 6, 1222b20-39) und in der *Magna Moralia* (I 10,
1187a29-b25) von Aristoteles. U.a. behandelt Aristoteles hier das Problem der menschlichen Freiheit. Das ist sicherlich ein festlicher Ort
und eine feierliche Stunde, da meiner Kenntnis nach hier die menschliche
Freiheit zum ersten Mal zum Gegenstand philosophischer Reflexion gesetzt wurde. Diese von dem politischen, sozialen und rechtlichen Frei-
heitsstatut radikal verschiedene metaphysische Freiheit des Menschen
ist von Aristoteles als eine Fähigkeit des reifen menschlichen Individuums definiert, sich in Abwesenheit äußerer Zwänge aufgrund einer
inneren Deliberation zwischen zwei entgegengesetzten, jedoch in seiner

Macht stehenden Möglichkeiten für die eine und gegen die andere zu entscheiden. Das Ziel der Entscheidung ist die Realisierung eines praktischen Zweckes, der vom Subjekt der Handlung als das Gute betrachtet
wird. Die Entscheidung ist deswegen frei, weil es mit logischen Mitteln
völlig unbeweisbar ist, welcher der beiden gleich-möglichen Wege zum
Guten und welcher zum Bösen führt. Wurde einmal eine Wahl getroffen,
folgt alles übrige aus dieser ersten Entscheidung mit unausweichlicher
Notwendigkeit. Da die altgriechische Sprache über keinen technischen
Terminus für die Bezeichnung des metaphysischen Freiheitsbegriffs verfügt,[32] versucht Aristoteles, wie er wiederholt betont, den Begriff
mit Hilfe eines plastischen Beispiels aus dem Bereich der Geometrie
deutlich zu machen: Die erste Entscheidung besitzt den Charakter eines
geometrischen Axioms, schreibt er. Der, der sich für einen von zwei
entgegengesetzten Wegen entscheidet, hält zwar diesen Weg für den einzig richtigen, ist sich selbst jedoch bewußt, daß die Richtigkeit, also
die Wahrheit seiner Entscheidung, nicht beweisbar ist. Die Handlungen,
die aus dieser ersten Entscheidung mit Notwendigkeit folgen, entsprechen
natürlicherweise den Theoremen als notwendiger Folgerungen der Axiome.
Als Parallele für die ethisch-politische Alternative *gut* und *böse* sugge-
riert Aristoteles die Alternative eines euklidischen und antieuklidischen
Dreiecks bzw. Vierecks. Nehmen wir als unbewiesenen Ausgangssatz, von
Aristoteles *arche*, *Prinzip*, oder im etymologischen Sinne *Hypothese* genannt - den Satz, der die Gleichheit der Winkelsumme mit zwei rechten
Winkeln behauptet, so folgen daraus mit Notwendigkeit bestimmte Theoreme. *Wandeln sich jedoch die axiomatischen Prinzipien, die Archai, ver-*
ändert sich das Dreieck so, daß die Summe seiner Innenwinkel nicht mehr
mit zwei rechten Winkeln gleich sein wird, so werden sich auch die Theo-
reme, die daraus folgen, mit Notwendigkeit und in entsprechender Weise
ändern.[33]

Merkwürdigerweise macht Aristoteles an dieser Stelle nicht einmal die
kleinste Anspielung darauf, ob in seiner Parallele die antieuklidischen
Sätze mit dem ethisch Guten oder im Gegenteil mit dem Bösen zu assimilieren wären. In den zweiten Analytiken gibt es jedoch eine Stelle, wo
er einen antieuklidischen Satz kategorisch mit dem ethisch Verdorbenen
und Bösen assimiliert (anal.poster. I 12, 77b16-28). Es ist übrigens
die einzige Stelle, wo er den einen antieuklidischen Satz mit einem
negativen, pejorativen, jedoch nicht logischen, sondern ethischen Terminus bezeichnet.[34] Dagegen betrachtet er im Problem 30.7 die beiden
entgegengesetzten Sätze sowohl ethisch als auch logisch à priori völlig
gleichberechtigt; (immerhin, ohne ihnen ein positives Prädikat zuzuschreiben).

Ohne es explizite auszusagen, suggerieren diese Passagen aus den Ethiken
eine eigenartige Auffassung. Der Bereich der axiomatisierten geometrischen Theorien wäre dementsprechend mit dem Bereich der freien Handlung
im ethischen Bereich gleichzusetzen; vor dem Menschen stünden also zwei
unbeweisbare und entgegengesetzte geometrische Sätze, und er kann und
muß sich frei, also ohne Beweis, für einen entscheiden. Diese Entscheidung kommt darin zum Ausdruck, daß die Wahrheit einem der beiden Sätze
zugeordnet wird; daraus folgt dann mit Notwendigkeit, daß dem entgegengesetzten Satz der Wert der Falschheit zugeschrieben werden muß. Beide
Möglichkeiten stehen jedoch vor dem menschlichen Wesen als gegeben und
als realisierbar, es steht in seiner Macht, beide der entgegengesetzten
Systeme effektiv auszubauen. Erst diese erste Entscheidung versieht den
gewählten Satz mit dem höheren Status einer *arche*, einer unbewiesenen
Grundwahrheit - also eines Axioms.

Die ganze griechische Philosophie ist von der Ablehnung der Möglichkeit
der Schöpfung charakterisiert.[35] Weder Mensch noch Gott kann etwas
völlig Neues, Noch-nicht-Dagewesenes erschaffen. Die Welt wurde von den
Göttern nicht erschaffen; schaffen hieße, etwas aus dem Nichtsein in
das Sein zu überführen, was unmöglich und absurd ist.[36] So besteht auch
die Freiheit des menschlichen Wesens immer nur in einer Wahl und in
einer Entscheidung für Möglichkeiten, die in der Welt gegeben und ihm
zugänglich, von ihm realisierbar sind. Die Freiheit des maimonidischen
Gottes besteht jedoch im Gegensatz zu der finiten menschlichen Handlungs-
freiheit und zu den Göttern des Olymp darin, daß es in seiner Allmacht
steht, eine Welt, ein Universum aus dem Nichts zu schaffen.

Auch wenn der Mensch theoretisch die Wahrheit des euklidischen Axioms
und die Falschheit der antieuklidischen Hypothese nicht beweisen kann,
wird daraus niemand schließen, daß er im Ernst vor einer unentscheid-
baren geometrischen Alternative stünde, und noch weniger, daß es mangels eines Beweises nur an seiner Freiheit liege, entweder den einen
oder den anderen Satz als Axiom zu wählen. Als letzte und entscheidende
Instanz ist die Erfahrung als korrumpierender Faktor noch immer da, um
die Entscheidung zugunsten der euklidischen Geometrie zu bestimmen. In
bezug auf den Menschen scheint es also müßig zu sein, diese Frage zu
stellen. Nicht jedoch in bezug auf den Allmächtigen. Die empirisch wahr-
nehmbare sinnliche Außenwelt ist noch nicht da, die ihn von der Unvermeidbarkeit und Wahrheit der euklidischen Geometrie überzeugen könnte,
sie entsteht erst als Ergebnis seiner freien Schöpfung. Was kann unter
diesen Umständen den geometrischen Gott dazu bestimmen, seine Welt eher
nach den Prinzipien der euklidischen als nach den Axiomen einer antieuklidischen Geometrie zu schaffen?

Maimonides Schriften waren dem ehemaligen Bischof von Regensburg, Albertus Magnus bekannt, und teilweise durch ihn, teilweise unmittelbar haben sie Thomas von Aquino beeinflußt.[37] Wahrscheinlich von Maimonides angeregt, fühlte sich Thomas verpflichtet, die Möglichkeit einer nichteuklidischen Welt in bezug auf die Schöpfung *ex nihilo* zu untersuchen. Sicher hat er auch einige einschlägige Stellen bei Aristoteles nachgelesen, da das antieuklidische Beispiel bei ihm deutlicher artikuliert und in jeder Hinsicht näher zum Wortlaut des *corpus aristotelicum* ist, als bei Maimonides. Gott ist nicht frei - lesen wir bei Thomas, - das Wunder eines Dreiecks hervorzubringen, in dem die Winkelsumme nicht gleich mit zwei rechten Winkeln wäre.[38] Im übrigen läßt Dante in seinem Paradies (XIII 91-104) den Hlg. Thomas erzählen, wie König Salomon, als Gott ihm einmal erlaubt hat, ihn zu fragen, was er wissen möchte, in seiner Weisheit nicht danach gefragt hat, ob ein Dreieck gemacht werden könnte, in dem die Summe der Winkel nicht mit zwei rechten Winkeln gleich wäre, sondern, wie er als Herrscher das Gute von dem Bösen unterscheiden könnte.[39]

Auch wenn das euklidische Theorem nicht immer explizit erwähnt wurde, gehörte seit Thomas die Geometrie zum Repertoire der Diskussionen über die göttliche Freiheit. Doch ist immerhin zu bemerken, daß die Geometrie in diesem Rahmen eine höchst untergeordnete Rolle spielte. Die Diskussionen um die göttliche Allmacht waren eher von politischer und sozialen Fragestellungen bestimmt. Es ging vor allem darum, ob die Welt so wie sie ist, die ethischen Normen, die juristischen Gesetze, ewige, notwendige, göttliche Wahrheiten darstellen oder nicht. In dem Hin und Her der Diskussionen war es wichtig festzustellen, daß es in der Welt mindestens ein Gebiet gibt, nämlich die Geometrie, wo die Sätze tatsächlich unerschütterliche, ewige und an-sich-notwendige Wahrheiten darstellen, - Wahrheiten, die zu ändern nicht nur Mensch, sondern auch Gott nicht imstande ist.

Es ist merkwürdig, daß eben die Gleichheit der Dreieckswinkelsumme mit zwei rechten Winkeln immer als paradigmatisches Beispiel für die Illustration der absoluten Notwendigkeit zitiert ist; ein Beispiel, das von Aristoteles, wie oben erwähnt, niemals zu diesem Zweck angeführt wurde. Diese Umwandlung erklärt sich daraus, daß Aristoteles den euklidischen Satz in die Diskussion der menschlichen Freiheit einbezogen hat. Von hier ist das Beispiel in den Kontext der Diskussionen über die Freiheit des jüdisch-christlichen Gottes gelangt. Der Allmächtige mußte zwangsläufig mit dem Wahrheitsgehalt des euklidischen Theorems konfrontiert werden. Um die subjektiv empfundene Unvorstellbarkeit eines nichteuklidischen Dreiecks objektiv begründen zu können, mußten die mittelalter-

lichen Theologen dem euklidischen Theorem eine a priori gegebene, imma-
nente und absolute Notwendigkeit zuschreiben, die nicht einmal die göttliche Allmacht zu brechen imstande sein durfte.

Wie es auch sei, in den nächsten Jahrhunderten hat der euklidische Satz
den Wert eines paradigmatischen Beispiels für eine Wahrheit gewonnen,
die eine absolute Notwendigkeit an sich besitzt. Die Wahrheit des Satzes
ist notwendig, d.h., das Dreieck, und damit das Universum der geometrischen Figuren und der Gesamtheit des physikalischen Raumes, kann nur
euklidisch und nicht anders ein. Die euklidische Geometrie ist die einzig mögliche Kosmologie, sie vertritt die einzig mögliche Wahrheit. Das
Sprechen von nichteuklidischen Räumen ist eine Rede, die zwar in unserer
euklidischen Welt geführt werden kann, betrifft jedoch Universa, die
für Gott selbst eine schlichte Unmöglichkeit darstellen. *Zweifeln Sie an
allem*, sagte Voltaire, *nur nicht an der Gleichheit der Winkelsumme des
Dreiecks mit zwei rechten Winkeln.* [40]

4. Die Cartesische Wende

Mit Descartes tritt jedoch eine entscheidende Wende[41] ein, sowohl in
der Behandlung der Frage der göttlichen Allmacht, als auch, damit zu-
sammenhängend, in der Charakterisierung der Notwendigkeit, die den euklidischen Wahrheiten zugeordnet wird. Nach Auffassung Descartes ist
Gottes Freiheit absolut unbeschränkt. Nichts, was für uns als solche
gilt, ist für ihn unmöglich. Für ihn gelten weder die Gesetze der aristotelischen Logik noch die Theoreme der euklidischen Geometrie als
verbindliche und an-und-für-sich-notwendige Wahrheiten. Hätte er ge-
wollt, so hätte er eine Welt schaffen können, wo das Gesetz des ausgeschlossenen Widerspruches nicht gilt (wie bereits Gabriel Biel aus
Tübingen, im XV. Jahdt. behauptete) wo die Summe von drei und zwei nicht
fünf ausmacht, wo die Winkelsumme des Dreiecks nicht gleich mit zwei
rechten Winkeln ist, wo der Pythagoräische Lehrsatz nicht gilt, usw.
Gott ist ebenso frei, den Wert der Wahrheit den Sätzen der euklidischen
oder, im Gegenteil, den Theoremen einer der euklidischen entgegengesetzten Geometrie zuzuordnen. Wir sind imstande, als möglich und wahr
zu erkennen, was Gott als möglich erscheinen läßt oder wirklich geschaffen hat. Wir sind aber nicht imstande, etwas als möglich zu betrachten, was Gott zwar hätte möglich machen können, es aber trotzdem
nicht gemacht hat und als unmöglich erscheinen lassen wollte. Und hätte
Gott das Universum mit anderen natürlichen oder geometrischen Sätzen
ausstatten wollen, als er es eben getan hat, so - kommentiert Spinoza[42]
- hätte er uns mit entsprechenden intellektuellen Eigenschaften ausge-
stattet, um uns die, von den heutigen abweichenden, Gesetze als wahr

empfinden zu lassen. In seiner von nichts beschränkten absoluten Freiheit hat Gott jedoch das Universum mit einer euklidischen Struktur versehen, und er hat diese von ihm geschaffenen Wahrheiten in unsere Seele
eingeprägt, so wie ein König sein Zeichen auf die Geldmünze geprägt hat.
Wir sind mit diesen Ideen geboren, und wir betrachten sie als notwendig
wahr, weil Gott sie so geschaffen hat. Es stellt sich jedoch die Frage,
wie erkennen wir die Wahrheit, die Gott geschaffen hat. Es hätte sein
können, daß ein ebenso mächtiger, listiger und böser Geist uns betrügen
wollte, und uns das Gegenteil der göttlichen Sätze als notwendige Wahrheiten vortäuscht. Hat etwa Gregorius de Rimini[43] nicht behauptet, daß
Gott auch lügen könnte? Wir können den Verführungen des Bösen Demiurgs
trotzdem entgehen, wenn wir an die Unizität und Existenz Gottes glauben,
und wenn wir überzeugt sind, daß der richtige Gott zwar *raffiniert, jedoch nicht bösartig* und so auch kein Betrüger ist. Wenn es so ist, dann
müssen wir in uns die von ihm geschaffenen Wahrheiten mit Klarheit und
Deutlichkeit als ewige und notwendige Wahrheiten erkennen. *Und so erkenne
ich* - schreibt er mit bekannter cartesischer Bescheidenheit, - *ich, der
ich immerhin etwas von Geometrie verstehe, daß die Winkelsumme des Dreiecks gleich mit zwei rechten Winkeln ist, und daß es anders überhaupt
nicht möglich sei, nämlich daß die Winkelsumme etwas nicht gleich mit,
sondern z.B. größer als zwei rechte Winkel wäre.*

Ein Atheist kann zwar den Satz über die Gleichheit der Dreieckswinkelsumme mit zwei rechten Winkeln korrekt beweisen, er wird jedoch niemals die Notwendigkeit dieser Wahrheit mit Gewißheit behaupten können;
der Grund, nämlich Gott, der die Wahrheit dieses Satzes garantiert, ent-
geht ihm. Unter diesen Umständen kann er die Wahrheit des euklidischen
Satzes bezweifeln, sogar das Gegenteil behaupten.[44]

Das Wichtigste, um die Notwendigkeit geometrischer Wahrheiten mit Gewißheit zu erkennen, ist, sich von der Existenz des einzigen Gottes
zu überzeugen. Die Existenz Gottes folgt jedoch mit derselben Notwendigkeit aus seinem Wesen, wie die Gleichheit der Winkelsumme mit zwei
rechten Winkeln aus dem Wesen des Dreiecks. Diese von Descartes (und
von Spinoza) so häufig wiederholte These steht zwar zu der cartesischen
Konzeption der Omnipotenz in keinem Widerspruch, konnte jedoch und hat
tatsächlich den Eindruck der Inkonsistenz erwecken. Die These selbst
wurde wahrscheinlich durch die oben bereits zitierte Stelle, anal.post.
II 2, 90a13-34, inspiriert, wo Aristoteles das Wesen des Dreiecks mit
seiner Winkelsumme identifiziert. In der Tat folgt, nach Aristoteles,
aus dem Wesen des Dreiecks, daß es eine bestimmte Winkelsumme hat, jedoch nicht, daß diese Summe zwei rechte Winkel ausmachen muß. Descartes
wäre sowohl mit seiner eigenen Theologie als auch mit der aristotelischen

Konzeption in besserem Einklang geblieben, hätte er behauptet: Gottes
Existenz folgt mit derselben Notwendigkeit aus seinem Wesen, wie die
Existenz einer Winkelsumme aus dem Wesen des Dreiecks; und genauso wie
Gottes Wesen ein euklidisches oder im Gegenteil ein nichteuklidisches
sein kann, kann auch die Winkelsumme des Dreiecks gleich oder nicht
gleich mit zwei rechten Winkeln sein.

Die cartesanische Konzeption hat sowohl in als auch außerhalb Frankreichs
heftige Diskussionen ausgelöst. Seine Gegener, so z.B. Arnauld,[45]
Gassendi[46] und andere haben ihm u.a. vorgehalten, daß es ihnen unvor-
stellbar wäre, wie man gegen die Verführung eines ebenso listig wie
bösartig *geometrisierenden Gottes* sich wehren könnte, wenn er uns eine
antieuklidische Geometrie als die wahre vortäuschen wollte. Wir würden
uns in diesem Fall ein antieuklidisches Dreieck oder die Wahrheit des
Satzes, der das euklidische Parallelenpostulat negiert, ebenso deutlich
und klar vorstellen, wie die anderen die Figuren und Theoreme der euklidischen Geometrie. Die Antworten von Descartes waren im allgemeinen
gereizt und kaum überzeugend. So. z.B. antwortete er Arnauld, eine mit
N rechten Winkeln gleiche Dreieckswinkelsumme wäre durchaus vorstell-
bar, jedoch dürfe und könne jeder dieser Fälle negiert werden, mit Ausnahme des Falles, wo die Summe der Winkel mit zwei rechten Winkeln gleich
ist.[47]

Die in den *Metaphysischen Meditationen* enthaltene cartesische Provoka-
tion hat in den nächsten Jahrhunderten die Auseinandersetzungen um das
Problem der göttlichen Allmacht und der Notwendigkeit der von Gott offenbarten ethischen Gesetze bestimmt. Aus einem untergeordneten Beispiel
ist der euklidische Satz in den Vordergrund des allgemeinen Interesses
getreten, er wurde überall erwähnt, wo von göttlicher Allmacht, Freiheit,
Schöpfung und von der Notwendigkeit der von Gott offenbarten ethischen
Gesetze die Rede ist. Wir finden das Beispiel bei Spinoza,[48] Malebranche,[49] Hobbes,[50] Hume, Locke und anderen. In seiner 1776 publizierten *Philosophy of Rhetoric* (ein Buch, das im XIX Jhrdt. 40 Auflagen
erlebt hat), empfiehlt Campbell den künftigen Rhetoren, in ihren Reden
das antieuklidische Dreieck als das wirksamste Beispiel für eine schlichte Unmöglichkeit zu verwenden. Im übrigen gilt für Fénelon die Anerkennung der Gleichheit der Winkelsumme des Dreiecks mit zwei rechten Winkeln von seiten der Griechen, Chinesen, Inder als Beweis göttlicher Allmacht: Auch die, die Gott nicht kennen und sogar die, die seine Existenz
kategorisch leugnen, sind überall auf dieselbe Idee der Gleichheit der
Winkelsumme mit zwei rechten Winkeln gekommen, und sie sind auch alle
des notwendigen Charakters ihrer Wahrheit gewiß. Es ist die Anwesenheit
und das Diktat des für sie verborgenen gebliebenen oder sogar von ihnen

geleugneten Gottes, dem sie nicht entgehen können.[51] Im Gegenteil, in
einer am Ende des 19. Jahrhunderts erschienenen Dissertation ist die
Existenz der nichteuklidischen Geometrie eben als ein mathematischer
oder sogar ontologischer Beweis für die Existenz Gottes hingestellt:
Da die Wahrheit des euklidischen Satzes aus den übrigen, also aus den
Axiomen der absoluten Geometrie Bolyais, nicht abgeleitet werden kann,
so bleibt ein Raum übrig, wo sich eine der Welt übergeordnete Freiheit
einsetzen kann, um dem Universum eine bestimmte geometrische Struktur
aufzuprägen. Ist unser physikalischer Raum euklidisch oder im Gegenteil
nichteuklidisch, so kann das Universum nur als Ergebnis der Tätigkeit
einer göttlichen Freiheit so sein, wie es ist. Die Tatsache, daß das
Universum eine konkrete euklidische oder nichteuklidische Struktur besitzt, deutet auf die Anwesenheit und auf die Wirkung eines transzendentalen, freien Agenten hin.[52]

Bereits Augustin hat die Theoreme der Mathematik als paradigmatisches
Beispiel für ewige und notwendige Wahrheiten betrachtet. Bezweifeln wir
diese, so können wir überhaupt an nichts mehr glauben, - schrieb er.
Mit Maimonides und Thomas, aber besonders mit Descartes ist jedoch eine
entscheidende Wende in dem Status dieses Beispiels eingetreten. Erstens,
anstatt im allgemeinen über mathematische Theoreme zu reden, hat sich
die Aufmerksamkeit auf einen einzigen euklidischen Satz, Elem. I 32.2,
konzentriert (nur sehr selten werden unbedeutende Varianten desselben
Satzes zitiert, wie z.B. der pythagoräische Lehrsatz). Das war jedoch
ein Satz, dessen Gewißheit im Bereich der geometrischen Grundlagenforschung ständig bezweifelt wurde, da ein allgemein akzeptierter Beweis
(unter Beweis verstand man immer einen absolut geometrischen Beweis)
fehlte. Da der beweistechnisch mangelhafte Status dieses Satzes schon
Aristoteles auffiel, konnte die Notwendigkeit, die ihm in spontaner und
intuitiver Weise zugeordnet wurde, von Maimonides nur damit begründet
und erklärt werden, daß nicht einmal Gott das Gegenteil erschaffen könnte. Descartes hat jedoch zum erstenmal erkannt und kategorisch behauptet,
daß umgekehrt die Gewißheit des euklidischen Satzes von der Entscheidung
eines absolut freien göttlichen Wesens abhängt. In dieser Sicht ist die
Gewißheit, die wir gegenüber diesem Satz empfinden, mit einem transzendentalen ethischen Argument - absolute Freiheit und Allmacht Gottes -
gerechtfertigt. Infolgedessen erscheinen die Bereiche der Geometrie und
der ethischen Handlung wieder wie einst bei Aristoteles in einer auf-
fallenden Parallele. Sowohl bei Aristoteles als auch bei Descartes steht
am Anfang der Ethik und der Geometrie eine freie Entscheidung. Jedoch
bei Aristoteles schafft diese Entscheidung nichts und kann nur zwischen
zwei Gegebenheiten die eine als das Gute wählen und die andere als das

Böse zurückweisen, ohne jedoch zur Gewißheit gelangen zu können, ohne
die Richtigkeit der Wahl begründen zu können. Bei Descartes dagegen entsteht Wahrheit erst durch einen absolut-freien, göttlichen Schöpfungsakt: nicht Gott ist derjenige, der in seinem Allwissen sich für das Wahre
und Gute entscheidet, sondern das, wofür sich Gott entschieden hat, wird
damit zum Guten und notwendig Wahren. Die Anerkennung der allgemeinen
Gültigkeit und Notwendigkeit ethischer Gesetze hängt genauso von der
Existenz eines einzigen Gottes ab, wie die Gewißheit des Glaubens an
die Ewigkeit und Notwendigkeit geometrischer Wahrheiten. Sowohl in der
Ethik als auch in der Geometrie kommt so alles auf die Anerkennung der
Existenz eines einzigen Gottes an. Wenn also die deduktiven Gedankengänge, die die Notwendigkeit der geometrischen Wahrheiten beweisen,
nichts als Vermittlungen der notwendigen Wahrheit sind, die von Gott
den geometrischen Axiomen frei aufgeprägt wurden, so muß auch die Notwendigkeit, die die ethische Intuition den Gesetzen der Moral zuordnet,
nichts als eine logische Folge der für jedes menschliche Wesen deutlichen, klaren und einleuchtenden Axiome sein, die das Wesen des einzigen
Gottes charakterisieren.

Diese Idee steht hinter der *ethica more geometrico demonstrata* von
Spinoza. Die Freiheit des Menschen besteht in der Anerkennung dieser
von Gott stammenden Notwendigkeit. Demzufolge besteht sein Weg zur Realisierung seiner Freiheit in dem Bewußtwerden der Notwendigkeit göttlicher Gesetze. Im Bereich der Geometrie besteht so seine Freiheit in der
Ansicht der göttlichen Notwendigkeit der Gleichheit der Dreieckswinkelsumme mit zwei rechten Winkeln. Der Mensch, der gegen die göttliche
Gesetze handelt, ist kein freier Mensch. Ebensowenig frei ist der, der
dem Dreieck eine Winkelsumme zuordnen würde, die nicht mit zwei rechten
Winkeln gleich ist. Die Immoralität widerspricht ebenso dem Wesen Gottes,
wie ein antieuklidisches Dreieck.

In diesem Sinne wird im 18. Jahrhundert das geometrische Beispiel wieder
und wieder zitiert: Die Notwendigkeit der ethischen Gesetze ist mit der
Notwendigkeit der Gleichheit der Winkelsumme des Dreiecks, und die ver-
meintliche Immoralität ethischer Handlungen mit der plausiblen Absurdität antieuklidischer Sätze in Parallele gesetzt. In einer merkwürdigen
Form erscheint diese Parallele bei Hobbes. In seinem *Leviathan* I 11
stellt er die folgende Frage: Weswegen streiten sich die Menschen mit
Feder und Schwert um die Wahrheit der ethischen und politischen Aussagen, und weswegen ist die Wahrheit der euklidischen Theoreme unbestrit-
ten. Seine Antwort lautet wie folgt: Falls das euklidische Theorem, das
die Gleichheit der Winkelsumme des Dreiecks mit zwei rechten Winkeln
behauptet, in Widerspruch geraten würde mit den Interessen derer, die

in Wirtschaft und Politik die Macht ausüben, so würden diese, insofern
es ihnen möglich wäre, alle Bücher, die die Geometrie von Euklid enthalten, vernichten. Und Malebranche sagt[53] in seiner Diskussion mit
Hobbes (seine Kritik zielt auf andere Passagen des Hobbeschen Werkes;
sie trifft jedoch auch auf die hier zitierte Stelle zu): Die Geometer
würden sich mit Ekel von einem Buch abwenden, das, wie Hobbes es gerne
möchte, dem Euklidischen entgegengesetzte Theoreme enthalten würden.

5. Parallelenproblem und Notwendigkeit geometrischer Wahrheiten

Am Anfang des 18. Jahrhunderts, unter Bedingungen, die mir noch nicht
klar sind, haben sich die beiden Problemkreise, das Parallelenproblem
und das Problem der absoluten Freiheit Gottes gekreuzt. Glaubten einige,
die Notwendigkeit ethischer Gesetze mit der Notwendigkeit der Wahrheit
euklidischer Theoreme begründen zu können, so ist ihnen schließlich klar
geworden, daß die Gewißheit der euklidischen Sätze selbst nicht gesichert
ist, solange das Parallelenproblem nicht im positiven, klassischen Sinn
gelöst wurde. Auf der anderen Seite wurde denen, die sich mit dem Parallelenproblem beschäftigten, klar: allein der Nachweis der logischen Unmöglichkeit eines antieuklidischen Systems ist befugt zu beweisen, daß
der Raum, gleichgültig ob reell oder ideal, nur euklidisch und nicht
anders sein kann; daß, mit anderen Worten, die Wahrheit, die den euklidischen Sätzen zugeschrieben wird, ihnen mit absoluter Notwendigkeit zukommt. Das Parallelenproblem erschien plötzlich unter einem ganz anderen
Gesichtspunkt. Es gewinnt an Relevanz, sein wissenschaftlicher Wert
steigt, hat auf einmal den Kreis der engbrüstigen, langweiligen und
müßigen Kommentare verlassen und sich in ganz anderen Dimensionen entfaltet. Selbständige Arbeiten, Doktordissertationen und Bücher werden
ihm gewidmet. Ihre gemeinsame Motivation, mehr oder weniger explizit
ausgedrückt, ist immer dieselbe: Die Existenz notwendiger und absoluter
Wahrheiten im Falle der euklidischen Geometrie streng zu beweisen.

Alle diese Tendenzen haben sich in der Philosophie von Kant fokalisiert
und unter ihrem Einfluß neue Intensität und Ausstrahlungskraft gewonnen.
Kant hat sich vorgenommen, ohne Gott die Gewißheit und die absolute Notwendigkeit der euklidischen Wahrheiten mit völlig neuartigen Argumenten
zu begründen.

In einer seiner vorkritischen Arbeiten, 1746, hat er noch zugegeben, daß
Gott zwar die Freiheit gehabt hätte, von den euklidischen verschiedene
Welten (so z.B. einen vierdimensionalen Raum) zu schaffen, und daß er
einen solchen Raum vielleicht irgendwo schon geschaffen hätte.[54] Ihm
ging es jedoch nicht um die Allmacht Gottes, und nicht einmal darum, ob
das Universum an sich euklidisch oder antieuklidisch strukturiert ist,

sondern um das menschliche Wesen, darum also, ob der Mensch die Freiheit oder die Fähigkeit besitzt, ein dem euklidischen entgegengesetztes
System als wahr anzunehmen oder nicht. Er hat so das Problem der Unizität und Notwendigkeit der euklidischen Geometrie aus dem theologischen
Rahmen herausgehoben und ihm einen säkularen Status gegeben. Er hat erkannt, daß die Notwendigkeit, die den euklidischen Theoremen zugeschrieben wird, weder mit Gott, noch mit Erfahrung, noch mit der Logik gerechtfertigt und begründet werden kann. Er selber war über den Stand des
Parallelenproblems auf dem laufenden, und hat im übrigen geglaubt, das
Problem hätte bereits die gewünschte, klassische Lösung gefunden.[55]
Die *Annahme* von asymptotisch-parallelen Geraden (also implizite einer
hyperbolischen Geometrie) bezeichnete er als eine schlichte *Unmöglich-*
keit und eben diese Unmöglichkeit ist es, die, seiner Meinung nach,
zwangsläufig zu einer *Theorie der Parallellinien leitet, welche nicht*
nur die wahre, sondern auch die einzig mögliche ist.[56] Mit unverhehl-
ter Verachtung äußerte er sich über ein aus einer solchen Annahme ab-
geleitetes antieuklidisches System: *Es gibt also jetzt eine licentia*
geometrica, so wie es längst eine licentiapoetica gegeben hat, - schrieb
er.[57] Er hatte sich vorgenommen, den richtigen Grund zu finden, aus dem
solche antieuklidischen Begriffsbildungen eliminiert werden könnten.
Seine Lösung ist bekannt: Nicht die Gesetze der Logik, sondern die transzendentale Intuition ist berufen, *Luftschlösser* und *Hirngespinste* aus
dem Bereich der Wissenschaft zu eliminieren, sie also gegenüber Literatur abzugrenzen und so Unizität des euklidischen Raumes sowie Notwendigkeit euklidischer Wahrheit mit Gewißheit zu begründen. Gottes Allmacht
wurde von Kant durch die Allmacht der transzendentalen Intuition ersetzt.

Mit seiner Philosophie wäre es eigentlich vereinbar gewesen, auf die
Lösung des Parallelenproblems zu verzichten und das Parallelenpostulat
als einen synthetischen Satz zu betrachten, dessen Notwendigkeit von der
transzendentalen Intuition a priori bestimmt ist. Das ist jedoch nicht
der Fall gewesen. Im Gegenteil. Die Kant'sche Philosophie hat die auf
die Lösung des Parallelenproblems gerichteten Versuche stärker gefördert,
als irgendeine andere Philosophie vorher. Die Anzahl solcher Versuche und
Dissertationen ist unter dem direkten Einfluß der Kant'schen Philosophie
sprungartig gestiegen. Viele seiner Anhänger gaben mit Ausnahme des Parallelenpostulats die a priori Notwendigkeit aller geometrischen Axiome
zu, und so war ihre Beschäftigung mit dem Parallelenproblem von der Hoffnung motiviert, durch einen logischen Beweis des euklidischen Fundamentalsatzes die a priori Notwendigkeit der gesamten euklidischen Geometrie
lückenlos zu beweisen. Diese Motivation hat am Anfang die Versuche zur

Begründung der nichteuklidischen Geometrie geleitet. Alle sind ursprünglich von der These der Unizität der euklidischen Geometrie ausgegangen
und alle haben sie versucht, diese durch die Unmöglichkeit einer antieuklidischen Geometrie zu beweisen. Als Ergebnis sind sie, und sie allein, zum Gegenteil, nämlich zur Begründung der nichteuklidischen Geometrie gelangt. Und die erste praktische Anwendung der neuen Entdeckung
sahen sie alle, vor allem Gauss,[58] eben in dem Zusammenbruch der ursprünglichen These, die den Wahrheiten der euklidischen Geometrie Allein-
gültigkeit und unbedingte Notwendigkeit zuschrieb. Die Kantische Philosophie ist mit der korrekten Lösung des Parallelenproblems durchaus vereinbar. Es ist nur die Pluralität (sowohl semantisch als auch syntak-
tisch) gleichbereichtigter Geometrien, die dem Unizitätsanspruch der
transzendentalen Intuition unversöhnbar gegenübersteht. Angeregt durch
die vielleicht müßigen, immerhin aber als harmlos erscheinenden dia-
lektischen Disputationen über die göttliche Allmacht wurden die Mathematiker des 19. Jahrhundert durch die seltsame innere Dialektik der
Ideengeschichte zur Entdeckung geführt, daß Omnipotenz und absolute Freiheit Merkmale sind, die den menschlichen Geist und allein diesen aus-
zeichnen, ihn im Bereich des Denkens zur *creatio continua* einer Pluralität geometrischer Universa bestimmen.

Ich möchte jedoch nochmals darauf hinweisen, daß die Entstehung der eigentlichen nichteuklidischen Geometrie auf keinen Fall als notwendige
Folge der Lösung des Parallelenproblems dargestellt werden kann. Die
korrekte Lösung des Parallelenproblems verlangt viel weniger. Für diese
ist es hinreichend, den Axiomencharakter des euklidischen Parallelenpostulats - d.h. seine logische Unabhängigkeit, seine Unentscheidbarkeit in bezug auf die absolute Geometrie Bolyais - zu erkennen. Dies
wurde im wesentlichen schon vor Euklid geleistet. Trotzdem führte diese
Erkenntnis nicht zu einer nichteuklidischen, sondern, mit Euklid, zur
axiomatischen Begründung der euklidischen Geometrie. Auf der anderen
Seite ist der strenge Beweis der logischen Unabhängigkeit des Parallelenaxioms für die Errichtung einer eigentlichen nichteuklidischen Geometrie nicht einmal notwendig. Die nichteuklidische Geometrie existierte
bereits seit 40 Jahren, als Beltrami den ersten, übrigens unvollständi-
gen Beweis dieses Metasatzes durchführte. Die relative Konsistenz des
euklidischen und des ihm entgegengesetzten Systems ist jedoch auch dann
und genau mit denselben Mitteln beweisbar, wenn das antieuklidische System als falsch angenommen wird. Die relative Konsistenz ist notwendig,
jedoch nicht hinreichend, um die Wahrheit den antieuklidischen Sätzen
zuordnen zu können; und sie ist noch weniger hinreichend, um die Wahrheit simultan dem euklidischen und dem nichteuklidischen Parallelenaxiom

zuzuordnen. In seiner 1713 publizierten *Clavis Universalis* hat Arthur Collier bereits darauf hingewiesen, daß die Idee einer materiellen Außenwelt an sich völlig widerspruchsfrei ist, daß sie jedoch, wie es ihm zu beweisen gelang, nicht existiert.

Wie gesagt, die korrekte Lösung des Parallelenproblems kann nicht als eine historische Folge des Entstehens der nichteuklidischen Geometrie betrachtet werden, da sie bereits vom Bestehen des euklidischen Systems impliziert ist. Jedoch nicht umgekehrt: Die korrekte Lösung des Parallenproblems impliziert weder die Existenz einer euklidischen, noch die Existenz einer nichteuklidischen Geometrie, da der Beweis der logischen Unabhängigkeit des euklidischen Parallelensatzes bereits in dem reduzierten Rahmen der absoluten Geometrie Bolyais durchgeführt werden kann. Die nichteuklidische Geometrie ist für die Lösung des Parallelenproblems nicht notwendig, die Lösung des Parallelenproblems für die nichteuklidische Geometrie nicht hinreichend.

Die nichteuklidische Geometrie erscheint in der Geschichte als ein Angebot ohne Nachfrage, als Ergebnis eines wahrhaftigen *acte gratuit* des wissenschaftlichen Denkens. In Form des antieuklidischen Systems erschien die nichteuklidische Geometrie als ein logisches Monster der geometrischen Fiktion. Das Ungeheuer war jedoch mindestens insofern nutzlos, als mit seiner Hilfe der Beweis der logischen Unabhängigkeit des euklidischen Parallelenaxioms bereits hätte gewährleistet werden können. Die nichteuklidische Geometrie, als wissenschaftliches System geometrischer Wahrheiten und räumlicher Wirklichkeit, erschien schlimmer als monströs: Sie war überflüssig. Dennoch, einmal erschienen, war es nicht mehr möglich, sie zu eliminieren oder wenigstens auf sie zu verzichten. Als wäre sie von einer *Invisible Hand* aufgezwungen, wurde ihre Existenz mit der Zeit als Notwendigkeit für die weitere Entfaltung der Mathematik erkannt. Unter diesen Umständen stellt sich die Frage, was für eine historische Kraft für die plötzliche Errichtung der nichteuklidischen Geometrie verantwortlich war. Meines Erachtens kann der Durchbruch vom antieuklidischen System zur eigentlichen nichteuklidischen Geometrie nur als Folge einer radikalen Umwälzung im allgemeinen Zustand des Bewußtseins verstanden und erklärt werden.

6.   Philosophische Implikationen der nichteuklidischen Geometrie

Bewußt geworden ist vor allem, daß den geometrischen Wahrheiten weder empirische noch göttliche oder transzendentale Notwendigkeit zugeordnet werden kann, sobald enantiomorphen Universa geometrischer Objekte Existenz, und entgegengesetzten geometrischen Axiomen Wahrheit simultan zugeordnet wird. Die wichtigste philosophische Folgerung dieser Umwälzung liegt auf wissenschaftstheoretischer Ebene. Die nichteuklidische Geometrie hat hier eine völlig neue Konzeption mit sich gebracht, die mit dem Ausdruck *Relativität der Wahrheit* bezeichnet werden kann. Der Terminus *relativ* darf in diesem Ausdruck jedoch nicht im gewöhnlichen, umgangssprachlichen, sondern nur im strengen Sinne der speziellen Relativitätstheorie Einsteins verstanden werden. Dem Begriff des Innertialsystems entspricht der Begriff des Axiomensystems. Den Koordinatenachsen entsprechen die Axiome; der linearen Unabhängigkeit der Koordinaten entspricht die logische Unabhängigkeit der Axiome. Dem Begriff Ruhe entspricht der Begriff Wahrheit; dem Begriff (uniforme lineare) Bewegung entspricht der Begriff Falschheit. Nur in bezug auf ein bestimmtes Innertialsystem hat der Begriff Ruhe einen Sinn; jedes Innertialsystem ist in Ruhe auf sich selbst bezogen und alles befindet sich im Zustand der Ruhe, was mit dem System starr verbunden ist. Ähnlich hat der Begriff Wahrheit nur auf ein bestimmtes Axiomensystem bezogen einen Sinn; jedes Axiomensystem ist auf sich selbst bezogen wahr. Wahr sind auch alle Sätze, die mit dem System der Axiome durch eine logische Kette "starr" verbunden sind. Es gibt kein privilegiertes Bezugssystem, es gibt also kein Bezugssystem, von dem aus behauptet werden könnte, es sei das einzige, das im Zustand der absoluten Ruhe sei. Es gibt kein privilegiertes Axiomensystem in der Geometrie; es gibt also keinen geometrischen Satz, von dem behauptet werden könnte, der Wert der Wahrheit käme allein und ausschließlich ihm zu. Es gibt keine absolute Wahrheit. Im Innern eines Inertialsystems läßt sich auf Grund eines mechanischen Experiments nicht entscheiden, ob es in Ruhe oder in (uniformer linearer) Bewegung ist. Im Innern eines deduktiven Systems läßt sich aufgrund der mit Mitteln dieses Systems durchgeführten Beweise nicht entscheiden, ob seine Axiome den Wert der Wahrheit besitzen oder nicht. Die nichteuklidische Geometrie hat so nicht nur das technische Instrumentarium für die künftige Relativitätstheorie vorbereitet und zur Verfügung gestellt, sondern sie hat zum ersten Mal das Prinzip einer erweiterten, im Sinne Einsteins aufgefaßten Relativität im Bereich der begrifflichen Bezugsysteme der Geometrie durchgesetzt, und so ihre mechanische Anwendung vorweggenommen.

Die Analogie hat jedoch ihre Grenzen. Dem Bewußtsein als dem Einbettungsraum des Denkens kann weder räumliche Extension noch zeitliche Dauer zugeschrieben werden. Jedem axiomatischen System kann je ein Universum zügeordnet werden, und, wie es aus einem von Saccheri bewiesenen Satz der absoluten Geometrie Bolyais folgt, jedes Universum enthält jeweils separat die Gesamtheit z.B. aller Dreiecke. Diese Universa existieren simultan als selbständige Totalitäten in dem Metakosmos des Denkens. Der Satz

von Saccheri verbietet jedoch, sie als komplementäre Teile eines umfassenden Ganzen aufzufassen.[59] Ähnliche Eigenschaften können den mechanischen Inertialsystemen kaum zugeschrieben werden. Auch schreiben
wir, zumal vorläufig, dem physikalischen Universum ohne Zögern Unizität zu. Demgegenüber ist es in der absoluten Geoemtrie Bolyais möglich,
isomorphe Modelle sowohl der euklidischen als auch der nichteuklidischen Geometrie zu konstruieren und mit ihrer Hilfe nachzuweisen, daß
dieselbe semantische Bedeutung, die den Grundbegriffen Punkt, Gerade,
Ebene usw. in der absoluten Geometrie Bolyais zukommt, ihnen sowohl in
der euklidischen als auch in der nichteuklidischen Geometrie zugeschrieben werden muß.[60] Unter diesen Umständen bezeichnen die undefinierten
Grundterme der euklidischen bzw. der nichteuklidischen Geometrie dieselben geometrischen Objekte. Die Schwierigkeiten, die aus dieser Simultanität entgegengesetzter axiomatischer Wahrheiten entstehen, können
leicht ausgeräumt werden, wenn die Axiome als Merkmale des Begriffes *Euklidische Geometrie* bzw. des Begriffes *Nichteuklidische Geometrie* betrachtet
werden. Die Wahrheit formal entgegengesetzter Axiome drückt in diesem Fall
die Tatsache aus, daß die Komprehension des Begriffes *Euklidische Geometrie* ein spezifisches Merkmal - das euklidische Parallelenaxiom, - enthält,
das dem spezifischen Merkmal des Begriffes *Nichteuklidische Geometrie* -
sc. dem hyperbolischen Parallelenaxiom - entgegengesetzt ist.

Die altgriechischen Geometer, Saccheri, Lambert, Taurinus sind keine
Vorläufer der nicht-euklidischen Geometrie, sondern Architekten eines
antieuklidischen Systems, dem sie einzig die Rolle des Zeugen für die
Unizität der euklidischen Geometrie zugewiesen haben. Die Konzeption
der Simultanität geometrischer Wahrheiten und Universa ist ein völliges
Novum, eine noch nie dagewesene Idee, die sich erst mit der Begründung
der nichteuklidischen Geometrie konstituiert hat. Diese Idee hat keine
Vorläufer. Sie konnte aus keiner bereits existierenden Philosophie abgeleitet werden: mit jeder blieb sie in unversöhnlichem Widerspruch.
Trotz des allgemeinen und äußerst heftigen Widerstandes aller Philoso-
phien, Ideologien, Religionen und nicht zuletzt der empirischen Erfahrung und des gesunden Menschenverstands - hat die nichteuklidische Geometrie ihre eigene Philosophie, eine radikal neue Denkungsart durchgesetzt. Mit ihr entsteht eine scharfe Diskontinuität in der Entwicklung
des Denkens - ein Bruch, eine *Katastrophe*. Sie hat keine Vorgeschichte,
keine Vergangenheit. Ihre explosionsartige Entfaltung wurde von keiner
langsamen Akkumulation vorbereitet. Bedeutet die Entstehung der nichteuklidischen Geometrie eine wissenschaftliche Revolution, so kann diese
in ihrem Ergebnis auf eine einzige Idee reduziert werden: die Koexsitenz
entgegengesetzter Welten und Wahrheiten im Bereich des sich selbst

wissenden Denkens. Diese und diese allein ist für die profunde und ir-
reversible Umwälzung verantwortlich, die das gesamte Bewußtsein von der
Ablehnung zu dem radikal entgegengesetzten Zustand der Akzeptierung der
Pluralität *aufgehoben* hat. Das Denken mußte sich selbst negieren.

Mit der Entstehung der nichteuklidischen Geometrie hat sich auf einmal
das klassische Verhältnis zwischen Ethik und Geometrie umgekehrt. Vorher waren alle Bestrebungen darauf gerichtet, die Notwendigkeit ethischer Gesetze mit geometrischer Gewißheit nach dem Muster der Geometrie
zu begründen. Kant selbst betrachtete noch die Geometrie und die Moral
als die einzigen Wissenschaften, deren Wahrheiten mit a priori Notwendigkeit begründet werden können und müssen.[61] Es hat sich jedoch heraus-
gestellt, daß eben die Geometrie eine Wissenschaft ist, wo es keine absoluten und notwendigen Wahrheiten gibt. Nicht die Ethik muß *more geometrico*, sondern im Gegenteil die Geometrie kann nur *more ethico* auf
Freiheit begründet werden.

Die für die nichteuklidische Geometrie konstitutive Freiheit unterscheidet sich jedoch grundsätzlich vom aristotelischen Freiheitsbegriff.
Bei Aristoteles ist der Bereich der Freiheit mit der Abwesenheit des
Zwanges, mit der Indifferenz, mit der Unentschiedenheit identisch. Zwang
und Freiheit stehen hier als zwei entgegengesetzte, dem Gesetz des ausgeschlossenen Dritten unterworfene Modalitäten, nebeneinander. Nicht
nur der Gegenstand, sondern auch die Tätigkeit der Freiheit selbst ist
dem Prinzip des ausgeschlossenen Dritten unterworfen, da sie in der
Entscheidung einer Alternative besteht. Diese Alternative selbst ist
einem Zwang unterworfen: Man muß mindestens, man kann jedoch höchstens
die eine der beiden gegebenen Möglichkeiten wählen.

Im Bereich der Geometrie ist die Abwesenheit des Zwanges mit der Unentscheidbarkeit des euklidischen bzw. des antieuklidischen Parallelenpostulats äquivalent. Der Akt, der die nichteuklidische Geometrie begründet, besteht jedoch nicht in einer von dieser Unentscheidbarkeit
ermöglichten Wahl zwischen den beiden entgegengesetzten Parallelenaxio-
men, sondern in der kategorischen Abschaffung dieses Unizitätszwanges
- mindestens und höchstens eines der beiden als wahr anzuerkennen. So
erweist sich Freiheit hier nicht als Abwesenheit des Zwanges, sondern
als eine dritte Modalität, als ein konstituiver Akt, der den bestehenden
Zwang der Unizität bricht und ihm gegenüber seine eigenen Konstruktionen
zur Existenz bringt und damit implizite die neue Idee der Pluralität
durchsetzt. Diese Freiheit besteht nicht in der Entscheidung einer Alternative und in der Wahl einer von zwei gegebenen Möglichkeiten, sondern eher in der Verwirklichung des Unmöglichen, in der simultanen

Wahl von zwei entgegengesetzten, relativ konsistenten Systemen wahrer
Sätze. Die nichteuklidische Freiheit ist nicht artistotelisch.

Dem ethischen Begriff der Freiheit entspricht auf ontologischer Ebene
der Begriff der Schöpfung. *Ich habe aus dem Nichts eine neue, eine
andere Welt geschaffen*, schrieb Johannes Bolyai 1823 seinem Vater. In
diesem Satz kommt zum ersten Mal das Bewußtsein der Schöpfung zum Ausdruck als unmittelbarer Reflex der eben durchgeführten Tat selbst. Die
Idee jedoch, daß der Mensch Schöpfer von neuen Wirklichkeiten, neuen
Wahrheiten und Welten ist, und daß die Mathematik auch zu diesem Bereich der menschlichen Schöpfung gehört, ist Ergebnis einer langen Entwicklung.

Diese Idee erscheint zum ersten Mal in einem besonders ausdrucksvollen
und schwer übersetzbaren Satz von Plotinos, (Enneade III 8.3):
*Ich bin es, der ich den Gegenstand meiner Anschauung durch die Tätig-
keit des Anschauens selbst hervorbringe, so, wie der Geometer den Gegenstand seiner Anschauung auch dadurch herstellt, daß er sie geometrisch
beweist.*[62] - Der Satz ist ein Wortspiel. Das Verb *betrachten, anschauen*
ist im Griechischem mit *theorein*, der Gegenstand der Betrachtung oder
der Anschauung mit der Vokabel *theorema* ausgedrückt. Da der letzte Terminus auch im Sinne von *mathematischer Satz* verwendet ist, kann das
Verb *theorein* die Tätigkeit bezeichnen, die das mathematische *Theorem*
hervorbringt. Dazu ist noch hinzuzufügen, daß das Verb *machen, hervor-
bringen* mit der Vokabel *poiein*, die auch mit *dichten* übersetzt werden
darf, ausgedrückt ist. Aus dem Neuplatonismus entspringt eine selbständige Gedankenströmung, in der der Mensch als *secundus Deus* definiert
wird. Nicolaus Cusanus interpretiert diese Definition in dem Sinn, daß
Gott den Menschen deswegen nach seinem Ebenbild geschaffen hat, weil
damit der Mensch zur Fortentwicklung des göttlichen Prärogativs der
Schöpfung von Gott selbst beauftragt wurde. Er ist fähig, Dinge zu erschaffen, die Gott am Tag der Schöpfung nicht gemacht hat: Produkte der
Technik, der Kunst oder auch der Mathematik.[63] Cardano behauptet zum
ersten Mal in ihrer Allgemeinheit die Idee, daß die ganze Geometrie nur
insofern eine Wissenschaft ist, als sie das Wissen dessen darstellt,
was das Denken selbst hervorbringt. Der Satz über die Gleichheit der
Winkelsumme des Dreiecks mit zwei rechten Winkeln ist nur deswegen notwendig wahr, besitzt nur deswegen eine absolute Gewißheit, weil das
Denken das Dreieck so gedacht, also gemacht hat.[64] Diese Idee entfaltet sich später bei Vico. Wir sind nicht imstande, sagt Vico, (1710) die
Natur zu erkennen, da wir sie nicht selber geschaffen haben. Das menschliche Denken ist imstande, richtig und vollständig nur das zu erkennen,
was er selbst geschaffen hat. Das menschliche Wissen ist eine Nachahmung

des göttlichen: *ad dei instar*, aus keinem gegebenen Ding, sondern quasi
aus dem Nichts hat das Denken die Grundbegriffe der Geometrie geschaffen.
Die Wahrheit der Sätze, die diese Dinge beschreiben, ist gewiß, notwendig
und absolut, und drückt nichts als das Einverständnis, die *adaequatio*
des Denkens mit sich selbst aus. Die Zuordnung des Satzprädikates 'wahr'
auf metasprachlicher Ebene drückt die Tatsache aus, daß das Denken das
wahrgenommen hat, was es selbst auf der Ebene der Objektsprache in Form
eines Satzes ausgesprochen hat. *Cogito ergo sum:* Ich denke ein Dreieck
mit einer Winkelsumme gleich mit zwei rechten Winkeln, also existiert
dieses Dreieck als solches.[65]

Merkwürdigerweise tauchen die Elemente dieser neuen Philosophie der
Mathematik Ende des 18., Anfang des 19. Jahrhunderts nicht bei Mathematikern, sondern bei vier Poeten und Schriftstellern der Romantik auf.
In England bei Coleridge (wahrscheinlich unter dem Einfluß Vico's), in
Deutschland bei Novalis, in Holland und Frankreich bei François Hemsterhuis[66] und in Amerika bei Edgar Allan Poe. Die reine Mathematik,
schreibt Novalis, ist die Anschauung des Verstandes *qua* Universum. Die
Mathematik ist *ächte Wissenschaft*, weil sie gemachte - also freige-
schaffene Kenntnisse enthält. Diese sind Produkte geistiger Selbsttätig-
keit, weil die Mathematik, wie Novalis sich ausdrückt, *methodisch ge-
nialisiert*. Das *Genie* ist nach Novalis *das transsubstantiierende Prinzip*,
da das Genie *das Unmögliche möglich macht. Die Algebra ist Poesie.*
(Poesie, Poet ist von Novalis im griechischen Sinne des Wortes als Macher,
als Schöpfer interpretiert). Der *poetische Philosoph*, schreibt er, *ist
en état de créateur absolu: Ein Triangel wird schon auf diese Art cre-
iert. Die sensibelste, frappanteste Wahrheit hat eine ihr entgegengesetzte, ebenso frappante Wahrheit.* Es ist auch merkwürdig, daß diese die
Mathematik betreffenden Bemerkungen den Titel *Postulat und Gegenpostu-
lat* tragen.[67]

Aber sicher ist niemend weitergegangen, als der Poet Edgar Allan Poe.
Für ihn sind die Kosmologien eigentlich poetische Werke in dem viel-
fältigen semantischen Sinn dieses Ausdrucks. Das freie Denken ist der
eigentliche Verfasser der Kosmologien, Schöpfer von unendlichen Universa, die alle in ihm wie in der Brust Gottes zustande kommen. Das einzige
Kriterium der Wahrheit der Sätze, die diese Universa beschreiben, ist
die Widerspruchslosigkeit, die Poe mit dem Ausdruck *consistency* bezeichnet. Die Axiome, in Poe's Bezeichnung (1848) *axiomatic principles*, sind
umwandelbar und mit ihrem Gegensatz ersetzbar. So entstehen eine unendliche Menge von Universa, die untereinander *symmetrisch* sind.[68] *Symmetrisch* im Sinne von Poe bedeutet die von entgegengesetzten Axiomen bestimmte Opposition; und das, was er unter dem *Prinzip der Symmetrie*

versteht, ist eigentlich identisch mit der auf den Rang eines Meta-
Prinzips gehobenen Idee der Negierbarkeit der Axiome. In der Tat, wenn
das Axiom im Mittelalter als eine *propositio fide digna quod negari non
potest* definiert wurde, muß es nach der Entstehung der nichteuklidischen
Geometrie als ein Satz definiert werden, der zwar *fide digna* ist, jedoch
den zu negieren möglich ist und in unserer Macht steht; ihn zu negieren
ist sogar eine moralische Pflicht, eine wissenschaftspolitische Aufgabe,
da nur durch seine Negation die freie Entfaltung der Wissenschaft realisiert werden kann. Diese *ungeheure Macht des Negativen*, die von Hegel
als *die Energie des Denkens, des reinen Ichs* bezeichnet wurde, ist dafür verantwortlich, daß das anti-euklidische System in der Gestalt der
nichteuklidischen Geometrie *ein eignes Daseyn und abgesonderte Freiheit*
gewinnen konnte. Die nichteuklidische Geometrie erscheint so als *der an
die Zeit entäußerte Geist.*[69]

Im übrigen kann der Gegensatz der Axiome tatsächlich als ein Symmetrie-
verhältnis gedeutet werden. Vor allem kann jedes Axiomensystem - als
logische Konjunktion von N+1 wahren, untereinander logisch unabhängi-
gen Sätzen - als ein N-dimensionaler Simplex aufgefaßt werden, dessen
Eckpunkte die einzelnen Axiome vertreten. Sind N Axiome eines anderen
Systems mit N Axiomen des ersten identisch, das (N+1)-te des einen dem
(N+1)-ten des anderen jedoch formal entgegengesetzt, so sind die beiden
N-dimensionalen Simplices zueinander - auf einen (N-1)-dimensionalen
Unterraum bezogen - *symmetrisch.*[70]

Keiner dieser Dichter hat von der nichteuklidischen Geometrie gehört.
Auch ist anzunehmen, daß die Schöpfer der nichteuklidischen Geometrie
nicht von den Texten eines Novalis oder Poe gewußt haben. Wir stehen
hier also vor einer allgemeinen Entwicklung des Denkens, die auf verschiedenen Ebenen dieselben Ergebnisse hervorgebracht hat. Die nichteuklidische Geometrie ist zweifelsohne das bedeutendste, deutlichste und
wirksamste Ergebnis dieser Entwicklung. Das Bewußtsein von dem, was das
Denken in Kunst und Mathematik tatsächlich macht, kommt in derselben Zeit
in verschiedenen Subjekten zum Ausdruck. Es ist der Ausdruck des neu
entstandenen Selbstbewußtseins des sich selbst betrachtenden Denkens.
Erst am Ende des 19., Anfang des 20. Jahrhunderts kommt dieser neue
Bewußtseinszustand bei Mathematikern wie Weierstrass, Kronecker, Dedekind, Cantor, Sylvester, Peirce zum Ausdruck. Vorwiegend in aphorisma-
tischen Formulierungen haben sie die Mathematik als eine Art Dichtung,
als Ergebnis der Freiheit oder der Schöpfung bezeichnet. Merkwürdigerweise haben sie jedoch ihre Äußerungen auf das Gebiet der verschiedenen
Zahlensysteme und Algebren oder auf den Bereich des Aktual-Unendlichen
begrenzt. In bezug auf die Geometrie war es Einstein, der als erster

mit notwendiger Klarheit und Kategorizität festgestellt hat: *Die Axiome der Geometrie sind freie Schöpfungen des menschlichen Geistes.*

In allen diesen Äußerungen blieb jedoch der nähere Sinn der Ausdrücke *Dichtung, Freiheit, Schöpfung* unerklärt und sie wirkten eher wie unverbindliche Metaphern. Die philosophische Relevanz der nichteuklidischen Geometrie besteht u.a. auch darin, daß die Explizierung ihrer genetischen Mechanismen eine scharfe Bestimmung der sachlichen Inhalte dieser Begriffe erlaubt und sie als unerläßliche, präzise, technische Grundtermini der strukturellen Geschichtsdarstellung auszeichnet.

Es muß vor allem festgehalten werden, daß die nichteuklidische Geometrie tatsächlich Ergebnis einer *Schöpfung* im eigentlichen Sinne des Wortes ist. Mit ihr ist eine neue Welt entstanden, die vorher nicht existierte. In der Tat ist jede Geometrie eine Welt. Der Existenzbereich dieser Welt ist die Totalität geometrischer Objekte, das Universum als Raum, als *res extensa*; in ihrem Satzsystem kommt eigentlich das Bewußtsein dieser geometrischen Welt als *res cogitans* zu sprachlichem Ausdruck. Der Seinsbereich der Welt ist geometrisch abgeschlossen: In dem euklidischen oder in dem nichteuklidischen System sind jeweils *alle* euklidischen und nichteuklidischen, und nur diese Objekte, anwesend. Das Satzsystem als Bewußtsein dieser Welt ist gespalten: Es enthält eine Untermenge von wahren Theoremen - diese bilden zusammen die *Wissenschaft* der Welt - und eine Untermenge in sich konsistenter, jedoch falscher Sätze; denen entspricht nichts im Bereich des Seins, sie sprechen von Dingen, die unmöglich sind. Die Schöpfung dieser neuen Welt geht jedoch nicht von dem leeren Nichts, sondern von einer schon gegebenen Welt aus. Die nichteuklidische Geometrie entsprang dem Innern der euklidischen Welt. Es ist leicht, eine Welt aus dem Nichts zu schaffen, da in bezug auf das Leere die konkrete Struktur der Welt nichts Unmögliches enthält. Es ist jedoch sehr schwer, aus einer schon gegebenen Welt ein in bezug auf das Gegebene unmögliches Universum von geometrischen Objekten hervorzubringen und dabei die negierte Welt als weiterhin wahr und reell aufzubewahren.

Das eigentliche Instrument dieser Schöpfung liegt auf logischer Ebene und besteht aus einem doppelten Negationsakt. Erst wird auf der Ebene der Objektsprache das euklidische Parallelenaxiom negiert. Als Ergebnis entsteht ein antieuklidischer Satz, der in trivialer Weise beweisbar falsch ist, und so ein eigentliches Theorem der euklidischen Geometrie darstellt. Bereits diese erste Negation erweist sich als Zeichen einer Freiheit: Ihr Ergebnis ist in der Tat völlig unabhängig von der Tyrannis der Wirklichkeit. In dem Bewußtsein der Falschheit, das diesen Satz

begleitet, drückt sich jedoch die Anerkennung der Tatsache aus, daß
eine andere Wirklichkeit, ein anderes Universum als das gegebene, nicht
existieren kann. Der eigentliche Schöpfungsakt ist durch eine zweite
Negation auf der Ebene der Metasprache vollzogen: Dem Negat des euklidischen Axioms wird ohne Beweis der Wert der Wahrheit zugeordnet. Damit
scheidet sich der antieuklidische Satz aus dem Netz der euklidischen
Theoreme aus, und konstituiert sich zu einer neuen, axiomatischen Wahrheit. Erst mit dieser zweiten Negation entfaltet sich die nichtaristo-
telische Freiheit als Schöpfung.

Freiheit, *Schöpfung*, *Negation* erscheinen so als drei Hypostasen eines
einzigen Denkens *in actu* . Die erste Negation kann noch als ein destruk-
tiver Akt bezeichnet werden, da sie einen Satz im Bewußtsein seiner
Falschheit produziert. Die zweite Negation ist jedoch ein durchaus positiver Akt. Sie zerstört nur die Hindernisse, die der freien Entfaltung
des Denkens, der Dilatation des geometrischen Metakosmos im Weg stehen.
Allein ihr Anspruch auf Alleinvertretung der Wahrheit und Wirklichkeit
wird der euklidischen Welt abgesprochen. Sie bewahrt jedoch ihre frühere
Wahrheit und Wirklichkeit auch nach der Schöpfung der nichteuklidischen
Welt. *Negare*, *tollere*, *conservare* sind die drei Momente dieses Schöp-
fungsaktes: Die bereits existierende euklidische Welt wird negiert, eine
zweite Negation vehikuliert die Entstehung einer neuen Welt, dabei wird
jedoch die alte Welt aufbewahrt. Diese zweifache Negation ist konstruk-
tiv und konstitutiv. Den eben angeführten trinären Ausdruck hat die
klassische deutsche Philosophie durch den plastischen Terminus *Aufheben*
ersetzt. Man kann also in einem Wort sagen, daß die nichteuklidische
Geometrie durch die *Aufhebung* der euklidischen Geometrie entstanden ist.
Da die Negation, die diese Aufhebung trägt, ein reiner innerer Akt des
Denkens ist, der ohne äußeren Anlaß entsteht, kann der Übergang von der
euklidischen zur eigentlichen nichteuklidischen Geometrie mit Recht als
eine von der Außenwelt völlig unabhängige, freie Selbstbewegung des
Denkens bezeichnet werden.

Der neue Zustand des Denkens, der diese Selbstbewegung steuerte und die
nichteuklidische Geometrie hervorbrachte, kann abschließend als Ergebnis
eines Selbstbewußtwerdens betrachtet werden: Das allgemeine Denken wird
sich dessen bewußt, daß sein Wesen in seiner nichtaristotelischen Freiheit besteht, und daß die Ausübung dieser Freiheit die notwendige und
hinreichende Bedingung ist für die unbehinderte Entfaltung aller seiner
Fähigkeiten.

Anmerkungen

1. A. Einstein, Johannes Kepler; in: Mein Weltbild (Hrsg. C. Seelig)
   Frankfurt/M-Berlin 1968, 150 - 151

2. Ibid. 119

3. Ibid. 151

4. Der surrealistische Akt wurde von André Breton als ein *Zusammentreffen zwischen einem Regenschirm und einer Nähmaschine auf
   einen Operationstisch* definiert.

5. *Die früheste Geometrie beschäftigte sich, wie uns die alte Über-
   lieferung lehrt, mit der Messung und Verteilung der Ländereien,
   woher sie* Feld-messung, γεω-μετρία, *genannt wird. Der Gedanke
   einer Messung nämlich war den Ägyptern an die Hand gegeben durch
   die Überschwemmung des Nil.* (Heronis Alex. geometric. reliqiae,
   Berlin 1864, hrsg. Hultsch, 138; vgl. auch Herodot, Reisebe-
   richte II 109, Strabo, Chrestom. XVII 787).

6. Im Kidrontal, in der Nähe von Jerusalem, befindet sich das sog.
   Grab von Abshalom, ein hellenistisches Baudenkmal (vgl. N. Avigad, The Ancient Monuments in the Kidron Valley, Fig. 50-53,
   Jerusalem 1967), das von einer ugf. 6 m hohen *Pseudosphäre* ge-
   krönt ist; (s. Figur). Auf der Oberfläche dieses Körpers mit
   konstanter negativer Krümmung kann ein (euklidisches) Modell
   der hyperbolischen Geometrie dargestellt werden. Diese Tatsache wurde jedoch erst 40 Jahre nach der Entstehung der nichteuklidischen Geometrie entdeckt und konnte erst mit Hilfe der
   schon ausgebauten hyperbolischen Geometrie nachgewiesen werden.

7. *Insofern sich die Sätze der Mathematik auf die Wirklichkeit beziehen, sind sie nicht sicher, und insofern sie sicher sind,
   beziehen sie sich nicht auf die Wirklichkeit;* (Einstein, Geometrie und Erfahrung, op. cit. 120).

8. G. Saccheri S. J., Euclides ab omni naevo vindicatus: sive conatus
   geometricus quo stabiliuntur prima ipsa universae Geometriae
   Principia, Mediolani/1733; Th. Reid, An Inquiry into the Human
   Mind on the Principles of Common Sense, (Ch. VI 9: Of the Geo-
   metry of Visibles), Edinburgh 1764; J. H. Lambert Theorie der
   Parallellinien (1776) in: Leipziger Magazin f. reine u. angew.
   Math., 1786, 137 - 164, 325 - 358; F. L. Wachter, Brief an
   Gauss, 12. Dez. 1816, in: Gauss' Werke VIII, 175 - 176; F. K.
   Schweikart, Über Astralgeometrie (1818), in: Gauss', Werke VIII
   180; F. A. Taurinus, Theorie der Parallellinien, Köln am Rhein
   1825; F. A. Taurinus, Geometriae prima elementa, Coloniae
   Agrippinae 1826.

9. W. Sartorius von Waltershausen (Gauss zum Gedächtnis, Leipzig 1856,
   81) schreibt, Gauss hat *mit dem Namen Antieuklidische Geometrie
   eine selbständige Geometrie bezeichnet, die er gelegentlich einmal verfolgt* hat; diese ist die Geometrie, die dann entsteht,
   wenn man das euklidische Parallelenaxiom *nicht zugeben* will. -
   Wachter an Gauss, 12. Dez. 1816: *Also die anti-Euklidische oder
   Ihre Geometrie wäre wahr; ... Das Resultat der Bisherigen wäre
   also so auszusprechen: Die Euklidische Geometrie ist falsch;*
   (vgl. Gauss, Werke VIII 175). - In seinen Erinnerungen schreibt
   Johannes Bolyai, sein Vater, Wolfgang, hat die *Absolute Wissenschaft des Raumes* im Febr. 1825 schmälernd und verachtend mit
   dem Argument zurückgewiesen, diese *sei NUR die Ausarbeitung des*

antieuklid'schen Systems; (vgl. P. Stäckel, Wolfgang und Johann Bolyai, Leipzig und Berlin 1913, I 87). - In seiner *Raumlehre* (1851) schreibt er: könnte eine *Entscheidung... nur zu Gnaden des Euklidischen Systems ausfallen,* so würde das *die Unwahrheit des Anti-Euklidischen, nämlich* des *auf der Voraussetzung der nicht-allgemeinen Wahrheit des XI. Euklidischen Axioms... gebauten Raumsystems* bedeuten; (Stäckel, op. cit. I 238).

10. Vgl. z.B. Chr. Richter, Ueber die fabelhaften Tiere, Gotha 1797; J. L. Borges-Margarita Guerrero, Manuel de zoologie phantastique, Paris 1970; Prof. Dr. Harald Stümpke, Bau und Leben der Rhinogradentia, Stuttgart 1975.

11. I. Toth, Das Parallelenproblem im Corpus Aristotelicum im Archive f. History of Exact Sciences III Nr. 4 - 5, Berlin-Heidelberg-New York 1967; Non-Euclidean Geometry before Euclid, in Scientific American 1969 Nov., 87 - 98; Aristoteles in der Entwicklungsgeschichte der geometrischen Axiomatik, Moskau 1971; Geometria more ethico; die Alternative: euklidische oder nicht-euklidische Geometrie bei Aristoteles und die axiomatische Grundlagen der euklidischen Geometrie, in: ΠΡΙΣΜΑΤΑ, Festschrift f. Willy Hartner (Hsg. Y. Maeyama und W. G. Saltzer) Wiesbaden 1977.

12. Anal. prior. II 17, 66a11-13; anal. Poster. I 12, 77b16-28; II 2, 90a13, 33-34; top. IX 10, 171a14-16; phys. II 9, 200a17-20, 30; de caelo I 12, 281b5-7; problem. XXX 7, 956a15-27; metaph. IX 10, 1052a4-11; eth. nicom. VI 5, 1140b13-15; magna moral. I 10, 1187a29-b4; eth. eudem. II 6, 1222b20-39.

13. Anal. prior. II 17, 66a11-13. Der Satz beweist die Unvereinbarkeit der elliptischen Geometrie mit der absoluten Geometrie Bolyai's, oder, - was auf dasselbe herauskommt, die Unmöglichkeit nicht-schneidender Geraden in der elliptischen Ebene.

14. Für eine ausführliche Analyse der beiden letzten Stellen (1140b13-15 und 956a15-27) vgl. I. Toth, Geometria more ethico (s. Anm. 11).

15. Das Parallelenpostulat ist in den Elementen, wie folgt, formuliert: für *alle* komplanaren Geradenpaare, *a*, *b*, gilt daß sie in einem Punkt, *P*, inzidieren, wenn sie mit einem gemeinsamen Scheidenden, *c*, in einer der beiden von *c* bestimmten Halbebenen Innenwinkel bilden, die zusammen kleiner als zwei rechte Winkel ausmachen. Kürzer: *alle* zueinander konvergierenden komplanaren Geradenpaare *a*, *b* sind inzident.

16. *Nicht alle* zueinander konvergierenden komplanaren Geradenpaare sind inzident; *es existiert* folglich mindestens ein konvergentes komplanares Geradenpaar, das nicht inzident ist.

17. Procli Diadochi in primum Euclidis Elementorum librum commentarii (ed. G. Friedlein), Lipsiae 1873, 175-177, 182-184, 191-193, 361-372.

18. Proclus, op. cit. 192

19. Proclus, ibid. 369-370. Für eine ausführliche Analyse dieses erstaunlichen Arguments vgl. Toth, Das Parallelenproblem (s. Anm. 11).

20. Nach einer vorläufigen Statistik haben sich mit der Lösung des Parallelenproblems beschäftigt eine Anzahl von ugf. 4 - 5 Autoren zwischen dem I. vorchristlichen und VI. nachchristlichen Jahrhundert; ugf. 6 zwischen IX-XIV., 3 - 4 im XVI., ugf. 11 (davon 5 in Italien) im XVII. Jahrhundert. Die Anzahl der Lösungsversuche war ugf. 20 (12 in Deutschland, 5 in Frankreich) zwischen 1700-1760 und ugf. 52 zwischen 1760-1800; davon wurden 36 in Deutschland, 6 in Italien, 3 in Frankreich, 3 in England publiziert.

21. Thabit b. Qurra (IX. Jhdt), Alhazen (XI. Jhdt), Omar Khayyam (XI-XII.), Nasreddin Tusi (XIII. Jhdt), Lewi b. Gerson (Südfrankreich, XIV.Jhdt). - Vgl. B. A. Rosenfel'd, Istorija neewklidowoj geometrii, Moskwa 1976, 42-89.

22. ...*ce seroit perdre le temps inutilement que de se rompre la teste pour la prouver* - schreibt Arnauld über die Euklidische Aussage (Nouveaux Elémens de Géométrie, 1683, 126-127).

23. Montucla, Histoire des mathématiques, Paris 1758, I 219-221.

24. Malebranche, La recherche de la vérité II 2.6

25. D'Alembert, Oeuvres I, 269, 279-280, 295

26. Lacroix widmet ganze Tiraden den *efforts inutiles pour démontrer dans toutes ses parties la théorie des parallèles*. Wozu alle diese Versuche gut sind, ist nichts als *alambiquer les notions les plus claires, obscurcir, par des preuves superflues, ce qui est évident par soi-même.* (Vgl. Lacroix, Essais sur l'enseignement des mathématiques Paris 1801, 174, 278, 283).

27. *Die meisten großen Mathematiker halten es nicht der Mühe werth, oder kümmern sich nicht darum, die Anfänge der Geometrie zu berichtigen* - so schildert ein Anonymer Verfasser einer Neuen Theorie der Parallellinien die Lage in 1830, (Quatember Zeitschrift Bd. 3, 35).

28. Moses ben Maimon, sein Leben, seine Werke und sein Einfluß (hsg. Bacher, Brann, Simonsen, Guttmann), Leipzig 1908

29. Moïse ben Maimoun, Le guide des égarés, I 73.3, 73.10, 74.6, 75.1, 75.5; II 13.2, 14.7 (hsg. S. Munk), Paris 1856. - Gegen Maimonides und Aristoteles behauptet Chasdai Crescas in seinem *Or Adonai* (1410) die *Möglichkeit der Existenz von mehreren Welten. Jedes Argument, das die Möglichkeit von mehreren Welten negiert*, - schreibt er, (propos. I 2), - ist *"eitel und ein Haschen nach Wind"*.

30. Darauf deutet eine Stelle aus Aristoteles, Metaphysik hin (1052a 6-7). - Vgl. auch Joannes Duns Scotus, Opera omnia, II 300: *tu cognoscis unumquemque triangulum; ergo omnem.*

31. Diese Stelle wird in allen Ausgaben, wie folgt, übersetzt: ich behaupte nämlich, daß es unmöglich ist für das Dreieck drei Winkel zu haben, die zusammen zwei rechte Winkel ausmachen, *wenn gewisse Prämissen gegeben sind* (εἰ τάδε, 281b6); und die Diagonale des Quadrates ist kommensurabel, *wenn gewisse andere Prämissen gegeben sind* (εἰ τάδε; b7). Diese Übersetzung folgt dem Text des Codex Vatic. Graec. 253 (fol. 17 rec., 19; 20) aus dem XIII.Jhdt. Die meisten Codizes enthalten jedoch das Demonstrativpronomen εἰ τάδε nur einmal, (in 281b6), - auf den ersten

Satz der Stelle bezogen; (so lautet der Text in allen Handschriften, die vor dem XIII. Jhdt. entstanden sind). Der zweite Satz der Stelle (...; und die Diagonale ist kommensurabel) verliert jedoch den sinnmäßigen Zusammenhang mit dem Kontext, wenn das Demonstrativpronomen nur dem ersten Satz hinzugefügt ist. Die Autopsie der Codizes hat jedoch zum Ergebnis geführt, daß die älteste Handschrift des de caelo, der Codex Vindobonensis phil.gr. 100, aus dem IX. Jhdt., anstatt dem Pronomen εἰ τάδε *(wenn das und das...)* in 281b6 den Ausdruck εἶτα δέ (fol.65 rec.6) enthält, dessen Sinn: *sodann ferner* (nach Liddell-Scott, Greek-English Lexicon, 1968: *and so, therefore, accordingly)* - eine klare implikative Bedeutung hat. (Derselbe Ausdruck befindet sich auch im Cod.Paris. gr. 2033, fol. 299 rec. 22; XV.Jhdt). Infolgedessen lautet die sinnvolle Übersetzung der Stelle 281b5-7: *ich behaupte nämlich, - ist es unmöglich für das Dreieck eine Winkelsumme gleich mit zwei rechten Winkeln zu haben; sodann ist auch die Quadratdiagonale* (mit der Seite) *kommensurabel.*

32. I. Toth, Die nichteuklidische Geometrie in der Phänomenologie des Geistes, Frankfurt/M 1972, 25, 69; (vgl. auch: D. Nestlé, Eleutheria, Tübingen 1967, bes. Kap. 6).

33. Magna moralia I 10 (1187a29-b4); ethica eudem. II 6 (1222b21-39); vgl. auch Toth, Geometria more ethico; (s. Anm. 11)

34. Diese Aussage ist zwar ein geometrischer Satz, schreibt Aristoteles, - (da sie in geometrischer Sprache formuliert ist), - sie ist jedoch *a-geometrisch*, in demselben Sinn wie ein musikalisches Werk *a-musikalisch* ist, da die Geometrizität in ihr (und vor allem in der ἀρχή, in dem Axiom also, aus dem sie abgeleitet ist), in einer *degenerierten, verdorbenen, schlechten Weise enthalten ist;* (φαύλως ἔχειν; 77b26).

35. I. Toth, Die nichteukl. Geom. in d. Phänom. d. Geistes (s. Anm. 32) 35, 109; vgl. Plato, Sympos. 205B; Soph. 236E-237A, 260D, 265B; Rep. 596B-597D, 602B-603B; Parm. 132D, sowie J. P. Vernant, Remarques sur les formes et les limites de la pensée technique chez les grecs, in: Rev. d'Hist. Sci., 1957, 205-255; B. Schweizer, Mimesis u. Phantasia, in: Neue Heidelb. Jhb. 1928, 28-132; Ella Birmelin, Die kunsttheoretischen Gedanken in Philostrats Appolonios, in: Philologie 1933, 149-180, 392-414; G. May, Schöpfung aus dem Nichts, Berlin 1978.

36. Plato, Sympos. 205B

37. J. Guttmann, Der Einfluß der maimonidischen Philisophie auf das christliche Abendland, in: Moses ben Maimon, Sein Leben etc. 175-204; (s. Anm. 28); Ders., Die Scholastik des XIII.Jhds. in ihren Beziehungen zum Judentum, Breslau 1902; M. Joel, Verhältnis Alberts des Grossen zu Moses Maimonides, Breslau 1876; W. Kluxen, Maimonides u. die Hochscholastik, in: Philos. Jhb. 1955, 151-165.

38. *Non potest fieri per miraculum (...): sicut quod triangulus non habeat tres angulos aequales duobus rectis;* (Summa theologiae, partis tertiae supplementum, Qu. 83 art. 3 ob. 2), vgl. auch Summa contra gentiles, lib. II cap. 25 *(Qualiter omnipotens dicatur quaedam non posse).* §1022: *Deus facere non possit: sicut (...) quod triangulus non habeat tres angulos aequales duobus rectis.*

82

39. *Questioni scholastice e vane di quel tempo*, kommentiert A. Cesari (Bellezze della Commedia di Dante III 226, Parma 1845); *notize inutili, benchè curiosissime* - vermerkt *ad locum* P. Venturi (Il Paradiso di Dante, 149; Lucca 1732).

40. Voltaire, Oeuvres, Paris 1855, VIII 465.

41. Auf die Eigenartigkeit und Originalität der cartesischen Konzeption wurde ich durch einen Vortrag von Harry Frankfurt, Descartes on the Creation of the Eternal Truths aufmerksam; (Princeton University 1976 März).

42. Spinoza, Cogitata metaphysica II 9

43. Gregorius Ariminensis Super Primum et Secundum Sententiarum, Venetiis 1522; fol. 166 - 167

44. Descartes, Oeuvres (hsg. Adam-Tannery), Paris 1973; VII 64-70, 379-384, I 152-153; III 383, 433; IV 118-120; V 223-224, 272-274.

45. Descartes, Oeuvres VII 201-202

46. Ibid. VII 278, 327; III 405-411.

47. Descartes, Discours de la méthode suivie des Méditations métaphysiques, Paris 1938, 224-225.

48. Spinoza, Ethica I 17 scholion, II 49; Epist. 21, 42, 56; Princ. philos. Cartes. I prolog, I 5 schol.

49. Malebranche, Oeuvres complètes III 31

50. Hobbes, Leviathan I 11

51. Fénelon, Traité de l'existence de Dieu, Paris 1880, 57.

52. G. Bellermann, Beweis aus der neueren Raumtheorie für die Realität von Zeit u. Raum u. für das Dasein Gottes, in: Progr. Realgymn. Königstadt; Berlin 1889.

53. Malebranche, loc. cit. (s. Anm. 49)

54. Kant, Werke (hsg. Weischedel) Darmstadt 1971, I 32-36, 502-503, 655; III 76

55. Kant's Nachlaß I 31-51

56. Nachlaß VII 421

57. Werke III 307; (vgl. Anm. 54)

58. Gauss an Wolfgang Bolyai 6. März 1932, in: P. Stäckel, op. cit. 94

59. Der Satz von Saccheri lautet: wenn in einem einzigen, so ist die Winkelsumme in jedem Dreieck entweder kleiner, oder gleich oder aber größer als zwei rechte Winkel. Die Hypothese, die Winkelsumme wäre in einem *Teil* der Menge *aller* Dreiecke gleich - in dem komplementären Teil jedoch nicht gleich mit zwei rechten Winkeln, steht in offensichtlichem Widerspruch zu Saccheris Theorem.

60.   I. Toth, An Absolute-Geometric Model of the Hyperbolic Plane and some Related Metamathematical Consequences, in den Akten des 6. Internat. Kongr. f. Logik, Philos u. Methodol. d. Wiss. Hannover 1979 (im Druck). Ders., Un problème de logique et de linguistique concernant le rapport entre la géom. euclidienne et non-euclidienne, in Actes du Coll. Internat. Pensée et Language Mathémat., Luxembourg 1978, 93-142.

61.   Kant, Kritik d. reinen Vernunft, Riga 1787, 508

62.   Τὸ θεωροῦν μου θεώρημα ποιεῖ, ὥσπερ οἱ γεωμέτραι θεωροῦντες γράφουσιν; Plotin, Ennéades (hsg. Bréhier) Paris 1925, III 157.

63.   Cusanus, Opera, Parisiis 1514, I fol. 26v, 82r, 165r, 184v, 189rv; II fol. 54v, 186r; III fol. 93-97; Brief an Nicolaus Albergati 1463, hrsg. Gerda v. Bredow (Sitzungsber. d. Akad. d. Wiss. Heidelberg 1955); vgl. auch H. Blumenberg, 'Nachahmung der Natur'; zur Vorgeschichte des schöpferischen Menschen, in: Studium Generale 1957, 266.

64.   Cardani Opera, Lugduni 1563, X 4.

65.   Vico, Opere, Napoli 1858, I 77, 117, 131.

66.   F. Hemsterhuis, Oeuvres compètes, Paris 1809, I 130, 247, II 50, 151, 178, 183.

67.   Novalis Schriften (hsg. Kluckhohn u. Samuel) Stuttgart 1960 - 1975; II 625, III 168, 175, 309, 415, 445, 473, 593-594.

68.   E. A. Poe, Complete Works (hsg. Harrison) New York 1902, XVI 196, 209, 241, 276, 302, 311.

69.   Hegel, Phänomenologie des Geistes, (hsg. Schulze), Berlin 1841, 25, 590. - Hegel war übrigens der Meinung (genau wie Diderot, Kant und die meisten der großen französischen Mathematiker des XVII. Jhdts), daß mit dem euklidischen System die Geometrie das Ende ihrer Geschichte erreicht hätte; (vgl. Phänomenologie 29-35; Sämtliche Werke, Stuttgart 1940, XXVII 38, XVIII 369).

70.   Vgl. I. Toth, Die nichteukl. Geom. in d. Phänom. d. Geistes, 86-91; Ders., Un probl. de logique, 138-142; (s. Anm. 32, 60).

UNITARY FIELD THEORY, GEOMETRIZATION OF PHYSICS OR PHYSICALIZATION OF GEOMETRY?*+

Peter G. Bergmann, Syracuse University, Syracuse, N.Y., USA

Almost from the day that Albert Einstein had completed the conceptual framework of the general theory of relativity, he began to search for an even more comprehensive theory that would deal uniformly with the forces of gravity, of electricity and magnetism, and of all the other force fields that might exist in nature. Many of his contemporaries, among them H. Weyl, Th. Kaluza, and E. Cartan, also involved themselves in this search, which became to be known as the program of unitary field theory (einheitliche Feldtheorie). That this search did not lead to definitive success in Einsteins's lifetime ist well known. Today this search has once more captured the interest and the imagination of theorists, with a somewhat different slant. As much of the emphasis of the talks scheduled at this symposium is on contemporary efforts, permit me to address myself both to past and to current activities.

When H. Minkowski cast Einstein's special theory of relativity into geometric language, Einstein is supposed to have commented: "Now the mathematicians have succeeded in casting my theory into such a form that I myself no longer understand it". We all know that this adverse reaction did not last long. Throughout his later life Einstein used geometric language freely. Nevertheless, late in his life Einstein appears to have retained his suspicions of geometry as a means for discovering new physical insights. I quote from his autobiographical notes (A l b e r t E i n s t e i n, *Philosopher-Scientist*, P. Schilpp, Ed., 1949):

> *Man darf aber die erwähnte Sünde nicht so weit legitimieren, daß man sich etwa vorstellt, daß Abstände physikalische Wesen besonderer Art seien, wesensverschieden von sonstigen physikalischen Größen ("Physik auf Geometrie zurückzuführen" etc.).*

I suppose that it would be possible to discover remarks of Einstein admitting of a contrary interpretation. But I believe that almost all of Einstein's explanations of his theories, in technical papers and in popular presentations, emphasize physical, not geometric reasoning.

Of course, geometric ideas have played an important role, in Einstein's

---

*   Summary of Talk Presented at the Einstein Symposium Berlin at the Technische Universität Berlin on March 27, 1979

+   Supported by the National Science Foundation (USA) Grant #Phy78-06721

search and in that of others. Can we identify the role of geometry in a physicist's search for the laws of nature? And more particularly, what is the role of geometry in the search for a unitary field theory?

Let me begin by recounting the perceived strengths of the general theory of relativity, strengths that ought to be retained in prospective unitary field theories. There is, first of all, the abandonment of the inertial frames of reference as privileged frames, and with it the replacement of the Poincaré group by the group of diffeomorphisms as the physically motivated invariance group of physics. Next there is the unification of the field equations with the ponderomotive laws, an accomplishment that in classical physics resembles the fusion of fields and particles in quantum field theory. There is, finally, the recognition that geometric laws may be a good way of formulating the dynamics of a physical field.

What about shortcomings of general relativity; There is, glaringly, a dichotomy between the gravitational field on one side, and all remaining fields on the other. Next, there is a real problem of fusing the essentially classical theory of gravitation with contemporary quantum physics. General relativity fails to incorporate any theory of elementary particles. And, perfectly reasonable conditions at one time may lead to field singularities at another. That looks as if general relativity carries within its conceptual belly the seeds of its own destruction. A unitary field theory should cure all of these shortcomings.

Let me begin by saying that all unitary field theories that I know have been based on one topological model, that of a manifold. They differ in the kind of structures that they superimpose on that basic framework.

Originally, physicists took it for granted that any model of space must be Euclidean. It was Minkowski who recognized the four-dimensional spacetime continuum as a pseudo-Euclidean manifold, one in which the relationship between two world points might be space-like, time-like, or lightlike. Einsteins's next step was to generalize Minkowski's spacetime to a pseudo-Riemannian manifold. The notion of a quadratic form as the basic geometric structure was retained as a local property of spacetime, instead of a global feature, as it had been with Minkowski. Levi-Civita, and the Italian school of differential geometers, had recognized the intimate relationship between the Riemannian metric and the affine connection, the possibility of identifying vectors at nearby points with each other, though only in a non-integrable fashion. With or without geometry, the Riemann-Christoffel curvature tensor turned out to be the repository of all local properties of the Riemannian geometry that involved at most second-order derivatives of the quadratic form (the metric tensor).

Indeed, Einstein's field equations directly involved the curvature tensor.

One might formulate the search for a unitary field theory as a search for a richer geometric structure than Riemann's, one that had room for all physical fields.

As a matter of fact, one of the earliest attempts, that by Weyl, did not so much enrich Riemannian geometry, it impoverished it. Weyl replaced the quadratic form, with ten components, by one that had but nine components, in that the determinant was to be set equal to -1 once and for all. The physical argument, if one can call it that, was that the magnitude of a vector should have no intrinsic significance, only its direction, and the ratios between magnitudes of different vectors at the same world point. This reduced metric of Weyl fails to determine the affine connection completely, one must introduce four components as an additional field, and these four additional field variables were then interpreted as the electromagnetic potentials. Weyl's geometry, in spite of valiant efforts, failed as a physical theory. Whether there are absolute universal standards of length in nature remains a contentious issue. Those who have been impressed by Dirac's hypothesis that the "universal standards" perceived are in reality slowly varying quantities, have come back to Weyl's geometry, or to conceptually similar models.

Incidentally, Weyl's geometry presents us with a paradigm for the study of fiber bundles. His geometry has been presented as a prime example of a fiber bundle, but there are equivalent presentations of that geometry whose invariance group is just that of diffeomorphisms, with no room left for an additional structure group.

Weyl's geometry permits one to interpret gravitation and electrodynamics as aspects of one reasonably cohesive geometry, but they occupy in that geometry quite different niches, so much so that their respective potentials belong to different orders of differentiation. In Kaluza's five-dimensional geometry this is no longer so. Gravitational and electromagnetic potentials form part of the five-dimensional pseudo-Riemannian metric. But Kaluza's theory involves two distinct geometric structures in the postulated five-dimensional manifold: its metric and its Killing vector field. Thus the unification is more apparent than real. In this respect the most popular generalization of Kaluza's construction, the so-called scalar-tensor theories, are worse: They add a third element, a scalar field, which is simply the norm of the Killing vector field; in the earlier version the Killing field had been a unit vector.

These earlier attempts at unitary field theory had either added no new physics to general relativity-cum-electrodynamics, or that new physics appeared to be contradicted by experiment and observation. In the fifties of our century the search for a unitary field theory gradually waned, because it was by necessity almost entirely speculative, and even the most promising beginnings seemed to be physically sterile.

Among current efforts directed at a unitary field theory I should like to mention three. The first is the resurrection of an ancient proposal by Cartan, to generalize Riemannian geometry by the introduction of an asymmetric affine connection. In Cartan's geometry it is still possible to construct locally a "free-falling frame", in which at one world point the metric is that of Minkowski, and all its first derivatives (and with them the Christoffel symbols) vanish. In such a frame the components of the affine connection do not vanish but induce a rotation of a parallel-displaced vector.

My misgiving with Cartan's geometry is that the geometry again decomposes into invariantly separable parts, one being just ordinary Riemannian metric, with all of its differential covariants, the other being the torsion, with 24 components, of which four form an invariantly separable substructure. Perhaps, from a different point of view things look more nearly like a whole.

The second direction involves the complexification of space-time. Formally, complexification involves the transition to an eight-dimensional manifold, with a complex structure in addition to the metric. Penrose's twistor theory belongs to this type of theory. The obvious question there is how the complex theory will relate to physical reality. At present there is no answer, but the answer may well be coming eventually.

The third approach involves the introduction of a hyper-complex number system, known as Grassman elements. Grassmann numbers afford the possibility of having anticommuting, rather than, or in addition to commuting numbers. The physical motivation here is the distinction between fermions and bosons, and the belief that among the elementary particles there are families including particles of both kinds. A Grassmann algebra makes it possible to combine commuting and anticommuting numbers in a single hypercomplex number system already at the (pseudo-) classical level. "Superfields" can be expanded into families of ordinary fields, but all these fields are mixed together in most algebraic operations. They really form an organic whole, which is masked by the resolution into ordinary fields.

Supersymmetry and supergravity recommend themselves by their ancestry, - from semi-phenomenological elementary particle physics, rather than from pure speculation, - and by their apparent unification of originally disparate fields. I have one adverse comment: Grassman algebra is homomorphic to ordinary (real or complex) algebra; hence the algebraic unification is less complete than it looks at first blush. But this criticism can be met by at least two possible stratagems: One is the transition to a Clifford algebra, the other the adoption of a non-associative algebra.

Permit me to come back to my original question: To what extent do geometric considerations play a role in these theories? I believe that the answer is largely heuristic. Any physical theory derives its justification from physical motives, and how well that theory helps us to understand the physical universe. The judgment as to desirable aspects of the invariance group of a theory must ultimately rest on the physical decision as to the equivalence of apparently distinct descriptions of a physical situation (choice of coordinate frame, choice of gauge frame, etc.). But given a conceptual framework, geometric realizations, and geometric imagery will often prove a powerful tool in elaborating the theory.

Komar and I published a paper several years ago in which we demonstrated that general relativity is invariant with respect to a much larger group than the group of four-dimensional diffeomorphisms. Our work, which orinally was intended to elucidate some of the puzzling aspects of Dirac's canonical formulation of general relativity, demonstrated that the theory is form-invariant with respect to canonical transformations of an infinite-dimensional symplectic manifold, which do not map a world point on a world point. Such transformations might point in a direction in which fields get unified under an invariance group in which the space-time manifold no longer plays a pre-eminent role. A world point would derive its identity from its dynamic environment, or it might possess no identity at all.

Whether such a theory can still be called geometric is a question of terminology, not of principle. Certainly, physicists will go in directions suggested to them by physical considerations, not by a mathematical discipline, and that is what geometry is. Consideration of an esthetic nature are important, to be sure; for that we have the testimony of such giants as Einstein and Dirac. But physics must not lose touch with its foundation, soil as it were: We must above all remain sensitive to what nature in its subtlety attempts to tell us.

TOWARDS A UNIFIED THEORY OF ELEMENTARY PARTICLE INTERACTIONS

John Iliopoulos, Laboratoire de Physique Théorique de l'Ecole Normale
Supérieure, Paris

I. Introduction

By a strange coincidence, this year marks the hundredth aniversary of
two events, both milestones in the long road towards a unified description of natural laws. On March 14, 1879 Albert Einstein was born in the
Southern German town of Ulm and on November 5 of the same year, James
Clerk Maxwell died in Cambridge. These are the two names that come first
to mind when one thinks about a unified theory. It was first Maxwell who,
in his famous work "A treatise on Electricity and Magnetism" achieved a
profound unification between these two, apparantly different, forces. It
is remarkable that the last chapter of his treatise is devoted to the
theories of action-at-a-distance in order to conclude that "the idea of
a medium cannot be got rid of". He writes: "Now we are unable to conceive of propagation in time, except either as the flight of a material
substance through space, or as the propagation of a condition of motion
or stress in a medium already existing in space." Several years later,
in 1905, the twentysix year old Einstein freed Maxwell's theory of this
burden and showed that no such material medium was needed. He did not
stop there. His effort to extend the principle of relativity to non-inertial frames led him to the equivalence principle and finally the general theory of relativity in 1916. From this point we can say that Einstein's efforts were concentrated to the formulation of what became known
as "the unified field theory". He pursued this goal undisturbed, with
stubborness and determination, essentially until his death in 1955. In
his own words "In the limited number of the mathematically existent simple
field types, and the simple equations possible between them lies the
theorist's hope of grasping the real in all its depth". In his mind such
a unified field theory would include, in a single framework, the general
relativity, *i.e.* the theory of gravitation, with the other forces in
nature. The elementary particles would emerge as special solutions of
these general equations.

Einstein never achieved this grandiose synthesis, neither did anybody
else after him. In fact, after the 1920's, physics has moved in a completely different direction. With the advent of quantum mechanics and
the subsequent research in atomic, nuclear and elementary particle phenomena, we were driven towards areas in which the effects of gravitational

forces are too weak to be observable. As a result, the study of gravitational interactions ceased for many years to be in the mainstream of
modern physics and the efforts were mainly concentrated on the unification of the other interactions. It is in this direction that remarkable
progress has been made during the last decade and it is this story that
I shall try to tell. However, let me say already that, as we shall see,
this strive towards a unification will bring us to distances such that
the gravitational interactions will become again important. We shall
come back to this point an the end.

II. Elemantary particles and their interactions

The structure of all known matter, from the stars and galaxies down to
the atoms and nuclei, as well as the living organisms, is due to four
types of interactions. (i) The strong interactions among nucleons build
the various nuclei. (ii) The electromagnetic interactions among charged
particles form the atoms and molecules. (iii) The weak interactions
are responsible for the β-decay of the neutron as well as the different
radioactive elements and play an important role in the evolution of the
stars. (iv) Finally we have the familiar gravitational interactions
with their well-known consequences in the macroscopic structure of the
universe.

Although our knowledge concerning the properties of these interactions
has been continuously expanding, our general notions about them have
been fairly stable. On the contrary, even today, the only working definition of an elementary particle can be stated as "an entry in the Table
of Elementary Particles" which is published every year by Rosenfeld and
co-workers. In other words, there is no absolute definition. It is a
time-dependent statement which reflects the state of our knowledge (or
ignorance). In recent times it went through the chain molecules → atoms
→ nuclei + electrons → protons + neutrons + electrons → ?

Table I shows the elementary particles known in 1932. We can make the
following remarks: (i) All "matter particles" (*i.e.* hadrons and leptons
out of which the world is made) have spin equal to 1/2. The spin of the
photon equals one. (ii) There is a manifest lepton-hadron symmetry
$(\nu; e^-) \longrightarrow (P; N)$. (iii) All elementary particles are "useful" *i.e.*
the role of each one in the structure of the Universe is essential and
clear.

TABLE I

Particles			Mass (Mev)	Life Time (sec)	Spin	Q	B	E
Matter particles / Hadrons		P	938	∞	1/2	1	1	0
Matter particles / Hadrons		N	939	918	1/2	0	1	0
Matter particles / Leptons		ν	0	∞	1/2	0	0	1
Matter particles / Leptons		e⁻	0,51	∞	1/2	-1	0	1
Quantum of Radiation		γ	0	∞	1	0	0	0

Table of elemenary particles of 1932. The neutrino had not yet been observed but its existence had been conjectured by Pauli.

The first attempt towards a partial unification in the lines we are following, can be traced back to Yukawa, the originator of the idea of the mesons. In his original proposal the same meson was supposed to mediate both strong and weak interactions by coupling strongly to the nucleons and weakly to the leptons. This idea failed when it was realized that muon decay, although it is a purely leptonic process, has the same strength as ß-decay rather than being much weaker. The whole program was abandoned with the discovery of a large number of "elementary particles", mainly hadrons, especially after the large accelerators were used. All three simple rules previously mentioned were violated. The new hadrons, both mesons and baryons, had different values of spin and made the distinction between "matter particles" and "radiation quanta" impossible. There was no obvious symmetry between leptons and hadrons and the role of all these particles in the structure of the world was, and still is, a mystery. Facing this complete disorder, the task of theoretical physics was clear and apparently hopeless: create order out of the chaos. We can distinguish two main lines of research: A classification effort in the continously growing jungle of elementary particles on the one hand and a study of the basic interactions on the other.

III.   Classification schemes for elementary particles

One of the most fruitful concepts in elementary particle physics has been that of a symmetry. Space-time symmetries (translations, rotations etc.) have been recognized and used since very many years, but here we shall deal mainly with the so-called "internal symmetries" *i.e.* transformations which leave the equations of motion invariant but they do not affect the space-time variables. An early example was provided in the early thirties by Heisenberg, just after the discovery of the neutron.

The apparent symmetry of the nuclear forces under the interchange of a proton with a neutron led him to conjecture that the strong interaction are invariant under a group of continuous transformations which can be viewed as rotations in a fictitious three dimensional space, the "isospin space". $180°$-rotations in this space, which change z to $-z$, are equivalent to a proton-neutron interchange. Mathematically, this group of transformations is equivalent to the familiar rotations in ordinary three-dimensional space. We know that, as a result of this invariance physical states are classified according to the value of their spin, which can take either integer or half-integer values. In exactly the same way, invariance under isospin rotations can be used to classify hadronic states according to the value of their "isospin", which again take integer or half-integer values. We thus put the proton and the neutron together in an iso-doublet $(I = 1/2)$, the three charge states of the π-meson $(\pi^+, \pi^-$ and $\pi°)$ in an iso-triplet $(I = 1)$ etc. After the discovery of the strange particles (*i.e.* hadrons carrying a new quantum number, conserved by the strong interactions) Gell-Mann and Ne'emann extended the concept of the internal symmetry from Heisenberg's group $SU(2)$ to $SU(3)$ (*i.e.* the group of three by three unitary matrices with determinant equal to one). Hadrons are now classified according to representations of $SU(3)$, for example the baryons (proton-neutron together with the Λ, Σ and Ξ hyperons) form an octet as do the pseudoscalar mesons π K and η.

This classification scheme was very successful not only in predicting new particles (the famous Ω^- hyperon was predicted and subsequently discovered in order to fill an empty space in the ten-dimensional representation) but also their strong interaction properties.

The ultimate form of this symmetry is expressed by the "quark" hypothesis (Gell-Mann and Zweig) which states that all hadrons are bound states of three, elementary, spin 1/2 constituents, named quarks (Gell-Mann found this name in a verse of James Joyce's "Finnegan's Wake"). Their assumed quantum numbers are given in Table II. Mesons are made out of a quark-antiquark pair and baryons out of three quarks. Table 3 gives some examples of hadron formation by quarks. Non-strange hadrons contain only p and n quarks while λ carries the strangeness quantum number. In the physicist's jargon the different quark types (p, n, λ or any further one may have to add) are often called "flavours".

In fact we are obliged to complicate a little the picture of the three quarks, sole constituents of all matter. In order to describe correctly the baryon wave functions, as well as to fit the experimental data on

TABLE II

Quark	Q	B	S	C
p	2/3	1/3	0	0
n	-1/3	1/3	0	0
λ	-1/3	1/3	-1	0
c	2/3	1/3	0	1

The quark quantum numbers. Q: electric charge; B: baryon number; S: strangeness. The first three quarks are the ones introduced by Gell-Mann and Zweig. The fourth (c) is the charmed one whose existence today is firmly established. We have evidence for a fifth one (b) with charge -1/3 and we believe that there is at least one more (t) with Q = 2/3.

TABLE III

Mesons	π^+	π^-	K^+	K^-	K^0	D^0
	$\bar{n}p$	$\bar{p}n$	$\bar{\lambda}p$	$\bar{p}\lambda$	$\bar{\lambda}n$	$\bar{p}c$
Baryons	P	N	$\Lambda^0, \bar{Z}^0$	Σ^+	Σ^-	P_c
	ppn	pnn	pnλ	ppλ	nnλ	cpn

Examples of hadron formation by quarks.

the π° life-time and the cross section of electron-positron annihilation, we must assume that each of the quarks comes in three distinct
types, named "colours". Thus one speaks of "blue", "white" and "red"
quarks. Therefore the number of quarks grows from three to nine. Actually, this scheme is not equivalent to a nine-flavour model because we
also assume that Nature is "colour-blind" *i.e.* all physically realizable
states contain equal proportions of the three colours and matter appears
colourless. In more technical terms, one can write the quarks in a 3 x 3
matrix where 1,2,3 mean "blue", "white", "red".

$$
\begin{pmatrix}
p_1 & p_2 & p_3 \\
n_1 & n_2 & n_3 \\
\lambda_1 & \lambda_2 & \lambda_3
\end{pmatrix}
$$

Therefore, there is a natural group of unitary transformations SU(3) x
SU(3). The first factor mixes the rows (flavours) but leaves the columns
(colours) of the matrix unchanged while the second does the opposite.
The flavour group is an approximate symmetry of strong interactions and
allows for a classification of physical states in multiplets of SU(3).
The second (colour group) is an exact symmetry but all physical states
are assumed to belong to the one-dimensional singlet representation.
According to this postulate, which still remains to be proven by the
dynamics of the theory, a single quark, which belongs to a colour triplet, cannot be a physically realizable state.

With the introduction of the quarks we succeeded in putting a little bit
of order in the jungle of elementary particles. Out of the three simple
rules of 1932,the first was restored. All matter-particles (quarks and
leptons) had spin 1/2. On the contrary, the other two remained violated.
There was no lepton-hadron symmetry (we had 4 leptons, *i.e.* the electron,
the muon and their respective neutrinos, but only three quark-flavours)
and, if the role of e^- , ν_e , p and n in the structure of matter was
clear, that of μ^- , ν_μ and λ was, and still remains, unknown.

IV.   Quantum field theory

1.   Q u a n t u m   e l e c t r o d y n a m i c s:

The first successful theory in elementary particles was provided by the
interaction among charged particles and the electromagnetic field. It is
the celebrated quantum electrodynamics (QED) which provided the arche-
type for any other physical theory. It is based on a very general technique called perturbation theory. Let H be the Hamiltonian describing

the dynamics of a physical system. In principle, we would know everything about the system, if we could solve the eigenvalue problem:

$$H\psi_n = E_n\psi_n \quad ,$$ (1)

where E_n are the eigenvalues and ψ_n the eigenfunctions of H. Unfortunately, the exact solution of (1) is unknown for all but some
very simple systems, and in practice we are obliged to use some kind of
approximation schemes. Perturbation theory amounts to splitting H into
two parts:

$$H = H_0 + \lambda H_I$$ (2)

where H_0 , called the "unperturbed part of H", is a Hamiltonian chosen
so that the solution of the eigenvalue problem is known exactly and
λH_I is called the "perturbation"; λ is some parameter which characterizes the strength of the perturbation. The idea is to find a splitting
such that λH_I is a relatively small part of H . Then, we solve the
eigenvalue problem of H_0 and calculate the corrections on the eigenvalues and eigenfunctions, induced by the presence of the perturbation,
as a series in powers of λ . Theoretically, there may be more than
one way to obtain such a splitting of the total Hamiltonian, leading to
more than one possible perturbation expansion, but in practice the choice
is very limited. In a relativistic quantum field theory in four space-
time dimensions, the only eigenvalue problem which is always exactly
solvable is that of a free field theory, *i.e.* if H_0 describes a system
of free particles. Consequently, we are obliged to include in the perturbation part λH_I , the entire interaction Hamiltonian. Furthermore,
the complexity of the calculations is such that we can only compute the
first few terms in the power series expansion. Obviously, such a scheme
has some chances to give sensible results only if the entire interaction
is "weak". This means, physically, that the energy due to the interactions must be small compared to the kinetic energies and masses of the
particles which are included in H_0 . Technically this is translated
to the requirement that λ , the "strength" of the interaction, is a
small number, $\lambda \ll 1$. In quantum electrodynamics this parameter is
the fine structure constant α which equals $1/137$. This small number
is responsible for the practical successes of QED because successive
terms in the perturbation expansion are proportional to increasing powers of α , so they get smaller and smaller, and a good approximation
is obtained by keeping only the first few of them. However, the computa-

tion of these higher order terms is very delicate and, if performed blindly, gives meaningless results because they turn out to be divergent. We remember from quantum mechanics that the formula for the higher order perturbation corrections involves a summation over the energies of a complete set of states and it is this summation which, in our case, diverges. The solution to this problem was obtained in 1947-1949 by Feynman, Schwinger, Tomonaga and Dyson with their famous "theory of renormalization" which allows one to extract meaningful finite answers out of the divergent expressions for all quantities which are physically measurable. All divergences are absorbed in non-physical quantities, such as the self-mass of the electron *i.e.* the difference between its physical mass and the mass it would have had, were there no electromagnetic interactions. The agreement with experiment of the "renormalized" QED is spectacular. For example, the experimental value of the anomalous magnetic movement of the electron is $(1\ 159\ 652\ 410 \pm 200) \times 10^{-12}$ while the value predicted by theory is $(1\ 159\ 652\ 359 \pm 282) \times 10^{-12}$.

Given this great success, we are naturally tempted to apply the same method to the other interactions. Unfortunately this is not straightforward. In strong interactions we can still, formally, write the analogue of Equation (2), but the corresponding parameter λ turns out to be large, $\lambda \gtrsim 1$, and the approximation scheme breaks down. This means physically that, for a system of hadrons, the energy due to their strong interactions is not a small part of their total energy. Then what about weak interactions? We know, experimentally, that they are indeed weaker than the electromagnetic ones and, therefore, we expect here perturbation theory to give even better results.

In fact, a phenomenological theory was proposed already in 1934 by Fermi which successfully described all weak interaction phenomena. It involved an operator $J_\lambda(x)$, the "weak current", which is the analogue of the familiar electromagnetic current.* The hamiltonian density describing the interaction energy of a system of weakly interacting particles was assumed, by analogy with the electromagnetic case, to be given by the product of two weak currents

$$H_F = \frac{G}{\sqrt{2}} J^\lambda(x) J^+_\lambda(x) \qquad (3)$$

where $\frac{G}{\sqrt{2}}$ is the Fermi coupling constant, which is equal to $10^{-5} m^{-2}_{proton}$,

---

* We have used a relativistic notation in which $J_\lambda(x)$ is a four-vector and λ takes the values 0,1,2,3. Similarly x represents the space-time point $\vec{x}$ and t.

the + represents the operation of hermitian conjugation and the summation convention over repeated indices is understood.

Through extensive experimental and theoretical work the properties of
$J_\lambda(x)$ were well-known. It is the sum of two parts, a leptonic part
$I_\lambda(x)$ and a hadronic one $h_\lambda(x)$. They are both superpositions of vector
and axial vectors and satisfy simple and elegant algebraic relations.
However, the crucial drawback of the Fermi theory is that the renormalization program does not apply, in other words one cannot remove the
infinities from the expansion. Several people have tried, without success, to invent a new algorithm adapted to the Fermi theory of weak
interactions, and finally the solution came from the opposite direction:
the algorithm remained the same, but the phenomenological theory was
replaced by a different one. The remarkable thing is that this new theory, which is far more beautiful from the aesthetic point of view, has
different experimental consequences, and it now seems that they are
verified. It looks like the old prejudice, that the search for internal
consistency and aesthetic beauty always leads to a deeper understanding
of the physical world, is once more confirmed.

2.   F i r s t   s t e p :   T h e   i n t e r m e d i a t e   v e c -
     t o r   b o s o n

As we said earlier, the model out of which we were getting our inspiration was QED. Therefore, the first step towards unification was the intermediate vector boson hypothesis. The Fermi theory (3) was replaced
by

$$H_W = gJ^\lambda(x) \; W_\lambda(x) + \text{hermitian conjugate} \tag{4}$$

where $W_\lambda(x)$ is the field of a charged spin one boson which is supposed
to mediate the weak interactions in exactly the same way as the photon
mediates the electromagnetic ones. It turns out that the hamiltonian
(4) gives, at low energies, the same results as (3) provided the constants G and g are related by $G/\sqrt{2} \sim g^2/m_W^2$ with m_W being the
W boson mass.

The hamiltonian (4) now looks very similar to QED since they both describe the interaction of a spin-one boson with an appropriate current.
However there are some important differences which cause the renormalization program to work for QED but not for (4). From the physical point
of view these differences are the following:

(i)   The electromagnetic interactions have a long range $\longrightarrow$ the pho-
       ton is massless.

Weak interactions give rise to short range forces ⟶ the intermediate vector bosons, if they exist at all, must be very massive
($m_W > 30$ GeV).

(ii)   The electromagnetic current is conserved, the weak one is not
($\partial_\lambda J^\lambda(x) \neq 0$).

(iii) The photon is neutral. The W's are charged.

The problem now is: Is it possible to modify (4) in such a way that we
obtain a renormalizable field theory, without upsetting its nice agreement with experiment? In order to answer it, we shall apply an old theoretical prejudice which says that the best theory is the most symmetric
one. The most symmetric way to couple spin-one bosons is given by the
"gauge theories" which we shall study next.

3.   S e c o n d   s t e p :   G a u g e   t h e o r i e s

Gauge transformations are known already in classical electrodynamics.
They correspond to the freedom we have to add to the vector potential
$A_\lambda(x)$ the gradiant of an arbitrary scalar function. $A_\lambda(x)$ ⟶
$A_\lambda(x) + \partial_\lambda \theta(x)$. The physical quantities, namely the electric and magnetic fields $\vec{E}(x)$ and $\vec{B}(x)$, are unaffected by this transformation. In
field theory we can express this invariance in the following way: consider, as an example, the Lagrangian density of a free Dirac field
$\psi(x)$:

$$L_0 = \overline{\psi}(x)\left(i\gamma_\lambda \frac{\partial}{\partial x_\lambda} - m\right)\psi(x) \tag{5}$$

where $\psi(x)$ is a four component complex Dirac spinor, γ_λ is a four
by four matrix and $\overline{\psi} = \psi^+ \gamma_0$. (5) is invariant under the $U(1)$ group
of the phase transformations:

$$\psi(x) \longrightarrow e^{i\theta} \psi(x) \tag{6}$$

where θ is an arbitrary, x-independent phase. This invariance means,
physically, that the phase of the field is not measurable and can be
chosen at will. On the other hand, since θ in (6) is x-independent,
it must be chosen the same over the entire universe and for all times.
This situation is somehow unsatisfactory on purely aesthetic grounds. We
would like instead to have a formalism which would allow us to fix the
phase locally in a region with the dimensions of our experiment, without
reference to far-away distances; in other words we would like to replace
(6) by

$$\psi(x) \longrightarrow e^{i\theta(x)} \psi(x) \tag{7}$$

where $\theta(x)$ is now a function of x. However, now

$$\frac{\partial}{\partial x_\lambda} \psi(x) \longrightarrow e^{i\theta(x)} \frac{\partial}{\partial x_\lambda} \psi(x) + ie^{i\theta(x)} \psi(x) \frac{\partial}{\partial x_\lambda} \theta(x) \tag{8}$$

i.e. the derivative of the field no longer transforms like the field itself and, as a result, the Lagrangian (5) is no longer invariant under (7). We shall call transformations of the form (7), *i.e.* with x-dependent parameters, "local" or "gauge" transformations.*

In differential geometry there is a standard way of restoring invariance under (7). Since the trouble arises from the derivative operator we must introduce a new "derivative" D_λ, called the "covariant derivative" which is again a first order differential operator, but with the property that it transforms under (7) like the field itself:

$$D_\lambda \psi(x) \longrightarrow e^{i\theta(x)} D_\lambda \psi(x) \tag{9}$$

In order to find such a D_λ we introduce the vector potential $A_\lambda(x)$, which plays the role of the afine connection of differential geometry, and which, by definition, transforms like:

$$A_\lambda(x) \longrightarrow A_\lambda(x) + \frac{1}{e} \frac{\partial}{\partial x^\lambda} \theta(x) \tag{10}$$

with e an arbitrary constant. Then D_λ is given by:

$$D_\lambda = \frac{\partial}{\partial x^\lambda} - ieA_\lambda(x) \tag{11}$$

and it is easy to verify that $D_\lambda \psi(x)$ transforms like (9) when $\psi(x)$ and $A_\lambda(x)$ transform according to (7) and (10). Invariance under gauge transformations is now restored if we replace everywhere in L_0 $\frac{\partial}{\partial x_\lambda}$ by D_λ:

$$L_0 \longrightarrow L_1 = \overline{\psi}(x) \left(i\gamma^\lambda D_\lambda - m\right) \psi(x) \tag{12}$$

L_1 contains the new field $A_\lambda(x)$. If we want to interpret the latter as the field representing the photon we must add to (12) a term corre-

---

* Sometimes they are called "gauge transformations of the second kind".

sponding to its kinetic energy. From classical electromagnetic theory we know that the energy of the electromagnetic field is given by $\vec{E}^2 + \vec{B}^2$ which, in our notation yields:

$$L_1 \longrightarrow L_2 = L_1 - \frac{1}{4} F_{\mu\lambda} F^{\mu\lambda} \tag{13}$$

where

$$F_{\mu\lambda}(x) = \frac{\partial}{\partial x^\mu} A_\lambda(x) - \frac{\partial}{\partial x^\lambda} A_\mu(x) \tag{14}$$

L_2 is nothing else but the Lagrangian density of QED. We have obtained it by imposing invariance under gauge transformations. A final remark is very important: L_2 does not contain a term proportional to $A_\lambda(x) A^\lambda(x)$ since, such a term is not invariant under (10). In other words, gauge invariance forces the photon to be massless.

This procedure has been generalized by Oscar Klein in 1938 and C.N. Yang and A. Mills and R. Shaw in 1954, to more complicated transformations than the phase transformations considered here. The results are similar and one obtains a massless spin-one field, like the photon, for every independent parameter of the transformation. For example, if one considers the isospin transformations mentioned earlier, one obtains a triplet of massless "photons" because one can perform three independent rotations in the fictitious three-dimensional isospin space. In such cases one speaks of "non-abelian" gauge transformations.* Since no massless spin-one bosons, other than the photon, are known in nature, it looks as if non-abelian gauge symmetries have nothing to do with physics in general and weak interactions in particular.

4. Spontaneously broken symmetries:

It was only in 1964 that the first formidable obstacle, namely the appearance of massless bosons, was removed from non-abelian gauge theories. This was achieved by Brout, Englert, Higgs and Kibble who studied the phenomena known as "spontaneous symmetry breaking".

We are used to look for symmetric solutions to symmetric problems. This considerably simplifies the construction of the solution, but there are cases in which the symmetric solutions are not the most interesting ones.

---

* The term "abelian" means that the transformations in this group commute with each other. Such was the case of the phase transformations studied earlier. "Non-Abelian" means that the different transformations do not commute. For example, the result of rotations about two different axes depends on the order in which they are carried out.

A classical example is provided by the bent rod. A cylindrical rod is compressed by a force F . The problem is obviously symmetric under rotations around the vertical axis but we all know, and we can verify it by solving the elasticity equations explicitly, that, if F is strong enough, the final state will be asymmetric with the rod bent in a random direction. In fact the symmetric solution (unbent but compressed rod) still exists but it turns out to be unstable. The non-symmetric solutions have lower energy. The original symmetry is still hidden in the sense that we cannot predict the direction in which the rod is going to bend. They all correspond to solutions with precisely the same energy. We call such a symmetry "spontaneously broken", and in this simple example we see all its characteristics: There exists a critical point, *i.e.* a critical value of some quantity (in this case the external force F ; in several physical systems it is the temperature) which determines whether spontaneous symmetry breaking will take place or not. Beyond this critical point:

(i)   the symmetric solutions becomes unstable;

(ii)   the ground state becomes degenerate.

There exist a great variety of physical systems, both in classical and quantum physics, exhibiting spontaneous symmetry breaking, but we will not describe any other one here. The Heisenberg ferromagnet is a good example to keep in mind, because we shall often use it as a guide, but no essentially new phenomenon appears outside the ones we saw already.

In elementary particles one can prove a theorem (Goldstone, Nambu) which states that to every generator (*i.e.* to every independent infinitesimal transformation) of a spontaneously broken symmetry there corresponds a massless particle, called the Goldstone particle. Let us now come to the field theory example studied by Brout *et al*. It is the gauge invariant Lagrangian of a complex scalar field known as "scalar electro-dynamics" because it describes the interaction of a charged spin-zero particle with the electromagnetic field. The Lagrangian density can be written as:

$$L = - \frac{1}{4} F_{\mu\lambda}(x) \ F^{\mu\lambda}(x)$$

$$+ [\left(\frac{\partial}{\partial x^{\lambda}} - ieA_{\lambda} \right) \ \phi(x)] \ [\left(\frac{\partial}{\partial x_{\lambda}} + ieA^{\lambda} \right) \ \phi^{*}(x)] \qquad (15)$$

$$- \mu^{2} \ \phi(x) \ \phi^{*}(x) - \lambda \left(\phi(x) \ \phi^{*}(x) \right)^{2}$$

where $\phi(x)$ is a scalar field and ϕ^{*} its complex conjugate. $F_{\mu\lambda}$ is given by (14) and we recognize the expression for the covariant derivative. μ^2 and λ are parameters which, in a quantum field theory, describe the mass of the scalar field and its self-coupling constant, respectively. (15) is invariant under the gauge transformation:

$$\phi(x) \longrightarrow e^{i\theta(x)} \phi(x) \quad ; \quad A_\mu \longrightarrow A_\mu + \frac{1}{e} \frac{\partial\,\theta(x)}{\partial x^\mu} \tag{16}$$

The equations of motion obtained from (15) will obey the same invariance. It is clear that $\phi(x) = 0$ is the only invariant solution. But let me assume for the moment that, for some dynamical reason, a spontaneous symmetry breaking occurs and a solution $\phi(x) = v \neq 0$, with v a constant, becomes energetically favourable. Indeed, it is easy to see in the above example, by minimizing the corresponding hamiltonian function, that this phenomenon happens if $\mu^2 < 0$ and $\lambda > 0$. In this case, replacing $\phi(x)$ by v in (15) we obtain, among other terms, one of the form: $e^2 v^2 A_\lambda(x) A^\lambda(x)$ which is precisely what corresponds to a mass term for the photon. A more careful analysis confirms this result. The spectrum of the theory consists of a massive spin-one particle and a massive spin-zero one. All massless particles (gauge boson as well as Goldstone) dissapear. What happens can be understood physically in the following way: A massless spin-one particle has two polarization states; a massive one has three. In the "Higgs phenomenon" the physical degree of freedom of the would-be Goldstone boson is absorbed by the massless gauge boson in order to allow it to increase the number of its polarization states from two to three and become massive.

5. C h a r m :

Application of the "Higgs mechanism" allowed us to give a mass to the gauge bosons without violating gauge invariance. But soon a second difficulty arose: As we said before, the weak current entering Equation (4) carried electric charge. We can represent it by using the field operators of known leptons for the leptonic part I_λ and those of the quark fields for the hadronic part h_λ . A simple form which fitted all experimental data, was:

$$J_\lambda^{(+)}(x) = I_\lambda^{(+)} + h_\lambda^{(+)}(x) \tag{17}$$

$$I_\lambda^{(+)}(x) = \bar{I}(x) \gamma_\lambda (1+\gamma_5)\, C_1^{(+)}\, 1(x) \quad ; \quad 1(x) = \begin{pmatrix} \nu_e(x) \\ \nu_\mu(x) \\ e^-(x) \\ \mu^-(x) \end{pmatrix} \tag{18}$$

$$h_\lambda^{(+)}(x) = \bar{q}^{(x)} \gamma_\lambda (1 + \gamma_5) C_q^{(+)} q(x) \ ; \ q(x) = \begin{pmatrix} p(x) \\ n(x) \\ \lambda(x) \end{pmatrix} \tag{19}$$

where we have used a spinor notation to write the four lepton fields as $l(x)$ and the three quark fields as $q(x)$. Each field $\nu_e(x), \ldots, \lambda(x)$ is by itself a four component complex Dirac spinor. The numerical matrices $C_l^{(+)}$ and $C_q^{(+)}$ are given by:

$$C_l^{(+)} = \begin{pmatrix} 0 & 0 & 1 & 0 \\ 0 & 0 & 0 & 1 \\ 0 & 0 & 0 & 0 \\ 0 & 0 & 0 & 0 \end{pmatrix} \quad ; \quad C_q^{(+)} = \begin{pmatrix} 0 & \cos\theta & \sin\theta \\ 0 & 0 & 0 \\ 0 & 0 & 0 \end{pmatrix} \tag{20}$$

θ is a phenomenological parameter called "the Cabibbo angle". The expressions (18) to (20) are just a shorthand notation for:

$$I_\lambda^{(+)}(x) = \bar{\nu}_e(x) \gamma_\lambda (1 + \gamma_5) e^-(x) + \bar{\nu}_\mu(x) \gamma_\lambda (1 + \gamma_5) \mu^-(x) \tag{21}$$

$$h_\lambda^{(+)}(x) = \cos\theta \, \bar{p}(x) \gamma_\lambda (1 + \gamma_5) n(x) + \sin\theta \, \bar{p}(x) \gamma_\lambda (1 + \gamma_5) \lambda(x) \tag{22}$$

(21) and (22) are the forms that were imposed upon us by experiment. Together with $J_\lambda^{(+)}(x)$ we need, in Equation (4) its hermitian conjugate which carries the opposite electric charge $J_\lambda^{(-)}(x)$. It is given by the same expressions (18) and (19) by replacing the matrices $C_l^{(+)}$ and $C_q^{(+)}$ by their transposed $C_l^{(-)} \ [C_l^{(+)}]^T$ and $C_q^{(-)} = [C_q^{(+)}]^T$. In a gauge theory however, one has to introduce a third current which is given by the commutator of $J_\lambda^{(+)}$ and $J_\lambda^{(-)}$ and carries no electric charge. This neutral current $J_\lambda^{(0)}$ will again be given by expressions like (18) and (19) but with the matrices $C_l^{(0)}$ and $C_q^{(0)}$ given by the commutators

$$C_l^{(0)} = [C_l^{(-)}, C_l^{(+)}] = \begin{pmatrix} 1 & 0 & 0 & 0 \\ 0 & 1 & 0 & 0 \\ 0 & 0 & -1 & 0 \\ 0 & 0 & 0 & -1 \end{pmatrix} \tag{23}$$

$$C_q^{(0)} = [C_q^{(-)}, C_q^{(+)}] = \begin{pmatrix} 1 & 0 & 0 \\ 0 & -\cos^2\theta & -\cos\theta\sin\theta \\ 0 & -\cos\theta\sin\theta & -\sin^2\theta \end{pmatrix} \tag{24}$$

Notice the asymmetry between leptons and quarks. The resulting leptonic neutral current is diagonal while the hadronic one is not.

In 1970, when this problem was tackled, experimental evidence for, or against, the existence of diagonal neutral currents was poor, however the non-diagonal ones were absolutely excluded. In fact a current of the form (24) couples $\bar{\lambda}n$ or $\bar{n}\lambda$ with $\mu^+\mu^-$, thus predicting decays like $K^0 \longrightarrow \mu^+\mu^-$ which were known to be absent at the 10^{-8} level. Staring for a minute at (23) and (24), one can find the remedy to this difficulty. One has to restore the old 1932 lepton-hadron symmetry by introducing a fourth quark. Then the hadronic current is given by (19) with

$$q(x) = \begin{pmatrix} c(x) \\ p(c) \\ n(x) \\ \lambda(x) \end{pmatrix} \quad ; \quad C_q^{(+)} = \begin{pmatrix} 0 & 0 & -\sin\theta & \cos\theta \\ 0 & 0 & \cos\theta & \sin\theta \\ 0 & 0 & 0 & 0 \\ 0 & 0 & 0 & 0 \end{pmatrix} \quad (25)$$

and it is easy to verify that the neutral current is, like the leptonic part, diagonal, since now the commutator $[C_q^{(+)}, C_q^{(-)}]$ equals the leptonic one of Equation (23).

The new quark c carries a new quantum number, which we called "charm". It is analogous to strangeness, carried by λ . It is conserved by the strong and electromagnetic interactions and violated by the weak ones. Bound states, in which c participates, produce a whole new class of hadrons, the charmed ones, and the classification group of strong interactions becomes SU(4).

V.   The breakthrough

The most important step in this long march was done in 1971 when G. 't Hooft proved that a spontaneously broken gauge theory, although it contains massive vector particles, is renormalizable. I cannot reproduce the proof here which, like all similar proofs in field theory, is quite technical, but it is easy to understand that this result opened the way to the study of realistic theories of weak and electromagnetic interactions.

1.   T h e   SU(2)xU(1)   m o d e l:

The theory which turned out to be the most successful was proposed already in 1961, before anything was known about renormalizability, by Glashow and completed with the Higgs mechanism in 1967 by Weinberg and in 1968 by Salam. Initially it described the weak and electromagnetic interactions of leptons but, it was extended to hadrons with the intro-

duction of charm. Furthermore, it was shown that renormalizability requires the introduction of coloured quarks, *i.e.* each quark flavour
c, p, n or λ should come in three colours the way we explained in
chapter II.

The model needs four vector bosons, two charged for ordinary weak interactions, one neutral for the weak neutral currents* and one for the
photon. It follows that the smallest possible group is SU(2) x U(1) .
The elementary fields of the model were the four known leptons (electron-
muon as well as their two neutrinos**) and four scalars necessary for
the spontaneous symmetry breaking. Before symmetry breaking all four
vector bosons are massless. Then the spontaneous breaking occurs but
it leaves unbroken the part of the symmetry which corresponds to the
exchange of the photon (a U(1) subgroup of SU(2) x U(1)). As a result
the two charged vector bosons W^+ and W^- as well as one neutral, called
Z, become massive while the photon remains massless. The strength of
the interactions is given by two arbitrary parameters (corresponding to
the two factor groups in SU(2) x U(1)) which can be chosen to be a coupling constant g and an angle θ_W . The Fermi coupling constant is
given by

$$\frac{G}{\sqrt{2}} = \frac{g^2}{8m_W^2} \tag{26}$$

and the electric charge by

$$e = g \sin\theta_W \tag{27}$$

By combining (26) and (27) we obtain a limit for the charged vector
boson mass $m_W \geq 37.5$ GeV while the mass of the neutral Z is given
by $m_Z = m_W/\cos\theta_W$. We see that the unification of weak and electromagnetic interactions means that the basic coupling constants of the
two forces are of the same order of magnitude. The apparent weakness of
weak interactions as well as their short range are due to the spontaneous
symmetry breaking which makes the intermediate vector bosons very heavy.
At very high energies (high with respect to m_W, *i.e.* higher than, say,
100 GeV) we expect both electromagnetic and weak forces to be of comparable strength.

---

* When the model was proposed the neutral currents had not yet been
  observed.
** We know that there exists at least one new lepton, the τ , with its
  own neutrino, so the model has to be enlarged.

2. The great experimental discoveries:

As we have mentioned already, the use of gauge field theories obliged us to change in several respects the original Fermi theory of weak interactions. These changes implied specific predictions which allow these theories to be tested experimentally. The most significant of these predictions are:

(i)   Gauge theories contain naturally neutral intermediate vector bosons. Their existence implies reactions of the type $\nu_\mu + e^- \longrightarrow \nu_\mu + e^-$ or $\nu_\mu + P(N) \longrightarrow \nu_\mu + X$ where   X   is some hadronic state, in other words it implies the existence of weak neutral currents.

In 1973 a European collaboration using the heavy liquid bubble chamber "Gargamelle" announced the observation of neutrino scattering reactions with no muon in the final state, *i.e.* reactions that show the existence of a weak neutral current. This discovery, one of the most important in recent years, opened a new chapter in the history of weak interactions. A great experimental effort has since been concentrated, both in Europe and the United States, in order to explore and study the properties of this newly discovered weak force. As a result we have measured today the parameters of this current, namely its overall strength as well as the relative importance of its vector and axial parts. These quantities are expressed, in the Weinberg-Salam model, through only one parameter, the angle θ_W .   The experimental results confirm this dependence and determine the value of θ_W to be $\sin^2\theta_W \approx 0.22$.

(ii)   We saw that we needed a new quantum number, "charm", and a whole new class of hadrons carrying it. Their masses were predicted to be light (less than ~ 10 GeV) and their decay products would involve strange particles. This last prediction is due to the particular form of the current of Equation (25). In fact we see that   c   is coupled to λ through $\cos\theta$ and to   n   through $\sin\theta$.   Since the experimental value of the Cabibbo angle is $\sin\theta$ ~ 0.2 ,   the $(c\lambda)$ coupling is much stronger.

In 1974 Ting from Brookhaven and Richter from Stanford independently discovered the J/ψ particle, a meson with the characteristic properties of a $c\bar{c}$ bound state. In the weeks and months that followed a whole rich spectroscopy was found which completely agrees with the charm scheme. The laboratories which shared the honours of these new discoveries were the $e^+ - e^-$ colliding beam machines of Stanford (SPEAR) and Hamburg (DORIS).

(iii)   We mentioned already that, for technical reasons, renormalizability requires the respect of lepton-hadron symmetry, with the hadrons represented by three-coloured quarks. This symmetry was restored by the introduction of charm. However, almost immediately after the discovery of J/ψ Perl at Stanford found a new lepton, named τ , with its own neutrino. It was the repetition of the story of the muon. The symmetry required now the discovery of two new quarks and, indeed, two years later Lederman, at FNAL found a new hadron, the Y , which seems to announce the existence of a new quark, in the same way that J/ψ showed the existence of charm. Both these discoveries have since been confirmed in Hamburg.

(iv) Maybe the most important prediction of gauge theories is the existence of the weak intermediate vector bosons $W^{\pm}$ and Z . Their masses are given in terms of θ_W . For $\sin^2\theta_W \approx 0.22$ we obtain $m_{W^{\pm}} \sim 84$ GeV and $m_Z \sim 94$ GeV. They are out of reach of the existing machines but a proton-antiproton colliding beam facility is beeing constructed at CERN with the purpose of producing and identifying them.
A further theoretical prediction is the particular form of self coupling among W's and Z given by the Yang-Mills theory. They will be harder to study experimentally and one should wait for at least a decade before a large $e^+ - e^-$ machine (LEP), at present under study in Europe, is built.

(v) Finally, the particular aspect of the Weinberg-Salam model, which distinguishes it from the 1961 Glashow theory and which ensures its renormalizability, is the existence of at least one physical, spin-zero, neutral particle, the remnant of the Higgs mechanism. However, at this point, the theory is not very precise. We know nothing about its mass and, in case we have more than one, we know next to nothing about their detailed properties. Several people believe, and I share this view, that the Higgs scheme is a convenient parametrization of our ignorance concerning the dynamics of spontaneous symmetry breaking and elementary scalar particles do not exist.

VI. Strong interactions (revisited)

1. A r e   s t r o n g   i n t e r a c t i o n s   s i m p l e ?

One of the main goals in particle physics has always been the understanding of the nature of strong interactions. For that we were using, most of the time, the results from hadronic collisions. The resulting picture invariably appeared to be too complicated to allow for a simple interpretation. We have by now good reasons to believe that this complexity should not be attributed to the fundamental interactions themselves, but is instead due to the fact that the objects we are dealing

with, namely the hadrons, are themselves too complicated. It is as if
we were trying to discover quantum electrodynamics by studying the interactions among complex molecules. The observed forces between hadrons
have the same relationship with the fundamental interactions, as the Van
der Waals forces with quantum electrodynamics. The great progress of the
last years was the realization that these fundamental interactions are
accessible to experimental observation. The main tool is the study of
high energy lepton-hadron scattering of the form lepton + nucleon $\longrightarrow$
lepton + hadronic system. We are particularly interested in the so-
called "deep inelastic region" in which both the incident lepton energy
and the momentum transfer between the initial and final leptons are very
large. These experiments were first performed at Stanford (SLAC) with
incident electrons but they have been repeated in Batavia (FNAL) and in
Geneva (CERN) with neutrino and muon beams. The results can be interpreted in the following way: (i) At the moment of the interaction the
target nucleon behaves like a conglomeration of elementary constituents
which may be identified with the quarks. (ii) Each of the constituent
quarks interacts with the incident lepton as if it were a free particle
despite the fact that it is bound inside the nucleon. (iii) The final
state consists of hadrons but never of free, unbound, quarks.

These results are very important for several reasons: First they confirm
the bound state nature of the nucleon. But also they teach us that
strong interactions at the more fundamental level of the quarks have
a very strange property: when we probe with leptons of very high energy
and large momentum transfer, *i.e.* at very short distances, strong interactions become very weak, the quarks behave like free particles. On the
contrary, when we try to extract a quark out of the nucleon we realize
that it is very strongly bound. In other words the strong interaction
intensity increases with distance, negligible at short distances it
becomes very strong, may be infinite, when the seperation becomes large.

2. Asymptotic freedom and infrared slavery

The fact that the effective strength of an interaction depends on the
distance can be understood by a simple classical argument. Let us take
electrodynamics. The coupling constant is the electric charge. Its magnitude is measured by its effects on surrounding charges. Let us assume
that we have a charge +Q inside a polarizable medium, for example, water.
We bring close to it another charge -Q. The water molecules between the
two are polarized by the electric field and they tend to screen the two
charges. The net result is that the charge -Q sees an effective value

of the positive charge, which depends on the distance between them. At
large distances (small momenta) the effect of the screening is very important and the effective value of the charge tends to zero. At small
distances (large momenta), on the other hand, this value gets larger.
In a quantum language the same effect occurs also in the vacuum because
of vacuum polarization. In a renormalizable field theory we introduce
a coupling constant g which is an arbitrary parameter whose value is
determined by experiment. The effective strength of the interaction is
a function $\bar{g}(\lambda, g)$ where λ is a parameter which scales momenta and
distances. Short distance means $\lambda \to \infty$. The dependence of $\bar{g}$ on λ
can be studied by using a technique known as "renormalization group".
It gives an equation of the form:

$$\lambda \frac{\partial \bar{g}}{\partial \lambda} = \beta(\bar{g}) \quad , \quad \bar{g}(1,g) = g \tag{28}$$

Where $\beta(x)$ is a function which characterizes the field theory but whose
knowledge requires the complete solution of the dynamical equations.
Equation (28) shows that if $\beta > 0$, $\bar{g}$ increases with increasing λ, and
it will continue to increase as long as β remains positive. The limit
of $\bar{g}$ when $\lambda \to \infty$ will be the first zero of β on the right of the initial
value g. If $\beta(x)$ has no zeros for x > g, then $\bar{g} \to \infty$ for $\lambda \to \infty$. Now let
us take the case when $\beta(g) < 0$. Then $\bar{g}$ decreases with increasing λ and
lim $\bar{g}(\lambda,g)$ = first zero of $\beta(x)$ for x < g. Finally, if $\beta(g) = 0$,
$\partial \bar{g}/\partial \lambda = 0$, and $\bar{g}$ is independent of λ.

This analysis shows that we can classify the zeros of β in two classes: Those

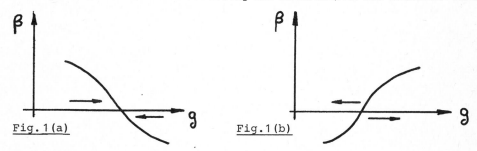

Fig. 1(a)   Fig. 1(b)

of Fig. 1a are "attractors", *i.e.* if we start somewhere in their neighbourhood, $\bar{g}$ approaches them for $\lambda \to \infty$. Those of Fig. 1b are "repulsors",
i.e. $\bar{g}$ goes further away from them when $\lambda \to \infty$. An attractor is always
followed by a repulsor (multiple zeros must be counted accordingly).
The conclusion is that the asymptotic behaviour of a field theory depends on the position and nature of the zeros of the function β.

As long as perturbation theory is our only guide, we cannot say anything

about the properties of β(x) for arbitrary x. We do not know whether it
has any zeros, let alone their nature. The only information that perturbation theory can hopefully provide is the behaviour of β(x) at the vi-
cinity of x = 0. We know that β(0) = 0 because g = 0 is a free field
theory. The nature of this zero (attractor or repulsor) will depend on
the sign of the first non-vanishing term in the expansion of β(g) in
powers of g. But this expansion is precisely perturbation theory. Therefore the properties of the zero of the β-function at the origin can be
extracted from perturbation. If β starts as in Fig. 2a, *i.e.* from positive values, the origin is a repulsor. The effective coupling constant
will be driven away to larger values as we go deeper and deeper into
the λ ⟶ ∞ region. On the contrary, if the first term of β is
negative the origin is an attractor. If we start somewhere between the

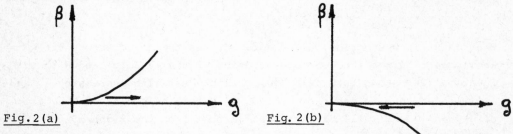

Fig.2(a)   Fig.2(b)

origin and the next zero of β, the effective coupling constant will become smaller and smaller and it will vanish in the limit. Such a theory
is called "asymptotically free".

And now we shall state the following, very important theorem:

Out of all renormalizable field theories only the gauge theories of non-
abelian groups are asymptotically free.

This theorem is proven simply by exhaustion. We calculate the first non-
vanishing term in the expansion of the β-function for all known, renormalizable field theories: ϕ^4 ; Yukawa, QED and non-abelian gauge theories. Only the latter have β negative.

3.   Q u a n t u m   c h r o m o d y n a m i c s

The message of this theorem is clear. The only way to understand the
deep inelastic lepton-nucleon scattering in a field theory framework
is to assume that strong interactions between quarks are described by
a non-abelian gauge theory. There are several ways to realize such a
theory and one, which is particularly simple and favoured by most theorists, is called "Quantum Chromodynamics" (QCD). It assumes that the
gauge group of strong interactions is the group SU(3) of the colour,
the one that mixes the three columns of blue, white and red quarks

leaving the rows unchanged. The vector gauge bosons are an octet of spin one, electrically neutral, but colour carrying mesons which we call "gluons". They are massless as required by the exact gauge symmetry. However, like the quarks, they never appear as free particles, they are permanently confined inside the hadrons.

This scheme has several experimental implications the most important of which is that it allows to make precise predictions for the deep inelastic experiments. The basic assumption is that we have already reached sufficiently high energies, so that the effective coupling constant of strong interactions is small and we can use the results of low-order perturbation theory. We can thus predict the results at FNAL or SPS energies by using, as input, the data from SLAC. We can also compute the cross sections for high energy electron-positron annihilation and explain the long life-times of the J/ψ resonances. They are essentially due to the fact that the heavy c and $\bar{c}$ quarks are very close together and the effective strength of strong interactions is small.

In all these applications QCD has been successful. However there still remains to prove rigorously that it possesses the property of confinement $i.e.$ that it binds permanently the quarks as well as the non-abelian gauge gluons inside the hadrons. This is one of the outstanding problems of elementary particle physics.

VII.   The Table of Elementary Particles

Table IV presents the assumed elementary particles according to our present ideas. It is more complicated than the one of 1932 but the order is partially restored. All matter-particles are spin-1/2 fermions and all radiation quanta are spin-one bosons. There is a lepton-quark symmetry. However the third point is still missing. If the role of (v_e, e^-) and (p, n) in the structure of matter is clear, nobody knows why the others (v_μ, μ^-), (v_τ, τ^-), (c, λ) and (t, b) exist. And if we don't know why they exist we have no reason to believe that the Table stops there. Maybe we are not yet at the end of the chain molecules → atoms → nuclei → nucleons → quarks.

		Useful			Useless		
Matter particles	Leptons	ν_e			ν_μ	ν_τ	...
		e^-			μ^-	τ^-	
	Hadrons	p_1	p_2	p_3	c_1	t_1	
		n_1	n_2	n_3	λ_1	b_1	...
Radiation Quanta	Strong	Eight Gluons					
	e.m.	Photon					
	Weak	W^+	W^-	Z^0			
	Gravit.	Graviton					

TABLE IV

This Table reflects mainly our theoretical prejudice. Only the leptons and the photon have been definitely identified. The quarks and the gluons are probably invisible; the W's and the graviton are, for the moment, beyond our reach.

VIII. Grand-unified theories

I hope that by now you are all convinced that gauge theories describe all interactions among elementary particles. The group structure seems to be SU(3) x SU(2) x U(1) . Although the scheme agrees with all available experimental data there are several reasons to believe that it is not the final step. One of the most important is that it is not a unified theory since one has three independent coupling strengths, one for each factor group. However, since they all describe gauge theories, it is natural to look for possible ways of obtaining a really unified picture in which all of them will be different manifestations of a single fundamental interaction. In other words one looks for a larger group G such that $G \supset$ SU(3) x SU(2) x U(1). If G is semi-simple it will only contain one coupling strength and a real unification will be achieved. However, how could one explain the apparent differences among the strengths of strong, electromagnetic and weak interactions? Here the notion of the effective coupling strength enters again. We assume that the true unification, involving only one coupling constant, happens at very high energies (Unification mass $H \sim 10^{15}$ GeV of Fig. 3). At this point a first spontaneous symmetry breaking occurs during which G breaks down to SU(3) x SU(2) x U(1) . The couplings of each group evolve according to their respective renormalization group equations of the form of Equation (28). Since the β-functions are different they follow different curves and this explains the observed differences at present energies ($\sim$ 10 GeV). At energies of the order of 100 GeV there is a second symmetry breaking, the one we described previously, which gives

masses to the W's and the Z leaving the photon and the gluons massless.
The symmetry breaks from SU(3) x SU(2) x U(1)   to SU(3) x U(1) .

FIG. 3

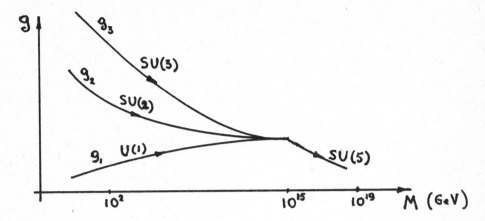

The simplest candidate for the group G is SU(5) introduced by Georgi
and Glashow, but higher groups, like SO(10) or E_6 have been also studied.
Leptons and quarks are assumed to be members of the same irreducible
representations of G and, as a result, all these schemes share the common
prediction that lepton as well as baryon numbers are not concerned. The
estimated life-time of the proton is $10^{31\pm2}$ years, quite close to the
present experimental limit of the order of 10^{29} to 10^{30} years. Several
experiments are planned for the near future in order to test this prediction.

We are still far from having a satisfactory grand-unified theory but one
thing is very suggestive: The energy scale of the grand-unification is
very large, of the order of 10^{15} GeV. This energy is quite close to the
so-called "Planck mass" $M_p \sim 10^{19}$ GeV above which gravitational effects
are expected to become important. If, on the other hand, we remember that
general relativity is also described by a non-abelian gauge theory, we
see that we have, probably gone a full circle and we arrive back at Einstein's original speculations about a unification of gravitation with
the other forces in nature. The problems are still formidable because
general relativity has resisted, so far, all attempts to describe it in
terms of a renormalizable field theory and in Professor Zumino's talk
we shall see all the problems involved and the progress which has been
made. However, the distance covered during the last ten years makes us
feel confident that we are on the right track. For the first time we
may hope that Einstein's dream about a unified field theory may not be
beyond our reach.

SUPERSYMMETRY: A WAY TO THE UNITARY FIELD THEORY

Bruno Zumino, CERN, Geneva, Switzerland

After his formulation of the general theory of relativity, which gave
a geometric interpretation to the gravitational force, Einstein devoted
most of his scientific efforts to the construction of a unified field
theory. He had particularly in mind the geometrization of electromagnetism through the introduction of a geometry more general than Riemannian
geometry [1] .

Today Einstein's dream of a unified field theory is again of great
actuality, although in a form somewhat different from that pursued
by him. In addition to gravitation and electromagnetism, we know now
that there exist other fundamental interactions: weak, strong and CP-
violating interactions. Furthermore, the natural framework for the
description of these interactions is relativistic quantum field theory.
Within this framework a number of theoretical schemes have been proposed,
which unify all elementary particle forces, with the exclusion of
gravitation [2]. These theories make use of the idea of non-abelian
vector gauge fields [3] and incorporate the mechanism of spontaneous
symmetry breaking through the vacuum expectation value of scalar fields
(Higgs effect [4]). First electromagnetic and weak interactions were
unified in the very successful SU(2) x U(1) model [5] . The strong
interactions were then also included in the reasonably successful SU(3)
color scheme of quantum chromodynamics [6] . Finally, all these interactions were put together in the so-called grand unified theories [7] ,
based on non-abelian gauge groups (the most popular example is SU(5))
which contain U(1) x SU(2) x SU(3). In a grand unified theory based on
a simple group there is only one gauge coupling constant (although there
are a number of interaction constants for the scalar fields) and the
difference in strength among weak, electromagnetic and strong interactions
is attributed to the variation of the interaction strength with energy
described by the equations of the renormalized group. Following this
variation with increasing energy one finds that all interactions become
of equal strength at the so-called grand unification energy [8] .
Although there are versions of the theory in which this energy is much
lower [9] in the more widely accepted schemes (for instance in the SU(5)
scheme), the grand unification energy is $\sim 10^{14}$ - 10^{16} GeV. Gravitation
is still neglected in this picture and this is usually justified by
recalling that the gravitational interaction does not become strong until

the so-called Planck energy $\sim 10^{19}$ GeV. It seems hardly reasonable to allow oneself the big leap forward from the energies typical of the SU(2) x U(1) model, 10^2 GeV, to the grand unification energy and to close ones eyes to gravity, which is just around the corner. Still, the grand unified theories have met with a certain measure of success [8] .Extrapolating back to the lower energies of SU(2) x U(1) they predict the weak mixing angle of this model. They also predict baryon number non-conservation and provide an explanation for the matter-antimatter asymmetry in the universe. A striking consequence of this baryon number non-conservation is the instability of the proton, its lifetime being predicted almost within reach of present experimental techniques, $\sim 10^{31\pm2}$ years.

Unification of non-abelian vector gauge theories with Einstein's theory of gravitation is the natural next step. A true unification does not consist in simply introducing an interaction between the fields of the two theories in a way which is covariant with respect to all relevant transformations. Nor does it help to observe that non-abelian gauge fields do have a geometric meaning as connections and that one can rewrite the same old equations in the more abstract mathematical language of differential forms and fibre bundles. A true unification will only be achieved if the theory possesses an invariance under transformations which transform the gravitational field into the matter fields and viceversa. Since the gravitational field has spin 2, while the matter fields have lower spins 1, $\frac{1}{2}$ and 0, this means transformations which change the spin. The only symmetry of this kind known today is Fermi-Bose supersymmetry and it is therefore the natural tool for the construction of unified theories including gravitation, or super unified theories. Observe that, in this sense, the grand unified theories are not truly unified even if one is willing to ignore gravitation, as reflected by the large number of arbitrary coupling constants appearing in them, about 20 to 25 depending on the scheme.

Supersymmetry [10] is a (graded) extension of the Poincaré algebra. In simple supersymmetry [11 - 13] the generators consist of the usual P_a, M_{ab}, of the Poincaré group (inhomogeneous Lorentz group) plus the fermionic generators Q_α (the index α is a four-component spinor index and the Q_α form a Majorana spinor; it is convenient to use a Majorana representation, in which the four Dirac gamma matrices are real). In addition to the usual commutation relations of the Poincaré algebra

$$\left[P_a, \ P_b \right] \qquad = \ 0$$

$$i \left[M_{ab}, P_c \right] \qquad = \ \eta_{ac} \, P_b - \eta_{bc} \, P_a \qquad\qquad (1)$$

$$i \left[M_{ab}, M_{cd} \right] \qquad = \ \eta_{ac} \, M_{bd} - \eta_{bc} \, M_{ad} - \eta_{ad} \, M_{bc} + \eta_{bd} \, M_{ac}$$

and the relations stating that Q is a translationally invariant spinor

$$\left[P_a Q_\alpha \right] \qquad = \ 0$$

$$i \left[M_{ab}, Q_\alpha \right] \qquad = \ (\sigma_{ab} \, Q)_\alpha \, , \qquad\qquad (2)$$

one has the basic anticommutation relation

$$\{ Q_\alpha, \ \bar{Q}_\beta \} = - 2 \, (\gamma^a)_{\alpha\beta} \, P_a, \qquad\qquad (3)$$

where $\bar{Q} = Q \gamma^\circ$. One knows many examples of local, renormalizable super-symmetric quantum field theories. In local field theory the fermionic generators can be represented as integrals over three-space of the time component of conserved vector-spinor currents.

The representation of the supersymmetry algebra in terms of one-particle states (supermultiplets) contain both integral and half-integral spins. For finite mass one can easily find their particle content by going to the rest frame, where (3) becomes essentially a Clifford algebra. One finds that the supermultiplet consists of four states of equal masses and of spins $s - \frac{1}{2}$, s, s, $s + \frac{1}{2}$ (s is a positive integer or half-integer; for $s = 0$ there are three states of spins $0, 0, \frac{1}{2}$). For zero mass it is convenient to work with helicity eigenstates and one finds that the supermuliplet consists of two states of spins s, $s + \frac{1}{2}$ (this again does not apply to $s = 0$, in which case one has three states of spins $0, 0, \frac{1}{2}$). If one applies a spinor generator Q_α to a one-particle state, one obtains another state of the super-multiplet having the same mass but spin (helicity) differing by one half. There are two especially interesting supermultiplets. The first is that with spins $\frac{1}{2}$ and 1 : it occurs in the widely studied supersymmetric Yang-Mills theories [14, 15] (it is certainly very tempting to imagine that the photon and the electron neutrino form a supermultiplet).The second, with spins $\frac{3}{2}$ and 2 is the supermultiplet of supergravity, where the graviton is taken to be the partner of a spin $\frac{3}{2}$ Rarita-Schwinger [16] field.

Supersymmetry can be broken either explicitly (but softly) [17]
or spontaneously [18] . In field theories with soft explicit or spontaneous breaking the particles of the supermultiplet no longer have the same
mass but satisfy instead certain mass sum rules. The various relations
among masses and coupling constants valid in the supersymmetric limit
are then modified by finite calculable corrections. The most appealing
alternative is that of spontaneous breaking, however, when supersymmetry
is spontaneously broken, a "Goldstone" massless particle emerges
[17, 19] which is a spin $\frac{1}{2}$ fermion, corresponding to the spinorial
nature of the generator Q_α which no longer leaves the vacuum invariant.
It has been suggested [12] that the (electron ?) neutrino may be this
Goldstone fermion, but the neutrino spectrum does not satisfy the low
energy theorems which must be valid for a Goldstone fermion [20] (see
below for a way out of this difficulty, the supersymmetric Higgs effect).

Supersymmetry can be combined with internal symmetry by attributing each
supermultiplet to some representation of the internal symmetry. The
generators Q_α are taken as singlets under the internal symmetry, just
as the generators of the Poincaré subalgebra. Most attempts at constructing relativistic supersymmetric field theories for particle physics are
based on this idea. A more interesting way of combining supersymmetry
with internal symmetry [21] attributes the spinorial generators to a
representation of the internal symmetry group, while naturally the
Poincaré generators are still singlets. Although in principle one could
imagine other possibilities, for various reasons one is led to an internal symmetry SO(N). One has N spinor charges Q^i (i = 1, ... N) belonging
to the vectorial representation of SO(N). Instead of (3) one writes

$$\left\{ Q_\alpha^i \ , \ \bar{Q}^j_\beta \right\} = -2\delta^{ij} (\gamma^a)_{\alpha\beta} P_a \ . \tag{4}$$

The supermultiplets are now richer. Let us consider the case of zero
mass. Starting with a state of maximum helicity λ (singlet under SO(N))
and applying to it the operators Q_α^i, one obtains N states of helicity
$\lambda - \frac{1}{2}$ (in the vector representation of SO(N)). Applying Q_α^i twice one goes
to helicity $\lambda - 1$ and multiplicity N(N − 1). Applying Q_α^i k times one
obtains N!/k!(N-k!) states of helicity $\lambda-k/2$. Finally, applying Q_α^i N
times one obtains one state (again a singlet) of helicity $\lambda-N/2$. If this
helicity equals − λ(which means that N = 4λ) the states form a set which
is complete under TCP and one can expect that a corresponding field
theory exists. Otherwise, one must adjoin the TCP conjugate states with
opposite helicities.

It is clear from the above that, if one fixes the maximum spin of the supermultiplet one cannot have an arbitrary number N of spinoral generators. In particular, if the maximum spin is 1, N cannot exceed 4 and, if the maximum spin is 2, N cannot exceed 8. The first case is realized in the SO(4) supersymmetric Yang-Mills theory [22, 23] . In this theory the group SO(4) is not gauged, but there is a gauge group which is to a large extent arbitrary and under which the spinor charges are singlets. All fields are in the adjoint representation of the gauge group and the number of adjoint representations is 1 for the vector, 4 for the spin $\frac{1}{2}$, 3 for the scalars and 3 for the pseudoscalars. This theory has the interesting property that the renormalization group function $\beta(g)$ vanishes in the one- and two-loop approximation [24,25] (it is not known what happens in higher orders).

If one assumes that there is only one graviton (a singlet under the SO(N) group), then the largest supergravity theory is that for N = 8. In any case, the existence of consistent interacting theories with fields of spin higher than 2 is very problematic. The particle content of supergravity theories (*i.e.*, maximum spin 2) for various values of N is given in the table. In these theories one achieves the unification of different spin fields with the graviton field in a single supermultiplet. In the theories with N from 4 to 8 even fields with spin zero are part of the supermultiplet. Observe that the theories with N = 7 and N = 8 have the same particle content. They are probably identical.

<u>Table</u>

Particle multiplicity of supergravity theories for various values of N

Spin	N = 1	2	3	4	5	6	7	8
2	1	1	1	1	1	1	1	1
3/2	1	2	3	4	5	6	8	8
1		1	3	6	10	16	28	28
1/2			1	4	11	26	56	56
0				2	10	30	70	70

The Lagrangians for the theories with $N \leq 4$ are explicitly known [26 - 33]. For the N = 8 theory (which contains all those for lower N) one knows [34] the lower order vertices in an expansion in the gravitational constant. A complete Lagrangian for N = 8 has also been written [35], but its invariance under local supersymmetry has not been fully verified as yet. It was obtained by first constructing the N = 1 supergravity in 11 space-time dimensions [36] and then applying the method of dimensional reduction [37] from 11 to 4 dimensions. In its simplest form this method, which is a generalization of that of Kaluza and Klein [38] , consists in assuming that all fields are independent of the additional space-time co-ordinates beyond our usual four. A higher dimensional field theory gives then rise to a four-dimensional one and higher dimensional conservation laws become four-dimensional conservation laws. Recently [39] a generalized method of dimensional reduction has been studied in which the fields are taken to depend upon the additional co-ordinates in a particular way. The resulting four-dimensional theory exhibits spontaneous symmetry breaking.

It is known [40] that the supergravity theories for various N are one-loop and two-loop finite (on the mass shell). This result should be compared with the situation in ordinary gravity, which is one-loop finite, but probably two-loop divergent, and with the situation with gravity coupled to spin 1, $\frac{1}{2}$ or 0 (or any combination of them) which is already one-loop divergent. Observe that these theories are contained as sub-theories in the supergravity theories; gravity coupled to spin 1 (Maxwell-Einstein) in the theories with $N \geqslant 2$, gravity coupled to spin $\frac{1}{2}$ (Dirac-Einstein) in the theories with $N \geqslant 3$, and gravity coupled with spin 0 in the theories with $N \geqslant 4$. Unfortunately, at the three-loop level there appear to arise divergences in supergravity [41] and the same is true for the higher loops. However disappointing this may be, the cancellation of divergences due to diagrams involving the spin 3/2 fields improves the convergence of the theories dramatically. It is essential for this that all fields belong to the same irreducible supermultiplet. For instance, it is known that simple supergravity coupled to spin $\frac{1}{2}$ - spin 1 supermatter is not even one-loop finite.

I shall now describe briefly simple supergravity [26, 27] , the gauge theory of simple supersymmetry. The Lagrangian can be formulated by introducing the gauge fields associated with the generators P_a, M_{ab}, Q_α of the supersymmetry algebra. We shall call them respectively $e_m{}^a$ (the vierbein field), $\omega_m{}^{ab}$ (the connection field), and $\psi_m{}^\alpha$ (the Rarita-Schwinger spin 3/2 field). These fields must be varied independently. One takes

$$L = L_E + L_{RS} , \tag{5}$$

where the Einstein Lagrangian is

$$L_E = -\frac{1}{2} \text{ Re} \tag{6}$$

and the Rarita-Schwinger Lagrangian is

$$L_{RS} = -\frac{i}{2} \varepsilon^{\ell mnr} \bar{\psi}_\ell \gamma_5 \gamma_m D_n \psi_r \ .$$ (7)

For simplicity we have chosen units in which the gravitational constant equals one. In (6) appear the determinant of the vierbein field

$$e = \det e_m{}^a$$ (8)

and the contracted curvature tensor

$$R = R_{mn,}{}^{ab} e_a{}^m e_b{}^n \ ,$$

$$R_{mn,}{}^{ab} = \partial_m \omega_n{}^{ab} - \partial_n \omega_m{}^{ab} - \omega_m{}^a{}_c \omega_n{}^{cb} + \omega_n{}^a{}_c \omega_m{}^{cb} \ .$$ (9)

The spinoral covariant derivative in (7) is defined as

$$D_n = \frac{\partial}{\partial x^n} - \frac{1}{2} \omega_{nab} \sigma^{ab} \ .$$ (10)

The expression is covariant in spite of the world index r carried by the spin 3/2 field because of the antisymmetrization in n and r due to the epsilon symbol. By construction (5) is invariant under general co-ordinate transformations and under local Lorentz transformations. The remarkable fact is that it is also invariant under the local supersymmetry transformation

$$\delta e_m{}^a = i\bar{\zeta}\gamma^a\psi_m \ .$$

$$\delta \psi_m = 2D_m\zeta \ ,$$ (11)

$$\delta \omega_m{}^{ab} = B_m{}^{ab} - \frac{1}{2} e_m{}^a B_c{}^{bc} + \frac{1}{2} e_m{}^b B_c{}^{ac} \ ,$$

where $\zeta(x)$ is a spinoral (Majorana) anticommuting infinitesimal parameter which can be an arbitrary of x and

$$B_a{}^{mn} = i\bar{\zeta}\gamma_5\gamma_a D_\ell \psi_r \varepsilon^{mn\ell r} \ .$$ (12)

The above is a first order formulation. Eliminating $\omega_m{}^{ab}$ in the Lagrangian by solving for it from its equation of motion, one obtains an equivalent second order form [26] which exhibits explicitly a quartic contact interaction of the spin 3/2 field with itself.

The transformation (11) on the Rarita-Schwinger field ψ_m is the generalization to curved space of the spinoral gauge invariance

$$\delta\psi_m = 2 \partial_m \zeta$$ (13)

oberserved by Rarita and Schwinger [16] for the free massless spin 3/2
Lagrangian , which would be given by (7) if one replaces the covariant
derivative by an ordinary derivative.

In checking the invariance of (5) under the local supersymmmetry transformation (11) one makes essential use of the fermionic nature of the
spin 3/2 field. Both the Rarita-Schwinger field ψ_m and the parameter
$\zeta(x)$ are treated as totally anticommuting (Grassmannian) quantities in
performing the necessary Fierz rearrangements.

If one calculates the commutator of two local supersymmetry transformations (11) of parameters ζ_1 and ζ_2, one finds a general co-ordinate transformation of parameter $\zeta^m = 2i\bar{\zeta}_1\gamma^a\zeta_2 e_a{}^m$ accompanied by a Lorentz transformation and by a supersymmetry transformation (both field dependent)
plus terms which vanish by use of the spin 3/2 equations of motion: the
algebra closes on the mass shell. In this well defined sense the general
co-ordinate invariance is a consequence of local supersymmetry, which
appears as the primary gauge principle.

The equations of motion derived from (5) were the first known example
of consistent equations for coupled higher spin fields and, together
with those of extended supergravity, probably the only consistent equations of this kind. The standard consistency checks proceed by deriving
constraints from the field equations by differentiation . Here the constraints turn out to vanish identically, provided one makes use of
the anticommuting nature of the spinor fields. The vanishing of the
constraints has also the consequence that no acausal wave propagation
occurs [27, 42] . This can be shown by computing the characteristic surfaces by the standard methods of the theory of partial differential
equations, as applied to theories with gauge groups, such as gravitation.
All these facts are direct consequences of the new gauge principle, local
supersymmetry.

There is an interesting generalization of (5) which also admits local
supersymmetry [42 - 44] . One adds to (5) a cosmological term and a spin
3/2 "mass" term in the particular combination

$$3\mu^2 e - \frac{i}{2} \mu \, \varepsilon^{\ell mnr} \, \bar{\psi}_\ell \gamma_5 \, \sigma_{mn} \, \psi_r \; . \tag{14}$$

The sum is invariant under a modified local supersymmetry

$$\delta \, e_m{}^a = i \, \bar{\zeta} \, \gamma^a \psi_m \; , \tag{15}$$

$$\delta \, \psi_m = 2D_m \, \zeta + \mu \, \gamma_m \, \zeta$$

(with a corresponding modification for $\delta\,\omega_m{}^{ab}$). The physical meaning of
this theory is not easy to extract, because the cosmological term requires quantization in a background de Sitter space. However, it is
interesting that a cosmological term with the opposite sign from that
in (14) is generated when supergravity is coupled to supersymmetric
matter with spontaneous breaking of supersymmetry. These two cosmological
terms can actually be cancelled against each other, by adjusting suitably
the constants in the theory. The resulting theory can be quantized in a
background Minkowski space.

When supergravity is coupled to supersymmetric matter with spontaneous
breaking of supersymmetry, a supersymmetric version of the Higgs effect
occurs [42] . The orginally massless spin 3/2 field (which is the gauge
field of supersymmetry) absorbs the degrees of freedom of the spin $\frac{1}{2}$
Goldstone fermion associated with the spontaneous breaking of supersymmetry and becomes massive. The absence in nature of the Goldstone
fermion can thus be understood. This is satisfactory because it leaves
one the freedom to use models of matter with spontaneous breaking of
supersymmetry.

For extended supergravity theories a generalization analogous to (14)
allows the gauging of the group SO(N). Oberserve that the number of spin
1 fields dictated by supersymmetry is exactly right for them to be the
gauge fields of SO(N). It is indeed possible [43] to introduce, besides
the gravitational couplings, a gauge coupling with constant g and pre-
serve local symmetry by adding also a spin 3/2 mass term and a cosmological term. This has been done explicitly for $N \leqslant 4$ and presumably can
also be done for the higher N's. Unfortunately, the resulting cosmological term is of the order of $(g/\kappa^2)^2$ (where we have introduced explicitly
the gravitational constant κ). If one takes g $\sim$1, it is enormous and in
obvious drastic disagreement with the known limits on the cosmological
constant.

To solve this difficulty, one would like to compensate the unwanted
cosmological term due to the introduction of the gauge coupling with
a cosmological term from the spontaneous breaking of supersymmetry,
in a manner similar to that described above for N = 1 supergravity
coupled to supersymmetric matter. This has not been done explicitly as
yet within an extended supergravity theory. However, it is encouraging
that some recent work [39] has provided an example of gauging (unfortunately abelian) and simultaneous symmetry breaking with vanishing cosmological term for the N = 8 theory.

Irrespective of this difficulty, even the gauged N = 8 theory does not seem to be sufficiently ample to describe the presently known elementary forces. As explained at the beginning of this lecture, according to the commonly accepted picture, one requires a gauge group containing U(1) x SU(2) x SU(3) (for flavor and color); SO(8) is too small. Another aspect of the same difficulty is that the basic supermultiplet of the SO(8) theory, however large it may appear, does not contain enough fields to accomodate the known vectors, leptons and quarks [45] .

This difficulty can perhaps be resolved as follows. It has been known for some time [46] that the equations of motion of supergravity theories admit a global U(N) (in some cases SU(N)) invariance which generalizes the orginal SO(N) invariance. This U(N) group is represented on the complex left-handed and right-handed components of spinors and on the complex scalar fields which have as real part the scalar and as imaginary part the pseudoscalar fields. On vectors, it is realized as a duality transformation which rotates the electric into the magnetic components of the field strength. Recently, it has been observed that, for $N \gtrless 4$, at least part of this unitary group can be made local [35] . In particular, the N = 8 theory is probably invariant under a local SU(8). Although the vectors of the basic supergravity multiplet are not the gauge fields of this SU(8), one can find [35] composite fields which transform exactly like the gauge vectors of SU(8). Presumably, these composite vector fields belong to a supermultiplet together with spin $\frac{1}{2}$ and spin 0 fields (as well as higher spins), belonging to (probably reducible) representations of SU(8). One sees here the possibility of avoiding the restrictions of the SO(N) basic multiplet, but at the composite level. In this picture, leptons and quarks would be composite objects, the elementary constituents (preons) being the fields of the basic super-gravity supermultiplet. One can imagine that, except for the graviton field, which is an SU(8) singlet, all other fields of the basic multi-plet are very strongly bound, with binding energies comparable to the Planck energy and therefore are experimentally irrelevant, while the composite fields behave essentially as point-like objects. At the same time, the problems and inconsistencies involved in the use of interacting elementary fields of spin higher than 2 are avoided. The gauge group SU(8) is certainly large enough (it contains SU(5) x SU(3)) and the idea that leptons and quarks are composite is perhaps no longer so extraordi-nary due to the experimental proliferation in the number of flavors. It could be especially acceptable here due to the uniqueness of the picture emerging in N = 8 supergravity. Also, the dynamical emergence of a supermultiplet containing composite gauge fields has been verified in simple models, the two-dimensional supersymmetric CP^{n-1} models [47] .

As it should have become apparent, supergravity for the time being is more a general framework than a concrete physical theory. The extended supergravity theories represent an obvious qualitative success. The gauge principle of extended local supersymmetry provides a way to unify the graviton with the lower spin fields and Einstein gauge invariance with internal symmetry. The improvement in convergence with respect to ordinary gravity coupled with matter is also remarkable. In spite of these positive results, some new input seems necessary for supergravity to become a physically relevant theory. We have mentioned the idea of using composite fields. Another possibility is to give up the strong requirement that all fields belong to the same irreducible supermultiplet. Indeed this requirement loses much of its motivation if, as seems to be the case, supergravity theories are not finite after all. Without this very strong restriction one has much more freedom in model building. For instance, within the framework of simple supergravity coupled to simply supersymmetric matter one can construct models which are almost realistic. Another case which deserves more study is that of $N = 4$ supergravity coupled to $N = 4$ Yang-Mills theory.

Does supergravity realize Einstein's dream of giving a geometric meaning to all elementary fields ? This depends on one's idea of what geometry is. The fact that fields of different spins, including the gravitational field, belong to the same irreducible representation (supermultiplet) of an algebra should satisfy those for whom geometry can be reduced to group theory. But if one is asking for an approach more along the lines of differential geometry, it is important to observe that supergravity can be described in terms of the differential geometry of a (super-) manifold, superspace.

Superspace [48, 49] is an extension of ordinary space-time in which one adjoins to the usual co-ordinates x^m a certain number of spinorial anti-commuting co-ordinates θ^μ. For simple supersymmetry and supergravity, for instance, the θ^μ form a single Majorana spinor. A field in superspace (superfield) is a function $V(x^m, \theta^\mu)$ of the superspace co-ordinates. Since the θ^μ anticommute, the square of each component vanishes. Therefore, since the number of components θ^μ is finite, an expansion of $V(x^m, \theta^\mu)$ in power series of θ^μ terminates after a finite number of terms. The coefficients of this polynomial are ordinary fields, functions of x^m alone, carrying different spins and satisfying different statistics. So, a superfield corresponds to a supermultiplet of fields. Global supersymmetry can be interpreted as the group of motions of a rigid superspace just like the Poincaré group is the group of motions of Minkowski space. In global supersymmetry superspace techniques have been extensively used not only for the construction of invariant Lagrangians but also for the study of perturbation theory, which can be formulated directly in super-space.

Just like Einstein's theory of gravitation can be obtained by general-izing flat Minkowski space to a curved space, similarly supergravity can be described geometrically by studying a suitable superspace, which generalizes the rigid superspace of global supersymmetry. General co-ordinate invariance plus local Lorentz covariance in superspace replace (and imply) local supersymmetry. For simple supergravity this approach has been fully developed and one has a complete method for the construc-tion of invariants, which permits in particular a simple discussion of the possible divergences arising in perturbation theory. I do not have the time here to enter into the formal details of the geometric theory of superspace which is highly technical and for which I refer to my own work in collaboration with J. Wess [50 - 54] and to J. Wess' lecture in these proceedings where other references can be found.

As we have seen, the study of supergravity has led to the development of two elegant mathematical techniques, both concerned with an enlarge-ment of the four-dimensional space-time of Einstein's theory: the super-space technique, in which the additional co-ordinates are fermionic, and the more traditional higher dimensional technique of the Kaluza-Klein variety, in which the additional co-ordinates are bosonic. These two approaches can easily be combined. Perhaps the additional dimensions have a real physical meaning and these mathematical developments will lead to new physical insights.

References

[1] For a review of the attempts, by Einstein and others, to construct a unitary field theory, see the lecture by P.Bergmann in these proceedings.

[2] See the lecture by J.Iliopoulos in these proceedings.

[3] C.N.Yang and F.Mills, Phys.Rev. 96 (1954) 191;
R.Shaw, Thesis, Cambridge, University (1954).

[4] P.W.Higgs, Phys.Letters 12 (1964) 132;
F.Englert and R. Brout, Phys.Rev.Letters 13 (1964) 321;
G.S.Guralnik, C.R.Hagen and T.W.Kibble, Phys.Rev.Letters 13 (1964) 585.

[5] S.L.Glashow, Nucl.Phys. 22 (1961) 579;
S.Weinberg, Phys.Rev.Letters 19 (1967) 1264;
A.Salam, in Proceedings of the Eighth Nobel Symposion, Stockholm 1968,edited by N.Svartholm (Almqvist and Wikell,Stockholm 1968).

[6] H.Fritzsch and M.Gell-Mann, Proc.XVI International Conference on High Energy Physics, Chicago 1972, Vol. 2; S.Weinberg, Phys.Rev.Letters 31 (1973) 494; H. Fritzsch, M.Gell-Mann and P.Minkowski, Phys.Letters 59B (1975) 257, and references therein.

[7] J.C.Pati and A.Salam, Phys.Rev. D8 (1973) 1240; D10 (1974) 275; H.Georgi and S.L. Glashow, Phys.Rev.Letters 32 (1974) 438.

[8] H.Georgi, H.R.Quinn and S.Weinberg, Phys.Rev.Letters 33 (1974) 451; A.J. Buras, J.Ellis, M.K. Gaillard and D.V.Nanopoulos, Nucl.Phys. B135 (1978) 66.

[9] V.Elias, J.C.Pati and A.Salam, Phys.Rev.Letters 40 (1978) 920.

[10] For reviews of supersymmetry with extensive references see, for instance, B. Zumino, Proc. 17th International Conference on High Energy Physics, London, 1974 (Science Research Council, Didcot, 1974) p. I-254; P.Fayet and S.Ferrara, Physics Reports 32 (1977) 251.

[11] Y.A. Golfand and E.P.Lichtman, JETP Letters 13 (1971) 323.

[12] D.V.Volkov and V.P.Akulov, Phys.Letters 46B (1973) 109.

[13] J.Wess and B.Zumino, Nucl.Phys. B70 (1974) 39.

[14] S.Ferrara and B.Zumino, Nucl.Phys. B79 (1974) 413.

[15] A.Salam and J.Strathdee, Phys.Letters 51B (1974) 353.

[16] W.Rarita and J.Schwinger, Phys.Rev. 60 (1941) 61.

[17] J.Iliopoulos and B.Zumino, Nucl.Phys. B76 (1974) 310.

[18] P.Fayet and J.Iliopoulos, Phys.Letters 51B (1974) 461; L.O'Raifeartaigh, Nucl.Phys. B96 (1975) 331.

[19] A.Salam and J.Strathdee, Phys.Letters 49B (1974) 465.

[20] W.Bardeen, unpublished; B.de Wit and D.Z.Freedman, Phys.Rev. D12 (1975) 2286.

[21] See R.Haag, J.T.Łopuszanski and M.Sohnius, Nucl.Phys. B88 (1975) 257 and references therein.

[22] L.Brink, J.H.Schwarz and J.Scherk, Nucl.Phys. B121 (1977) 77.

[23] F.Gliozzi, J.Scherk and D.Olive, Nucl.Phys. B122 (1977) 253.

[24] D.R.T.Jones, Phys.Letters 72B (1977) 199.

[25] E.C.Poggio and H.N.Pendleton, Phys.Letters 72B (1977) 200.

[26] D.Z.Freedman, P.van Nieuwenhuizen and S.Ferrara, Phys.Rev. D13 (1976) 3214.

[27] S.Deser and B.Zumino, Phys.Letters B62 (1976) 335.

[28] S.Ferrara and P.van Nieuwenhuizen, Phys.Rev. Letters 37 (1976) 1669.

[29] S.Ferrara, J.Scherk and B.Zumino, Phys.Letters 66B (1977) 35.

[30] D.Z.Freedman, Phys.Rev.Letters 38 (1977) 105.

[31] A.Das, Phys.Rev. D15 (1977) 2805.

[32] E.Cremmer, J.Scherk and S.Ferrara, Phys.Letters 68B (1977) 234.

[33] E.Cremmer and J.Scherk, Nucl.Phys. B127 (1977) 259.

[34] B.de Wit and D.Z.Freedman, Nucl.Phys. B130 (1977) 105.

[35] E.Cremmer and B.Julia, Phys.Letters 80B (1978) 48;
      Ecole Normale preprint LPTENS 79/6.

[36] E.Cremmer, B.Julia and J.Scherk, Phys.Letters 76B (1978) 409.

[37] J.Scherk and J.H.Schwarz, Phys.Letters 57B (1975) 463;
      E.Cremmer and J.Scherk, Nucl.Phys. B103 (1976) 399.

[38] T.Kaluza, Sitzungsberichte Preuss.Akad.Wiss.Berlin, Math.Phys.KA
      (1921) 1966;
      O.Klein, Z.Physik 37 (1926) 895.

[39] J.Scherk and J.H.Schwarz, Ecole Normale preprints LPTENS 78/28
      and 79/2.

[40] See the review by M.Grisaru and P.van Nieuwenhuizen in Deeper
      Pathways in High-energy Physics, Proc.Orbis Scientiae, Coral
      Gabels 1977 (Plenum Press, New York, 1977).

[41] S.Deser, J.H.Kay and K.S.Stelle, Phys.Rev.Letters 38 (1977) 527;
      S.Ferrara and B.Zumino, Nucl.Phys. B134 (1978) 301.

[42] S.Deser and B.Zumino, Phys.Rev.Letters 38 (1977) 1433.

[43] D.Z.Freedman and A.Das, Nucl.Phys. B120 (1977) 221.

[44] P.K.Townsend, Phys.Rev. D15 (1977) 2802.

[45] M.Gell-Mann, unpublished.

[46] S.Ferrara, J.Scherk and B.Zumino, Nucl.Phys. B121 (1977) 393.

[47] A.D'Adda, P.Di Vecchia and M.Lüscher, DESY preprint 78/75.

[48] A.Salam and J.Strathdee, Nucl.Phys. B76 (1974) 477.

[49] S.Ferrara, J.Wess and B.Zumino, Phys.Letters B51 (1974) 239.

[50] J.Wess and B.Zumino, Phys.Letters 66B (1977) 361; 74B (1978) 51.

[51] R.Grimm, J.Wess and B.Zumino, Phys.Letters 73B (1978) 415.

[52] J.Wess, Supersymmetry-Supergravity, Lectures at the VIII G.I.F.T.
      Seminar (Salamanca 1977), Karlsruhe University preprint, to be
      published in Springer tracts.

[53] J.Wess and B.Zumino, Phys.Letters 79B (1978) 394.

[54] R.Grimm, J.Wess and B.Zumino, Nucl.Phys. B152 (1979) 255.

BOLTZMANN UND PLANCK: DIE KRISE DES ATOMISMUS UM DIE JAHRHUNDERTWENDE UND IHRE ÜBERWINDUNG DURCH EINSTEIN

Res Jost, Eidgenössische Technische Hochschule, Zürich, Schweiz

Meine Damen und Herren, ich weiß, daß A l b e r t   E i n s t e i n selbst H e n d r i k   A n t o o n   L o r e n t z als den für seine eigene Entwicklung bestimmendsten Physiker bezeichnet hat. Dessen ungeachtet sind für die Gegenstände, die der Diskussion des heutigen Tages zugrunde liegen: nämlich Einsteins Arbeiten zur B r o w n s c h e n B e w e g u n g und zur Q u a n t e n t h e o r i e, B o l t z - m a n n und P l a n c k die eigentlichen Vorläufer und Wegbereiter. An Boltzmann schließen Einsteins frühe Arbeiten zur Statistischen Mechanik an, mit welchen er die Grundlage seiner souveränen Meisterschaft über die Schwankungserscheinungen schafft; und die Strahlungsformel von Max Planck wird ihm der Ausgangspunkt für die Entdeckung des verborgendsten Elementarteilchens, des sichtbaren Photons. Das Quantenrätsel aber ist der "cantus firmus" in Einsteins wissenschaftlichem Leben.

So verschmelzen im jungen Einstein die Bestrebungen der beiden verbitterten Streiter des ausgehenden 19. Jahrhunderts, des Österreichers Ludwig Boltzmann und des Deutschen Max Planck. Als dritter erscheint in diesem Kampf E r n s t   M a c h, der grosse Unabhängige, dessen Einwirkung auf den jungen Einstein hinreichend bekannt ist.

Max Planck und Ernst Mach

Die Behauptung, daß Mach im Kampf zwischen Planck und Boltzmann eine Rolle gespielt habe, bedarf einer Rechtfertigung. Vor der Jahrhundertwende finde ich Ernst Mach in Plancks *Physikalischen Abhandlungen und Vorträgen* [PAV] nur einmal erwähnt, nämlich im Vortrag über *"Die Maxwell'sche Theorie der Elektricität von der mathematischen Seite betrachtet"* aus den Jahresberichten der Deutschen Mathematischen Vereinigung von 1899. Ich zitiere:

> *Alles zusammengefaßt möchte ich also sagen: die M a x w e l l'sche Theorie zeichnet sich vor den älteren Theorien aus nicht durch größere Richtigkeit, sondern durch größere Einfachheit, oder mit anderen Worten: es ist im letzten Grunde nichts anderes als das Prinzip der Ökonomie im Sinne von M a c h gesprochen, welches in der Durchführung der M a x w e l l'schen Elektricitätstheorie einen seiner schönsten Triumphe gefeiert hat.* [1]

Im Dezember 1908, vor den Studenten der Universität Leiden tönt es dann allerdings anders.[2] Der IV. Abschnitt dieses Vortrages ist der Beginn einer Polemik des im Zenith seiner Laufbahn stehenden Max Planck gegen einen seit Jahren leidenden, gelähmten Ernst Mach. Hier eine bekannte Stelle - und ich zitiere - :

Zum Schluß noch ein Argument, das vielleicht auf diejenigen, welche trotz alledem den menschlich-ökonomischen Gesichtspunkt als den eigentlich ausschlaggebenden hinzustellen geneigt sind, mehr Eindruck macht als alle bisherigen sachlichen Überlegungen. Als die großen Meister der exakten Naturforschung ihre Ideen in die Wissenschaft warfen: als N i k o l a u s K o p e r n i k u s die Erde aus dem Zentrum der Welt entfernte, als J o h a n n e s K e p l e r die nach ihm benannten Gesetze formulierte, als I s a a c N e w t o n die allgemeine Gravitation entdeckte, (...), als M i c h a e l F a r a d a y die Grundlagen der Elektrodynamik schuf – die Reihe wäre noch lange fortzusetzen –, da waren ökonomische Gesichtspunkte sicherlich die allerletzten, welche diese Männer in ihrem Kampfe gegen überlieferte Anschauungen und gegen überragende Autoritäten stählten. Nein – es war ihr felsenfester, sei es auf künstlerischer, sei es auf religiöser Basis ruhender Glaube an die Realität ihres Weltbildes. Angesichts dieser doch gewiß unanfechtbaren Tatsache läßt sich die Vermutung nicht von der Hand weisen, daß falls das M a c h sche Prinzip der Ökonomie wirklich einmal in den Mittelpunkt der Erkenntnistheorie gerückt werden sollte, die Gedankengänge solcher führender Geister gestört, der Flug ihrer Phantasie gelähmt und dadurch der Fortschritt der Wissenschaft vielleicht in verhängnisvoller Weise gehemmt werden würde.[3]

Man sieht, Planck bekennt sich jetzt (und damit ist er nicht allein) zu einer idealistisch-romantischen Auffassung über die Wurzeln des Fortschrittes in der Physik. Es ist vielleicht die Anmerkung angebracht, daß mit der Abkehr von der Machschen Nüchternheit die Physik in immer größere Abhängigkeit von Wirtschaft und Ökonomie getrieben wurde und schließlich mit der nackten Gewalt (nämlich dem Militär) akkordiert hat. Das mag eine zufällige Koinzidenz sein.

Wichtig für uns ist Plancks scharfe Duplik *Zur M a c h schen Theorie der physikalischen Erkenntnislehre. Eine Erwiderung* [12] ; denn sie enthält die aufschlußreiche Rechtfertigung:

Die Berechtigung zu einer Meinungsäußerung über die M a c h sche Theorie der physikalischen Erkenntnis glaube ich aus dem Umstand ableiten zu dürfen, daß ich mich mit dieser Theorie seit Jahren eingehend beschäftigt habe. Zählte ich mich doch in meiner Kieler Zeit (1885 - 1889) zu den entschiedenen Anhängern der M a c h schen Philosophie, die wie ich gerne anerkenne, eine starke Wirkung auf mein physikalisches Denken ausgeübt hat.[4]

Das ist das Bekenntnis, das wir suchten: P l a n c k ist als M a c h i s t (um einen Ausdruck von W l a d i m i r I l j i t s c h U l j a n o w zu verwenden) gegen Boltzmann angetreten. Und nach der schmerzlichen Niederlage glaubte er aus Recht und Pflicht vor allem die junge Generation vor Mach und seiner Erkenntnistheorie warnen zu müssen. Allein dies geschah in verletzender Form und auch Einstein, der zu heilen suchte, wo er konnte, war unfähig den Schaden zu bessern. Das Unrecht war geschehen.[5]

Wir aber glauben den Einfluß Machs auf Planck schon in der Münchner Zeit und auch später in der ersten Berliner Dekade erkennen zu können. Mach war bekanntlich ein Anti-Atomist, Planck war hinsichtlich der Existenz der Atome ein Skeptiker bis zu seiner Bekehrung im Herbst 1900. Um den

"Glauben an die Realität der Atome"[6] geht es denn auch im Streit zwischen Mach und Planck. Woher aber diese Feindschaft gegen die Atome? Hören wir ein Zeugnis Machs aus seinem Frühwerk *Die Geschichte und die Wurzel des Satzes von der Erhaltung der Arbeit*, Prag 1872, eine Schrift, die Planck wahrscheinlich in seiner Münchner-Zeit studiert hat. Mach schreibt mit Hinblick auf den 1. Hauptsatz der Thermodynamik:

> *Auf dieses Verschwinden der Wärme bei Leistung von Arbeit und die Bildung von Wärme bei Verbrauch von mechanischer Arbeit (...) hat man nun ein besonderes Gewicht gelegt. Man schloß daraus: Wenn Wärme sich in mechanische Arbeit verwandeln kann, so wird Wärme in mechanischen Vorgängen (,) in Bewegung bestehn.*
>
> *Dieser Schluß, der sich wie ein Lauffeuer über die ganze cultivirte Erde verbreitete, rief nun eine Masse von Schriften über diesen Gegenstand hervor und man ist nun allerorten bemüht, die Wärme durch Bewegungen zu erklären; man bestimmt die Geschwindigkeiten, die mittleren Wege und die Bahnen der Molecüle und es gibt fast kein Räthsel mehr, das nicht auf diesem Wege mit Hilfe hinreichend langer Rechnungen und verschiedener Annahmen vollständig erledigt würde.*[7]

Zu den in Machs Augen unkritischen Rechnern gehören natürlich R u d o l f C l a u s i u s und Ludwig Boltzmann, aber auch G u s t a v K i r c h h o f f und wohl H e r m a n n v. H e l m h o l t z *"das Viergestirn"* der Theoretischen Physik in Deutschland um 1880, wie Planck es nannte.[8]

Es kennzeichnet die anti-autoritäre Haltung des jungen Planck, daß er die Meinung von Ernst Mach zuneigte. In seiner Preisschrift über *Das Prinzip der Erhaltung der Energie* [7] aus dem Jahr 1887 sekundiert er vorsichtig und ohne Namensnennung mit den Worten:

> *Der Ansicht aber, die wohl auch manchmal geäußert wird, daß man die mechanische Theorie (der Atome) als ein a priori Postulat der physikalischen Forschung zu akzeptieren habe, müssen wir mit aller Entschiedenheit entgegentreten; dieselbe kann nicht von der Verpflichtung befreien, jene Theorie auf legalem Wege zu begründen.*[9]

Und er fährt - auch hier Mach folgend[10] - fort:

> *Die Naturwissenschaft kennt überhaupt nur ein Postulat: das Kausalitätsprinzip; denn dasselbe ist ihre Existenzbedingung. Ob dies Prinzip selber erst aus der Erfahrung geschöpft ist, oder ob es eine notwendige Form unseres Denkens bildet, brauchen wir hier nicht zu untersuchen.*[11]

Plancks Haltung ist eine grundsätzlich-axiomatische. Für ihn ist die Allgemeingültigkeit des Energiesatzes eine Tatsache, gesichert wie wenig andere in der Naturwissenschaft. Er ist A u s g a n g s p u n k t, die mechanische Naturauffassung mögliches oder sogar wahrscheinliches Z i e l der Forschung.

Planck arbeitete an der Preisschrift über die Erhaltung der Energie nach eigener Angabe im Frühjahr 1885, also noch in München.[12]

Planck als Entropiker

Planck war anti-autoritär und fand daher im deutschen Sprachgebiet, das
um 1880 die eigentliche Wallfahrtsstätte für Studenten der theoretischen
Physik war, keinen Lehrer im eigentlichen Sinne des Wortes. Er sagt selbst
in seiner Antrittsrede als Mitglied der Preußischen Akademie:

> *Mir ist nicht das Glück zuteil geworden, daß ein hervorragender Forscher*
> *oder Lehrer in persönlichem Verkehr auf die spezielle Richtung meines*
> *Bildungsganges Einfluß genommen hat. Was ich gelernt habe, entstammt*
> *ausschließlich dem Studium der Schriften unserer Meister, (....).* [13]

Seine Dissertation trägt den Titel *Über den zweiten Hauptsatz der mechanischen*
Wärmetheorie.[14] Die mechanische Wärmetheorie von damals ist unsere heutige
phänomenologische Thermodynamik. Planck glaubte an die absolute Gültigkeit des Satzes von der Vermehrung der Entropie: er war ein   E n t r o -
p i k e r. Das machte ihn schon früh zum Anti-Atomisten. Erstaunt lesen
wir beim erst 23jährigen:

> *Zum Schluß möchte ich hier noch auf eine allerdings schon bekannte Tat-*
> *sache ausdrücklich hinweisen. Der zweite Hauptsatz der mechanischen Wär-*
> *metheorie consequent durchgeführt, ist unverträglich mit der Annahme end-*
> *licher Atome.* Es ist daher vorauszusehen, daß es im Laufe der weiteren*
> *Entwicklung der Theorie zu einem Kampfe zwischen diesen beiden Hypothesen*
> *kommen wird, der einer von ihnen das Leben kostet. Das Resultat dieses*
> *Kampfes jetzt schon mit Bestimmtheit voraussagen zu wollen, wäre aller-*
> *dings verfrüht, indeß scheinen mir augenblicklich verschiedenartige An-*
> *zeichen darauf hinzudeuten, daß man trotz der großen bisherigen Erfolge*
> *der atomistischen Theorie sich schließlich doch noch einmal zu einer*
> *Aufgabe derselben und zur Annahme einer continuirlichen Materie wird*
> *entschließen müssen.*
>
> *München, Dec. 1881*
>
> ** Vgl. J. Clerk-Maxwell, Theory of Heat. Deutsche Übersetzung (...) von*
> *F. Neesen, P 373 1878.*[15]**

Will man der Fußnote des Textes voll vertrauen, dann waren es also
J a m e s   C l e r k - M a x w e l l s   Ausführungen über die *"Begrenzung*
des zweiten Hauptsatzes der Thermodynamik", welche Plancks Widerstand gegen die
Existenz von Atomen wachriefen. Es ist dies der Abschnitt im zweiundzwan-
zigsten Kapitel der Theorie der Wärme, die vom Maxwellschen Dämon handelt
und der die Warnung enthält, daß sich unsere Erfahrung an makroskopischen
Körpern nicht mehr auf die feineren Beobachtungen und Versuche anwenden
lassen, von welchen nur wenige Moleküle betroffen sind.[17] Planck konnte
damals keine Beschränkung des Gültigkeitsbereichs des Entropieprinzips
dulden. Für ihn waren die Hauptsätze, wie schon erwähnt, Axiome, Wahrheiten, erhärtet durch jahrhundertlange Erfahrung. Auch teilte er die

---

** Die Bedenken Plancks gegen den Atomismus sind also älter als seine
   Theorie der ideal verdünnten Lösungen, die er in späteren Jahren
   auch als Ursache für seinen frühen Antiatomimus angibt.[16]

menschliche Schwäche, jede erfolgreiche Anwendung der Thermodynamik als zusätzliche Stütze der Hauptsätze aufzufassen. Daher die häufigen Beteuerungen, in zutreffenden aber auch in unzutreffenden Fällen, die Gültigkeit eines speziellen Resultates stehe und falle mit der Gültigkeit des Entropiesatzes. Man ist versucht, dem Entropiesatz im Denken von Max Planck die Rolle einer autonomen Instanz zuzuschreiben.

Es ist der absolute Glaube an die b e i d e n Hauptsätze, der Planck von Mach trennte und ihn zu den Energetikern (wie Georg Helm und Wilhelm Ostwald) in Gegensatz brachte. Mit B o l t z m a n n war er natürlich verfeindet.

Denn so wie Planck ein "Entropiker" war, für den die Welt einem Ziel zustrebte - dem Wärmetod, so war Ludwig Boltzmann ein Atomist, für den die Richtung des Zeitablaufs, ähnlich wie die Richtung der Schwerkraft, nur lokal - und dazu in weit auseinanderliegenden Inseln - definiert war.[18]

Diese unklaren Andeutungen lassen sich nicht durch die bekannten Klebezettel "fortschrittlich" und "reaktionär" verdeutlichen. Zum Lebenslauf der beiden Großen passen eher die Adjektive "optimistisch" und "fatalistisch" oder "pessimistisch". Der Optimist ist offenbar auf ein Fortschreiten, auf ein Ziel angewiesen; wer grundsätzlich an keine Entwicklung glaubt, wird auf das Leben gern verzichten, wenn er genug hat.

Wir verlassen an dieser Stelle unser gegensätzliches Paar, um uns für kurze Zeit mit einem ungemein liebenswerten Mann aus der untergegangenen Stadt der Vernunft zu unterhalten.

Der Atomismus in der Elektrizität

Uns ist heute selbstverständlich, daß das Faradaysche Grundgesetz der Elektrolyse zusammen mit dem Atomismus der Materie notwendig zur Existenz einer elektrischen Elementarladung führt. Es ist daher erstaunlich, daß dieser Sachverhalt erst sein H. v. Helmholtz' Faraday Lecture vom 5. April 1881 allgemein anerkannt wird.[19] Vor allem verwunderlich ist Clerk Maxwells ambivalente A b l e h n u n g des Atomismus der Elektrizität, wie sie im § 260 des Treatise on Electricity and Magnetism zum Ausdruck kommt.[20] Nachdem er dort den Begriff der elektrischen Elementarladung *(one molecule of electricity)* eingeführt hat, schreibt er:

> *This phrase, gross as it is, and out of harmony with the rest of this treatise, will enable us at least to state clearly what is known about electrolysis, and to appreciate the outstanding difficulties.*[21]

Und später:

> *This theory of molecular charges may serve as a method by which we may remember a good many facts about electrolysis. It is extremely improbable however that when we come to understand the true nature of electro-*

lysis we shall retain in any form the theory of molecular charges, for then we shall have obtained a secure basis on which to form a true theory of electric currents, and so become independent of these provisional theories.[22]

Es verbergen sich hinter diesen Vorbehalten alte, wertlose Berzeliussche Ideen und wohl auch das, noch bei Einstein nachwirkende, Unbehagen über das Eindringen eines fremden, atomistischen Elementes in die Feldtheorie des Elektromagnetismus: das Elektron als Fremdling in der Maxwellschen Theorie.

Noch bei Helmholtz lesen wir den Satz:

Now the most startling result of F a r a d a y's law is perhaps this. If we accept the hypothesis that the elementary substances are composed of atoms, we cannot avoid concluding that electricity also, positive as well as negative, is divided into definite elementary portions, which behave as atoms of electricity.[23]

"Startling", überraschend, bestürzend nennt Helholtz dieses Resultat.

Willkommen heißt es   E m i l   W i e c h e r t   in Königsberg; freilich erst ein gutes Dutzend Jahre später. Damit bin ich bei dem Mann angekommen, dessen theoretische und experimentelle Verdienste um die Elektronentheorie ich für einen Augenblick der Vergessenheit entreißen möchte. Seine Verdienste in der Geophysik sind ja aus Göttingen bekannt. Ihm gebührt, soweit ich sehe, vor   J. J.   T h o m s o n   die Priorität für die Entdeckung des Elektrons. Vor allem aber hat er unabhängig von H. A. Lorentz in den Jahren 1894 - 96 eine Elektronentheorie skizziert. Ich zitiere aus seiner Abhandlung *Die Theorie der Elekrtodynamik und die Röntgen'sche Entdeckung* [13] :

Da über die Realität der naturwissenschaftlichen molekularen Hypothese heute nicht der mindeste Zweifel mehr herrschen kann, bleiben wir auf sicherem Boden, wenn wir sie auch für die Elektrodynamik unumwunden anerkennen. Es ist eine hohe Freude zu sehen, wie schön sich dann die M a x w e l l 'schen Ideen den älteren Theorien einfügen, wie leicht und klar sich Alles gestaltet. [24]

Das, was uns als >Materie< den eigentlichen Inhalt der Welt auszumachen scheint, ist zusammengesetzt aus sehr kleinen, selbständigen Bausteinen, den chemischen A t o m e n. - Es kann nicht oft genug betont werden, daß man heutzutage bei dem Wort >Atom< durchaus nicht an irgendwelche der alten philosophischen Spekulationen denkt: Wir wissen ganz genau, daß die Atome, um die es sich für uns handelt, keineswegs die denkbar einfachsten Urelemente der Welt sind; ja, eine Reihe von Erscheinungen, vor Allem die der Spektralanalyse, führen zu dem Schluß, daß die Atome sehr komplicirt gebaute Dinge sind. Angesichts der heutigen Naturwissenschaft müssen wir wohl überhaupt den Gedanken aufgeben, ins Kleine gehend irgend einmal auf die letzten Fundamente der Welt zu stoßen, und ich glaube, wir können es leichten Herzens thun. Die Welt ist eben nach allen Richtungen >unendlich<, nicht nur nach oben, ins Große, sondern auch nach unten, ins Kleine hinein. Verfolgen wir von unserem menschlichen Standpunkt ausgehend den Inhalt der Welt weiter und weiter, so gelangen wir in beiden Richtungen schließlich zu nebelhaften Fernen, in welchen uns erst *die Sinne, und dann auch die Gedanken im Stich lassen.*[25]

Natürlich bildet für Wiechert der Satz *"Die Elektricität erscheint atomistisch gebaut, gerade so wie die Materie"*[26] einen Angelpunkt seiner Theorie.

Max Planck gegen Ludwig Boltzmann

Nach diesem Abstecher in ein Märchenland der Vernunft, zurück zur Haupthandlung. Das Verhältnis zwischen Boltzmann und Planck blieb belastet. Seitdem Boltzmann 1888 als Nachfolger des verstorbenen G. Kirchhoff nach Berlin berufen worden war, fühlte der Österreicher bis an sein Lebensende eine gewisse Verpflichtung, Berlin gegenüber kritisch zu sein, wozu er aus seiner Mitgliedschaft in der Preußischen Akademie der Wissenschaften weitere Veranlassung zog. Das war für Planck, der unterdessen Nachfolger von Kirchhoff geworden war, gewiß unbequem aber kein Grund, von seinen Überzeugungen abzustehen. Hören wir aus seinem Vortrag an der 64. Versammlung der Naturforscher und Ärzte am 24. September 1891 in Halle:

> *Indessen scheinen nach den ersten glänzenden Resultaten der kinetischen Gastheorie ihre neueren Fortschritte den daran geknüpften Erwartungen nicht zu entsprechen; bei jedem Versuch, diese Theorie sorgfältiger auszubauen, haben sich die Schwierigkeiten in bedenklicher Weise gehäuft. Jeder, der die Arbeiten derjenigen beiden Forscher studiert, die wohl am tiefsten in die Analyse der Molekularbewegungen eingedrungen sind: M a x w e l l und B o l t z m a n n, wird sich des Eindrucks nicht erwehren können, daß der bei der Bewältigung dieser Probleme zu Tage getretene bewundernswürdige Aufwand von physikalischem Scharfsinn und mathematischer Geschicklichkeit nicht im wünschenswerten Verhältnis steht zu der Fruchtbarkeit der gewonnen Resultate.*[27]

Diese Worte erinnern an das, was Ernst Mach 20 Jahre früher in Prag gesagt hatte. Der Mut jedoch zu dieser Kritik fließt Planck aus seiner Theorie der verdünnten Lösungen, die allerdings neben den Hauptsätzen noch allerlei "Zaubermittel" (W. Pauli) verwendet, wie etwa das reversible Überführen der verdünnten Lösung durch Temperaturerhöhung allein und ohne Änderung der Chemie in eine Mischung idealer Gase.

Die Kritik traf Boltzmann wohl besonders empfindlich, weil er sie auf seine monumentalen Rechnungen *"Zur Theorie der Gasreibung"*[28] beziehen mußte. Nun kämpfte vielleicht die kinetische Theorie der verdünnten Lösungen mit gewissen Anfangsschwierigkeiten, aber der Plancksche Purismus der Hauptsätze war der Transporttheorie, etwa der Gasreibung, gegenüber im Quantitativen völlig ohnmächtig: alles was er zu liefern hatte war ein Vorzeichen + oder - ; und das wußte auch Planck.

Es kann nun nicht unsere Aufgabe sein, die Scharmützel der folgenden Jahre zwischen Planck und Boltzmann einzeln aufzuzählen. Wir konzentrieren uns auf die letzten 6 Jahre des vergangenen Jahrhunderts, die Zeit, während welcher Planck fast ausschließlich mit dem Problem der Hohlraumstrahlung gerungen hat - und wir bleiben auch dann noch oberflächlich.

Irreversibilität und Hohlraumstrahlung

Am 11. Juni 1894 wird Max Planck Mitglied der Preußischen Akademie der Wissenschaften. Mit der Erreichung dieses höchsten äußeren Ziels scheinen sein Mut und seine Kräfte zu wachsen. Am 28. Juni hält er seine Antrittsrede. Als das für alle Zeiten unverrückbar feststehende Ziel der Forschung bezeichnet er (sehr modern scheint uns) die *"Herstellung des einen großen Zusammenhangs aller Naturkräfte."*[29] Als einen Schritt in dieser Richtung muß ihm die Aufklärung des Ursprungs der Irreversibilität erschienen sein. Daß hier die Mechanik wegen ihrer Zeitumkehr-Invarianz machtlos ist, wußte er seit mehr als 13 Jahren. Das war ja der Grund, weshalb er die *"endlichen"* Atome der kinetischen Theorie abgelehnt hatte.[30] Aber es gab die Elektrodynamik, eine Kontinuumtheorie, und es gab die Hohlraumstrahlung als Kreuzungspunkt zwischen Thermodynamik und Elektrodynamik. War von hier aus vielleicht eine *"elektromagnetische Theorie der Wärme"* zu gewinnen, *"-nicht mit Hülfe besonderer neuer Hypothesen, sondern einfach in consequenter Fortbildung der M a x w e l l 'schen Ideen von dem Zusammenhang zwischen Licht und Elektricität?"*[31]

Das ist das Problem, welchem Planck nun rund ein Dutzend Arbeiten widmet. Es war von vornherein klar, daß in einem Hohlraum mit ideal spiegelnden Wänden von einer Annäherung an ein universelles Strahlungsgleichgewicht nicht die Rede sein kann. Die Wechselwirkung des elektromagnetischen Feld mit der Materie in anderen Worten: Emissions- und Absorptionsprozesse müssen eine wesentliche Rolle spielen. Planck wählt, wie das damals und auch noch viel später nahe lag, als >Strahlungsumwandler< einen Hertz' schen Dipol, einen geladenen Oszillator also, dessen Ausdehnung gegenüber der Wellenlänge vernachlässigt werden kann.

Fällt ein Wellenzug endlicher Ausdehnung und geeigneter Frequenz auf einen zunächst ruhenden Hertz'schen Oszillator, so beginnt dieser zu schwingen, indem er dem Wellenzug Energie entzieht. Der schwingende Oszillator strahlt und strahlt weiter, nachdem der Wellenzug vorbeigezogen ist. Der Oszillator verändert also den Wellenzug im Sinne eines Ausgleichs der räumlichen und zeitlichen Inhomogenität. Das ist das Bild, welches Planck in seinen Vorarbeiten entwickelt. Dabei entdeckt er, ein Jahr vor J. L a r m o r , dem diese Erfindung allgemein zugeschrieben wird, die Strahlungsdämpfungskraft.[32] Sie wird zu einer Quelle der Inspiration; denn hier hat man es mit einer neuen Art der Dämpfung zu tun, bei der nicht Arbeit in Wärme verwandelt wird, die also nicht "verzehrend" *(consumptiv)* sondern "arbeitserhaltend" *(conservativ)* ist. Daher Plancks Worte:

Das Studium der conservativen Dämpfung scheint mir deshalb von hoher

Wichtigkeit zu sein, weil sich durch sie ein Ausblick eröffnet auf
die Möglichkeit einer allgemeinen Erklärung irreversibler Prozesse
durch conservative Wirkungen - ein Problem, welches sich der theo-
retisch-physikalischen Forschung täglich drängender entgegenstellt.[33]

Nun fühlt sich Planck gerüstet, die irreversiblen Strahlungsvorgänge im
Strahlungshohlraum zu untersuchen. Man hat sich dabei einen durch ideale
Leiter begrenzten Hohlraum vorzustellen, in welchem sich Hertz'sche Oszillatoren befinden. Über die Natur dieser Oszillatoren wird nichts vorausgesetzt. Planck legt manchmal nahe, daß sie molekulare Systeme sein
können, insistiert aber, daß man dabei durchaus a u c h an verlustlose
(ideale) makroskopische Oszillatoren denken kann. Solche makroskopische
Oszillatoren besitzen dann natürlich auch eine Entropie.

Es ist wissenschafts-psychologisch interessant, für die Entwicklung der
Physik allentscheidend, daß Planck an seinem Programm, an der absoluten
Gültigkeit des Entropiesatzes, so überaus zäh festhält. Aber die Boltz-
mannsche Auffassung des Entropiesatzes als eine Wahrscheinlichkeitsaus-
sage war ihm und seiner Umgebung so zuwider, daß mit seiner Billigung,
sein Assistent E r n s t   Z e r m e l o, der sich später als Mathematiker einen sehr bedeutenden Namen gemacht hat, gegen Boltzmann mit
jugendlicher Unverschämtheit und großer mathematischer Strenge zu Felde
zog [14, 15] . Großartig und viel zuwenig bekannt ist Zermelos Beweis
von Poincaré's Wiederkehrsatz.[34] Physikalisch aber geht sein Angriff
völlig daneben. Aber Planck ist auf   s e i n e r   Seite, wie u.a. aus
einem Brief an seinen Freund und Münchner Kollegen   L e o G r a e t z
vom 23. März 1897 hervorgeht, dessen Kenntnis wir dem verstorbenen Hans
Kangro verdanken. Planck schreibt:

> *In dem Hauptpunkt der Frage stehe ich auf   Z e r m e l o's Seite, indem*
> *auch ich der Ansicht bin, daß es principiell ganz aussichtslos ist, die*
> *Geschwindigkeit irreversibler Prozesse, z.B. der Reibung oder Wärmelei-*
> *tung, in Gasen, auf wirklich strengem Wege aus der gegenwärtigen Gas-*
> *theorie abzuleiten. Denn da Boltzmann selber zugibt, daß sogar die*
> *R i c h t u n g, in der die Reibung und Wärmeleitung wirkt, nur aus*
> *Wahrscheinlichkeitsbetrachtungen zu folgern ist, so wäre völlig unver-*
> *ständlich, woher es denn kommt, daß unter allen Umständen auch die*
> *G r ö ß e   dieser Wirkungen einen ganz bestimmten Betrag darstellt.*[35]

Das also liegt hinter der Kritik, die Planck 1891 gegen den*"bewundernswür-*
digen Aufwand von physikalischem Scharfsinn und mathematischer Geschicklichkeit, der
nicht im wünschenswerten Verhältnis zu der Fruchtbarkeit der gewonnenen Resultate"
steht,[36] vorgebracht hatte.

Boltzmann verteidigte sich souverän gegen Zermelo, was die Physik an-
geht,[37] seine Attacke auf Zermelos Mathematik schießt allerdings mit
plumper Artillerie neben dem eleganten, schlanken Zermeloschen Beweis des
Wiederkehrsatzes vorbei.[38] Dann nimmt er den Meister selbst an und

schlägt ihn, scheinbar vernichtend, mit dem expliziten Beweis der Zeitumkehrinvarianz der Elektrodynamik.[39] Aber er ist müde und seine Verzweiflung geht aus dem Vorwort des zweiten Teils seiner *Vorlesungen über Gastheorie* (1898) hervor:

> *Es wäre daher meines Erachtens ein Schaden für die Wissenschaft, wenn die Gastheorie durch die augenblicklich herrschende ihr feindseligen Stimmung zeitweilig in Vergessenheit geriethe, (...)*

schreibt er, und:

> *Wie ohnmächtig der Einzelne gegen Zeitströmungen bleibt, ist mir bewußt. Um aber doch, was in meinen Kräften steht, dazu beizutragen, daß, wenn man wieder zur Gastheorie zurückgreift, nicht allzuviel noch einmal entdeckt werden muß, nahm ich in das vorliegende Buch nun auch die schwierigsten dem Mißverständnisse am meisten ausgesetzten Theile der Gastheorie auf (....).*[40]

Nicht so Planck. Scheinbar wenig beirrt setzt er seine Mitteilungen über irreversible Strahlungsvorgänge fort.[41] Durch die Beschränkung auf *"natürliche Strahlung"* (d.h. passende Mitteilung über Phasen) entzieht er sich der Boltzmannschen Kritik und gelangt schließlich zwar methodisch aber undurchsichtig zu einem ihn vorläufig befriedigenden Beweis der Irreversibilität.* Wenn in der fünften Mitteilung Planck schließlich im Abschnitt über *"Thermodynamische Folgerungen"* sein eigentliches Gebiet der Meisterschaft betritt, dann lichten sich die Wolken: er erkennt als erster klar, daß das Wiensche Strahlungsgesetz zwei neue u n i v e r s e l l e   K o n s t a n t e n (er nennt sie a und b) enthält, die zusammen mit der Lichtgeschwindigkeit c und der Gravitationskonstanten f natürliche Einheiten der Zeit, der Länge, der Masse und der Temperatur ergeben. Er schließt triumphierend-rätselhaft mit der Periode:

> *Diese Größen behalten ihre natürliche Bedeutung so lange bei, als die Gesetze der Gravitation, der Lichtfortpflanzung im Vacuum und die beiden Hauptsätze der Wärmetheorie in Gültigkeit bleiben, sie müssen also, von den verschiedensten Intelligenzen nach den verschiedendsten Methoden gemessen, sich immer wieder als die nämlichen ergeben.*[42]

Hier ist das Neuland in Sicht, das Planck ein Jahr später, dank seiner Standhaftigkeit betreten sollte.

Im übrigen lesen wir erneut die Beschwörungsformel: *"Ich glaube hieraus schließen zu müssen, daß (...) das   W i e n sche Energievertheilungsgesetz eine nothwendige Folge der Anwendung des Princips der Vermehrung der Entropie auf die elektromagnetische Strahlungstheorie ist und daß daher die Grenzen der Gültigkeit dieses Gesetzes, falls solche überhaupt existieren, mit denen des zweiten Hauptsatzes der*

---

* Es ist interessant, worauf vor allem Hans Kangro hinweist, daß in der vierten Mitteilung, wenn auch nur als Beispiel, für die Entropie eines Oszillators (in Abhängigkeit von der Energie U) der Ausdruck $S(U) = \log U$ auftritt. Dieser Ausdruck führt zur Äquipartition - also zum Rayleigh-Jeanschen Gesetz, dem klassischen Strahlungsgesetz.

Wärmetheorie zusammenfallen."[43]

Diese Ausführungen vom 1. Juni 1899 finden eine Bestätigung und Amplifikation am 7. November desselben Jahres.[44] Neu und bedeutungsvoll ist die, allerdings nur unter Vorbehalten durchgeführte, Analogie zwischen der Hypothese der *natürlichen Strahlung* und der Boltzmannschen Hypothese der molekularen Unordnung - als eines Hilfsmittels zur Herleitung des H-Theorems, das Planck ja ständig als Beispiel vor Augen hat.[45]

Das dramatische Finale ist wohlbekannt. Zunächst zeigen sich experimentelle Diskrepanzen zum Wienschen Strahlungsgesetz, die Planck zu einem erneuten Überdenken seines Ansatzes für die Entropie eines Oszillators führen. Die alten Argumente werden verworfen, ein neues wird zur Rettung der alten Formel beigebracht.[46] Frische Messungen von O. Lummer und E. Pringsheim einerseits, H. Rubens und F. Kurlbaum andererseits, beide Gruppen an der Physikalisch-Technischen Reichsanstalt, beweisen endgültig die Unhaltbarkeit des Wienschen Strahlungsgesetzes. Nur als Grenzgesetz für kurze Wellenlängen bzw. tiefe Temperaturen hat es Gültigkeit. Plötzlich ist man herausgetreten aus den akademisch-philosophischen Betrachtungen über die Irreversibilität der natürlichen Prozesse i n d i e Z w ä n g e d e r P r a x i s. Aber die gesammelte Erfahrung ist jetzt unbezahlbar. Aus dem Meer von zweifelhaften Ausdrücken und Aussagen greift Planck die entscheidende zuverlässige Formel heraus, welche die spektrale Energiedichte mit der Thermodynamik eines linearen Oszillators verbindet. Immernoch ist es von Vorteil, daß dieser Oszillator durchaus als ideale makroskopische Maschine betrachtet werden kann, auf die man die Thermodynamik anwenden m u ß. Gesucht ist die Abhängigkeit der Entropie von der Energie des Oszillators. Besonders einfach ist die zweite Ableitung dieser Funktion. Hier ergibt sich eine naheliegende Modifikation des alten Ausdrucks, und das Glück ist mit dem Tüchtigen! Die neue Strahlungsformel hält der Prüfung durch das Experiment stand![47]

Die geheimnisvolle Parenthese in der Note vom 19. November 1900 im Zusammenhang mit der logarithmischen Abhängigkeit der Entropie von der Energie *(was anzunehmen die Wahrscheinlichkeitsrechnung nahe legt)*[48] kann nur bedeuten, daß Planck den Boltzmannschen Zusammenhang zwischen Entropie und Wahrscheinlichkeit übernommen hat.

Acht Wochen später, am 14. Dezember, beginnt der zweite Abschnitt des Vortrages von Planck in der Deutschen Physikalischen Gesellschaft lapidar

"Entropie bedingt Unordnung (....)" [49]

Planck hat vor Boltzmann kapituliert. Jetzt triumphiert er durch die erste Präzisionsbestimmung der Loschmidtschen Zahl und der elektrischen

Elementarladung. Die e i n e Strahlungskonstante k von Planck
trägt heute Boltzmanns Name, das Plancksche Wirkungsquantum h beherrscht seit 1905 unsere Physik.

==================

Blicken wir zurück: Planck suchte den Grund für die absolute Irreversibilität der natürlichen Prozesse. Diese erwies sich als Trugbild. Statt
ihrer fand er das Wirkungsquantum.

Die älteren unter uns kennen wohl noch aus dem 1. Buch Samuel Kap. 9 die
Geschichte von Saul, dem Sohn von Kis, der auszog, seines Vaters Eselin-
nen zu suchen, der Samuel, den Mann Gottes traf und ein Königreich fand.

Ähnlich ging es Planck.
Die Irreversibilität fand er nicht, dafür gewann er die Einsicht, daß die
natürlichen Vorgänge mit zeitumkehrinvarianten fundamentalen Naturgeset-
zen sich vertragen: er erkannte eine v e r b o r g e n e S y m m e -
t r i e der Naturgesetze.

Sie erinnern sich: Planck bezeichnete in seiner Preisschrift das Kausali-
tätsprinzip als eigentliche Existenzbedingung der Naturwissenschaften.
Heute wissen wir aus Quantenmechanik und Relativitätstheorie, daß eine
Theorie nur dann kausal sein kann, wenn sie eine Symmetrie besitzt, wel-
che auch den Zeitsinn umkehrt. Die handgreiflichste Konsequenz dieses
Faktums ist die Existenz des Antiteilchens zu jedem Teilchen.

Bezüglich Zukunft und Vergangenheit zeigen sich die Naturgesetze merkwürdig indifferent. Boltzmann hat wohl recht: die beobachtete Irreversibilität beruht auf einem singulären Zustand in der Vergangenheit.

Einstein widerlegt den Entropiesatz

Wir erinnern uns an Plancks äußerstes Widerstreben, eine Beschränkung der
Gültigkeit des Entropiesatzes für möglich zu halten; wir erinnern uns an
die Warnung Clerk Maxwells im Anschluß an die Beschreibung "seines" Dä-
mons:

> *This is only one of the instances in which conclusions which we have
> drawn from our experience of bodies consisting of an immense number
> of molecules may be found not to be applicable to the more delicate
> observations and experiments which we may suppose made by one who
> can perceive and handle the individual molecules which we deal with
> only in large masses.*[50]

Wir erinnern uns an die eigentümliche Auffassung Boltzmanns über die Richtung der Zeit und seine fundamentale Einsicht, daß allem gesunden Menschenverstand zum Trotz, die grundlegenden Naturgesetze zeitumkehrinva-

riant seien. Aber diese Einsicht konnte nur sehr indirekt und mit kühnem Vorgriff in die qualitative Dynamik unserer Tage begründet werden.

Nun kommt der 26jährige Albert Einstein, ein kleiner Beamter aus Bern, und zeigt, daß Abweichungen vom zweiten Hauptsatz von jedem (angefangen bei Antoon von Leeuwenhoek), der schon einmal durch ein Mikroskop geschaut hat, beobachtet worden sind. Nur, das Gesehene muß man auch erkennen, muß es auch quantitativ beschreiben können - dann lernt man die absolute Größe der Atome messen.

Doch hören wir Einstein selbst:

> *In dieser Arbeit soll gezeigt werden, daß nach der molekularkinetischen Theorie der Wärme in Flüssigkeiten suspendierte Körper von mikroskopisch sichtbarer Größe infolge der Molekularbewegung der Wärme Bewegungen von solcher Größe ausführen müssen, daß diese Bewegungen leicht mit dem Mikroskop nachgewiesen werden können. (....)*
>
> *Wenn sich die hier zu behandelnde Bewegung samt den für sie zu erwartenden Gesetzmäßigkeiten wirklich beobachten läßt, so ist die klassische Thermodynamik schon für mikroskopisch unterscheidbare Räume nicht mehr als genau gültig anzusehen und es ist dann eine exakte Bestimmung der wahren Atomgröße möglich. Erwiese sich umgekehrt die Voraussage dieser Bewegung als unzutreffend, so wäre damit ein schwerwiegendes Argument gegen die molekularkinetische Auffassung der Wärme gegeben.* [51]

Mit andern Worten, Einstein proponiert in diesem heiklen Gebiet, wo Entscheidungen bisher nach Geschmack und Neigung gefällt worden waren, ein e x p e r i m e n t u m   c r u c i s   direktester und einfachster Art. Ausgangspunkt und wohl Keimzelle der Untersuchung ist der Satz:

> *Nach dieser* (molekularkinetischen Wärme-) *Theorie unterscheidet sich ein gelöstes Molekül von einem suspendierten Körper l e d i g - l i c h   durch die Größe, und man sieht nicht ein, warum einer Anzahl suspendierter Körper nicht derselbe osmotische Druck entsprechen sollte, wie der nämlichen Anzahl gelöster Moleküle.* [52]

Aber was soll man an einem solchen Riesenmolekül, das wir uns der Einfachheit halber kugelförmig vorstellen, beobachten? Natürlich folgt aus der kinetischen Wärmetheorie für dessen mittlere Translationsenergie der Wert $3/2 \cdot (R/L)T$ (R die Universelle Gaskonstante, L die Loschmidtsche Zahl, T die Absolute Temperatur). Aber die Bewegung des Schwerpunktes ist derart erratisch, daß eine Geschwindigkeit unmöglich gemessen werden kann. Einstein gibt uns die Lösung: sei Δx die Verrückung des Teilchens in der x-Richtung im Zeitintervall t, mittelt man $(\Delta x)^2$ über viele solche Zeitintervalle so ist dieses Mittel proportional zu t und die Proportionalitätskonstante ist das doppelte der Diffusionskonstante D der suspendierten Teilchen: in Formeln angedeutet

$$<(\Delta x)^2>_t \ = \ 2Dt.$$

Das Wahrscheinlichkeitsgesetz für Δx ist eine Gauss'sche Normalverteilung, *"was"* nach Einstein *"zu vermuten war."*[53] In der angegebenen Formel steckt der eigentliche probabilistische Aspekt der Einsteinschen Theorie.

Wie aber berechnet man D ? Behandelt man das suspendierte Kügelchen als makroskopischen Körper, so kann man leicht die stationäre Geschwindigkeit angeben, die ihm eine konstante Kraft K erteilt. Sie beträgt $K/6\,\pi\cdot\eta\cdot\varrho$, wobei η die Zähigkeit der Flüssigkeit und ϱ der Radius des Kügelchens ist. Unter der Wirkung der äußeren Kraft K stellt sich aber eine solche Dichteverteilung der suspendierten Körper ein, daß der von der äußeren Kraft erzeugte Teilchenstrom (der nach der strikt thermodynamischen Auffassung allein vorhanden wäre) durch den Diffusionsstrom ausgeglichen wird. In diesem Zustand werden die Kräfte aus dem inhomogenen osmotischen Druck durch die äußeren Kräfte gerade aufgehoben. Eliminiert man aus den angedeuteten Beziehungen die äußere Kraft, so erhält man den Ausdruck

$$D = \frac{RT}{L} \cdot \frac{1}{6\,\pi\,\eta\,\varrho}$$

für die Diffusionskonstante. Kombiniert man dies Resultat mit dem mittleren Verschiebungsquadrat $\langle(\Delta X)^2\rangle_t$, so findet man die Einsteinsche Formel

$$\langle(\Delta X)^2\rangle_t = \frac{RT}{L} \cdot \frac{t}{3\,\pi\,\eta\,\varrho}$$

In dieser Formel sind alle Größen außer L d i r e k t m e ß b a r.[54] L , die Zahl der Moleküle im Mol gibt aber die absolute Masse der Moleküle und Atome.

Ein eindrückliches qualitatives Merkmal der von Einstein 1905 theoretisch entdeckten, von Robert Brown schon 1828 eingehend beschriebenen "Brownschen Bewegung" ist ihre zeitliche Persistenz. Man hat hier ein echtes sich ewig Bewegendes, ein mobile perpetuum vor sich. Noch merkwürdiger ist ihre Zeitumkehrinvarianz. Hören wir Einstein ein Menschenalter nach seiner Entdeckung in einem Brief an seinen Freund M i c h e l e B e s s o vom 29. Juli 1953:

> *Denke Dir die Brown'sche Bewegung eines Teilchens kinematographisch aufgenommen und die Bilder genau in der zeitlichen Folge konserviert, was die Benachbartheit der Bilder anlangt; nur ist nicht notiert worden ob die richtige zeitliche Folge von A bis Z oder von Z bis A ist. Der pfiffigste Mann wird aus dem ganzen Material den Zeit-Pfeil nicht ermitteln können.* [55]

Am großartigsten aber ist, daß Einstein dieselbe Methode der Brownschen Bewegung später auf die Untersuchung der Hohlraumstrahlung ausgedehnt hat: sie wird ihm zum Zauberstab, mit welchem er das Rätsel von Welle und Korpuskel für Strahlung und Materie aufdeckt und einer Lösung nahebringt. Doch davon handeln andere Vorträge.

Anmerkungen

1. [PAV] Bd. I, 601 - 613 bes. p. 604

2. Die Einheit des physikalischen Weltbildes,
    [PAV] Bd. III, 6 - 29

3. l.c. 2. p. 28

4. [12] p. 1187

5. Zum Verhältnis Mach-Planck-Einstein siehe [16].

6. [11] p. 603

7. [3] p. 16 f

8. Theoretische Physik (1930),
    [PAV] Bd. III, 209 - 218, p. 209

9. [7] p. 155

10. [3] p. 50

11. [7] p. 155

12. Wissenschaftliche Selbstbiographie,
    [PAV] Bd. III, 374 - 401, p. 379

13. Antrittsrede zur Aufnahme in die Akademie vom 28. Juni 1894,
    [PAV] Bd. III, 1 - 5, p. 4 f

14. [PAV] Bd. I, 1 - 61

15. Verdampfen, Schmelzen und Sublimiren,
    [PAV] Bd. I, 134 - 163, p. 162 f

    Es ist dies diejenige Arbeit, die Hermann v. Helmholtz in seinem
    Wahlvorschlag in die Akademie ausdrücklich von seiner Laudatio
    ausnimmt.

16. Die Stellung der neueren Physik zur mechanischen Naturanschauung
    (1910),
    [PAV] Bd. III, 30 - 46, p. 32 f
    aber auch schon

    Allgemeines zur neueren Entwicklung der Wärmetheorie (1891)
    [PAV] Bd. I, 372 - 381, p. 373

17. [5] p. 373 ff

18. z.B. Zu Hrn. Zermelos Abhandlung "Über die mechanische Erklärung
    irreversibler Vorgänge",
    [BWA] Bd. III, 579 - 586, § 4

19. On the Modern Development of Faraday's Conception of Electricity
    (1881),
    [HWA] Bd. III, 52 - 87

20. [6] vol. I, p. 379 ff

21. l.c. 20. p. 380

22. l.c. 20. p. 381

23. l.c. 19. p. 69

24. [13] p. 2

25. [13] p. 3

26. [13] p. 17

27. Allgemeines zur neueren Entwicklung der Wärmetheorie (1891),
    [PAV] Bd. I, 372 - 381, p. 372 f

28. Zur Theorie der Gasreibung I, II, III (1880/1881),
    [BWA] Bd. II, 388 - 556

29. l.c. 13. p. 4

30. l.c. 15.

31. Die Maxwell'sche Theorie der Elektricität von der mathematischen
    Seite betrachtet (1899),
    [PAV] Bd. I, 601 - 613, p. 613

    Es ist bemerkenswert, daß diese Arbeit aus dem Jahr 1899 die ein-
    zige bedingungslose Anerkennung für Mach und sein "Princip der
    Ökonomie" enthält (l.c. 1.).

32. Über electrische Schwingungen, welche durch Resonanz erregt und
    durch Strahlung gedämpft werden (1896),
    [PAV] Bd. I, 466 - 488, p. 480 Gleichung (25)

33. l.c. (32) p. 469 f

34. [14] p. 486 - 488

35. [2] p. 131

36. l.c. 27.

37. Entgegnung auf die wärmetheoretischen Betrachtungen des Hrn.
    E. Zermelo (1896),
    [BWA] Bd. III, 567 - 578
    und l.c. 18.

38. Über einen mechanischen Satz Poincaré's (1897),
    [BWA] Bd. III, 587 - 595

39. Über irreversible Strahlungsvorgänge I (1897),
    [BWA] Bd. III, 615 - 617
    Über irreversible Strahlungsvorgänge II (1897),
    [BWA] Bd. III, 618 - 621
    Über vermeintlich irreversible Strahlungsvorgänge (1898),
    [BWA] Bd. III, 622 - 628

40. [1] p. VI Vorwort.

41. Über irreversible Strahlungsvorgänge, 5 Mitteilungen (1897 - 1899),
    [PAV] Bd. I, 493 - 600

42. l.c. 41. p. 600

43. l.c. 41. p. 597

44. Über irreversible Strahlungsvorgänge (1899),
[PAV] Bd. I, 614 - 667

45. l.c. 44. p. 619 - 621

46. Entropie und Temperatur strahlender Wärme (1900),
[PAV] Bd. I, 668 - 686, p. 679 - 682

47. Über eine Verbesserung der Wien'schen Spectralgleichung (1900),
[PAV] Bd. I, 687 - 689

48. l.c. 47. p. 689

49. Zur Theorie des Gesetzes der Energieverteilung im Normalspektrum
(1900),
[PAV] Bd. I, 698 - 706, p. 698

50. [4] p. 309

51. [10] p. 549

52. [10] p. 550

53. [10] p. 558/559

54. Die Formel steht so nicht bei Einstein. Einstein schreibt
x, N, k, P   für ΔX , L, η , $\wp$
respektive. Man gewinnt die angegebene Formel aus der letzten
Formel § 5 und der letzten Formel § 6
l.c. 53.

55. [8] p. 499

Literatur

[BWA] Ludwig Boltzmann, Wissenschaftliche Abhandlungen, 3 Bde.,
J. A. Barth, Leipzig 1909

[HWA] Hermann von Helmholtz, Wissenschaftliche Abhandlungen, 3 Bde.,
J. A. Barth, Leipzig 1895

[PAV] Max Planck, Physikalische Abhandlungen und Vorträge, 3 Bde.,
F. Vieweg & Sohn, Braunschweig 1958

[1] Ludwig Boltzmann, Vorlesungen über Gastheorie, II. Theil,
J. A. Barth, Leipzig 1898

[2] Hans Kangro, Vorgeschichte des Planckschen Strahlungsgesetzes,
Steiner, Wiesbaden 1970

[3] Ernst Mach, Die Geschichte und die Wurzel des Satzes von der
Erhaltung der Arbeit, Calve, Prag 1872

[4] J. Clerk Maxwell, Theory of Heat, 2nd ed., Longmans, Green and Co.,
London 1872

[5] J. Clerk Maxwell, Theorie der Wärme übersetzt von F. Neesen,
F. Vieweg, Braunschweig 1878

[6]   J. Clerk Maxwell, A Treatise on Electricity and Magnetism, 2 volumes, 3rd edition republished by Dover Publications, Inc., New York

[7]   Max Planck, Das Prinzip der Erhaltung der Energie, 2. Aufl., B. G. Teubner, Leipzig und Berlin 1908

[8]   Albert Einstein, Michele Besso, Correspondence 1903 - 1955, Pierre Speziali ed., Hermann, Paris 1972

[9]   Robert Brown, A brief Account of Microscopic Observations (...), Phil. Mag. N.S. 4 (1828) 161 - 173

[10]   Albert Einstein, Über die von der molekularkinetischen Theorie der Wärme geforderte Bewegung von in ruhenden Flüssigkeiten suspendierten Teilchen, Ann. d. Phys. (4) 17 (1905) 549 - 560

[11]   Ernst Mach, Leitgedanken meiner naturwissenschaftlichen Erkenntnislehre und ihre Aufnahme durch die Zeitgenossen, Phys. Z. S. 11 (1910) 599 - 606

[12]   Max Planck, Zur M a c h schen Theorie der physikalischen Erkenntnis. Eine Erwiderung, Phys. Z. S. 11 (1910) 1186 - 1190

[13]   Emil Wiechert, Die Theorie der Elektrodynamik und die Röntgensche Entdeckung. Schriften der Physikalisch-Ökonomischen Gesellschaft zu Königsberg in Pr., 37 (1896) 1 - 48

[14]   Ernst Zermelo, Über einen Satz der Dynamik und die mechanische Wärmetheorie, Wied. Ann. 57 (1896) 485 - 494

[15]   -, Über mechanische Erklärungen irreversibler Vorgänge. Eine Antwort auf Hrn. Boltzmann's "Entgegnung", Wied. Ann. 59 (1896), 793 - 801

[16]   Erwin N. Hiebert, The Conception of Thermodynamics in the Scientific Thought of Mach and Planck. Wissenschaftlicher Bericht Nr. 5/68, Ernst-Mach-Institut, Freiburg i. Br.

ALBERT EINSTEIN UND DAS QUANTENPROBLEM

Max Jammer, Bar-Ilan University, Ramat-Gan, Israel

"Albert Einstein und das Quantenproblem" ist ein in vielfacher Hinsicht beachtenswertes Thema.

Rein menschlich betrachtet, behandelt es das heroisch-tragische Ringen eines der grössten Denker der Menschheit, unbeirrt einen einsamen Weg zu gehen, den ihm Wissen, Gewissen und Intuition als den einzig gangbaren zum Erreichen eines wissenschaftlichen Zieles anwiesen.

Historisch betrachtet, zeichnet sich Einsteins Stellung zur Quantentheorie dadurch aus, daß es in der gesamten Geschichte der Wissenschaft wohl kaum einen zweiten Forscher gegeben hat, der so entscheidend zum begrifflichen Aufbau einer Theorie beigetragen hat, um dieselbe, nach Abschluß ihrer Entwicklung, nicht weniger entscheidend abzulehnen.

Philosophisch betrachtet, greift das Thema in die Tiefe der philosophischen Problematik der Quantenmechanik hinein und kann daher, wenn nur oberflächlich behandelt, zu argen Mißverständnissen führen, wie es tatsächlich oft der Fall ist.

Die Grundlehren der modernen Physik sind die Theorien der Relativität und der Quanten. Daß die Relativitätstheorie, spezielle wie allgemeine, vor allem dem Genie Einsteins zu verdanken ist, wird heute allgemein anerkannt. Viel weniger ist man sich aber bewußt, daß der Name Einsteins auch mit der Begründung und Entwicklung der Quantentheorie aufs engste verbunden ist. Zwar kennt jeder Physiker Einsteins klassische Arbeiten zu dieser Theorie. In einer neueren Einstein-Biographie heißt es sogar:

> *No physicist had more to do with the creation of quantum mechanics than Einstein. His work in this field would have been in and of itself a full scientific career for any other physicist.*[1]

Diese Behauptung wird jedoch weder in dieser Biographie noch in anderen physik-geschichtlichen Werken überzeugend bewiesen.

Im ersten Teil meiner Ausführungen werde ich daher zeigen, in welch entscheidendem Maße Einsteins Gedanken für die Begriffsbildung der Quantentheorie wegweisend und für ihre Entwicklung bahnbrechend waren.

Im zweiten Teil werde ich auf die Motive eingehen, die Einstein veranlaßten, gegen die allgemein anerkannte Theorie Stellung zu nehmen. Im Gegensatz zur herrschenden Meinung werde ich dabei die These vertreten, daß Einstein die orthodoxe Quantentheorie ablehnte, nicht weil *er*,

Einstein, vielleicht aus intellektueller Trägheit oder Senilität zu
konservativ war, sich gedanklich umstellen zu können, sondern weil, ganz
im Gegenteil, *die Theorie* für ihn zu konservativ war, um den neuen empirischen Tatbeständen gerecht werden zu können.

Einstein begann seine wissenschaftliche Laufbahn mit einem gründlichen
Studium der Thermodynamik, in der er einen verlässlichen Führer zum
Lösen aktueller physikalischer Probleme sah. Auch beschäftigte er sich
eingehend mit statistischer Mechanik, wobei sich sein Studium statisti-
scher Schwankungen besonders fruchtbar erwies, und zwar nicht nur in
seinen Untersuchungen zur Brownschen Bewegung, sondern auch in seiner
Behandlung des Strahlungsproblemes, das er gerade dadurch zu einem
Quantenproblem verwandelte.

Im Frühling 1905, seines *annus mirabilis*, schrieb er an seinen Freund
Conrad Habicht, der kurz vorher in Mathematik promoviert hatte:

> *..Warum haben Sie mir Ihre Dissertation immer noch nicht
> geschickt?.. Ich verspreche Ihnen vier Arbeiten dafür, von
> denen ich die erste in Bälde schicken könnte, da ich die Frei-
> exemplare baldigst erhalten werde. Sie handelt über die Strah-
> lung und die energetischen Eigenschaften des Lichtes und ist
> sehr revolutionär, wie Sie sehen werden...* [2]

Mit diesen Worten beschrieb er seine Arbeit *"Über einen die Erzeugung und
Verwandlung des Lichtes betreffenden heuristischen Gesichtspunkt"* [3], die
er am 17. März 1905 als erste seiner drei epochemachenden, im Band 17
der *Annalen der Physik* veröffentlichten, Arbeiten an die Schriftleitung
dieser Zeitschrift absandte.

Über die letzte und berühmteste dieser drei Arbeiten, seine Arbeit zur
Relativitätstheorie "Zur Elektrodynamik bewegter Körper",[4] schrieb er
in jenem Brief:

> *Diese Arbeit liegt im Konzept vor und ist eine Elektrodynamik
> bewegter Körper unter Benützung einer Modifikation der Lehre
> von Raum und Zeit; der rein kinematische Teil dieser Arbeit
> wird Sie sicher interessieren.* [5]

Es ist erstaunlich, daß Einstein schon damals die Arbeit zu den Lichtquanten und nicht die zur Relativitätstheorie, *"revolutionär"* nannte,
ein Urteil das uns heute völlig berechtigt erscheint. Denn trotz ihrer
weitreichenden Revision in unseren Anschauungen über Raum und Zeit ist
die spezielle Relativitätstheorie eine rein klassische Theorie, ja im
Grunde genommen, wenn man so sagen darf, "klassischer" als die Galileo-
Newtonsche Theorie, von der sie sich im Prinzip nur dadurch unterscheidet, daß sie das *"Leibniz Postulat"* betreffs intra-systemischer Gleichzeitigkeit durch das *"Einstein Postulat"* ersetzt, welches, kurz ausgedrückt, der eben genannten Relation die Transitivität abspricht und erst

gerade dadurch den logisch-konsequenten Aufbau einer kausalen Raum-Zeit
Topologie und Metrik, und damit des Minkowski Raum-Zeit Kontinuums, ermöglicht;[6] während die Idee der Lichtquanten, eigentlich ein quanten-
feld-theoretischer Begriff, mit dem Begriffsschema der klassischen Physik
unvereinbar ist.

Wie kam nun Einstein zu dieser "revolutionären" Begriffsbildung? Für den
Wienschen Grenzfall des Planckschen Strahlungsgesetzes berechnete er die
Entropie der Strahlung, die ein Volumen V einnimmt. Mittels der Boltz-
mannschen Beziehung zeigte er dann, daß die Wahrscheinlichkeit, in einem
beliebig herausgegriffenen Zeitmoment diese gesamte Strahlungsenergie
E in einem Teilvolumen V' von V anzutreffen, durch

$$W = (V'/V)^{E/h\nu}$$

gegeben ist. Da im Falle von n unabhängigen *Teilchen* für eine solche
Verteilung dieselbe Formel gilt, wenn nur der Exponent durch n ersetzt
wird, folgerte Einstein, daß die Strahlung sich so verhält *"wie wenn sie
aus voneinander unabhängigen Energiequanten von der Größe* hν *bestünde."*[7]

Zum Abschluß wies er darauf hin, daß durch diese Hypothese zahlreiche
Erfahrungs-Tatsachen, wie die Stokesche Regel oder der photoelektrische
Effekt, einfach und zwanglos ihre Erklärung erhalten.

Einsteins Hypothese übertraf in ihrer Kühnheit bei weitem Plancks Einführung des Wirkungsquantums [in seiner Quantisierung des harmonischen
Oszillators]. Denn obgleich dieser Schritt Plancks mit Recht als der Auf-
takt der Quantentheorie gelten muß, (da man ja diese als die Lehre von
der Rolle, die h in der Natur spielt, definieren kann,) so beruhte er
doch letzten Endes auf einer Interpolation zwischen zwei Ausdrücken, von
denen der eine dem Wienschen, der andere dem Rayleigh-Jeansschen Strahlungsgesetz entsprachen. Erst Einstein erkannte, daß dieser Kompromiß,
der für Planck, wie er es einmal nannte, ein "Akt der Verzweiflung" gewesen war, eine Verschmelzung von Welle und Teilchen bedeutete. Im Gegensatz zu Planck, der sich ständig, wenn auch erfolglos, bemühte, das
Wirkungsquantum "irgendwie" mit der klassischen Physik zu versöhnen, bekannte sich Einstein bereit, klassische Begriffe aufzugeben, sobald er
sich von den Grenzen ihrer Anwendbarkeit überzeugt hatte.

Wie seine Lichtquanten-Arbeit bezeugt, hatte er schon 1905 eine Stellung
eingenommen, die er später mit folgenden Worten charakterisierte:

> *Begriffe, welche sich bei der Ordnung der Dinge als nützlich
> erwiesen haben, erlangen über uns leicht eine solche Autori-
> tät, daß wir ihren irdischen Ursprung vergessen und sie als
> unabänderliche Gegebenheiten hinnehmen. Sie werden dann zu
> 'Denknotwendigkeiten', 'Gegebenen a priori', usw. gestempelt.
> Der Weg des wissenschaftlichen Fortschritts wird durch solche*

Irrtümer oft für lange Zeit ungangbar gemacht. Es ist deshalb
durchaus keine müßige Spielerei, wenn wir darin geübt werden,
die längst geläufigen Begriffe zu analysieren und zu zeigen,
von welchen Umständen ihre Berechtigung und Brauchbarkeit ab-
hängt, wie sie im einzelnen aus den Gegebenheiten der Erfah-
rung herausgewachsen sind. Dadurch wird ihre allzu große
Autorität gebrochen. Sie werden entfernt, wenn sie sich nicht
legitimieren können, korrigiert, wenn ihre Zuordnung zu den
gegebenen Dingen allzu nachlässig war, durch andere ersetzt,
wenn sich ein neues System aufstellen läßt, das wir aus irgend-
welchen Gründen vorziehen.[8]

Die Bedingung, die Einstein [wie er es bei anderen Gelegenheiten betonte]
den neuen Begriffen auferlegte, war die, daß sie mit den alten Begriffen,
die sie ersetzten, innerhalb deren Anwendungsbereiche wenn nicht kon-
zeptuell so doch wenigstens operativ annähernd übereinstimmen sollen.

Schon hier zeigt es sich, wie paradoxal es wäre, Einstein, der solche
Gedanken proklamierte und in dessen gesamten Lebenswerk Begriffsanalysen
und Begriffs-Verfeinerungen eine ausschlaggebende Rolle spielten, Mut
oder Fähigkeit abzusprechen, sich den Gedankengängen der modernen Quantenmechanik anzupassen.

Daß Einstein, in der Tat, eine derartige Begriffsanalyse schon in seiner
Lichtquanten-Arbeit vollzogen hat, geht aus seinen Bemerkungen hervor,
in denen er zwar einräumt, daß die kontinuierliche Wellentheorie sich
für rein optische Effekte, wie Interferenz oder Beugung, vortrefflich
bewährt, daß sie dabei aber nur auf zeitliche Mittelwerte und nicht auf
Momentanwerte Bezug nimmt und daher, wenn auf Erscheinungen der Licht-
erzeugung oder Lichtverwandlung bezogen, wie etwa beim photoelektrischen
Effekt, mit der Erfahrung zu Widersprüchen führt. Auch wies er, unab-
hängig von Rayleigh, auf die Schwierigkeiten hin, in die die Maxwellsche
Theorie gerät, wenn sie mit dem Gleichverteilungsgesetz vereinigt wird,
Schwierigkeiten, die nach Ehrenfest mit dem Schlagwort "ultraviolette
Katastrophe" bezeichnet werden.

Noch zwei Bemerkungen zur Lichtquanten-Arbeit. Es war dies die erste
Arbeit, in der Einstein einen für die Theorie grundlegenden und daher
folgenschweren Schluß aus einem einfachen Gedankenexperiment ableitete,
eine Technik, die er bekanntlich später auch in anderen Zusammenhängen
mit großem Erfolg benutzte.

Außerdem beruhte die Einsteinsche Beweisführung, rein logisch betrachtet,
abgesehen von den physikalischen Prämissen, auf einem Analogie-Schluß,
nämlich auf jenem Vergleich der erwähnten Wahrscheinlichkeits-Ausdrücke.

Wegen dieses Analogie-Argumentes, das übrigens, wie Jon Dorling[9] 1971
zeigte, eliminierbar ist, war Einsteins Schlußfolgerung, wie jeder Analogie-Schluß, dem Einwand ausgesetzt, logisch nicht zwingend zu sein.

Kein Wunder, daß Robert Millikan, der neun Jahre später die Einsteinsche Formel für den photoelektrischen Effekt experimentell verifizierte, Einsteins Argument eine "kühne, wenn nicht waghalsige Hypothese" ("a bold, not to say reckless hypothesis") nannte.

In der Tat, bis Arthur Compton 1923 den nach ihm genannten Effekt entdeckte, hatten fast alle namhaften Physiker "ernste Bedenken gegen" diese Hypothese. In dem von Planck, Nernst, Warburg und Rubens 1913 unterzeichneten Empfehlungsschreiben an das Preußische Unterrichtsministerium zur Ernennung Einsteins als ordentliches Mitglied der Berliner Akademie der Wissenschaften hieß es zum Beispiel:

> *Daß er in seinen Spekulationen auch einmal über das Ziel hinausgeschossen haben mag, wie zum Beispiel in seiner Hypothese der Lichtquanten, wird man ihm nicht allzusehr anrechnen dürfen. Denn ohne einmal ein Risiko zu wagen, läßt sich auch in der exaktesten Naturwissenschaft keine wirkliche Neuerung einführen.* [10]

Daß auch Planck dieses Dokument unterzeichnete, ist besonders bedeutungsvoll. Denn in seiner Suche nach weiteren Indizien für die Existenz der Lichtquanten hatte Einstein schon 1906, in seiner zweiten Arbeit zu diesem Thema *"Zur Theorie der Lichterzeugung und Lichtabsorption"* [11] diese - in Plancks eigener Strahlungstheorie gefunden. So schrieb dort Einstein:

> *Die vorstehenden Überlegungen scheinen mir zu zeigen, daß Hr. Planck in seiner Strahlungstheorie ein neues hypothetisches Element, die Lichtquantenhypothese, in die Physik eingeführt hat.* [12]

1909, in seiner Arbeit *"Zum gegenwärtigen Stand des Strahlungsproblems"*, [13] verstärkte er die Beweiskraft seiner Argumentation durch Abschwächung der Bedingungen jenes Gedankenexperimentes. Statt auf die höchst unwahrscheinliche Verteilung der Gesamtenergie in einem Teilvolumen Bezug zu nehmen, berechnete er einfach das mittlere Quadrat der Energieschwankungen und erhielt einen **zweigliedrigen** additiven Ausdruck, in dem das eine Glied auf Grund der Wellentheorie und das zweite, im Geltungsbereich des Wienschen Gesetzes dominierende, Glied nur auf Grund der Lichtquantenhypothese verständlich war.

Die Lichtquantenhypothese wählte Einstein auch zum Hauptthema für seinen Vortrag auf der Salzburger Versammlung Deutscher Naturforscher und Ärzte, wo er, Ende September 1909, zum ersten Mal an einer Physiker-Tagung teilnahm. In seinem Vortrag *"Über die Entwicklung unserer Anschauungen über das Wesen und die Konstitution der Strahlung"* [den Pauli [14] "ein Landmark in der Entwicklung der theoretischen Physik" nannte] erklärte Einstein:

*Die nächste Phase der Entwicklung der theoretischen Physik
wird uns eine Theorie des Lichtes bringen, welche sich als
eine Art Verschmelzung von Undulations- und Emissionstheorie
des Lichtes auffassen läßt.*

Um zu zeigen, daß diese beiden Struktureigenschaften nicht unvereinbar
sind, skizzierte er zum Schluß seiner Ausführungen ein Modell, das die
Energie des Feldes in Singularitäten lokalisiert und mit einem Super-
positionsprinzip operierte, so daß sich ein Wellenfeld ergibt, das dem
der Maxwellschen Theorie sehr ähnlich ist.

Dieses nur in Umrissen skizzierte Modell bildete den Anfang einer langen
Reihe von Gedanken-Konstruktionen, mit denen sich Einstein von nun ab
unabläßlich beschäftigte, um ein   theoretisches Modell zu finden,
das die Quantenphänomene einfach und zwanglos erklärt.

Es wäre ein fundamentaler Irrtum in Einsteins "Verschmelzung", wie er
es nannte, von Wellen und Teilchenaspekten etwa eine Antizipation von
der Bohrschen Dualität im Sinne des Komplementaritäts-Prinzips zu sehen.
Für Einstein, im Gegensatz zu Bohr, haben beide Aspekte gleichzeitig und
in Beobachter-unabhängiger Weise Realität. Die sogenannten Bohr-Ein-
steinschen Gleichungen $E = h\nu$ und $p = hk$ werden zwar von beiden als
vollgültig anerkannt, aber nicht im gleichen Sinne, eine Tatsache, die
durch das mehrdeutige Gleichheitszeichen verschleiert wird. Übrigens
gibt es wohl in der gesamten Physik keine zweite Gleichung, die mit zwei
Namen bezeichnet wird, deren Träger über den Sinn der Gleichung von so
grundlegend verschiedener Meinungen sind. Was Einstein anbelangt, so
erinnere ich nur an das, was er einmal an Cornelius Lanczos geschrieben
hat:

*Wenn jemand Ihnen sagt, er verstehe, was $E = h\nu$ bedeute,
sagen Sie ihm, er ist ein Lügner.*[15]

Einsteins Vorschlag, Energiequanten als Singularitäten eines kontinuier-
lichen Feldes aufzufassen, hatte, wie wir sehen werden, einen großen
Einfluß auf De Broglie, Born und andere, und damit auf die gesamte Entwicklung der Quantenmechanik, ausgeübt.

Doch bevor ich auf diesen Punkt näher eingehe, erlauben Sie mir, meine
Damen und Herren, kurz Einsteins Arbeiten zur spezifischen Wärme zu erwähnen. Denn diesen Arbeiten ist es vor allem zu verdanken, daß die
Quanten*hypothese* den Status einer *Theorie* und erst damit auch allgemeine Anerkennung gewonnen hat.

Um dies zu verstehen, darf man nicht vergessen, daß anfangs viele Physiker die Arbeiten zur Quantenphysik der Strahlung nicht allzu ernst
nahmen und sie als eine voraussichtlich bald vorübergehende *ad hoc*

Hypothese betrachteten.

Es war daher von entscheidender Bedeutung, daß Einstein schon 1907 erkannte, daß der Begriff der Quanten auch in der molekular-kinetischen
Theorie der spezifischen Wärmen eine wichtige Rolle·spielen müßte. So
schrieb er 1907:

> [Es drängt sich ... die Frage auf:] Wenn sich die in der Theorie
> des Energieaustausches zwischen Strahlung und Materie anzu-
> nehmenden Elementargebilde nicht im Sinne der gegenwärtigen
> molekular-kinetischen Theorie auffassen lassen, müssen wir dann
> nicht auch die Theorie modifizieren für die anderen periodisch
> schwingenden Gebilde, welche die molekulare Theorie der Wärme
> heranzieht? Die Antwort ist nach meiner Meinung nicht zweifel-
> haft. Wenn die Plancksche Theorie der Strahlung den Kern der
> Sache trifft, so müssen wir erwarten, auch auf anderen Gebie-
> ten der Wärmetheorie Widersprüche zwischen der gegenwärtigen
> molekular-kinetischen Theorie und der Erfahrung zu finden, die
> sich auf dem eingeschlagenen Wege heben lassen.[16]

Ich brauche nicht näher auszuführen, wie es Einstein gelang, diese Widersprüche, d.h. die Abweichungen spezifischer Wärmen vom Dulong-Petitschen
Standardwert   (besonders bei tiefen Temperaturen, wie sie schon seit
Jahrzehnten bei den leichten Elementen wie Bor, Beryllium, Silizium oder
beim Diamanten bekannt gewesen waren und denen die klassische Physik
mit ihrem Satz der Energie-Gleichverteilung ratlos gegenüber stand)
mittels Quantisierung der Atomschwingungen aufzuklären, wobei sich die
Dulong-Petitsche Formel, genau wie die Rayleighsche Strahlungsformel,
als ein "Grenzgesetz" für kleine $h\nu/kT$ herausstellte.

Wissenschaftstheoretiker schreiben einer wissenschaftlichen Hypothese
den Status einer Theorie erst dann zu, wenn sie ihren *ad hoc* Charakter
dadurch verliert, daß sie auch Erscheinungen von ganz anderer Art erklärt als diejenigen, für die sie ursprünglich aufgestellt wurde. In
diesem Sinne muß man der Einsteinschen Arbeit aus dem Jahre 1907 *"Die
Plancksche Theorie der Strahlung und die Theorie der spezifischen Wärme"* [17]
die Rolle zusprechen, die Plancksche *Hypothese* zum Status einer *Theorie*
erhoben zu haben.

Es ist aber mehr als zweifelhaft, ob man sich seinerzeit der wissenschaftstheoretischen Bedeutung dieser Einsteinschen Arbeiten bewußt war.

Nichtsdestoweniger führten dieselben Einsteinschen Arbeiten zum gleichen
Ziel, nicht auf Grund ihrer eben genannten *inneren* Logik, sondern durch
die Mitwirkung äußerer historischer Umstände.

Auf ganz anderem Wege, nämlich von der phänomenologischen Thermodynamik,
war unterdessen Walther Nernst, der Berliner Professor für physikalische
Chemie, auf Grund seines nach ihm benannten Theorems, des Dritten Hauptsatzes der Thermodynamik, zu dem Ergebnis gekommen, daß die Atomwärme

kondensierter Stoffe bei Annäherung an den absoluten Nullpunkt äußerst klein werden sollte. Als die mit seinen Mitarbeitern Lindemann und Magnus ausgeführten Messungen ergaben, daß der Abfall der Atomwärme mit sinkender Temperatur mit der Einsteinschen Formel gut übereinstimmt, begann Nernst die Quantentheorie ernst zu nehmen[18] und sogar als ihr Fürsprecher aufzutreten.[19] So erklärte er vor der Berliner Akademie im Januar 1911:

> *Zur Zeit ist die Quantentheorie wesentlich eine Rechnungsregel, und zwar eine solche, wie man wohl sagen kann, sehr seltsamer, ja grotesker Beschaffenheit; sie hat aber in den Händen von Planck, was die Strahlung, in den Händen von Einstein, was die Molekularmechanik anlangt, so reiche Früchte gebracht und ist, wie ich im vorstehendem angedeutet habe, auch so vieler sonstiger Anwendungen fähig, daß der Forschung die Pflicht erwächst, möglichst vielseitig dazu Stellung zu nehmen und sie der experimentellen Prüfung zu unterziehen.*[20]

Solche Worte von Geheimrat Prof. Dr. Walther Nernst, der vielleicht einflußreichsten Persönlichkeit in der damaligen wissenschaftlichen Hierarchie Deutschlands, verfehlten nicht ihre Wirkung.

Nernst war ein Mann von Charakter. Wenn er von einer "Pflicht der Forschung" sprach, so richtete er diesen Appell zuerst an sich selbst. Mit Hilfe des Belgischen Industriellen Ernest Solvay arrangierte er Ende Oktober 1911 einen Kongress in Brüssel, auf dem die prominentesten Physiker jener Zeit fünf Tage lang zum Quantenproblem Stellung nahmen.

Die Bedeutung dieses sogenannten "Ersten Solvay Kongresses" für die Entwicklung der Quantentheorie kann kaum überschätzt werden, nicht nur weil er mit seinem offiziellen Titel *"Die Theorie der Strahlung und der Quanten"*, und damit als erster Kongress der Quantentheorie, demonstrativ die Anerkennung dieser Theorie förderte. Niels Bohrs Interesse an der Quantentheorie wurde erweckt, als er von Rutherford einen Bericht über die Brüsseler Diskussionen erhielt. Louis de Broglie berichtet in seiner Autobiographie, mit welcher Begeisterung er die von seinem Bruder Maurice redigierten Brüsseler Protokolle studierte und wie ihn dies zum Beschluß brachte:

> *Je m'étais promis de consacrer tous mes efforts à comprendre la véritable nature des mystérieux quanta... dont on n'apercevait pas encore la significance profonde...*[21]

Wie wir sehen, all dies waren Folgen, wenn auch auf Umwegen, von Einsteins Anwendung des Quantenbegriffes auf die molekularkinetische Theorie.

Aber auch ein direkter Einfluß blieb nicht aus. So versuchte Nernst, wie er es ausdrückte, "eine Darstellung der Quantentheorie zu geben, die

nach dem Vorgang Einsteins die Strahlungserscheinungen nur als beglei-
tende Umstände auffaßt". Da, streng genommen, die Strahlungstheorie ein
Kapitel der Quanten-Feldtheorie ist, hat es prinzipielle Bedeutung festzustellen, daß Einsteins Gedanken eine ganze Reihe von Versuchen inspirierte, die Quantentheorie logisch unabhängig von Strahlungsphänomenen
aufzubauen. Derartige Versuche, zu denen noch Otto Sterns 1962 veröffentlichte Arbeit *"On a proposal to base wave mechanics on Nernst's theorem"* [22] gehört, sind ein ziemlich unbekanntes Kapitel in der Geschichte
der Quantenphysik.

Wie bekannt, geht die moderne Quantentheorie in ihrer Form als Matrizenmechanik auf Heisenbergs zum ersten Male streng gültige und verall-
gemeinerungsfähige Quantisierung des anharmonischen Oszillators zurück,
die er Ende Mai 1925 während seines Erholungsaufenthalts in Helgoland
ausführte. Die dabei mit Erfolg angewandte Methode, nur beobachtbare
Größen zu benutzen, hatte er, wie er selbst wiederholt erklärte, von
Einstein übernommen.

Als Beispiel erwähne ich nur Heisenbergs Bericht[23] über seinen Besuch im
Frühling 1926 bei Einstein in dessen Wohnung in der Haberlandstraße 5,
Berlin, wo Einstein mit Bezug auf Heisenbergs Kolloquium-Vortrag den
Einwand machte:

> *Aber sie glauben doch nicht im Ernst, daß man in eine
> physikalische Theorie nur beobachtbare Größen aufnehmen
> kann.*

wozu Heisenberg erwiderte:

> *Ich dachte, daß gerade Sie diesen Gedanken zur Grundlage
> Ihrer Relativitätstheorie gemacht hätten...*

Daß Einstein schon seit langem die positivistische Einstellung, die gerade Heisenberg zu seiner Entdeckung der Matrizenmechanik inspirierte,
aufgegeben hatte und daher jetzt Heisenberg antworten konnte:

> *Erst die Theorie entscheidet, was man beobachten kann.*

ist natürlich in unserem gegenwärtigen Zusammenhang irrelevant.

Auch die Geburt der Wellenmechanik wurde durch Einsteinsche Gedanken be-
fruchtet, wobei es sich hier mehr um den physikalischen Inhalt als um
die Methode von Einsteins Arbeiten handelte.

In Bezug auf seine Entdeckung der Wellenmechanik schrieb Schrödinger am
23. April 1926 an Einstein:

> *Die ganze Sache wäre sicherlich nicht jetzt und vielleicht
> nie entstanden (ich meine, nicht von meiner Seite), wenn mir
> nicht durch Ihre zweite Gasentartungsarbeit auf die Wichtig-
> keit der de Broglie'schen Ideen die Nase gestossen wäre.*[24]

Auf die Frage, warum es gerade Schrödinger war, der die de Broglie'schen Ideen über Materiewellen zur Wellenmechanik entwickelte, gibt die historische Forschung mehrere Antworten. Daß Schrödinger, nach dem eben erwähnten Zitat, durch Einsteins Hinweis auf de Broglies These darauf geführt wurde, ist eine Möglichkeit. Eine andere Antwort beruht auf der Tatsache, daß schon 1922 Schrödinger in seiner Arbeit *"Über eine bemerkenswerte Eigenschaft der Quantenbahnen eines einzelnen Elektrons"*,[25] in der er die Weylsche Erweiterung der Einsteinschen Feldtheorie auf quantisierte Elektronenbahnen anwandte, zu einem Ergebnis kam, bei dem er - wie es Fritz London einmal ausdrückte - "sogar den Resonanzcharakter der Quantenforderung lange vor de Broglie in den Händen gehabt hätte". An dritter Stelle wäre der Bericht von Edmond Bauer[26] zu erwähnen, nach dem Schrödinger durch Vermittlung von Paul Langevin und Victor Henri ein Exemplar der de Broglieschen These direkt erhalten haben soll. Wenn wir uns erinnern, daß seinerseits de Broglie in seiner These *"Recherches sur la Théorie des Quanta"* (1924) die Zuordnung von periodischen Phänomenen mit dem Teilchenbegriff nur auf Grund der speziellen Relativitätstheorie erreicht hatte, so ist es ganz klar, daß - was immer die Antwort - die Wellenmechanik nicht entstanden wäre, wie sie entstanden ist, ohne den Einfluß Einsteinschen Gedankengutes.

Nicht nur die Entwicklung der Formalismen der Quantenmechanik, auch das Entstehen ihrer Interpretationen wurde durch Einstein beeinflusst.

Selbst eine so fantastische Deutung wie die Everett-Wheeler-Graham-DeWitt'sche "Many Worlds Interpretation", nach der sich bei jedem Quantenprozess die Welt in eine Unmenge neuer Welten spaltet, geht auf Einsteinsche Gedanken zurück, da sie ja aus der Forderung erwachsen ist, die Quantentheorie auf kosmologische Konsequenzen der allgemeinen Relativitätstheorie anwenden zu können. Schon der Umstand, daß sie ein Produkt der Princeton Relativisten-Schule ist, bestätigt meine Behauptung.

Bei der allgemein verbreiteten Bornschen probabilistischen Interpretation, nach der $|\Psi|^2$ als eine Wahrscheinlichkeitsdichte gedeutet wird, kann der Einsteinsche Einfluß sogar dokumentarisch belegt werden.

Als Born 1926/27 in seinen Arbeiten zur "Quantenmechanik der Stoßvorgänge"[27] diese Interpretation vorschlug, glaubte er in den Fußstapfen Einsteins gegangen zu sein. Dies hatte Born in seiner Nobel-Preis-Rede und bei anderen Gelegenheiten wiederholt ausdrücklich betont, wie zum Beispiel in seiner Rede vor der Physikalischen Gesellschaft zu Berlin am 18. März 1955,[28] hier in der Technischen Universität Berlin,

> *... Einstein hält die statistische Interpretation für unbefriedigend und hat immer wieder versucht, sie zu widerlegen.*

> *Dabei geht die Deutung des Quadrats der Wellenfunktion als*
> *Wahrscheinlichkeit auf Einstein selbst zurück. Er hat aus-*
> *gesprochen, daß die mittlere Dichte der Photonen in einem*
> *Lichtstrahl übereinstimmen muß mit der Energiedichte der*
> *elektromagnetischen Wellen, die diesen Strahl beschreiben.*
> *Es ist dieser selbe Gedanke, den ich 1927 zur Deutung der*
> *Schrödingerschen Wellenfunktion vorschlug...*

Auch andere Interpretationen, wie die stochastischen oder die Theorien der verborgenen Parameter, wurden durch Einsteinsche Gedanken beein-flußt.

Aus Zeitmangel kann ich nicht in Einzelheiten eingehen. Ich glaube je-doch, daß die schon erwähnten Argumente genügen zum Beweis der Behaup-tung:

> *No physicist had more to do with the creation of quantum*
> *mechanics than Einstein.*

Wenn ich nun, meine Damen und Herren, im zweiten Teil meiner Ausführun-gen erklären möchte, warum Einstein die Theorie, die - wie wir sahen - ihm so viel verdankt, kategorisch ablehnte, so müsste ich, der Voll-ständigkeit halber mit einer eingehenden Analyse seiner philosophisch-methodologischen Schriften über die Struktur und Funktion physikalischer Theorien beginnen. Von besonderer Bedeutung für unser Thema wären seine Aufsätze *"Physik und Realität"*[29] (1936), *"Quantenmechanik und Wirklich-keit"*[30] (1948) und seine autobiographisch-kritischen Bemerkungen in dem bekannten *Schilpp-Buch*.

Aus Zeitmangel kann ich Einsteins wissenschaftstheoretisches *Credo* nur schlagwortartig wie folgt zusammenfassen:

Physik ist ein sich ständig entwickelndes Gedankensystem, dessen Grund-lagen *nicht* induktiv aus unmittelbaren Erlebnissen eindeutig herausde-stilliert, sondern nur durch *freie Erfindung* gewonnen werden können, dessen logische Folgerungen sich aber an den Sinneserlebnissen bewähren müssen, dabei vollzieht sich diese Entwicklung in Richtung wachsender Einfachheit der logischen Fundamente.

Auf Grund dieser methodologischen Erwägungen und gewisser metaphysischer Voraussetzungen, auf die ich noch eingehen werde, forderte Einstein von der Theorie (1) eine möglichst einheitliche Erklärung aller physikali-schen Naturerscheinungen, (2) eine kausal-deterministische Erfassung der letzteren und (3) - meiner Meinung nach die wichtigste Forderung - eine theoretische Ausdrückbarkeit einer diesen Erscheinungen zu Grunde liegenden physikalischen Realität.

Sein Streben nach theoretischer Vereinheitlichkeit zeigte sich schon 1909 in der erwähnten Arbeit *"Zum gegenwärtigen Stand des Strahlungspro-*

blems",[31] in der er versuchte, das Plancksche Wirkungsquantum h und
das Elementarquantum   e der elektrischen Ladung auf eine gemeinsame
Wurzel zurückzuführen. [So schrieb er:]

> *Es scheint mir aus der Beziehung h e^2/c hervorzugehen, daß*
> *die gleiche Modifikation der Theorie, welche das Elementar-*
> *quantum e als Konsequenz enthält, auch die Quantenstruktur*
> *der Strahlung als Konsequenz enthalten wird.*

Obgleich er zugeben musste, daß es ihm nicht gelungen ist, ein zur Konstruktion beider Quanten geeignetes Gleichungssystem zu finden, schien
ihm dennoch die Mannigfaltigkeit der Möglichkeiten nicht zu groß, um
vor einer solchen Aufgabe zurückschrecken zu müssen.

Als zehn Jahre später die allgemeine Relativitätstheorie ihre ersten
Erfolge verzeichnen konnte, sah Einstein in ihr ein Vorbild für seine
Quantentheorie, die den ihr auferlegten Forderungen genügen könnte. Mit
der feldtheoretischen Methode partieller Differentialgleichungen, die
die zeitliche Fortsetzung des Geschehens eindeutig bestimmen, könnte
man dem Kausalgesetz gerecht werden. Da aber, wie es zum Beispiel die
Bohrsche Theorie des Wasserstoffspektrums zeigte, in der Quantenphysik,
im Gegensatz zur klassischen Physik, die Anfangsbedingungen eines Systems
nicht frei wählbar sind, würde man gezwungen sein, die Feldvariablen
durch Gleichungen zu "überbestimmen", das heißt, mit einer größeren Anzahl von Differentialgleichungen zu arbeiten als die Zahl der durch sie
bestimmten Feldvariablen.

Diese Idee der "Überbestimmung" lag ihm, meines Erachtens, deshalb so
nahe, da man sie schon in der allgemeinen Relativitätstheorie antrifft.
Verlangt man nämlich, daß alle Komponenten des Riemannschen Krümmungstensors verschwinden, stellt man also im vier-dimensionalen Kontinuum
20 algebraische unabhängige Differentialgleichungen für die 10 Koef-
fizienten g_{mn} der quadratischen metrischen Grundform auf, so ist die
Raum-Zeit-Mannigfaltigkeit euklidisch und läßt daher überhaupt keine
"Anfangsbedingungen" zu.

Auf diese Weise hoffte Einstein damals, das Quantenproblem zu lösen. Am
3. März 1920 schrieb er an Born:

> *Ich brüte in meiner freien Zeit immer über das Quantenproblem*
> *vom Standpunkt der Relativität. Ich glaube nicht, daß die Theo-*
> *rie das Kontinuum wird entbehren können. Es will mir aber*
> *nicht gelingen, meiner Lieblingsidee, die Quantenstruktur aus*
> *einer Überbestimmung durch Differentialgleichungen zu verstehen,*
> *greifbare Gestalt zu geben.*[32]

Mit welchem Eifer und mit welcher Zuversicht er sich dieser Aufgabe
widmete, zeigt sein 1923 veröffentlichter Aufsatz: *"Bietet die Feldtheo-*
rie Möglichkeiten für die Lösung des Quantenproblems?",[33] in dem er sogar

an die Mathematiker für Mitarbeit appellierte, da er überzeugt war, daß der eingeschlagene Weg verfolgbar ist und unbedingt zu Ende gedacht werden muß.

Von besonderem Interesse ist in diesem Zusammenhang ein Brief, den er 1924 an seinen Freund Michele Besso geschrieben hat:

> *Die Idee, mit der ich mich herumschlage, betrifft das Verstehen der Quantenthatsachen und heißt: Überbestimmung der Gesetze durch mehr Differentialgleichungen als Feldvariable. So soll die Nichtwillkürlichkeit der Anfangsbedingungen begriffen werden, ohne die Feldtheorie zu verlassen. Natürlich kann der Weg ganz falsch sein, aber er muß versucht werden und ist jedenfalls logisch möglich. Die Bewegungsgleichung materieller Punkte (Elektronen) wird ganz aufgegeben; das motorische Verhalten der letzteren soll durch die Feldgesetze mitbestimmt werden... Das Mathematische ist enorm schwierig, der Zusammenhang mit dem Erfahrbaren wird leider immer indirekter. Aber es ist doch eine logische Möglichkeit, um ohne sacrificium intellectus der Wirklichkeit gerecht zu werden.*[34]

Was an diesem Briefe besonders interessant ist, ist nicht nur die Tatsache, daß Einstein in ihm zum ersten Male, und zwar gerade in Verbindung mit dem Quantenproblem, den Gedanken aussprach, Bewegungsgesetze aus Feldgleichungen abzuleiten - ein Programm, das 14 Jahre später mit Leopold Infeld und Banesh Hoffmann in der Relativitätstheorie verwirklicht wurde - sondern, in unserem Zusammenhang, die Schlußworte. Denn, wie wir bald sehen werden, war für Einstein die orthodoxe Auffassung der Quantentheorie, und besonders die Komplementaritäts-Interpretation, ein *sacrificium intellectus*, *ohne* der Wirklichkeit gerecht zu werden.

Vorläufig jedoch, 1924, also drei Jahre bevor Bohr auf dem Como Congress seine Komplementaritäts-Interpretation präsentierte, war es in der Hauptsache der Indeterminismus der offiziellen Quantentheorie, den Einstein scharf verurteilte. Als 1924 Bohr, Kramers und Slater ihre seinerzeit Aufsehen erregende Arbeit[35] erschienen ließen, in der es hieß:

> *die kontinuierlichen Strahlungserscheinungen werden mit den diskontinuierlichen Atomprozessen durch Wahrscheinlichkeitsgesetze nach dem Vorgang von Einstein verbunden...*

schrieb Einstein an Born:

> *Bohrs Meinung über die Strahlung interessiert mich sehr. Aber zu einem Verzicht auf die strenge Kausalität möchte ich mich nicht treiben lassen, bevor man sich nicht noch ganz anders dagegen gewehrt hat als bisher. Der Gedanke, daß ein einem Strahl ausgesetztes Elektron aus freiem Entschluß den Augenblick und die Richtung wählt, in der es fortspringen will, ist mir unerträglich. Wenn schon, dann möchte ich lieber Schuster oder gar Angestellter in einer Spielbank sein als*

Physiker.[36]

Als Heisenberg Anfang 1927 seine Unbestimmtheitsbeziehungen aufstellte, deren grundlegende Bedeutung von führenden Physikern sofort erkannt wurde, glaubte Einstein, irrtümlich wie er später selbst einsah, daß sie zu logischen Widersprüchen führen müssten. So versuchte er schon im Oktober 1927, auf dem 5. Solvay Kongress, wie auch auf dem folgendem im Oktober 1930, die logische Unhaltbarkeit dieser Relationen durch raffiniert ausgedachte Gedankenexperimente zu beweisen. Bohr hat bekanntlich in seinem Beitrag zum Schilpp'schen Einstein Buch diese Diskussionen mit unübertrefflicher Klarheit beschrieben.

Erlauben Sie mir nur, einen Höhepunkt dieser Debatte zu erwähnen, nämlich wie Einstein mit dem Gedankenexperiment eines Photonen-Kastens die Energie-Zeit Unbestimmtheitsbeziehung zu widerlegen versuchte und wie Bohr Einsteins Angriff mit Einsteins eigener Theorie von der Abhängigkeit eines Uhrenganges vom Gravitationspotential abwehren konnte. Einstein wurde widerlegt aber nicht überzeugt. Denn intuitiv fühlte er, daß dieses Gedankenexperiment begrifflich noch nicht voll ausgeschöpft worden ist. In der Tat erkannte er bald, daß dieses Gedankenexperiment ihm einen Anhaltspunkt zu geben vermag, zwar nicht die *Widerspruchsfreiheit* der Quantenmechanik, wohl aber ihre *Vollständigkeit* in Frage zu stellen.

Ich kann hier nicht auf die interessanten Einzelheiten eingehen, wie Einstein durch eine schrittweise Modifikation dieses Gedankenexperimentes zu seinem Unvollständigkeitsbeweis geführt wurde, den er mit Podolsky und Rosen 1935 unter dem Titel *"Can quantum-mechanical description of physical reality be considered complete?"* im *Physical Review* (vol. 47)[37] veröffentlichte. Ich möchte nur erwähnen, daß Einstein schon am 4. November 1931 hier in Berlin auf einem Kolloquium eine solche Modifikation besprach, als er zeigte, daß man *entweder* die Energie *oder* den Emissionszeitmoment des Photons, aber nicht beide, genau voraussagen kann, und zwar sogar auch nachdem das Photon schon längst den Kasten verlassen hat.[38]

Einen wichtigen Teil des EPR-Argumentes hatte Einstein schon hier zum Ausdruck gebracht, indem er zeigte, daß die Möglichkeit, die eine oder die andere von zwei quantenmechanisch inkompatibilen Eigenschaften dem Photon zusprechen zu können, davon abhängt, welche Messung man am Kasten auszuführen beschließt, und dies sogar wenn das Photon vom Kasten willkürlich weit entfernt ist.

Ein wesentlicher Punkt des EPR-Argumentes - und damit, wie wir sehen werden, der eigentliche Kernpunkt des Konfliktes zwischen Einstein und den Vertretern der orthodoxen Theorie - kam allerdings in jenem Berliner

Kolloquium-Vortrag noch nicht zum vollen Ausdruck: daß es eine physikalische Realität gibt, die in der Theorie ihre Repräsentation haben muß.

Betreffs des jetzt eine zentrale Rolle spielenden Realitätsbegriffes
stellte Einstein, wie bekannt, die folgende hinreichende Bedingung auf:

> If, without in any way disturbing a system, we can predict with
> certainty the value of a physical quantity, then there exists an
> element of physical reality corresponding to this physical quantity.

Nach Einstein ist nun eine physikaische Theorie nur dann *vollständig*,
wenn jedem Element der physikalischen Realität ein Gegenstück (counter-
part) in der Theorie entspricht. Da nun zwei inkompatibile Größen, wie
Lage und Impuls, wie das Gedankenexperiment zeigt, einerseits die erwähnte Realitätsbedingung erfüllen können, andererseits aber als durch
nicht-vertauschbare Operatoren repräsentiert, nicht gleichzeitig in der
Theorie gegeben sein können, ist die Unvollständigkeit der Quantenmechanik erwiesen.

Ich kann nicht auf die Bohrsche Erwiderung noch auf die zahlreichen
anderen Einwände eingehen, die gegen das Argument erhoben wurden.

Wie bekannt, erhielt Einstein eine große Anzahl kritischer Briefe von
Lesern dieses Artikels. Was ihn besonders amüsierte, war der Umstand,
daß jeder dieser Leute von fester Überzeugung war, den Fehler gefunden
zu haben, aber jeweils an einer anderen Stelle im Argument.

Worauf ich aber kurz eingehen möchte, ist die weit verbreitete, aber
unberechtigte Behauptung, Einstein hätte mit dem Unvollständigkeitsbeweis die Einführung verborgener Parameter befürwortet.

So heißt es zum Beispiel in John Stewart Bell's einflußreicher Arbeit
von 1964 *"On the Einstein-Podolsky-Rosen Paradox"*:[39]

> The paradox ... was advanced as an argument that quantum
> mechanics ... should be supplemented by additional variables.

Oft beruft man sich auch auf die folgende Stelle in Einsteins Epilog
zum *Schilpp'schen Buch*:

> Assuming the success of efforts to accomplish a complete
> description, the statistical quantum theory would, within
> the framework of future physics, take an approximately
> analogous position to the statistical mechanics within
> the framework of classical mechanics. I am rather firmly
> convinced that the development of theoretical physics
> will be of this type. [40]

Wir besitzen nur wenige Äußerungen von Einstein zu Theorien verborgener
Parameter, da diese erst 1952 mit David Bohms erster Arbeit, also nur
drei Jahre vor Einsteins Tod, allgemeines Interesse erweckten.

Am 3. Mai 1953 schrieb Einstein an Mauritius Renninger über Bohms Hypothese:

> *Ich glaube nicht, daß diese Theorie sich halten läßt.* [41]

In einem Brief an den (damals an der Notre Dame University, jetzt am Caltech tätigen) Physiker Aron Kuppermann schrieb er:

> *Ich denke, daß man zu einer Beschreibung der individuellen Systeme überhaupt nicht durch bloße Ergänzung der gegenwärtigen statistischen Quantentheorie gelangen kann.* [42]

Es wäre also ein Irrtum, wenn man, sagen wir, auf Grund der Resultate von Kochen und Specker behaupten wollte, Einstein wäre bewiesenermaßen auf einem Fehlweg gegangen. Oder wenn man, wie vielleicht Herr Ehlers vorgestern andeutete, behauptete, daß die Experimente von Clauser, Holt, Shimony etc. Einstein's Stellung widerlegten. Die Idee, durch Einführung verborgener Parameter die probabilistische Quantenmechanik zu vervollständigen, hatte Einstein nie vertreten.

In seiner letzten Arbeit zum Quantenproblem und der einzigen, in der er ausdrücklich auf Bohms Theorie eingegangen ist, *"Elementare Überlegung zur Interpretation der Grundlagen der Quanten-Mechanik"*, die in dem Sammlerband *"Scientific Papers Presented to Max Born"* 1953 erschienen ist, zeigte Einstein, daß die Theorie der verborgenen Parameter genau so wenig wie die orthodoxe Theorie befriedigend ist. Er betrachtete, im folgenden Gedankenexperiment, eine kleine, aber noch als Makro-Objekt ansehbare Kugel, die zwischen zwei parallelen elastisch reflektierenden Wänden hin und her geht, und zeigte, daß keine dieser Theorien für den Impuls dieser Kugel einen Wert liefert, der auch nur angenähert mit dem aus der klassischen Mechanik berechneten Werte übereinstimmt. Da aber die Kugel, als Makro-Objekt, zu jeder Zeit in einem Real-Zustand ist, der durch die klassische Mechanik wenigstens angenähert richtig beschrieben wird, folgerte Einstein, daß beide Theorien nicht in der Lage sind, Real-Beschreibungen eines individuellen Falles zu liefern.

Wie wir sehen, lag der Schwerpunkt der Einsteinschen Kritik an der orthodoxen Interpretation in seiner Erkenntnis, daß diese nicht in der Lage ist, eine Real-Beschreibung eines individuellen Systems zu liefern.

So schrieb er am 15. April 1950 an Besso:

> *Die Frage der 'Kausalität' steht nicht eigentlich im Mittelpunkt sondern die Frage des realen Existierens...* [43]

Und zwei Jahre später schrieb er an Besso:

> *Ein Realzustand läßt sich in der gegenwärtigen Quantentheorie überhaupt nicht beschreiben, sondern nur ein (unvollständiges) Wissen in bezug auf einen Realzustand. Die 'orthodoxen'*

*Quantentheoretiker verbieten überhaupt den Begriff des Real-
zustandes. Man gerät so in eine Situation, die ziemlich genau
der des guten Bischofs Berkeley entspricht.*[44]

Ich gebe zu, Einstein war Realist und forderte von der Theorie, die
Existenz eines absolut-objektiv gegebenen Tatbestandes anzuerkennen.
Aber ich behaupte auch, daß man ihm daraus, daß   seine epistemologi-
sche Einstellung auf metaphysischen Voraussetzungen basierte, keinen
Vorwurf machen darf. Auch die Bohrsche oder die Heisenbergsche Philosophie hatte metaphysische Voraussetzungen. Was E. A. Burtt in Bezug
auf die klassische Physik einmal sagte, gilt gewiß auch   für die
moderne Physik:

> *There is no escape from metaphysics ... the only way to avoid
> becoming a metaphysician is to say nothing.*[45]

Einsteins Überzeugung von der realen Existenz physikalischer Gegebenheiten erklärt nicht nur seine ablehnende Kritik der orthodoxen Theorie,
sie führte ihn auch, wie ich jetzt zeigen möchte, zu der Erkenntnis,
daß *diese Theorie nicht radikal genug ist*, um das Quantenproblem lösen
zu können.

Die orthodoxe Interpretation beruht auf einer Erkenntnis, die Einstein
vollauf akzeptierte. Sie wurde schon 1927 in seinem Como Vortrag von
Bohr wie folgt formuliert:

> *Charakteristisch für die Quantentheorie ist die Erkenntnis einer
> fundamentalen Begrenzung der klassischen physikalischen Begriffe,
> wenn sie auf atomare Phänomene angewandt werden.* [46]

Die aus dieser Erkenntnis erwachsenden Schwierigkeiten versucht nun die
orthodoxe Auffassung in zwei Schritten zu überwinden: einerseits dadurch,
daß sie zwar die klassischen Begriffe aufrecht erhält; andererseits dadurch, daß sie dennoch der Anwendung solcher Begriffe Begrenzungen auf-
erlegt, indem sie untersagt, komplementäre Begriffe gleichzeitig einem
Mikro-Objekt zuzuschreiben.

Der erste Schritt, die Beibehaltung klassischer Begriffe, wurde, wie
bekannt, durch die Behauptung gerechtfertigt, daß eine Deutung des Er-
fahrungsmaterials klassische Begriffe benötigte, da nur mit diesen eine
eindeutige intersubjektive Kommunikation und Verständigung über experimentelle Erfahrung ermöglicht wird.

Der zweite Schritt, verbunden mit dem Quantenpostulat, demzufolge (in
Bohrs Worten) "jede Beobachtung atomarer Phänomene eine nicht zu ver-
nachlässigende Wechselwirkung mit dem Messungsmittel fordert", führt
zu der logischen Konsequenz, daß einem atomaren Phänomen *keine* selbständige physikalische Realität zugeschrieben werden kann.

Beide Schritte waren für Einstein ungangbar. Eine Theorie, wie die orthodoxe, die prinzipiell die physikalische Realität theoretisch nicht erfassen kann, musste er als zumindest "unvollständig" ablehnen. Dies
brachte er genau in dem EPR-Argument zum Ausdruck, mit dem er aber auch
die Schlußfolgerung aus dem Quantenpostulat entkräftigen zu können glaubte; denn in verschränkten Systemen könnte man ja für ein System einen genauen Messwert erhalten,ohne überhaupt auf dieses einwirken zu müssen.
Die Bohrsche Behauptung von der Unvermeidbarkeit von einer "nicht zu
vernachlässigenden Wechselwirkung mit dem Messungsmittel" wäre daher
widerlegt,und alle auf dieser Behauptung beruhenden Schlußfolgerungen
wären als unbegründet erwiesen.

Wie anfangs erwähnt, hielt Einstein es für nötig, geläufige Begriffe zu
analysieren, um zu überprüfen, von welchen Umständen "ihre Berechtigung
und Brauchbarkeit abhängt." Berechtigung und Brauchbarkeit waren für
Einstein untrennbar. Begriffe, die durch neue Erfahrungen sich als un-
berechtigt erweisen, verlieren dadurch ihre Brauchbarkeit.

Im Mai 1928 erhielt Einstein ein Schreiben, in dem Schrödinger die
Komplementaritätstheorie scharf kritisierte und dem er eine Kopie von
Bohrs Brief an Schrödinger beifügte, in dem Bohr sein Beibehalten klas-
sischer Begriffe verteidigte.

In seiner Antwort vom 31. Mai 1928 schrieb Einstein an Schrödinger:

> *Ich denke, daß Sie den Nagel auf den Kopf getroffen haben...*
> *Ihr Verlangen, daß die Begriffe p, q (Impuls, Lage) ver-*
> *lassen werden müssen, wenn sie nur so eine 'Wackelbedeutung'*
> *beanspruchen können, scheint mir ganz berechtigt. Die Heisen-*
> *berg-Bohr'sche Beruhigungsphilosophie - oder Religion? - ist*
> *so fein ausgeheckt, daß sie dem Gläubigen einstweilen ein*
> *sanftes Ruhekissen liefert, von dem er nicht so leicht sich*
> *aufscheuchen lässt. (Also lasse man ihn liegen).*[47]

Einstein verwarf die orthodoxe Auffassung, weil sie nach seiner Meinung
nur eine "Beruhigungsphilosophie", ein Palliativ oder, wie er es auch
einmal ausdrückte, eine "Vogel-Strauß-Philosophie" war, die die Schwierigkeiten nur hinwegtäuschte, sie aber nicht beseitigte. Im Gegensatz zu
Bohr war Einstein bereit, wie wir sehen, die klassischen Begriffe wenn
nötig aufzugeben, um sie durch neue Konzeptionen zu ersetzen, die dem
Erfahrungsmaterial entsprechen.

Einstein verwarf die orthodoxe Theorie, da sie auf Grund ihrer These von
der Beschränktheit des menschlichen Begriffsbildungsvermögens an klassischen Begriffen festhalten zu müssen behauptete, diese aber wegen ihrer
empirisch erwiesenen Anwendungsbeschränktheit dem Komplementaritäts-
prinzip unterwarf und dadurch die Beschreibung und selbst den Begriff
einer objektiv bestehenden, Beobachter-unabhängigen physikalischen

Realität aufgeben musste.

Einsteins Stellung zum Quantenproblem war radikaler als die der Anhänger der orthodoxen Theorie. Denn er war bereit, die klassischen Begriffe durch grundsätzlich neue Konzeptionen zu ersetzen, um das Erfahrungsmaterial zu erfassen,ohne auf die Realitätsforderung zu verzichten.

Die Tatsache, daß Einstein sich beim Suchen nach solchen Konzeptionen feldtheoretischer Begriffsbildungen bediente, darf nicht als eine konservative oder reaktionäre Weigerung gegen die Einführung neuer Denkstrukturen gedeutet werden.

Im Epilog zum Schilpp-Buch schrieb er:

> *Festhalten am Kontinuum stammt bei mir nicht aus einem Vorurteil,*
> *sondern aus dem Umstand, daß ich daneben nichts Organisches aus-*
> *denken konnte.*[48]

> *Adhering to the continuum originates with me not in a prejudice,*
> *but arises out of the fact that I have been unable to think up*
> *anything organic to take its place.*

In seinem Vorwort zu dem Buche *"Albert Einstein - Max Born - Brief-wechsel"* schrieb Werner Heisenberg:

> *Fast jeder Forscher ist ... bereit, neue Erfahrungsinhalte aufzu-*
> *nehmen oder neue Ergebnisse anzuerkennen, wenn sie in den Rahmen*
> *seiner philosophischen Einstellung passen. Es kann aber im Fort-*
> *schritt der Wissenschaft vorkommen, daß ein neuer Erfahrungsbe-*
> *reich nur dann voll verständlich wird, wenn die enorme Anstren-*
> *gung geleistet wird, diesen Rahmen zu erweitern und die Struk-*
> *tur des Denkens selbst zu ändern. Einstein war offenbar im Falle*
> *der Quantentheorie nicht mehr bereit oder nicht mehr in der Lage,*
> *diesen Schritt zu tun.*[49]

Wahrlich, Einstein hatte einen solchen Schritt nicht getan - aber nicht weil er zu ihm "nicht mehr bereit" war, sondern weil er ihm nicht gelungen war.

Im Alter von 38 Jahren, am 19. März 1917, in einem Briefe an Besso, stellte Einstein die Frage:

> *... ich fühle, daß der eigentliche Witz, den uns der ewige*
> *Rätselaufgeber da vorgelegt hat, absolut noch nicht begriffen*
> *ist. Ob wir die rettende Idee erleben werden?* [50]

38 Jahre später, am 3. Februar 1955, zehn Wochen vor seinem Tode, in einem Briefe an von Laue, gab Einstein die Antwort:

> *Wenn ich in den Grübeleien eines langen Lebens etwas gelernt*
> *habe, so ist es dies, daß wir von einer tiefen Einsicht in*
> *die elementaren Vorgänge viel weiter entfernt sind, als die*
> *meisten Zeitgenossen glauben..*[51]

Obgleich es Einstein nicht gelang, eine ihn befriedigende Lösung des Quantenproblems zu finden, war seine Arbeit nicht umsonst.

Selbst Niels Bohr erklärte 1961, daß sich die Quantenmechanik ohne die kritischen Herausforderungen Einsteins nie so schnell entwickelt haben würde.

Der Fall "Einstein und das Quantenproblem" ist einer jener Fälle, von denen das antike Werk "Über das Erhabene"[52] sagt:

$$\mu\varepsilon\gamma\acute{\alpha}\lambda\omega\nu \quad \overset{'}{\alpha}\pi o\lambda\iota\sigma\theta\alpha\acute{\iota}\nu\varepsilon\iota\nu \quad \overset{\iota}{o}\mu\omega\varsigma \quad \varepsilon\overset{'}{\upsilon}\gamma\varepsilon\nu\overset{'}{\varepsilon}\varsigma \quad \overset{\iota}{\alpha}\mu\acute{\alpha}\rho\tau\eta\mu\alpha$$

oder frei übersetzt:

Wenn es um große, tiefe Probleme geht, kann auch ein Streben nach Lösung ohne Erfolg etwas hoch Wertvolles sein.

Anmerkungen

1. J. Bernstein, *Einstein* (New York, 1973) S. 141.

2. C. Seelig, *Albert Einstein. Eine dokumentarische Biographie* (Zürich, Stuttgart, Wien, 1960), S. 124-125.

3. *Annalen der Physik 17* (1905), 132-148.

4. *Annalen der Physik 17* (1905), 891-921.

5. C. Seelig, *Op. cit.*

6. Vgl. z.B. J. A. Winnie, "The causal theory of space-time", in: *Minnesota Studies in the Philosophy of Science*, Band 8 (Minneapolis, 1977), 134-205.

7. *Op. cit.* (Anm. 3), S. 143.

8. *Physikalische Zeitschrift 17* (1916), 101.

9. *The British Journal of the Philosophy of Science 22* (1971), 1-8.

10. C. Seelig, *Op. cit.*, S. 174-175.

11. *Annalen der Physik 20* (1906), 199-206.

12. *A.a.O.*, S. 203.

13. *Physikalische Zeitschrift 10* (1909), 185-193.

14. In seinem Beitrag zu dem von P. A. Schilpp herausgegebenen Buch: *Albert Einstein als Philosoph und Naturforscher* (Stuttgart, 1955).

15.   C. Lanczos, *The Einstein Decade* (London, 1973), S. 62.

16.   *Annalen der Physik 22* (1907), 180-190.

17.   *Anm. 16.*

18.   Siehe seine Arbeit "Zur Theorie der spezifischen Wärme und über die Anwendung der Lehre von den Energiequanten auf physikalisch chemische Fragen überhaupt," *Elektrochemie 17* (1911), 265-275.

19.   Siehe seine Arbeit "Spezifische Wärme und Quantentheorie," *Elektrochemie 17* (1911), 817-827.

20.   *Berliner Berichte 1911*, S. 86.

21.   Siehe seine Selbstbiographie in: *Louis de Broglie - Physicien et Penseur* (Paris, 1953), S. 458.

22.   *Helvetica Physica Acta 35* (1962), 367-368.

23.   W. Heisenberg, *Der Teil und das Ganze* (München, 1969), S. 91.

24.   K. Przibram, *Briefe zur Wellenmechanik* (Wien, 1963), S. 24.

25.   *Zeitschrift für Physik 12* (1922), 13-23.

26.   Siehe M. Jammer, *The Conceptual Development of Quantum Mechanics* (New York, 1966), S. 258.

27.   *Zeitschrift für Physik 38* (1926), 803-827.

28.   *Die Naturwissenschaften 11* (1955), 425-436.

29.   *Journal of the Franklin Institute 221* (1936), 349-382.

30.   *Dialectica 2* (1948), 320-324.

31.   *Anm. 13.*

32.   *Albert Einstein - Max Born, Briefwechsel* (München, 1969), S.48-49.

33.   *Berliner Berichte 1923*, S. 359-364.

34.   *Albert Einstein - Michele Besso, Correspondance* (Paris, 1972), S. 197.

35.   *Philosophical Magazine 47* (1924), 785-802; *Zeitschrift für Physik 24* (1924), 69-87.

36.   *Anm. 32*, S. 118.

37.   *Physical Review 47* (1935), 777-780.

38.   Für weitere Einzelheiten siehe M. Jammer, *The Philosophy of Quantum Mechanics* (New York, 1974), S. 166-197.

39.   *Physics 1* (1964), 195-200.

40.   *Anm. 14.*

41.   *Anm. 38*, S. 254.

42. *Ibid.*

43. *Anm. 34*, S. 129.

44. *Ibid.*, S. 48.

45. E. A. Burtt, *The Metaphysical Foundations of Modern Physical Science* (Garden City, New York, 1953), S. 227.

46. *Die Naturwissenschaften 16* (1928), 245-257.

47. *Anm. 24*, S. 29.

48. *Anm. 14*, S. 510.

49. *Anm. 32.*

50. *Anm. 34*, S. 103.

51. *Anm. 2*, S. 397.

52. Longinus, *Peri hypsous* 3, 4.

CONNECTION BETWEEN BROWNIAN MOTION AND QUANTUM MECHANICS

Edward Nelson, Princeton University, Princeton, N.J. , USA

It is a great pleasure to address the Einstein Symposium in Berlin and to represent Princeton and mathematics at this event. I came to Princeton a few years after Einstein's death and never saw him, but his presence is a permanent part of Princeton, and his work has deeply affected the course of mathematics. Let us not be afraid to state the obvious: Albert Einstein was like no one else. It is a joy to celebrate his extraordinary depth as a thinker and his remarkable human spirit.

Dissipative diffusion

At the beginning of this century there were many scientists, including Mach, who did not believe in the reality of atoms and molecules. They were regarded as a useful mathematical device for deriving the laws of thermodynamics, a kind of hidden variables. Einstein set out to find physically observable effects of molecular motion.

Now picture, with Einstein, a drop of water with molecules moving rapidly about. Picture this as occuring in reality. It seems to me that Einstein's strong intuition of objective reality was a key ingredient in all of his major discoveries, and also in his rejection of quantum mechanics in the form which it came to assume. Most physicists do not work this way: the common approach in this century (and it has been very fruitful) is to develop an abstract formalism and then to look for verifiable predictions. Few physicists try to intuit reality directly, and many believe that any attempt to go beyond describing the connections between experiments is unscientific. In 1905 this positivistic attitude took the form: molecules cannot be seen, therefore they are not real.

The rapidly moving molecules in a drop of water cannot be seen, but consider a colloidal particle in the water which is large enough to be seen under a microscope. It is really being bombarded by the molecules of water. This bombardment should cause the colloidal particle to move along an irregular trajectory. This motion should be observable, and should yield definite information about the size of the water molecules. In this way Einstein was led to predict the phenomenon of Brownian motion, without initially being aware that the phenomenon was well known and had been intensively studied since 1827.

Before discussing Einstein's work on Brownian motion let us describe the kinematics of diffusion. If $x(t)$ is the position at time t of a particle diffusing in n-dimensional Euclidean space then its differential $dx = dx(t) = x(t+dt) - x(t)$, where $dt > 0$, is given by the Langevin equation

(1)
$$dx = bdt + dw.$$

Here $b = b(x(t),t)$, called the drift or mean forward velocity, is given by a smooth function of position and time, and $dw = dw(t)$ is the fluctuation term. The random variables $dw(t)$ are independent of the past history of the process (independent of the $x(s)$ for all $s \leq t$), are of mean 0 and have the covariance

$$Edw^i dw^j = 2v\delta^{ij} dt$$

where E denotes the expectation, v is a constant called the diffusion coefficient, and δ^{ij} is 1 if $i = j$ and is 0 otherwise. Also, the higher moments of dw are $o(dt)$; that is, they tend to 0 faster than dt . We always neglect terms which are $o(dt)$ in our equations. There is a unique stochastic process w having these properties. It is called the Wiener process because in 1920 Norber Wiener [7], inspired by Einstein's work, constructed the probability measure on the space of continuous trajectories having the above properties. The fact that $dw(t)$ is of order $\sqrt{dt}$ accounts for many peculiarities of diffusion. In particular, the trajectory of a diffusing particle is very rough (velocities do not exist).

Let f be a smooth function of position and time, and use E_t to denote the conditional expectation given the position $x(t)$ at time t .

Then

$$E_t df(x(t),t) = E_t(\tfrac{\partial}{\partial t}fdt + \frac{\partial}{\partial x^i}fdx^i + \frac{1}{2}\frac{\partial^2}{\partial x^i \partial x^j}fdx^i dx^j) = (\tfrac{\partial}{\partial t} + b \cdot \nabla + v\Delta)fdt.$$

Since this is true for arbitrary f , the probability density ρ satisfies the adjoint equation, which is the Fokker-Planck equation

(2)
$$\frac{\partial \rho}{\partial t} = v\Delta\rho - \text{div}(b\rho).$$

We may also describe the same process with the direction of time reversed. Keeping the convention that $dt > 0$ we have $dx = b_* dt + dw_*$. Here b_* is the mean backward velocity and the $dw_*(t)$ are of mean 0 and are independent of the future (independent of the $x(s)$ for all $s \geq t + dt$). Since $Edw^i dw^j = Edx^i dx^j = Edw_*^i dw_*^j$ we also have $Edw_*^i dw_*^j = 2v\delta^{ij} dt$.

We have $E_{t+dt} df(x(t),t) = (-\frac{\partial \rho}{\partial t} - b_* \cdot \nabla + v\Delta) f dt$ and so we have the backward Fokker-Planck equation

(3) $$\frac{\partial \rho}{\partial t} = -v\Delta \rho - \text{div}(b_* \rho) .$$

Let $v = (b+b_*)/2$, called the current velosity. Adding (2) and (3) we have the equation of continuity

(4) $$\frac{\partial \rho}{\partial t} = -\text{div}(v\rho).$$

Let $u = (b-b_*)/2$, called the osmotic velocity. Subtracting (3) from (2) we obtain $v\Delta \rho - \text{div}(u\rho) = 0$, so that $\text{div}(v \, \text{grad} \, \rho - u\rho) = 0$. A closer look, using the fact that the adjoint with respect to ρ of the operator $(\frac{\partial}{\partial t} + b \cdot \nabla + v\Delta)$ is the operator $-(\frac{\partial}{\partial t} + b_* \cdot \nabla - v\Delta)$, shows that in fact $v \, \text{grad} \, \rho - u\rho = 0$; that is,

(5) $$u = v \frac{\text{grad} \, \rho}{\rho} .$$

If $b = 0$ then the Fokker-Planck equation (2) has the fundamental solution

(6) $$\frac{1}{(4\pi vt)^{3/2}} e^{-\frac{x^2}{4vt}}$$

giving the probability density at time t of a particle starting at the origin at time 0 and diffusing in three dimensional space without drift.

The formula (6) was derived by Einstein [1], not using partial differential equations but rather by an argument quite similar to that used by de Moivre in 1733 to establish the central limit theorem for the symmetric random walk . Using (6) one can measure the diffusion coefficient by observations of diffusing particles.

Einstein was concerned to know the size of v , for this is crucial if the effect is to give visible evidence of the existence of molecules. He supposed the particle to be moving in an external force field F . Now if we throw some pellets into a tank of water they soon lose their initial velocity, due to friction, and fall at a constant speed under the force of gravity. That is, if the fluctuation effects of Brownian motion are negligible, the effect of an external force F on a particle in a frictional medium is to impart a velocity $F/m\beta$ proportional to the force. Here m is the mass of the particle, $m\beta$ is the coefficient of friction, and β^{-1} is a relaxation time. Thus the dynamics of the

limiting theory (negligible fluctuation effects) is dissipative Aristotelian dynamics.

As we have seen, a solution to (1) does not have a finite velocity, so
one cannot set $F/m\beta = dx/dt$. Einstein, and also Smoluchowski in his
investigations of Brownian motion, in effect set

(7)
$$\frac{F}{m\beta} = b$$

in (1), thus converting it from a kinematical description to a dynamical equation, called the Einstein-Smoluchowski equation.

Next Einstein supposed that instead of a single diffusing particle there
is a suspension of many identical diffusing particles with density ρ.
He argued that in equilibrium the force F would be balanced by the osmotic pressure forces of the suspension, so that

(8)
$$F = kT \frac{\text{grad } \rho}{\rho} \quad ,$$

where T is the absolute temperature and k is Boltzmann's constant.
In equilibrium, ρ is the probability density for any given particle and
the current velocity is 0 by invariance under time reversal, so that
in equilibrium $b = u$. By (5), (7) with $b = u$, and (8) we obtain

$$\frac{kT}{m\beta} \frac{\text{grad } \rho}{\rho} = \frac{F}{m\beta} = v \frac{\text{grad } \rho}{\rho} \quad ,$$

so that

(9)
$$v = \frac{kT}{m\beta} \; .$$

This is Einstein's formula for the diffusion coefficient.

The above is in essence Einstein's derivation [1] of the formula. It is
an amazing argument because both the force F and the suspension are
purely virtual: the result also holds for a single diffusing particle
with no external force. The formula has stood the test of time. It was
confirmed experimentally by Perrin, subject to an unavoidably large experimental error. It is also confirmed by a more refined theory of
Brownian motion due to Wiener [8] and Ornstein-Uhlenbeck [5].

Is it surprising that so probabilistic a theory as Einstein's theory of
Brownian motion was developed by someone who repeatedly and emphatically

said "The Lord does not play dice with the universe"? Not at all. It is clear in principle how probability enters the description of Brownian motion in terms of ignorance of initial conditions, and it is more or less understood how the Einstein - Smoluckowski equation arises from a coupling of the particle to a classical mechanical system.

Conservative diffusion

Next I want to celebrate another aspect of Einstein's scientific career: his stubborn refusal to accept the Born interpretation of quantum mechanics. This interpretation violated his understanding of objective reality, and so almost alone he maintained throughout his life that it was only a provisional theory.

I will tell a lengthy fable describing how modern physics might have developed. Although the tone of the fable is perhaps somewhat polemical, for which I apologize, my serious purpose in presenting it is the hope of providing a fresh perspective on some much debated foundational problems.

There once was a planet called Lagado, far out in the complement of our past and future light cones. The physicists turned their attention to constructing a theory of microprocesses to explain such puzzles as the existence of discrete energy levels of atomic systems. It seemed clear that fluctuations play a large role at this scale, so being well versed in diffusion theory they attempted to explain the phenomena in terms of diffusion.

The Hamiltonian of a free system is quadratic in the momenta, of the form $\frac{1}{2}g^{ij}p_ip_j$ where g^{ij} is a symmetric positive-definite tensor. The covariance matrix of a diffusion process is of the form $2v^{ij}$ where v^{ij} is also a symmetric positive-definite tensor. This suggested associating a diffusion process with the system by setting v^{ij} proportional to g^{ij}. The idea was that no system is isolated, but interacts with a medium of infinitely many degrees of freedom, the physical cause of the diffusion. For example, a charged particle interacts with the electromagnetic field. The dimensions of g^{ij} are inverse mass and the dimensions of v^{ij} are length squared per time, and so the equation $2v^{ij} = \hbar g^{ij}$ requires $\hbar$ to have the dimensions of action. In this way Planck's constant entered the theory phenomenologically. This procedure the physicists called stochastic quantization, although the origin of the term "quantization" is obscure. For simplicity of exposition I will deal henceforth with the system consisting of a single nonrelativistic particle of mass m , so that $g^{ij} = \frac{1}{m}\delta^{ij}$ and the covariance matrix is $2v\,\delta^{ij}$ with diffusion coefficient

(10)
$$v = \frac{\hbar}{2m} \ .$$

The covariance matrix does not fully describe a diffusion; there is also the drift. The kinetics of any diffusion process are given by the Langevin equation (1). In the theory of dissipative diffusion the limiting dynamics (when fluctuation effects are negligible) was Aristotelian, $dx/dt = F/m\beta$. The dynamics of conservative diffusion must be such that the limiting dynamics is Newtonian.

The guiding idea in discovering such a dynamics was energy conservation. Now $\frac{1}{2}m(\frac{dx}{dt})^2$ is infinite. In the theory of dissipative diffusion the infinite quantity $\frac{dx}{dt}$ is replaced by b in the dynamical equation; in the theory of conservative diffusion the infinite quantity $\frac{1}{2}m(\frac{dx}{dt})^2$ is replaced by the average of $\frac{1}{2}mb^2$ and $\frac{1}{2}mb_*^2$, which may also be written as $\frac{1}{2}mu^2 + \frac{1}{2}mv^2$. For a particle moving in the presence of a time-independent potential V the basic hypothesis was that the expected value of the energy $\frac{1}{2}mu^2 + \frac{1}{2}mv^2 + V$ is constant in time:

(11)
$$\frac{d}{dt} \int (\frac{1}{2}mu^2 + \frac{1}{2}mv^2 + V)\rho = 0 \ .$$

To calculate the left hand side of (11) all that is needed is (4), (5), and integration by parts: it is $m(I_1 + I_2 + I_3)$ where

$$I_1 = \frac{d}{dt} \int \frac{1}{2} u^2 \rho = \int (u \cdot \frac{\partial u}{\partial t} \rho + \frac{1}{2}u^2 \frac{\partial \rho}{\partial t})$$

$$= \int ((u \cdot \frac{\partial}{\partial t} v \frac{\mathrm{grad}\ \rho}{\rho})\rho - \frac{1}{2}u^2 \mathrm{div}(b\rho))$$

$$= \int ((u \cdot v \frac{\mathrm{grad}\ \rho}{\rho^2}) \mathrm{div}(v\rho)\rho - u \cdot v\ \mathrm{grad}\ \mathrm{div}(v\rho) - \frac{1}{2}u^2\ \mathrm{div}(v\rho))$$

$$= \int (\frac{1}{2}u^2\ \mathrm{div}(v\rho) - u \cdot v\ \mathrm{grad}\ \mathrm{div}(v\rho))$$

$$= -\int (v\Delta u + u \cdot \nabla u) \cdot v\rho \ ,$$

$$I_2 = \frac{d}{dt} \int \frac{1}{2} v^2 \rho = \int (v \cdot \frac{\partial v}{\partial t} \rho + \frac{1}{2}v^2 \frac{\partial \rho}{\partial t})$$

$$= \int (v \cdot \frac{\partial v}{\partial t} - \frac{1}{2}v^2\ \mathrm{div}(v\rho)) = \int (\frac{\partial v}{\partial t} + v \cdot \nabla v) \cdot v\rho \ ,$$
(where we have assumed that v is a gradient),

$$I_3 = \frac{d}{dt} \int \frac{V}{m} \rho = - \int \frac{V}{m} \, \text{div}(v\rho) = \int \frac{\text{grad } V}{m} \cdot v\rho .$$

Therefore

$$\frac{d}{dt} \int (\tfrac{1}{2}mu^2 + \tfrac{1}{2}mv^2 + V)\rho = m \int \left[-(v\Delta u + u \cdot \nabla u) + (\frac{\partial v}{\partial t} + v \cdot \nabla v) + \frac{\text{grad } V}{m} \right] \cdot v\rho .$$

If this is to hold for general v and ρ the term in square brackets must be 0 :

(12) $$-(v\Delta u + u \cdot \nabla u) + (\frac{\partial v}{\partial t} + v \cdot \nabla v) + \frac{\text{grad } V}{m} = 0 .$$

Also,

(13) $$\frac{\partial u}{\partial t} = -v \, \text{grad div } v - \text{grad}(v \cdot u)$$

since

$$\frac{\partial u}{\partial t} = \frac{\partial}{\partial t} v \frac{\text{grad } \rho}{\rho} = v \, \text{grad} \frac{\partial}{\partial t} \log \rho = -v \, \text{grad} \frac{\text{div}(v\rho)}{\rho} =$$

$$= -v \, \text{grad}(\text{div } v + v \cdot \frac{\text{grad } \rho}{\rho}) = -v \, \text{grad div } v - \text{grad}(v \cdot u) .$$

Let $R = \tfrac{1}{2} \log \rho$, so that $\text{grad } R = 2v \, u$. Assume that v is also a gradient and let S be such that $\text{grad } S = 2vv$, and let $\psi = e^{R+iS}$. Recalling that $v = \hbar/2m$ and choosing the arbitrary additive constant in S properly, one sees by a simple calculation that (12) and (13) are equivalent to the Schrödinger equation

(14) $$\frac{\partial \psi}{\partial t} = i \frac{\hbar}{2m} \Delta \psi - \frac{i}{\hbar} V \psi .$$

If $(\frac{\hbar}{2m} \Delta + \frac{1}{\hbar} V)\psi = \lambda \psi$ then λ is the energy

$$\lambda = \int (\tfrac{1}{2}mu^2 + \tfrac{1}{2}mv^2 + V) \rho ,$$

so the theory gave the observed energy levels of bound systems.

This is not the place to recount the many other triumphs of the Schrödinger equation. Rather let us describe the interpretation which it received on Lagado.

The Schrödinger equation is not an equation of motion for the state of a physical system, any more than is the Fokker-Planck equation. It is a linear partial differential equation which enables one to determine the drift b in the Langevin equation (1). The latter is the equation of motion: solutions of it give the trajectories of the diffusing particle (or system of interacting diffusing particles).

Let $a = -(v\Delta u + u\cdot\nabla u) + (\frac{\partial v}{\partial t} + v\cdot\nabla v)$, called the mean acceleration. Then (12) may be written as $F = ma$ where $F = -\text{grad } V$ is the external force. The two constituents of a are called the osmotic accelera- tion $-(v\Delta u + u\cdot\nabla u)$ and current acceleration $(\frac{\partial v}{\partial t} + v\cdot\nabla v)$. The osmotic acceleration is of a form familiar from hydrodynamics whereas the current acceleration is simply the acceleration of a stream of particles moving with velocity v . Thus in the kinematics of diffusion the mean accele- ration combines aspects of wave motion and particle motion. It was not too surprising that solutions of (12) and (13), or equivalently of the Schrödinger equation (14), show interference effects. A particle dif- fusing through one of two slits in a screen arrives at a later screen with a probability density characteristic of diffraction, but the par- ticle travels in a continuous trajectory and passes through one or the other of the two slits.

The fullest description of the state of a system of particles which the theory provides is given by the probability density ρ and drift b at a given time (or equivalently by the function ψ at a given time) to- gether with the positions of the particles at that time. Then if the forces are known the transition probabilities for the future motion of the particles are determined.

So far there is little that is surprising in our fable. What we have re- counted is the development of the physics of microprocesses as it normal- ly occurs on planets inhabited in part by physicists. However, the next development on Lagado is of particular interest to us.

A mathematician observed that certain random variables α , for example functions of the position $x(t)$ and linear combinations of such for different times t , have the property that the expected value of α is a Hermitean form in ψ , so that $E_\alpha = (\psi, A\psi)$. The correspondence between α and A is linear. The mathematician described a method for multiplying two Hermitean forms, subject to certain technicalities about domains of operators. The product $A_1 A_2$ is in general not a Hermitean form and has no relation to the expected value of $\alpha_1 \alpha_2$. He proposed to regard the state of a physical system as being completely given by ψ , without including the positions of the particles, and to regard the Hermitean form A not as giving the expected value of a physical ob- servable but as itself being the physical observable. The probability

distribution of a (hypermaximal) Hermitean form A was given by a
mathematical construct called its spectral resolution. The new theory
provided no new predictions, since it and the old theory yield the
same probability distribution for all particles at any given time.

Some objections were made to the new theory. First, consider a free
particle and let X(t) be the Hermitean form corresponding to the
position random variable x(t) . Then

(15)   X(t) = X(0) + tX'

were X' is a time-independent Hermitean form interpreted as the velocity. This relation was of couse well-known for the expected values of
the position of a free particle, but in the new theory (15) was asserted
to hold for the position observables themselves. This suggested a picture in which free particles travel in a straight line with constant
velocity, very much at variance with the picture of diffusing particles.
In particular, it was hard to see how such a picture could account for
interference effects. This objection turned out to be due simply to the
unfamiliarity of the new theory. The linear relations (15) for position
observables at different times were not open to experimental observation
and so could logically be asserted in the new theory, and interference
effects are a mathematical consequence of the Schrödinger equation regardless of its interpretation.

Second, it was objected that while a hypermaximal Hermitean form A
has a well-defined probability distribution, for a given ψ , this is
no longer true in general for a pair of such forms. What sort of properties of a physical system were these which could only be considered one
at a time? This objection was answered by developing a principle, called
the principle of complementarity, which asserts that if two properties
of a physical system are complementary then they can only be considered
one at a time.

Third, it was objected that since the new theory regarded ψ as a
complete description of the state of a physical system, and since ψ
develops in time according to the Schrödinger equation, the new theory
could not give an account of the state of a physical system after an
observation is made. This objection was answered by pointing out that
the state of a physical system changes in two ways: continuously according to the equation of motion and abruptly when information about it
enters the consciousness of the observer.

Now the inhabitants of Lagado are creatures who look at a theory on its merits without regard to established notions. Although the new theory could not be distinguished from the old by experiment, the appeal of the non-commutative algebra of partially defined linear transformations on an infinite dimensional vector space, together with the principle of complementarity and the role of consciousness in affecting the states of physical systems, was so great that they soon adopted the new theory. In this way they came to share the understanding of microprocesses enjoyed on our own planet.[*]

Outlook

What we have done is to give a new derivation, based on energy conservation, of the stochastic interpretation of the Schrödinger equation due to Féynes [2] (see also [3]). Since the interpretation of the Schrödinger equation in terms of diffusion gives no new predictions, what if anything does it contain which is of interest to physics? What it gives us is some challenges.

1.   The stochastic picture describes fluctuation effects beyond those described by quantum mechanics. For example, for the ground state of the hydrogen atom one can compute not only the probability density for the position of the electron but also a relaxation time such that for two times which are separated by a greater amount the positions at the two times are not strongly correlated. The analogy with fluctuations predicted by kinetic theory versus thermodynamics is striking. If the electron is diffusing in reality, are such fluctuation effects forever and in principle beyond the reach of experimental observation? Perhaps so and perhaps not, but conventional quantum mechanics precludes us from even asking the question, which may be unwise.

2.   Are there physical instances of conservative diffusion other than elementary processes taking place in the vacuum, instances with a diffusion coefficient much bigger than $\hbar/2m$? I am completely ignorant of how colloidal particles diffuse in liquid helium, but presumably the analysis in terms of friction and dissipative diffusion breaks down. One can observe individual continuous trajectories; is it possible that one can also see diffraction patterns in probability densities?

3.   The realms of mathematical validity of stochastic mechanics and quantum mechanics overlap, but neither contains the other. In particular, it is possible to apply the method of stochastic quantization to systems with nonholonomic constraints. Is it possible that this would be helpful in quantizing Einstein's gravitational theory?

---

[*] For further information on the science of Lagado see [6].

4. Does classical mechanics really give rise to the type of stochastic process we have described? Is it possible to realize Einstein's desire to derive the effects described by quantum theory from purely classical considerations? The classical electrodynamics of point charges is not yet a finished theory. Does this interaction give rise to the type of diffusion described by the Schrödinger equation, and if so is the diffusion coefficient $\hbar/2m$ determined by the parameters e, c, and m of the theory?

There is a growing literature on stochastic interpretations of quantum mechanics (see the references in [4]). I shall not attempt to review it. There is much of interest in this literature, but it is fair to say that much of it is in a preliminary state. There are questions here which need the same kind of precision that mathematical physicists have brought to the study of quantum field theory, statistical mechanics and general relativity. It is time, in March 1979, to declare this field of research to be respectable.

Perhaps Einstein was right. Perhaps the intuition of objective reality is still a useful guide to a deeper understanding of nature, and perhaps there is a theory of microprocesses in which probability is merely an expression of ignorance of the initial conditions. It seems to me that Einstein's view of nature has after all withstood the attack of Born, Heisenberg, and Bohr, and if this is the case, would it not be odd if his program of a classical field theory of nature holds no promise for the future? In Einstein's words, carved in the fireplace of the mathematics building at Princeton in which he worked, "Raffiniert ist der Herr Gott, aber boshaft ist er nicht."

References

[1]   A. Einstein, Die von der molekularkinetischen Theorie der Wärme geforderte Bewegung von in ruhenden Flüssigkeiten suspendier- ten Teilchen, Annalen der Physik 17 (1905) 549 - 560

[2]   Imre Féynes, Eine wahrscheinlichkeitstheoretische Begründung und Interpretation der Quantenmechanik, z. Phys. 132 (1952) 81-106

[3]   Edward Nelson, Dynamical theories of Brownian motion, Princeton University Press, Princeton, NJ (1967)

[4]   Steven M. Moore, Can stochastic physics be a complete theory of nature?, Foundations of Physics, to appear.

[5]   G. E. Uhlenbeck and L. S. Ornstein, On the theory of Brownian motion, Physical Review 36 (1930) 823-841

[6]   Jonathan Swift, Gulliver's Travels, The Modern Library, Random
       House, New York (1958)

[7]   Norbert Wiener, The mean of a functional of arbitrary elements,
       Ann. of Math. 22 (1920) 66-72

[8]   -, The average of an analytical functional and the Brownian
       movement, Proc. Nat. Acad. Sci. USA 7 (1921) 294-298

The idea of studying energy conservation in the stochastic theory was
suggested to me by David Shucker.

QUANTA IN CONTEXT*

Joseph Agassi, Boston University, USA, and Tel-Aviv University, Israel

The context of a scientific theory can be epistemological and methodological. Or it can be metaphysical, relating to the intellectual framework within which we cast it. Or it can be intertheoretical, both syn-
chronically and diachronically. My concern here will be mainly dia-
chronical - the historical context of quantum theory, what is required
of it vis-à-vis that context and how well it fulfills this requirement.
But I shall come to this only at the later part of this essay. I shall
have to clear the ground by discussing the epistemic and metaphysical
contexts first.

1.   Epistemic

The claims for the foundation of a scientific theory are made on the
basis of experience or intuition or both. Not only is there no foundation for any theory; also looking at anything as possible foundation
distorts it.

When experience is looked at as foundation, only the success of a theory
is considered; failure is overlooked. This makes it impossible to look
at a theory in its historical context since the initial success of a
theory is all too often the overcoming of the failures of its predecessor. It is often claimed that the predecessor was never meant to explain
or predict data it failed to describe correctly. This is often historically false and heuristically confusing since the problem a new theory
comes to solve often stems from false predictions based on old theories.
To claim that old theories are not refuted since they are still in use
is to confuse theory with practice. To claim that they are not refutable
since they are only tools of predictions is to make science a part of
technology devoid of any intellectual value, which is absurd. To claim
that the old theory is still asserted with qualifications may be true, yet
the knowledge of the qualifications, when available, is the corollary of
the new theory. This point will prove crucial in the later part of this
essay, so let me elaborate with an example.

Newtonian mechanics is putatively considered as still true within the
limits of small velocities and weak gravitation if not tested too accurately. This is not true, since, however small the velocity of a nucleus
is, if it disintegrates it violates both Newtonian conservation of mass
and Newtonian conservation of energy, and the violation may be of as

large an order of magnitude as that of a nuclear explosion. Nevertheless, suppose we have managed to specify the qualifications within which New- tonian mechanics is held to be true. Not only did the Newtonians deny any such qualification - they explicitly denied all qualifications! - they could not possibly know all those which we make, since this know- ledge is part and parcel of later theories. Often the success of stating a qualification of a theory opens the road to its replacement. For ex- ample, after special relativity was established it became clear that Newtonian gravity that acts at a distance had to be replaced with a gra- vity that is propagated at the speed of light at most, thus showing the road to general relativity.

The same holds for intuition: the attempt to see it as foundations di- rects our attention to success, not to failure. Thus, those who attempted to found Newtonian mechanics on intuition had to declare it perfectly intuitive despite the historical fact that when it appeared it was deemed highly counter-intuitive, as noticed by Imre Lakatos in his classic paper on infinite regress in the foundations of mathematics.

What is intuitive and what is not is hard to decide. We all agree that pre-Copernican astronomy is so very intuitive that even the most ad- vanced astronomers still use it when they permit the sun to rise in the east and set in the west. It is, however, quite unproblematic because the newer theories explain the success of older theories, both empirical and intuitive. For many people the Newtonian replacement of Galilean gravity with the interaction of the falling apple with the earth is counter-intuitive. Consequently there is the opinion extant that those who find Newtonianism intuitive do so out of sheer habit. This is a typi- cal foundationist attitude: either intuition is genuine or a mere habit and so fake. In truth, as Lakatos has indicated, we educate our intui- tions.

The education of our intuitions need not progress on a unique path. Newton has suggested one possibility: extend Galilean gravity from the falling apple to the moon: make the moon fall. Make then the moon inter- act with the earth. Transfer then the interaction to the apple. Newton has also suggested that this is how he developed his own ideas; which may indeed be true. No doubt, the intuitions of discoverers are more developed than those of their contemporaries, and grow faster. But they are no guarantee, no proper foundations.

Niels Bohr's central contention was that classical physics is intuitive but not quantum mechanics. He concluded that classical physics is here to stay - which, in the sense in which even pre-Copernican astronomy is here to stay, surely is true. He raised the question, how can classical and quantum theories coexist? The answer should be, by having quantum physics explain the success of classical physics. Barring this possibility, we may wish to have a theory that should explain the success of both classical and quantum physics. Bohr's answer was his principle of complementarity, which differs from the above two alternatives. Hence, whatever precisely it says, it will not be endorsed here. (See note 5 below.)

To conclude, both intuition and experiment are at time successful, at times they fail. Progress is achieved when a new step explains the success of the old and avoids its failures. This, however, need not always be so. Newtonian mechanics explained the success of both Kepler's theory and Galileo's. The special theory of relativity failed to explain the success of Newtonian gravity and so had to be replaced - by the general theory of relativity, as it turned out.

2. Metaphysical

By metaphysics I mean the general presuppositions, the intellectual framework, within which a scientific theory is or should be couched. When discussing quanta two interconnected metaphysical points are raised at once: realism and determinism. Einstein endorsed both; Bohr rejected both. It makes little sense to endorse determinism and reject realism, but the other way round is possible and proposed by Karl Popper, Alfred Landé, Mario Bunge, and many others. It will be endorsed here.

Realism is proposed here as naive, not in the sense of naive relaism (that declares true all careful observations) but as the proposal to take scientific theories fairly literally. Also it may be deemed as a desideratum, as the demand of any theory to be an attempt to describe reality. Finally it may be deemed as the demand that science be truly explanatory. All this may prove excessive: we may fail in our attempt to describe reality, to offer true explanations, etc. The way to avoid the excess is not to forget that what we want and what we get are different things. We then can view Daltonian atomism, for example, realistically, and assess its success and failure, and explain them both by the high but not absolute stability of the nucleus. Tentative realism thus permits pictures of science that are more historically accurate, more heuristically satisfying, more dialectical and more logically coherent. Tentative realism, then, is the same as the attempt to view a theory in its historical context.

Determinism was, for Einstein, realism pushed all the way: the attempt to describe and explain everything is the attempt to have an all-embracing deterministic theory. This was his grand argument for determinism. He saw no need to elaborate on it. Many have attributed to him a deterministic scientific theory rather than a deterministic metaphysics and program. In his famous "Replies to Criticisms" he protested: he had no scientific theory, he stressed, only a program toward it. Those who did not see the difference he could not argue with.

One who did see the difference between having a scientific theory and the program to develop one was Wolfgang Pauli. The return to determinism, he said in a prophetic remark, was neither possible nor desirable.[1] He did not elaborate. The following elaboration is due to Alfred Landé and Karl Popper mainly.

The alleged determinism of classical mechanics was based on the supposition that particle dynamics fully determine trajectories out of any permissible initial conditions on any permissible equations of force and motion. Statistics, then, was deemed as rooted in ignorance or nescience. Einstein endorsed all this. He was doubtlessly in error: classical particle theory could not avoid singularities or discontinuities, and trying to handle these sooner or later led to quantum mechanics. History aside, ignorance cannot explain facts like the second law of thermodynamics. What Einstein assumed, of course, is not that ignorance gives us laws of distribution, but that we postulate laws of distribution in ignorance of the initial conditions of the system which include all details distributed.

Yet even if we count heads and observe the distribution, we cannot yet explain its stability without postulating it or some other stable distribution. The variational principle on probabilities says, no matter in what distribution a system is, it will get to the equilibrium distribution in the shortest way. This is postulated as a law of nature.

Suppose that we knew all systems and in all desired detail and predicted them all to move along classical trajectories toward equilibrium. Would that make statistics redundant? Certainly not; it would make it a general fact, a law of nature. Hence statistics is not eliminable by knowledge. Nescience has nothing to do with it.

We have here a pressing question: the systems that are subject to strict particle mechanics and also obey laws of distribution of statistical mechanics, are they not over-determined? If they are not over-determined then, perhaps, the laws of distribution can be deduced from the equations of particle mechanics. Even so, the laws will be deduced for each system separately. We could not possibly deduce them for all systems, since it is not deducible from general particle mechanics. Therefore it needs an explanation which does not belong to particle mechanics. The fact of the logical independence of thermodynamics - and hence of statistical mechanics - from particle mechanics was discovered by Maxwell. He said, for thermodynamics to belong to particle mechanics, it should follow from a Hamiltonian. But Hamiltonians are time-reversible and the second law of thermodynamics is not. A more pictorial proof of the independence is Maxwell's demon, who violates thermodynamics without violating particle mechanics. A number of authors, expecially Brillouin, have tried to outlaw Maxwell's demon, but he has already done his duty. One can argue against all this by saying, the systems that are permissible by general particle mechanics but not by statistical mechanics, *e.g.* one of particles all moving in parallel to and fro between two perfect mirrors, are of the measure zero. This is true, yet not deducible from particle mechanics. To show this all we need is to view the improbable systems as models to establish independence in the classical Hilbertian way.

Thus, perfect knowledge of all the initial conditions (and computations) of all the systems in the universe will not yield the laws of statistical mechanics.

The question arises, at once, is not a system with both particle mechanics and statistical mechanics over-determined? Most likely it is. Does this make the system inconsistent? We do not know. Perhaps it simply excludes certain sets of initial conditions (like the above model of the two mirrors), perhaps it excludes full determinism in particle dynamics - an exclusion which can be blamed on multiple collisions, discontinuities in elastic collisions of inelastic bodies (*i.e.* energy and momentum conserving collisions between absolutely rigid bodies), and other imperfections of classical particle dynamics.

My discussion of classical dynamics is not motivated by a desire to grant it any special position, intellectual or physical. I was questioning Einstein's claim that an attempt at a complete description and explanation of nature has to be deterministic. Perhaps so. Perhaps, however, far from having to be deterministic it simply cannot be.

So much for realism and determinism. Metaphysics also concerns relations between scientific theories, since we may be looking for a framework to accommodate diverse, coexistent, yet very different ones. I will not enter this discussion here: I have done so elsewhere. But at least one more topic is traditionally within the domain of metaphysics, the principles of the theory of the ultimate structure of matter. Evidently both the bootstrap theory and chromodynamics pertain to this topic. I will not enter this topic except for a brief casual remark, and from want of proficiency.

3.   Quantum mechanical

The two preceding discussions are preparatory to the discussion of the
position quantum theory has vis-à-vis its predecessors. To complete the
preparation we may want to present a coherent picture of quantum theory
itself.

When we talk of quantum mechanics, then, what exactly do we have in
mind? It is three quarters of a century since quantum mechanics was in-
augurated, half a century since the debate about it began, and a quarter
of a century since the end of the contributions of and debates between
Bohr and Einstein. The debate was fluid, in content and even in reference: there was no canonic version of quantum theory. This did not
trouble Bohr. He felt that as long as a theory includes the Heisenberg
principle of uncertainity then what he had to say in this discussion of
his own principle at complementarity stood fast. This did not trouble
Einstein either. He liked fluidity since it stimulates heuristic. Canonic forms, he felt, are reached at the end of a road, perhaps after ages,
he noticed, as was the case with Euclidean geometry. Without insisting
on canonic forms, however, one may want to reach a narrowing down of the
reference, since the very existence of variants, especially in time
series, presents changing challenges to analysis.

No participant in the debate makes much of the absence of a canonic
definitive version of quantum mechanics but the absence still is something adumbrated in the literature, perhaps as a serious problem, at
least as an added technical difficulty. Nearest comes, I feel, Carl
Friedrich von Weizsäcker, who is, by the way, as establishment as they
come, these days. His essay "Probability and Quantum Mechanics", in the
British Journal for the Philosophy of Science of 1973, presents as a
problem for discussion the problem, what in quantum mechanics is meant
as a universal theory, what as specific to quanta? Clearly, the logic,
the mathematics, the probability, all these are general. "Bohr's concept of complementarity was never understood", he says (p. 323), "because it was misinterpreted as a generalization of a particular empirical concept of physics, while Bohr intended it to indicate a universal
structure of all human experience which could be particularly well exemplified only in quantum theory." Similarly, he says of quantum logic
(loc.cit), that though its discovery "was induced by experience", once
it was discovered "it can be understood without reference to experience".
Let me overlook the unfortunate foundationist expression "induced from
experience" as well as the historical fact that the principle of com-

plementarity and quantum logic came to overcome difficulties within quantum mechanics. My aim is to reproduce the problem-situation as presented by Weizsäcker: wanting to distinguish within quantum theory what is specific to it and what not he found the need to be specific about the theory: the desire to be at times context-independent made him want to specify the context exactly first.

Weizsäcker chooses as his context second quantization. It, too, was traditionally misconstrued, he notices: "it was never quite clear", he tells us (p. 334), "what the iteration of the quantization process really meant." It is, he adds, quite generally, "a process of ensemble building according to the peculiar rules of probability that are characteristic of quantum theory. And this is exactly the thesis of the present paper: quantum theory is nothing but a general theory of probability, *i.e.*, of expectation values of relative frequencies in ensembles." This last sentence is crucial.

There is a program here, and it is executed by Weizsäcker at once in two quick steps. First the formulation of quantum theory in general, and second the manifestation of its character as a theory of expectation values of relative frequencies. The formulation is Feynman's formulation of quantum field theory, so called. I think things can be made more specific: not merely Feynman's theory, but in the presentation of Julian Schwinger as it is sketched, say, in his *Particles and Sources* of 1969. As Schwinger explains in his preface there, his formulation overcomes traditional difficulties and is more parsimonious in its assumptions; he calls it phenomenological in the sense that it avoids speculations not intrinsic to the more formal part of the theory. It is not phenomenological in any sense used in philosophy; for example it employs Feynman's virtual particles.[2]

The second step, then, exhibiting the phenomenological part of the theory as that of distributions of expected values, is also a matter of a careful development of the formal theory and also already executed by Schwinger. What Weizsäcker adds is precisely what we are looking for: the inter-theoretical context: classical mechanics, he says, is the limiting and special case of quantum mechanics. It obtains for classical ensembles when the quantum phases are not crucial so that quantum probabilities can be replaced by classical probabilities. And it obtains for classical single particles when many quantum states come close enough to look like a single and highly probable state. This is Weizsäcker's point, his addition to the picture, and the reason I refer to his paper here.

Before coming to examine the truth or falsity of Weizsäcker's point we may want to know more about the content and the meaning of the diverse parts of the theory. It is not a question of sympathies but of reading. For, some of the physicists who share Weizsäcker's reading but not his sympathies, conclude that anyone who is not pleased with the present situation must get out of Hilbert's space. A few people have undertook such adventurous journeys, including heretics like Mendel Sachs and orthodox like Leon Rosenfeld, but this is a different matter. The matter at hand is quantum theory as understood by the reading mentioned here, and the classical difficulties it encountered, the so-called quantum paradoxes. What has happened to them? What can one expect from a statistical reading of quantum mechanics about the quantum paradoxes?

The less one can expect from the theory, the less paradoxial it should seem, and quite a priori. For example, in the case of the two slit experiment, this supports (a priori) authors like Niels Bohr and P. A. M. Dirac, who have declared the crucial central bothersome question quite meaningless, on the simple ground that the question is quite unrelated to the job at hand. The question, we remember, is, through which slit did the particle go, and how did it decide that the other slit is open or closed, as the case may be? To begin with, does quantum theory cover the two slit experiment? Let us take this question a bit slowly. Do we have a quantum equation for the experiment whose solution fits the observed facts of the matter? Of course, not. But, quantum physicists hasten to add, this is only small technicality. May be; why are they so convinced that in principle there is no problem here? Which principle applies here? The answer is very simple and in two steps. First, the classical equation, the variation principle, classically applies just to a case like this. And when we take it as the limiting case of some unknown quantum equation such that should take the phases as probabilities and replace them with some vector out of Hilbert space, then, without deciding which equation it be, the facts as observed make sense and only the facts unobserved become paradoxical![3] Query: can we have a quantum theory that should offer expected values of quanta passing through this slit of that? Yes. Could these be compared with observations? Yes. Could the observations of the slits and the observations of the screen be superimposed? Yes, but this would not have any specific physical meaning. For this we can take the superposed calculated results - meaning, not really but presumably calculated, we may remember - and compare them with superposed observed results. Now, says Bohr, there is no such physical superposition to compare with any theoretical superposition of this sort. This, in essence, is Bohr's reply to Einstein, Podolsky,

and Rosen. It always strikes me as odd that Bohr has elucidated his
reply to the old objection of the two slit experiment by discussing a
new one. But these things happen.

Why should it be presumed impossible to observe a particle twice without interfering with its business, which is the business of contribut-
ing to an interference pattern? The answer was, this will violate the
Heisenberg principle. This answer may be true, but it does not belong
to quantum mechanics: Bohr and Dirac declared the question outside
quantum mechanics, and all those who read it statistically must do so
too. And if the question does not belong, then the answer to it does
not belong either. And the Heisenberg formula, being statistical, cannot stop us from observing one particle's trajectory, even if Nature
does stop us from doing so. But suppose we cannot observe a particle's
trajectory. Why can we not envisage it? After all, we cannot see a
classical wave front or wave train, yet we can and do envisage it. The
quantum paradoxes derive, as Bohr was the one to tirelessly emphasize,
from our wish to envisage electrons either as particles or as waves,
which can be done only partially and complementarily. No one contests
this point of Bohr. The question is, what is the proper quantum description or envision of the electron's trajectory? Bohr says, it is
complementary. Others say, it does not exist. Here is the dispute. If
quantum theory is statistical and the quantum paradoxes not, then the
two do not impinge on each other, and complementarity is redundant at
best.

Quantum mechanics has no trajectories of individual electrons. This is
by no means paradoxical or disastrous, nor even unusual. New theories
often replace answers given by older theories, but often they only reject old answers without replacing them. Spontaneous emissions, quantum
jumps, relativistic simultaneity, even action-at-a-distance, all these
leave gaps in previously filled cases. The gaps are at times permitted
to be filled by additional assumption, at times not - *e.g.* yes in the
case of action-at-a-distance and no in the case of simultaneity. In
1927 Heisenberg tried to show the impossibility of quantum trajectory
by the use of his microscope thought experiment. And in 1928 Bohr added
that what can never be empirically decided should be left outside science
for good. Yet Heisenberg's microscope thought experiment employed not
quantum electrons but the arbitrary mixture of classical wave and particle presentations of it. And Bohr confused what is outside the domain
of quantum mechanics with what is outside science at large - a confusion
known in the philosophical jargon as hypostatization. Clearly von Neumann

tried to prove the impossibility of hidden variables in 1932 because
he was not fully satisfied with Heisenberg's and Bohr's discussions.
Yet hidden variables proved possible even if admittedly hideous. Mean-
while, heavy beam microscopes either had the Heisenberg-Bohr claim that
we can never see atoms refuted, or shown it too vague for a proper
debate. With this, much of the force of Bohr's thought experiments was
gone.[4]

The question is, then, what is the proper domain of the quantum paradoxes? Whom should they worry and why? But first we may better ask, do
they exist at all, and what makes them problematic? This, I contend,
is a problem for the inter-theoretical considerations.

4.   Inter-theoretical

Let me explain briefly how a theory should relate to its predecessor.
I speak of the general requirement, due to Einstein and well formulated
by Popper, that a new theory replacing a once-successful theory should
yield its predecessor as an approximation or as a special case and preferably both. The domain of approximation is that where the new theory
refines results of the old theory. The domain where the old is a special
case only is that where the new theory introduces new parameters, and
new domains of facts to explain and predict. Once a theory does that,
a crucial experiment between the old and the new takes place and if
the new succeeds it is here to stay even though we expect it to be
superseded.

The previous paragraph included claims that will be taken as self-under-
stood by some people, such as Einstein or Schrödinger, Popper or Bunge,
and as obviously unsatisfactory and misleading at best by others, such
as Poincaré and Duhem, Heisenberg, Weisskopf, perhaps also Bohr.[5]   I
know many say there is no crucial experiment in science; I think the
presence of crucial experiments is an obvious fact and will not discuss
all this here. I did it in many other works of mine and I always find
the discussion on whether crucial experiments exist rather tedious:
what else was Eddington's observation?

Let me repeat: the idea is of a requirement from a theory; not a matter
of historical fact. This means that when the theory in question does
not fulfill the requirement we may well be dissatisfied with it. How
then, if at all, does quantum mechanics conform to this requirement?

Historically, relating quanta to classical matters always was a troublesome affair. In 1900 Planck postulated that when his constant was viewed
as zero classical theory came back intact. Einstein in 1905 saw this

as defective; he offered a different rule of approximation: when a field is very weak it appears quantized, but when strong enough it is classical. In 1913 Bohr came with a different rule: distances from the nucleus that are small enough (compared with Bohr's radius) are the domain of quantum orbits; larger distances permit electrons to behave classically. In 1918 His celebrated correspondence principle closed the gap a bit: even for quantum orbits classical calculations for the intensities of spectral lines may be good approximations. It was clearly all in a fluid state. In 1924 Bohr, Kramers, and Slater published a paper offering almost nothing but a new revolutionary rule of approximation: energy is conserved only statistically. And they referred to Einstein as their inspiration. (The rule was refuted at once in one of the best known experiments. The importance of this episode for the history of physics is thus very great and is constantly underestimated because foundationist bias[6] make us overlook refuted ideas.) With the advent of classical wave and matrix mechanics things became less clear, and even Schrödinger's proof of their overlap did not help. Soon Dirac showed that quantum fields and double quantization were the same. The situation became increasingly exciting and people awaited a picture to emerge. After about fifty years it is still open enough to be debated.

Perhaps a word of caution is in order here. The question, what are the rules of approximation between quanta and classic theories are fundamentally different from the question, by what rules classic equation become quantized. The question how to quantize refers to future new theories and their expected explanatory function. The question of approximation pertains to the old theories and their explanatory success. The question was, as with many cases, publicly suppressed yet privately studied. One root of the trouble is the lack of a clear identification of the canonical theory. There are many specific theories with quantum characteristics, yielding bits and pieces of classical theories as approximations. Now that a canonic version seems to be emerging, the question can be repeated. Taking quantum mechanics to be quantum field theory (including a variant of Dirac's equation, quantum statistics, weak and strong interaction), how much of the success of the classical theories is explained? There is no need to go into the failures of classical theories here, especially since there is no question about the superiority of quantum mechanics each time it competes with classical mechanics. The question is not, what is included, but rather, is anything left out?

Quantum field theory is a theory of both scatter and interactions by creation and annihilation of particles and virtual particles. It includes

the theory of electrons in their orbits, and of the propagation of free
particles, including photons and electrons. As to scatter, clearly it
is a peculiarly quantum mechanical effect, and, considering Compton's
effect, it is naturally relativistic - it belongs to quantum field
theory. The only scatter that may be left out - I cannot say - is the
Rayleigh dispersion so-called, which is classical, and so should follow
from a quantum formula. The picture is much more problematic with the
accelerated propagation of classical particles in accord with Lorentz
force. Here we have classical trajectories of electrons and if there is
no quantum trajectory at all, precise or not, then the classical electron trajectory is not covered by quantum theory. Yet quantum theoreti-
cians have no hesitance in using Lorentz force, in J. J. Thomson or
cathode ray tubes such as oscillographs and boob tubes, in accelerators,
such as cyclotrones and linear accelerators, in tracking down charged
particles in Wilson chambers and in bubble chambers, and, most distur-
bingly, in plasma physics in general.

We may remember that Weizsäcker has declared classical particle states
to be derivative of quantum statistics. Does this include trajectories?
If yes, how? Nor is it hard to see the connection between this and the
quantum paradoxes: the J. J. Thomson electron has a path, and the electron's path leads to the quantum paradoxes. Heisenberg, in his debate
of his uncertainty principle, tried to soften the path by making it im-
precise. This very imprecision may very well do the job of covering
the J. J. Thomson path as a good approximation. Yet it is precisely
this kind of Heisenbergian fuzzy path that is hit by the Einstein, Podolsky, and Rosen paradox. If quantum theory is statistical it is not
hit by the paradox and it fails to account for the J. J. Thomson path.
If it does account for it then it grants the electron its path and then
it is possibly hit by that special paradox. It is a clear choice; and
both options seem unpleasant.

This seems to me to be the best way to present the quantum paradoxes,
in the abstract and pertaining to approximation rules. Thus, the two
slit experiment is not a matter to envisage intuitively but the claim
that there is a possible experimental arrangement with conflicting results: by Einstein's approximation rule (weak fields are quantum mechanical, strong ones are classical) it is quantal but by Bohr's rule
(small distances are quantal, large distances classical) it is classical. The Heisenberg microscope, too, uses two different approximations,
the one to wave representation the one to particle representation, of
the same case. Schwinger claims, perhaps, that slow electrons follow
classical orbits, but his presentation admittedly fails because of the

classical difficulty: the accelerated electron should radiate. In classical considerations of an electron we overlook the problem of radiation. In quantum field theory, where emission and absorption are the means of describing interactions, the situation seems to me to be seriously troublesome.[7] This is another way to say, what could be said with no reference to any special version of quantum mechanics: returning to Weizsäcker's point, we can say, there is no difficulty to imagine a quantum wave looking fairly much like a classical wave in limiting cases or a cluster of quantum states looking like a classical particle state in limiting cases. This will not do: we have a classical path but not a cluster of quantum paths. Why? Can these be supplemented?

I shall turn to this question presently. Let me notice, however, that quantum theory does become much easier to comprehend - even to intuit - and much simpler, once we reject both the wave and the particle presentation. This does not mean that the two presentations are taboo: even the pre-Copernican sunrise is not taboo, we remember, much less classical waves and particles. But the quantum thing, the quanton, as Bunge calls it, is simply neither.

And thus the proposal of the present essay is to reverse the concern about relations between classical and quantum mechanics: a part of the old theory which covers a part of the new theory may thereby win survival value, but this is unproblematic. The problematic is whether the new theory covers all the valuable parts of the old theory, so that the replacement incurs no loss. Yet to this effect we should be able to consider the new theory on its own. And here is another problem lurking: what do the Planck and de Broglie formulas say? Seemingly they are translation rules between the wave presentation and the particle presentation. But admitting this takes us back to the wave-particle duality of the old quantum theory, and to the demand that the (semi-)classical theories cover quantum theory and to the recognition of the partiality of this coverage and the complementarity principle as an attempt to make the best of a bad job. What else than can the Planck and de Broglie formulas mean?

De Broglie insisted all his life: these formulas quantize wave packets. Schrödinger insisted, they are derivative from quantum resonance laws which say that energy exchanges are in whole multiples of Planck's constant. Schrödinger could not account for the seeming localization of the quanton and was always bothered by it. He tended to accept de Broglie's view of it as a wave packet. But the wave packet should be stable. De Broglie hoped to discover stable wave packets, perhaps as solutions to nonlinear equations, preferably derived from equations linear in the

space of general relativity and made nonlinear by transfer to a localized space of special relativity. And, no doubt, had there been any
equations offering as solutions in addition to the current solutions,
also wave packets, some stable some not, with agreement with known facts,
what a joy that would be. Moreover, we do have the phonon, which is a
bundle of elastic energy that fails to dissipate due to quantum restrictions that prevent it from splitting. It is possible to look at the
rules for quarks as such restrictions thereby explaining their contain-
ment and perhaps also allowing them to appear as series analogous to
the spectral series of the early quantum theory that were explained by
the early quantum restrictions. If quantons were quantum wave packets
then their interference and localization would be intuitively comprehended with ease, though they would not be anything like classical waves
since the cause of their stability would be quantum selection or exclusion rules. Not only that. Classical optics postulates ad hoc the
requirement that diffraction grids be of the order of magnitude of the
wave length of the diffracting waves. Were quantons wave packets, this
would then be a most obvious requirement and the causal anomaly debate
about the particle moving through one slit but being influenced by
another would then be seen as a gross exaggeration.

The aim of the last paragraph is not to advocate these speculations.
Nor is it to lull the sense of discomfort by reference to possible
speculative resolutions of a difficulty. My aim in presenting these
possible resolutions is precisely the opposite. By showing such speculations as appealing we also indicate the current troubles, the difficulties that these speculative possible solutions might repair, though
not very likely. By showing what speculation may be effective we learn
what it is that we are after. And once we notice what is our program
we may also rehabilitate Einstein's program, though only as a program
which one may but need not endorse, and with no deterministic basis or
pretension to it.

5.   Einstein's Program

Einstein never objected to quantum mechanics, contrary to almost every
physicist's understanding of what he said: quantum theory in Born's
interpretation, as a statistical theory, is quite satisfactory. Yet,
he added, as statistical theory it wants supplementation of a particle
mechanics. Briefly he wanted an X that would stand to quantum mechanics roughly as classical particle dynamics stands to classical
statistical mechanics. I say roughly, because Einstein did not want
total differential equations, only partial ones with proper boundary

conditions and quasi-singularities, *i.e.*, a field theory with particle-like parts having proper trajectories. Obviously, what formal apparatus involved should not be prejudged. And, obviously, contrary to Einstein's (deterministic) view, we need not insist on a classical trajectory: all we can suggest is the program of seeking quantons whose propagation should be so described or envisaged as to yield classical - forced and free - propagations of particles as special cases and as approximations to the propagations of quantons. It is easy to see that classical propagations cannot be the general case, since quantum scatter and creation-annihilation processes are there. But we may want to overcome the absence of the special case if it is indeed the case that there are no quantum mechanical formulas yielding classical accelerated charges particles as approximate.

All this is different from the way Einstein discussed the situation, but only because his discussions were couched in a deterministic framework while arguing against opponents who rejected realism. He could not, then, go into further detail, particularly because his idea of the limits of quanta as he held in 1905 was erroneous and yet not replaced to his satisfaction by any other. And for Einstein, clearly, size was of no import. His astonishing ideas about statistics were the application of statistics to visible particles that exhibit Brownian motion and to galaxies. His induced radiation theory has as its major step increasing temperatures and radiation intensity beyond any limits.[8] The only experiment carrying his name, the Einstein-de-Haas effect, exhibits the micro-particle spin as a macro-phenomenon. His drawing attention to Bose's statistic and to the Davisson and Germer experiment with material waves, again, have to do with the recognition of the applicability of quantum ideas to broader domains of heavy particles. And he noticed at once that Bose's statistics deprive the proton of its sharp edges. In his "Replies to Criticisms" he noticed that were it possible to slow down billiard balls far enough to make their de Broglie wave-lengths comparable to their radii and hit a grid they would exhibit an interference pattern. The experiment would last as long a time as the age of the universe, but it proves that the two-slit experiment holds for billiard balls no less than for photons. This is a logical fact, and so the unperformability of the experiment it describes is immaterial.

It seems that Einstein conceded here too much to Bohr. It is clear that there are two different aspects of quanta here that Bohr took together and Einstein conceded. The Einstein-Podolsky-Rosen paradox is not a quantum paradox. It does not put into question quantum mechanics proper, whether as understood by Einstein or by Weizsäcker:

it was meant as a paradox for those discussing quantum mechanics for individual particles, not quantum mechanics as statistics. Not so with the soft edge of the proton, with the phenomena that exhibit properties that are neither wave-like nor corpuscle-like (since the Bose-Einstein condensation is a statistical matter). There is little difficulty having something that is not quite classical, and the difficulty is of the sort of having a simultaneity not quite classical, gravity not quite classical, etc., etc. The difficulty with quanta is greater: we know when relativistic simultaneity looks Galilean, when Einsteinian gravity looks Galilean or even Aristotelian. We do not quite know when quantum electrons become classical Lorentz or J. J. Thomson ones. If they do not, then we want a theory unifying the two. If they do, then all we need notice, as Weizsäcker rightly points out, is that in the paradoxically seeming experiments the paradox vanishes upon the observation that the electron is not supposed to act classically but quantum mechanically. But we do not even know what exactly this is. Einstein's program of having a quantum theory of single particle propagation should be useful, and allowing the theory to be statistical rather than deterministic may be the modification it wants, and is what followers of the statistical reading of quantum theory, Weizsäcker for example, should support. But it seems to be more of a program then physicists are willing to admit.

Notes

*   Paper written while an Alexander von Humboldt senior fellow resident at the Zentrum für Interdisziplinäre Forschung, Universität Bielefeld, and read at the Einstein Symposium, Berlin, on the 28th March, 1979. Professor E. Scheibe has read the final version.

1.   Pauli was not interested in the question whether classical physics is or is not deterministic or even causal, since he was convinced that the future of physics lies in a still less deterministic region. He ended his editorial introduction to the 1948 *Dialaectica* issue devoted to the philosophy of quantum mechanics saying (p. 331), "We are here in the very beginning of a new development of physics which will certainly lead to still further generalising revisions of the ideals underlying the particular description of nature which we today call the classical one." By contradistinction, Einstein and Bohr both appreciated the importance for their debate of the question, is classical physics deterministic? This was shown in their discussion with Karl Popper, in Princeton, after his lecture there on the topic, whose content was published in 1950. (See Bibliography.) I have refuted Popper's argument to my own satisfaction in a paper read in the Fifth International Conference on Logic, Methodology, and Philosophy of Science in London, Ontario. This, however, does not detract from the

importance of his claim and his approach: the argument was
simply replaced by one developed by Landé and Popper later on.

2.   The ontological status of Feynmann's virtual particles is con-
     tested among physicists. The naive scientific realism as ad-
     vocated here should take their existence as a matter of course.
     Yet it is far from clear what real existence is there to vir-
     tual existence. That a proton is virtually a neutron plus is
     a fact, yet there is no virtual neutron analogous to a virtual
     pion in the theory; but then, had one found use for it perhaps
     it would have been brought into action. What virtual particles
     do is reduce interaction between particle and field into that
     between particle and virtual particle. Hence virtual particle
     is field action under quantum constraints, which, the theory
     postulates, may be released as a particle proper. The fact
     that the virtual particle has this dual role all the way is
     what distinguishes it, and gives it more reality, than the
     virtual neutron that the proton contains in it. All this is
     acceptable to all parties within quantum field theory and needs
     further separate eleaboration within each of the different sub-
     theories.

3.   Hans Reichenbach, in his *Philosophical Foundations of Quantum Me-
     chanics*, presents the quantum paradoxes as the wave-particle
     duality discussed by Einstein early in the days of quanta, and
     observes that Heisenberg's principle stops us short of checking
     the electron for wave or particle nature. He concluded that we
     should confine the theory to the observed facts and thus have
     no problem left. And his proposal does work. He views the prob-
     lem, especially of the two-slit experiment, as that of a causal
     anomaly. And he abolishes all causality. Quite generally, Popper
     has observed (see his 1963 book), depriving a theory of its
     realistic pretence solves all its problems. It is like the use
     of strong poison as medicine on the true ground that it stops
     all complaint and all ailment. All that remains, then, is a
     rigid corpse instead of live science. Yet some do like their
     science dead. Strange but true.

     As Bunge has noticed, the positivistic fashions of the day left
     their impact on the early literature of quantum mechanics.
     Even the vague and useless term "observable" testifies to that.
     Clearly, not all observables are observable - not even in prin-
     ciple, *e.g.*, ground levels, which the new quantum theory but
     not the old one makes different from zero. Nor are all quantum
     transitions observable - at least not adiabatic ones, *i.e.*
     those between states of equal energy levels. This may be dis-
     missed as irrelevant on the ground that quantum statistics
     does not distinguish two such states and Pauli's principle may
     even identify them. And this seems a victory for positivism,
     a profit accrued from the use of Occum's razor. Yet the in-
     applicability of Pauli's principle to bosons may suffice to
     cheat positivism of its victory: why are there more bosons
     but not more fermions with the same quantum characteristics?
     Postitivist may say, this is an empirical matter: we can count
     quantum particles with the same characteristics but not dis-
     tinguish between them, and so quantum statistics takes account
     of the number but not of the combination of bosons in the same
     state, whereas we can distinguish large particles and so clas-
     sical statistics does take account of their combinations. If
     this were true, then the limit between classical and quantum
     statistics will depend on our tools of observation! Moreover,
     Pauli's principle identifies two particles with the same

quantum characteristics, not two with the same energy level,
such as the two electrons in an orthohelium atom who can exchange spins with no loss or gain of energy; the two states,
the one before and the one after the isoenergetic transitions
are quantum statistically indistinguishable yet quantum mechanically not identical. Hence quantum theory does not endorse the Leibnizian principle of identity of indescernibles,
contrary to what many textbooks say in the twilight of positivistic fervor.

There is still the question, why is quantum statistics so
different from classical statistics? This question is very
different from the question handled in the present essay since,
no doubt, quantum statistics does yield classical statistics
as a good approximation. Hence the question is not a matter of
methodology but of metaphysics and pertains to the fact that,
precise numerical values aside, we prefer to stick to classical statistics on the macro level. Schrödinger has claimed that
on the macro level we would not distinguish between combinations of states, only of things, and hence bosons, and more so
fermions, are not things but states. This idea goes well with
the speculations presented here in the wake of de Broglie and
Schrödinger. But I am not here advocating these speculations
and there may be other explanations for the difference between
classical and quantum statistics. Let me only mention, however,
that Schrödinger used his idea to solve Gibbs' paradox - see
his 1946 book - namely the fact that uniting two containers of
equal pressure and volume of gas will or will not alter the
entropy level depending on whether the gases in the two are
the same or not. What this means is that only quantum statistics, not Boltzmannian one should account for ordinary thermodynamics! This is a far reaching claim. It also amounts to
saying that any two quantum statistically indistinguishable
states are indeed identical (though not for positivistic reasons
but for realistic ones). I cannot judge all this, except that
it sounds suspicious, especially since the two vessels in Gibbs'
paradox are large enough to count as things.

4. The fact that the same quantum thought experiments were viewed so
    very differently by different people is disturbing: it shows
    that the game is played without precise rules. It seems to me
    now, in retrospect, that there is no harm in this, on condition
    that it is made clear. Yet it was not made clear, chiefly since
    sycophants of the Copenhagen school both repressed differences
    of opinions within the school - and even Bohr confessed
    disagreement with Heisenberg only in private conversations -
    and claimed utter victory over Einstein - though Bohr never did
    and always stayed troubled, as it is well-known.

    Yet there must be a limit to the looseness of any rules or
    else the game becomes pointless fast. It seems to me that the
    proper rule is, idealization is either a part of the theory,
    or a supposition that opponents should be invited to contest,
    or concessions to opponents. And possible deviations from such
    rules are better noted during the debate. I think Heisenberg
    violated the rules most, especially when he said, having ob-
    served one particle's position precisely twice may give it
    a trajectory but only in the past, which is uninteresting. In
    essence thought experiments do not sit well with such a cava-
    lier attitude to all that is not predictive. And he said, Bohm's
    hidden variables are untestable and so do not count. This is
    cheating: the game was logical, not empirical, as understood
    by von Neumann and as is still understood by all students of
    hidden variables today, including those who claim to have

tested and refuted the assumption of hidden variables. Their claim, incidentally, is much more restrictive than it sounds.

Bohr's use of general relativity to neutralize Einstein's argument from the weighing of a photon was deemed a tour de force. I never understood why. On the contrary, I found it logically fantastic that such a remote theory should be dragged into the debate. I have discussed this with a number of physicists and found little sensitivity to this. I was fortunate in having an occasion to discuss this with Schrödinger, but he too was unimpressed, saying, if Einstein brought in gravity, Bohr was at liberty to bring in the best theory of gravity available. To my surprise Schrödinger lost patience and would not have my response to this. Karl Popper did me the honour of publishing my views on the matter. (See Bibliography for his 1959 publication.) Max Jammer criticizes my point while implicitly conceding it. (See Bibliography, his 1976 publication.) He says, Einstein's argument can be neutralized without the use of general relativity. Jammer is more concerned with the outcome – neutralizing Einstein's argument – than with the rule.
For me, however, it remains the case that it is not who wins but how the game is played. It is no accident that the game has lost popularity; it can only regain it by making it better played, *i.e.* played more in accord with the rules.

5. It is hard to judge what was Bohr's methodological position, on account of its idiosyncracy, fluidity and notorious difficulty to comprehend. In his contribution to the *Dialectica* issue of 1948 edited by Pauli, "Causality and Complementarity", he said (p. 316), "In presenting a generalization of classical mechanics suited to allow for the existence of the quantum of action quantum mechanics offers a frame sufficiently wide to account [also] for empirical regularities which cannot be comprised in the classical way of description." Putting aside the fact that he viewed the rule of approximation to be, quantum mechanics has classical mechanics for the limiting and special case when Planck's constant is equated with zero, Bohr's view expressed here is the one endorsed in the present essay. Pauli's understanding of Bohr, cited in Note 1 above, is more pronounced an expression of the same view, as he speaks there of "generalizing revisions". Yet, Pauli also endorses there Heisenberg's theory of science as of "closed theories", adding that it accords well with the dialectical view of science. It does not, though it may be viewed, dialectically, as an approximation and a special case, perhaps.

6. I may be overestimating the importance of foundationism. The apologetic oversight of the Bohr-Kramers-Slater theory (observed by B. L. van der Waerden) may be closer to contemporary physics. The denial of strict conservation laws had its import in its glorious denial with the history of the theory of the neutrino: though every conjecture about the neutrino was refuted, the rationale of introducing it, namely the defence of strict conservation laws in the face of evidence from beta decay, was amply empirically vindicated. Nevertheless, and protestations to the contrary notwithstanding, the fact remains: almost all physicists reject strict conservation, and even high-handedly. Mario Bunge is almost the only one explicitly and systematically endorsing it. Others often enough declare the tunnel effect to be an empirical refutation of strict conservation. The view that the law of strict conservation of energy is decidedly violated, but for periods of time short enough to guard the

violation against detection by the Heisenberg time-energy
uncertainty, this view is metaphysical and irrefutable and
unempirical in the extreme, yet it is endorsed unhesitatingly
by most physicists, including those who viciously ridicule
much lesser violations of empiricism.Bunge denies even the
validity of the Heisenberg formula for time and energy. Also
Bohr was consistent here. He said, since energy and momentum
conserve strictly, once a particle is permitted to have precise initial conditions (regardless of our knowledge or ignorance of them) it is thereby doomed to precise path all the
way. At heart, it seems, Bohr was committed to classical physics in its classical interpretation, and he thus found a most
important function for the uncertainty principle: it had to
make room for quanta! But this is no longer the only view open
to us. Once we recognize that both Bohr and Einstein were too
impressed with classical arguments, once we see the tunnel
effect as a violation of strict conservation (regardless of
our assuming that strict conservation holds for energy transfer), then we have to decide again on the large issues, and in
a manner that will decidedly put the heroic Bohr-Einstein debate well into the background. What stops physicists from this
move is their apologetic mood. And the louder one criticizes
them the more apologetic they become. Pity.

7. Julian Schwinger's *Particles, Sources, and Fields*, 1970, intro-
    duces Lorentz force (p. 11) under the strange title of Galilean
    relativity, commenting (p. 12) that the systems described there
    "give a simple description of the behavior of a particle that
    is influenced by a macroscopic, controllable environment."
    Next comes a crucial sentence, quite out of tune with the whole
    volume and its tensor: "Since a classical theory of such inter-
    actions underlies the measurement of free particle properties,
    a test of self-consistency is also involved." What is bother-
    some is that the classical theory underlies the measurements
    without quantum theory underlying it. Schwinger's presentation
    is not clear to me. He derives the Coulomb and Ampère energies
    for the charge and current interactions for very slowly moving
    photon source (p. 77); the emission of the slow electron is
    taken up again later on and the radiation proves to be infinite
    (p. 274), and this impediment is then removed. The overall re-
    sultant picture is not clear to me.

8. See B. L. van der Waerden's thoughtfull book on the sources of
    quantum mechanics for the fact that Einstein's radiation theory
    embarrassed the establishment. It is a historical fact that
    soon after the new quantum theory and quantum field theory were
    established this theory was neglected and not even mentioned
    in many textbooks, not even in those which introduced the topic
    in a historical manner, as is quite ususal. The advent of la-
    sers sent many a physicist back to school to study his Ein-
    steinian A's and B's.

Bibliography

Joseph Agassi, "Between Micro and Macro", *Brit. J. Phil. Sci.*, 14 (1963)
    26 - 31.

   Towards an Historiography of Science, History and Theory, Beiheft
    2; facsimile reprint, Weslyan University Press, Middletown, 1967.

   "The Kirchhoff-Planck Radiation Law", *Science*, 157, April 7, 1967,
    30 - 37.

"The Correspondence Principle Revisited", *Science*, 157, August 18, 1967, 794 - 5.

Faraday as a Natural Philosopher, Chicago University Press, 1971.

"The Interface Between Philosophy and Physics", *Philosophy of Science*, 39, 1972, 263 - 5.

Science in Flux, Reidel, Dordrecht and Boston, 1975.

Y. Aharonov and D. Bohm, "The Significance of Electromagnetic Potentials in Quantum Theory", *Phys. Rev.*, 115, 1959, 485 - 491.

Y. Aharonov, H. Pendelton and A. Peterson, "A Deterministic Quantum Indeterminacy Experiment", *Int. J. Theoretical Physics*, 3, 1970, 443 - 448.

L. E. Ballantine, "The Statistical Interpretation of Quantum Mechanics", *Rev. Mod. Phys.*, 42, 1970, 358 - 81.

Niels Bohr, *Atomic Theory and the Description of Nature*, Cambridge University Press, 1934.

"Can Quantum Mechanical Description of Physical Reality be Considered Complete?", *Phys. Rev.*, 48, 1935, 696 - 702.

"Causality and Complementarity", *Dialectica*, 2, 1948, 312 - 324.

Atomic Physics and Human Knowledge, Cambridge University Press, 1958.

Max Born, *Natural Philosophy of Cause and Chance*, Clarendon Press, Oxford, 1949.

L. Brillouin, *Science and Information Theory*, N.Y., 1956.

Science, Uncertainty and Information, N.Y., 1964.

Mario Bunge, *Foundations of Physics*, Springer, N.Y., 1967.

Scientific Research, Springer, N.Y., 1967.

(ed.), *The Delaware Seminar in the Foundations of Physics*, Springer, N.Y., 1967.

(ed.), *Quantum Theory and Reality*, Springer, N.Y., 1967.

Philosophy of Physics, Reidel, Dordrecht, 1973.

"Quantum Mechanics and Measurement", *Int. J. Quantum Chemistry*, Vol. 12, Sppl. 1, 1977, 1 - 13.

M. Bunge and A. Kálnay, "Welches sind die Besonderheiten der Quantenphysik gegenüber der klassischen Physik?", in R. Haller und J. Götschl, *Philosophie und Physik*, Vieweg, Braunschweig, 1977.

R. G. Chambers, "Shifts of Electron Interference Patterns by Enclosed Magnetic Flux", *Phys. Rev. Letters*, 5, 1960, 3.

L. de Broglie, *Non Linear Wave Mechanics, A Causal Interpretation*, trans. A. J. Knodel and J. C. Miller, Elsevire, Amsterdam, 1960.

Introduction to the Vigier Theory of Elementary Particles, Elsevire, Amsterdam, 1963.

The Current Interpretations of Wave Mechanics, A Critical Study, Elsevire, Amsterdam, 1965.

P. A. M. Dirac, *Principles of Quantum Mechanics*, 4th ed., Clarendon Press, Oxford, 1958.

Albert Einstein, "Autobiographical Notes" and "Replies to Criticisms", in P. A. Schilpp, ed., *Albert Einstein, Philosopher Scientist*, Open Court, La Salle, Ill., 1949.

A. Einstein, B. Podolsky and N. Rosen, "Can Quantum Mechanical Description of Reality be considered Complete?", *Phys. Rev.*, 47, 1935, 777 - 80.

R. P. Feynmann, *The Feynmann Lectures on Physics*, Vol. 3, Addison Wesley, Reading/Mass., 1965.

Michael R. Gardner, "Two Deviant Logics for Quantum Theory: Bohr and Reichenbach", *Brit. J. Phil. Sci.*, 23, 1972, 89 - 109.

Werner Heisenberg, *The Physical Principles of Quantum Theory*, transl., Carl Eckart and Frank C. Hoyt, Chicago University Press, Chicago, 1930; Dover, New York, 1949.

"Der Begriff 'Abgeschlossene Theorie' in der moderne Naturwissenschaft', *Dialectica*, 2, 1948, 331 - 336.

Physics and Philosophy, Harper, New York, 1958, Allen and Unwin, London, 1959.

"The Development of the Interpretation of Quantum Theory" in Wolfgang Pauli, *Bohr Festschrift*, New York, McGraw Hill, 1955, 12 - 29.

Max Jammer, *The Conceptual Development of Quantum Mechanics*, McGraw Hill, N.Y., 1966.

The Philosophy of Quantum Mechanics, John Wiley, N.Y., 1976.

Imre Lakatos, "Infinite Regress and the Foundations of Mathematics", *Arist. Soc. Suppl. Vol.*, 36, 1962, 155 - 84.

Alfred Landé, *Foundations of Quantum Theory*, Yale University Press, New Haven, 1955.

From Dualism to Unity in Quantum Physics, Cambridge University Press, London, 1960.

New Foundations of Quantum Mechanics, Cambridge University Press, London, 1965.

Quantum Mechanics in a New Key, Exposition Press, Jericho, N.Y., 1973.

G. Ludwig, *Wave Mechanics*, Pergamon Press, Oxford, 1968.

"A Theoretical Description of Single Microscopic Systems", in: W. C. Price and S. S. Chiswick, eds., *The Uncertainty Principle and the Foundations of Quantum Mechanics*, John Wiley, N.Y., 1977.

J. von Neumann, *Mathematical Foundations of Quantum Mechanics*, Princeton University Press, Princeton, 1955.

Wolfgang Pauli, Editorial, *Dialectica*, 2, 1948, 307 - 311.

Wolfgang Pauli, *Niels Bohr and the Development of Physics*, Essays dictated to Niels Bohr on the Occasion of his Seventieth Birthday, McGraw Hill, New York, 1955, Pergamon Press, Oxford, 1962.

    Aufsätze und Vorträge über Physik und Erkenntnistheorie, Vieweg, Braunschweig, 1961.

Karl R. Popper, "Indeterminism in Quantum Physics and in Classical Physics", *Brit. J. Phil. Sci.*, 1, 1950, 117 - 33, 173 - 95.

    The Logic of Scientific Discovery, Huthcinsin, London, 1959.

    Conjectures and Refutations, Routledge and Kegan Paul, London, 1963.

    "Quantum Mechanics without the 'Observer'", in M. Bunge, ed., *Quantum Theory and Relativity*, Springer, N.Y., 1967.

W. C. Price and S. S. Chiswick, eds., *The Uncertainty Principle and the Foundations of Quantum Mechanics*, John Wiley, N.Y., 1977.

M. L. G. Readhead, "Wave-Particle Duality", *Brit. J. Phil. Sci.*, 28, 1977, 65 - 80.

Leon Rosenfeld, "On Quantization of Fields", *Nuclear Physics*, 40, 1963, 353 - 6.

    "The Macroscopic Level of Quantum Mechanics", in C. George, I. Prigogin and L. Rosenfeld, *Mathematisk-physiske meddelelsar*, Copenhagen, 1972.

    "Statistical Causality in Atomic Theory", in Y. Elkana, ed., *The Interaction Between Science and Philosophy*, Humanities, N.Y., 1975, 469 - 480.

Mendel Sachs, "A New Theory of Elementary Matter", *Int. J. Theoretical Physics*, 4, 1971, 433 - 51, 453 - 76; 5, 1972, 35 - 53, 161 - 197.

Erwin Schrödinger, *Statistical Thermodynamics*, Cambridge University Press, London 1946.

    "Are There Quantum Jumps?", *Brit. J. Phil. Sci.*, 3, 1953, 109 - 123, 233 - 242.

    (Reprinted in his *What is Life and Other Essays*, Anchor, Doubleday, New York.)

E. Schrödinger, M. Planck and A. H. Lorentz, *Briefe zur Wellenmechanik*, Springer, Wien, 1963.

Julian Schwinger, *Particles and Sources*, Notes by Tung-mow Yan, Gordon and Breach, N.Y., 1969.

    Particles, Sources, and Fields, Addison Wesley, Reading/Mass., 1970.

Abner Shimony, "Metaphysical Problems in the Foundations of Quantum Mechanics", *Int. Phil. Quarterly*, 18, 1978, 3 - 17.

B. L. van der Waerden, *Sources of Quantum Mechanics*, North Holland, Amsterdam, 1967.

Victor Weisskopf, "Niels Bohr", *New York Review of Books*, April, 20, 1967.

C. F. von Weizsäcker, "Probability and Quantum Mechanics", *Brit. J. Phil. Sci.*, 24, 1973, 321 - 337.

THE EINSTEIN-BOHR DEBATE OVER QUANTUM MECHANICS: WHO WAS RIGHT ABOUT WHAT?

Mario Bunge, McGill University, Montreal, Canada

1.   The debate of the century

Einstein's work on Brownian motion, the photoelectric effect, and the specific heat of solids contributed vigorously to the birth of contemporary atomic physics and its central theory, quantum mechanics. So did his active and fascinating discussions with Bohr, Born, de Broglie, Schrödinger and others over two decades at Solvay conferences, in physics journals, and by letter.

However, when the new theory was finally on its feet, Einstein turned against it. What did Einstein have to object to in quantum mechanics? Was his one more case of conservatism in the face of a scientific upheaval or did he have any solid objections? I submit that Einstein did raise genuine difficulties which were unjustly ignored by most physicists; that they have not lost their scientific and philosophic interest, and that they can be met without giving up quantum mechanics.

Einstein leveled criticisms of three kinds against quantum mechanics. In a first batch we find technical objections such as the EPR paradox and the far simpler paradox that a micro-object in a stationary state (*e.g.* an electron in a box) may not move, whereas it should if quantum mechanics were in fact a generalization of classical mechanics. A second set of objections concerned indeterminacy: Einstein could not accept the idea that there are primary or irreducible probabilities -- *i.e.* that chance is objective -- but took probability to be a temporary device hiding our ignorance of causes. Therefore he hoped that quantum mechanics would eventually prove to be derivable from a deeper nonprobabilistic theory. The third set of objections concerned objectivity: Einstein was a realist who wanted physical theories to represent reality rather than facts such as they appear to some observer.

All three objections of Einstein's to quantum mechanics surfaced in his debate with Bohr (Einstein *et al.* 1935, Bohr 1935) and in related publications of his (Einstein 1934, 1936, 1948, 1949, 1950, 1953) and his adversaries (Bohr 1934, 1937, 1948, 1949, Born 1956, Heisenberg 1947, 1958, Pauli 1953, 1961, Rosenfeld 1953) as well as in numerous letters (*e.g.* Born 1971, Przibram ed. 1963).

While most physicists seem to believe that Bohr won the intellectual duel, I happen to believe that neither he nor Einstein reaped a full victory but that each won a round. This paper will argue that Bohr was right in holding that quantum mechanics is probabilistic, has a lasting value, and is unlikely to be replaced by, or derived from, a classical or neoclassical theory. And it will also argue that Einstein was right in demanding that all subjectivistic components be extruded from physical theory.

However, I will also argue that Bohr was wrong in regarding the general schema of quantum mechanics as complete and therefore final, as well as in upholding his own interpretation (the so-called Copenhagen interpretation). And I will submit that Einstein was mistaken in sharing Bohr's belief that that interpretation was unavoidable: in fact it can be replaced with a strictly objectivistic (though not fully causal) interpretation without altering the mathematical formalism. I will also argue that quantum mechanics, though probabilistic, has a causal component. As for the technical difficulties, I will suggest that while some are non-existent others can be circumvented in ways that neither Bohr nor Einstein might have approved of.

2.   Value of quantum mechanics

Bohr and Einstein disagreed on the value of quantum mechanics. Whereas to the former the general theory was complete, hence final, to Einstein it was only a temporary lesser evil: he hoped that one day it would be replaced with a field theory in the style of Maxwell's theory of the electromagnetic field and his own of the gravitational field (Einstein 1934). While recognizing that on the whole quantum mechanics enjoyed a strong empirical support, Einstein gave several reasons for his dissatisfaction with it. One of his reasons was that the theory gave certain paradoxical results.

One of the paradoxes is this. If one computes the quantum mechanical average of the momentum of a "particle" in certain stationary states, such as an atomic "orbit" with zero angular momentum, or an electron in a box, he finds that it is nil. This result contradicts classical mechanics, which ought to hold to a first approximation. Hence there seems to be something wrong with quantum mechanics.

In my view this result is paradoxical only if one views quantum mechanics as a mechanics, hence as containing a kinematics that should approach classical kinematics in some limit (*e.g.* when Planck's constant can be neglected). But quantum mechanics is no such theory: it was not built for solving kinematical problems but for handling atomic spectra, "particle" scatterings, the Compton effect, and the like, none of which

require a knowledge of trajectories. No wonder then that the theory
should fail to yield definite particle trajectories. (In particular,
Dirac's relativistic quantum mechanics is hardly of any use in designing
betatrons, for it does not allow one to compute the trajectory of an
electron in an electromagnetic field.) In sum there is no paradox about
the zero velocity of an electron in certain stationary states provided
one realizes that 'quantum mechanics' is a misnommer, for the theory is
only vaguely analogous to classical mechanics -- but then also to wave
optics.

A more serious objection was the celebrated Einstein-Podolsky-Rosen
(or EPR) paradox (Einstein *et al*. 1935) -- so serious in fact that it
continues to elicit a spate of articles every year. The paradox consists
in the following. If two things interact for a while and then come far
apart, a measurement on one of them seems to affect the result of a
measurement on the other even though they have long ceased to interact.
Einstein, Pcdolsky and Rosen (1935) held rightly that this result is not
acceptable, for it contradicts experiment as well as the very idea of
physical independence.

The culprit of the EPR paradox is von Neumann's projection postulate, or
axiom of the collapse of the wave packet (or state function) to an eigenfunction of the operator representing the dynamical variable that is
being measured. Luckily the only use of this postulate is to make trouble: it can be given up without in any way curtailing the explanatory
and predictive power of quantum mechanics (Margenau 1936).

In other words, the EPR paradox dissolves if von Neumann's quantum mechanical theory of measurement is relinquished and the general axioms of
the theory are formulated without reference to measurements (Bunge 1967).
Nothing is thereby sacrificed, as this measurement theory is never used
to design or predict any real experiments, and consequently has never
been put to experimental test. (One may go as far as to assert that the
only purposes that the theory has served are to be the subject of doc-
toral dissertations and a cause of promotions.)

Another objection of Einstein's to quantum mechanics was the apparent
impossibility of joining it with his own theory of gravitation -- which,
incidentally, has proved to be far richer and more accurate than suspec-
ted during Einstein's lifetime. The difficulty is triple. Firstly, general relativity is not self-sufficient (like Maxwell's electrodynamics
and quantum electrodynamics) but must borrow the matter tensor from
other theories. (Likewise Newtonian mechanics is not self-reliant, for
is must get the force law from some other theory.) Secondly, non-rela-

tivistic quantum mechanics does not have a matter tensor, so it cannot
feed the ever hungry theory of gravitation. Thirdly, relativistic quantum mechanics does have such a tensor (namely Tetrodes's) but, if grafted
onto general relativity, the gravitational field potentials turn out to
be probabilistic. (This does not seem to have been done but I can think
of no obstacle to this enterprise.) I submit that the first difficulty
can hardly be blamed on quantum mechanics, that the second is irrelevant, and that the third should not be a cause for concern -- unless of
course one clings to the ideal of a nonprobabilistic theory.

In sum, Einstein's technical objections to quantum mechanics can be met,
though in a manner that would not have satisfied him -- or Bohr. This
does not entail that there is nothing wrong with quantum mechanics and
quantum electrodynamics. To begin with there is much wrong with the usual
interpretation of these theories -- but this too can be remedied in the
manner indicated elsewhere (Bunge 1967, 1973). Then there are well-known
mathematical problems such as the infinities associated with continuous
spectra and those occurring in quantum electrodynamics -- not to mention
virtual particles and virtual processes, and the failure to yield classical (*i.e.* continuum) mechanics in the classical limit.

However, classical physics too is defective, and this was the rationale
for inventing the quantum theories. (Recall among others the runaway
solutions of classical electrodynamics, the electron self-acceleration,
and the wrong solution to the black-body problem.) Not that this could
serve as an excuse for the real shortcomings of the quantum theories, but
it is a reminder that the old times were no better, so adopting a nostal-
gic attitude won't help much. Here as elsewhere progress can be made by
looking ahead rather than by embellishing the past.

3.   Hidden variables

Einstein had employed probabilistic notions in his papers on Brownian
motion (1904, 1905) and on the specific heat of solids (1914). However,
like most other scientists, he disbelieved that chance (or randomness)
is an objective mode of being and becoming. (Remember his famous dictum:
"God does not play with dice", reminiscent of Spinoza's assertion that
nothing in nature is random: that "A thing appears random only through
the incompleteness of our knowledge".)

Einstein held the subjectivistic interpretation of probability inherited
from the 18th century. According to this interpretation, a formula such
as "The probability of fact x equals y", or "$Pr(x) = y$", should be read
as "The degree of rational belief or certainty about fact x equals y".

At that time only a minority sided with the statistical interpretation introduced by Venn and worked out by von Mises, according to which probabilities are definable as long run frequencies of observed events. And still fewer dared assert that the probability of a state or an event is just as objective a property of a physical thing as its size or its energy.

The unspoken and undoubted presupposition of the subjectivistic interpretation of probability is that chance is not objective: that probability is an artifact rather than the quantitation of real possibility. (In short, the ontological presupposition of the subjectivistic interpretation of probability is necessarianism or classical determinism of the Laplacian type.) Only a handful of scientists -- notably Cournot, Peirce, and Smoluchowski, and later on the mathematician Fréchet -- held what we now call the propensity interpretation of probability. According to this view a physical probability is a measure of real possibility on the same footing as electric charge and pressure. (See Fréchet 1946.)

However, objective probability is not in one piece: it may be reducible or irreducible, according as it is, or is not, the result of the encounter of causal lines. Classical physics knew the former, *i.e.* it computed or measured the probability that two independent, or nearly independent, causal lines would meet somewhere or within a certain time interval. Typically, in classical mechanics the individual entities (atoms, Brownian particles, etc.) were assumed to move according to Newtonian mechanics and to be subjected to random disturbances by other equally deterministic entities, either on the same level (as in gases) or on deeper levels (as in Brownian motion). The idea was that, if left alone, every physical entity, whether corpuscular or field-like, would behave deterministically, not probabilistically. (Oddly enough Dirac kept this belief in his famous textbook.) Consequently a really fundamental theory should contain no probabilities: the latter should always be derivable.

As we know, the quantum theory changed all this by introducing irreducible probabilities, or probabilities that, far from resulting from the encounter of nonrandom processes, were primary. According to that theory even a single free electron behaves probabilistically: *i.e.* its states are represented by the values of a probability amplitude, and its dynamical properties (*e.g.* linear momentum, angular momentum, and spin) have probability distributions. Moreover, whereas a classical dynamical variable has a meaning by itself, a quantum-mechanical one does not. Indeed, a quantum-mechanical dynamical variable (or "observable") A must always be taken together with some state function ψ to form a density of the form $\overline{\psi} \, A \, \psi$. Consequently in general the mean standard deviation of A won't be nil -- *i.e.* the values of A will spread.

To be sure, for a microthing in an eigenstate of A, *i.e.* such that $\psi \in eif \, A$, the mean standard deviation of A will be nought -- but in return for this definiteness the spread of the canonical conjugate variable will be infinite. So, the occasional sharpness in A is paid for by a total smudging in its canonical conjugate. And such inderminacy is not a result of incomplete knowledge: rather on the contrary, "complete" knowledge is impossible because of the objective blurredness or indeterminacy. Better: there *is* nothing to be known about a dynamical variable that is thoroughly blurred, such as the position of an electron with a precise momentum value.

Einstein could not accept this situation: he regarded quantum mechanics as a statistical theory, not as one concerning individual things. And, since for him statistics was just a stop-gap, he regarded quantum mechanics as a temporary expedient utilizable as long as a deterministic theory did not come along to take its place. (In this regard he shared Planck's viewpoint: see Planck 1933.)

In 1951 Einstein summoned a brilliant theoretical physicist teaching at nearby Princeton University who had just published what Einstein regarded as the best exposition of the worst formulation of quantum mechanics, namely the Copenhagen (Bohm 1951). The old master explained to the young scientist what his misgivings about quantum mechanics were, and encouraged him to try and find an alternative. Bohm responded quickly by enriching the theory with a couple of "hidden variables", *i.e.* dynamical variables with zero statistical spread (Bohm 1952).

The additional (hidden) variables were a position and a momentum. The latter was defined by "$p = \nabla S$", where S is the phase of the state function. By integrating this equation one obtains the particle trajectory. Since the particle is now attributed both a definite (hidden) position and a definite (hidden) momentum at every instant, the Heisenberg inequalites do not hold for them. But the trajectory turns out to be rather wild even in the absence of external forces. This result was explained as the effect of the ψ-field on the particle.

There are several problems with this enrichment of quantum mechanics. One is that it does not eliminate any of the features of the theory that the determinist regards as objectionable: in fact the old theory becomes embedded in the new one. Another is that the ψ-field is regarded as real yet not as part of the particle but as external to it: so much so that it exerts a (quantum) force on the particle. A third difficulty is that it forces one to regard every quantum mechanical object, even a complex biomolecule, as a point particle with a zero-width trajectory -- which is certainly not a classical feature. A fourth difficulty is that the new ideas cannot be carried over to quantum field theory, where the notion of trajectory makes little sense (except of course in the case of a radiation field as the analog of the eiconal). Fifthly and perhaps most decisively, the new theory predicts just the same facts as the old one, so whatever confirms or weakens the latter confirms or weakens the former. For these and possibly further reasons the search for hidden variables was not pursued with enthusiasm. (See however de la Peña and Cetto 1977, and Claverie and Diner 1977, for some of the recent work on the so-called stochastic quantum theory, based on classical electrodynamics.)

The search for hidden variables is well-nigh over. Yet the adventure did yield two results. One was to show that, while von Neumann was right in claiming that there are no hidden variables in quantum mechanics, he was wrong in suggesting that none could be introduced therein. Another was to draw the attention of many physicists and philosophers to the extreme dogmatism with which the usual version of quantum mechanics had been defended, and to cause many to wonder whether the Copenhagen interpretation was ineed inevitable. In particular, the problem of indeterminacy could now be posed anew.

4.   Determinacy

Strictly speaking, the Copenhagen school did not embrace indeterminism: it did not state that nature is random, and this for the simple reason that it held that quantum mechanics is not about nature. Indeed, Bohr (1937), Heisenberg (1947), Frank (1936) and others explained carefully -- but did not prove -- that the theory makes no assertions concerning autonomous, *i.e.* observer-independent, things: that all its statements are about experimental situations. (This is why Bohr, and initially also Rosenfeld, stated that no special theory of measurement was necessary: they believed that quantum mechanics was already a theory of measurement.)

However, experimental situations (what Bohr called 'phenomena') are under the (direct or indirect) control of an experimenter, and the latter is supposed to make his decisions freely. Hence one could argue that the Copenhagen school is indeterministic after all, for it claims that quantum mechanics concerns phenomena freely "conjured up" by the experimenter (Rosenfeld 1953). Still, beyond the smoke curtain of such philosophical declarations one can see the quantum theoretician generating or applying formulas representing things -- electrons, photons, nuclei, atoms, molecules, etc. -- that are not being molested in any way by experimental devices. In particular, the calculations of the eigenvalues and eigenfunctions of a dynamical variable (such as the energy) of an atom or a molecule do not presuppose that a measuring device is exerting any influence on the system.

Regardless of the sophisticated position that the Copenhagen school adopted towards the problem of determinism -- namely that it was a metaphysical question -- the quantum theory is usually taken to be indeterministic. Is it? This is a loaded question, for the term 'determinism' is ambiguous. If indeterminism is equated with probabilism, then quantum mechanics is obviously indeterministic and basically so because probabilities (or rather probability distributions) occur in it as fundamental or irreducible. But since quantum-mechanical probability may (wrongly) be interpreted in a subjectivistic manner, the previous answer is unsatisfactory: it does not tell us whether nature itself is chancy. To obtain a sensible answer to the original question we should adopt an objectivist interpretation of quantum-mechanical probabilities (namely as real dispositions or propensities) and should refine the concept of determinacy to include probabilistic laws.

More precisely, it seems convenient to demand of every form of determinism that it abide only by the following principles (Bunge 1959): (a) *lawfulness* ("Every event is lawful, none is chaotic") and (b) *transformationism* or *non-magic* ("Nothing comes out of nothing or goes into nothing: every thing and every state of a thing have precursors and leave descendants"). If this redefinition of 'determinism' is accepted, quantum mechanics proves to be deterministic -- though not fully causal. More precisely, one may argue (Bunge 1977) that quantum mechanics is half-stochastic and half-causal.

Consider an arbitrary eigenvalue equation of the form "$A_{op}u_k = a_k u_k$" for a dynamical variable ("observable") A represented by an operator A_{op} in the state space of a physical thing ("particle", field, or what have you). Assume, for the sake of simplicity, that all the eigenvalues a_k are non-degenerate. According to the Copenhagen interpretation, a_k is one of the values that an observer is bound to find when measuring A with a suitable instrument -- never mind which one. Einstein did not challenge this interpretation and so was unhappy with quantum mechanics, which seemed concerned with human operations rather than with nature.

However, the given formula can be reinterpreted in a manner consistent with Einstein's realistic epistemology, by observing that it makes no reference to any measuring devices, let alone their readings by an observer. Indeed, the formula is assumed to hold for any of the dynamical variables of a nondescript quantum-mechanical object. Moreover it concerns such an object rather than a thing composed of it and some macroobject. Therefore we should interpret a_k as one of the really possible values of A, whether or not somebody happens to measure A. As for the eigenfunctions u_k of A_{op}, they may be regarded as mathematical auxiliaries serving only to compute the probability that the A of the microobject has value a_k when in a given state ψ. (To compute this probability expand the state function in eigenfunctions of A_{op}. The probability that A takes on value a_k equals the square of the k-th coefficient in the expansion.)

The previously mentioned law, or rather law schema, is definitely stochastic for, in conjunction with the state function, iit yields the probability distribution of each dynamical variable A. However, such probability distributions are not unchangeable: they can be influenced by external circumstances such as obstacles (represented by boundary conditions), fields, etc. Changing external circumstances will distort the original probability distribution, now in one direction, now in another: there will be a shift of the center towards lower or higher values, or a skewness will be introduced (or eliminated), or a flattening (or bunching up) will happen, and so on.

In other words, the quantum-mechanical object is not immune to causal agencies. And such influences are occasionally so strong that the distribution of certain dynamical variables becomes peaked, so that in practice they degenerate to a single value at a time. (Think of the instantaneous position distribution of a charged particle in a high energy accelerator.) So, the eigenvalue equations are a mixture of causality and randomness (lawful chance).

What holds for an arbitrary dynamical variable holds of course for the energy operator or hamiltonian H. Taken by itself H is nonprobabilistic and in general it contains potentials from which the force acting on the thing concerned can be derived. (These forces may be internal to the system, as in the case of an electron subjected to electromagnetic radiation.)

Quantum mechanics, in short, does not ignore the notion of a force, which quantitates that of cause. But it does modify the classical concept of force, for according to it same forces (causes) need not have same effects. Indeed, one and the same force may have different effects, though each with a fixed probability given by the law of evolution of states.

I.e. quantum mechanics retains the concepts of cause and effect but modifies their relation, which is now probabilistic rather than causal. (For example, if an electron collides with another "particle", it has a definite probability of being scattered within a given solid angle.) It is only when no causes are acting, as is the case with the free particle and the free photon, that the entity concerned evolves by itself in a fully random fashion -- though of course lawfully and keeping intact all of its constants of motion (notably its momentum, total angular momentum, and energy).

In sum, when interpreted in realistic terms quantum mechanics is seen to be deterministic in an important large sense. Moreover it is seen to contain important causal ingredients, namely interactions and environmental agents. Unfortunately considerations such as these played no role in the Einstein-Bohr debate, which was moreover marred by the confusions of blurredness (or absence of sharp value) with indeterminacy, and the latter with uncertainty.

5. Objectivity

Bohr (1934, 1937, 1948, 1949, 1958) held repeatedly that quantum mechanics had produced an epistemological revolution, namely the replacement of the classical picture of a subject-free nature with a new vision centered on the observer. (In short, whereas classical physicists depicted still lifes, quantum physicists would paint self-portraits.) At one time Heisenberg (1947, 1958) endorsed this view holding that, whereas classical physicists strived to discover laws of nature, quantum physicists talked about such laws from the point of view of the active observer. (In his later years he gave up this anthropomorphic world view and adopted a sort of Platonic objective idealism: Heisenberg 1969.) And von Neumann (1932) stated that the demarcations between micro-object, measuring device and observer were arbitrary, since the experimenter had to be counted in anyway. In short, the Copenhagen school rejected the traditional distinction between subject and object and moreover preached a return to anthropocentrism.

Einstein believed this to be the case and, for this very reason, could not accept quantum mechanics with a light heart. Presumably to him, as to many other scientists brought up in the scientific tradition and not much swayed by the fashionable positivistic philosophy of the day, the quantum physicists had betrayed the glorious legacy of Galilei, Kepler, Huyghens, Newton, Euler, Faraday, Maxwell, Helmholtz, Boltzmann, Lorentz, Kelvin, and Rutherford -- all of whom had contributed to the collapse of the age-old anthropocentric world view.

Einstein tried to undermine the semisubjectivistic interpretation of quantum mechanics by devising ingenious imaginary experiments that purported to show that the theory failed at certain crucial places. Bohr replied by inventing further no less ingenious gedankenexperiments purporting to show that quantum mechanics was demanded by experiment. With hindsight one cannot help smile at this tournament and compare it to the rivalry between those eminent cyberneticians Trurl and Klapaucius in Stanislaw Lem's *Cyberiad:* theirs, too, were imaginary contraptions.

One of the devices invented by Bohr was purported to illustrate his own "fourth indeterminancy relation", which would hold between energy and time. But of course it did nothing of the sort, because in quantum mechanics time is a "c number", hence one with zero spread. (That is, the quantum-mechanical average of t equals t , hence the mean standard deviation of t is exactly nil for all possible states of a quantum-mechanical entity.) Nor did any of the other devices prove anything, for gedankenexperiments have at most a heuristic value and no proving virtue whatsoever.

Unfortunately neither Einstein nor Bohr seems to have realized how pointless it was to continue along this course. Nor did Einstein realize that he had been lured into the positivist trap when trying to find in experiment the meaning of theoretical formulas. (Experiments may test formulas, not endow them with meaning. Meaning is prior to test: nobody would know how to go about putting a meaningless formula to the test.)

Curiously enough, Einstein might have found an ally in von Neumann, whose influential book (1932) was wrongly assumed to support the Copenhagen school. This assumption was false because von Neumann stated explicitly that the theory centered on Schrödinger's equation (or some equivalent of it) concerns things that fail to be under experimental control. So, Einstein's apprehensions should have been allayed: here was an eminent practitioner of quantum mechanics who parted company with Bohr, Born, Heisenberg, Pauli and the other defenders of the semisubjectivistic interpretation of the theory. But somehow nobody seems to have noticed the incompatibility between von Neumann's and Bohr's views.

As soon as the experimenter steps in, von Neumann wrote, a different theory is to be used, namely the measurement theory based on his projection postulate, mentioned in Sec. 2 as responsible for the EPR paradox. Remember the gist of the postulate: Under the action of an A-measuring apparatus, the state function of the system will not evolve according

to the Schrödinger equation but will collapse onto one of the eigenfunctions of the operator representing A . The observer is deemed to be
free to produce the collaps but impotent to guide it·to a definitive
eigenfunction or even a narrow wave packet. Margenau (1963) has argued
persuasively that the postulate may describe the preparation of some
states but not an experiment proper. But here we are interested in the
philosophical aspect of the situation.

The von Neumann formulation of quantum mechanics has been used as an
argument for both indeterminism and subjectivism: for the former because the precise collapse is not predictable, for the latter because
the collapse can be triggered at will by the experimenter. Yet one may
argue the other way around, shifting one's attention from type I processes (measurement, projection postulate) to type II processes (no
measurement, Schrödinger equation), which according to von Neumann himself follow their course without being influenced by any observer. As
for experimental situations, one may try and describe them with the help
of the Schrödinger equation, $i.e.$ just the same as type II processes --
as London and Bauer (1939) did -- or one may argue that no general theory of measurement is possible because there are no all-purpose measurement devices. In either case one can refuse to accept the subjectivistic
ingredient attached to the projection postulate without thereby dimini-
shing the power of quantum mechanics.

But of course rejecting subjectivism is not enough: one must offer a
substitute for it. I submit that this is easily obtained by focusing on
the key concepts and formulas of the quantum theory, forgetting about
the philosophical arguments and the didactic props, and concentrating
instead on the realistic successful applications of the theory. This
procedure should yield an objectivistic interpretation of the theory,
$i.e.$ one exhibiting it as being concerned with physical entities rather
than conscious abservers. (For an axiomatic formulation of elementary
quantum mechanics couched in strictly objective terms see Bunge 1967.)

Einstein might have been pleased with a realistic interpretation of
quantum mechanics but he would have objected to its keeping primary or
irreducible probabilities -- unless he could have been persuaded that
probabilities are not just a cloak of ignorance but are a representative
of part of the very marrow of reality. However, such gedankenexperiments
on ideas are no less misleading than gedankenexperiments on things.

6.   Conclusions

The Einstein-Bohr debate over quantum mechanics touched on both technical and philosophical aspects of the quantum revolution. We have not

examined it in the traditional hermeneutic fashion, which seeks to extract the truth from the words of the discussants. Nor have we attempted to perform an accurate historical reconstruction of the celebrated duel. We have adopted instead the approach consisting in letting the quantum theory speak for itself, *i.e.* in analyzing it with a minimum of philosophical bias in order to ferret the philosophy out of it. Our conclusions are as follows.

(i)   The issues discussed by Einstein and Bohr are *still topical* in the sense that they are still research subjects. However, some of them -- notable the possibility of adding hidden variables and the EPR paradox -- may be regarded as solved.

(ii)   On the whole, *Bohr was right scientifically:* quantum mechanics is a powerful and deep theory not a makeshift, and is unlikely to be superseded by a classical (or neoclassical) theory. But Bohr was wrong in claiming that quantum mechanics is complete and therefore final. First, because -- as we know since Gödel -- no theory containing number theory can be both complete and consistent. Second, because in factual science we do not want complete theories but, instead, theories that can be enriched with hypotheses and data concerning details of the system of interest. Third, because the quantum theory does have some blemishes that will have to be corrected -- such as the divergences and the virtual particles. Fourth, because the Copenhagen interpretation of the theory is the prisoner of an obsolete philosophy and contains unnecessary classical metaphors, in particular those of particle and wave, that are responsible for a number of paradoxes and even contradictions. But then that interpretation is not mandatory.

(iii)   On the whole, *Einstein was right philosophically* in demanding that physical theories represent reality rather than subjective experience. He was also right in noting that the quantum theory presents certain technical problems such as the EPR paradox. Fortunately, though, the latter paradox is avoided by relinquishing the von Neumann theory of measuring centered on the projection postulate (Margenau 1936). Other difficulties, such as the interpretation of perturbation calculations and the Feynman diagrams in terms of virtual particles and virtual processes, can likewise be avoided by abstaining from foisting a physical interpretation on every symbol -- particularly when the interpretation violates basic scientifc principles, such as "Thou shalt not explain existents in terms of nonexistents" (Bunge 1955, 1970). In short, Einstein's realism has emerged unscathed.

(iv)   *Bohr and Einstein were both wrong in the matter of determinism*, for both interpreted the mean standard deviations occuring, *e.g.*, in Heisenberg's inequalities, as indeterminacies and even uncertainties. They would indeed be indeterminacies if quantum-mechanical objects were classical point particles; and they would be uncertainties if the theory actually referred to physicists rather than to physical entities. As it is, the mean standard deviations can be interpreted differently -- *e.g.* as objective latitudes or spreads in the properties of the thing of interest, which would acquire sharp values only in exceptional circumstances. With this shift, the quantum theory is seen not to be indeterministic -- particularly if determinism is not equated with causalism (or Laplacian determinism) but is taken in the minimal sense of lawfulness together with non-magic.

(v)   *The Einstein-Bohr debate was a model* of a passionate yet not acrimonious argument between two men who sought the truth rather than celebrity. (Both had been awarded the Nobel prize before they engaged in the debate.) It was a very instructive debate conducted in reputable publications. Why are there no more debates of that kind in physics? Is is because we have reached consensus about all important scientific-philosophic issues? Because people have no time to waste on scientific debates? Because discussion is being systematically discouraged or even silenced? Whatever the reason, the absence of debate over foundational matters is just as disquieting a sign of decadence as would be the concentration on polemics at the expense of the production of new results.

Bibliography

David Bohm, *Quantum Theory*, New York: Prentice-Hall, 1951

-,   *Phys. Rev.* **85** (1952) 166, 180

Niels Bohr, *Atomic Theory and the Description of Nature*, Cambridge: Cambridge University Press (1934)

-,   *Phys. Rev.* **48** (1935) 696

-,   *Erkenntnis* **6** (1937) 293

-,   *Dialectica* **1** (1948) 312

-,   Discussion with Einstein on epistemological problems in atomic physics, in: Schilpp, ed.   (1949)

-,   *Atomic Physics and Human Knowledge*, New York: John Wiley (1958)

Max Born, *Natural Philosophy of Cause and Chance*, Oxford: Clarendon Press (1949)

—, *Physics in my Generation*, London: Pergamon Press (1956)

—, *The Born-Einstein Letters*, New York: Walker (1971)

Mario Bunge, *Methodos* 7 (1955) 295

—, *Brit. J. Phil. Sci.* 6 (1955) 1, 141

—, *Causality*, Cambridge, Mass.: Harvard Univeristy Press (1959), Rev. ed.: *Causality in Modern Science*, New York, Dover, 1979

—, *Foundations of Physics*, Berlin-Heidelberg-New York, Springer Verlag (1967)

—, The so-called fourth indeterminancy relation, *Can. J. Phys.* 48 (1970) 1410

—, *Intern. J. Theor. Phys.* 3 (1970) 507

—, *Philosophy of Physics*, Dordrecht: Reidel (1973)

—, *Intern. J. Quantum Chem.* XII, Suppl. (1977) 1

P. Claverie and S. Diner, *Intern. J. Quantum Chem.* XII, Suppl. (1977) 41

De la Peña-Auerbach, L. and A. M. Cetto, *Intern. J. Quantum Chem.* XII, Suppl. (1977) 39

Albert Einstein, Preface to Planck (1933)

—, *Mein Weltbild*, Amsterdam: Querido Verlag (1934)

—, *J. Franklin Institute* 221 (1936) 349

—, *Dialectica* 2 (1948) 320

—, Autobiography, in: Schilpp, ed. (1949)

—, *Out of my Later Years*, New York: Philosophical Library (1950)

—, in: *Scientific Papers Presented to Max Born*, New York: Hafner (1953)

Albert Einstein, B. Podolsky and N. Rosen, *Phys. Rev.* 47 (1935) 777

Philipp Frank, *Erkenntnis* 6 (1936) 303

Maurice Fréchet, *Les mathématiques et le concret*, Paris: Presses Universitaires de France (1946)

André George, ed., *Louis de Broglie, physicien et penseur*, Paris: Albin Michel (1953)

Werner Heisenberg, *Wandlungen in den Grundlagen der Naturwissenschaften*, 7th ed. Zürich: Hirzel (1947)

—, *Physics and Philosophy*, New York: Harper & Brothers (1958)

—, *Der Teil und das Ganze*, München: R. Piper (1969)

F. London and E. Bauer, *La théorie de l'observation en mécanique quantique*, Paris: Hermann (1939)

Henry Margenau, *Phys. Rev.* <u>49</u> (1936) 240

-,   *Annals of Phys.* <u>23</u> (1963) 469

Johann von Neumann, *Mathematische Grundlagen der Quantenmechanik*, Berlin: Julius Springer (1932), New York: Dover, 1943

Wolfgang Pauli, in: A. George ed. (1953)

-,   *Aufsätze und Vorträge über Physik und Erkenntnistheorie*, Braunschweig: Vieweg (1961)

Max Planck, *Where is Science Going?* London: George Allan & Unwin (1933)

K. Przibram, *Schrödinger, Planck, Einstein, Lorentz: Briefe zur Wellenmechanik*, Wien: Springer-Verlag (1963)

Leon Rosenfeld, *Science Progress*, No. 163 (1953) 393

Paul A. Schilpp, ed., *Albert Einstein: Philosopher-Scientist*, Evanston, Ill.: The Library of Living Philosophers (1949)

AUF DEM WEG ZU EINER RELATIVISTISCHEN QUANTENFELDTHEORIE

Konrad Osterwalder, Eidgenössische Technische Hochschule, Zürich,
                     Schweiz

I. Einleitung

Die Erfolge der Quantenelektrodynamik und neuerdings auch der nichtabelschen Eichtheorien begründen die weitverbreitete Meinung oder Hoffnung, daß eine relativistische Quantenfeldtheorie schließlich einmal ein
befriedigendes Modell für die Welt der Elementarteilchen abgeben würde.
Viele Leute gehen sogar so weit zu glauben, es handle sich lediglich noch
darum, die "richtige Wechselwirkung" zu finden, um die Elementarteilchen-
physik völlig zu verstehen. Diese Meinung muß uns aber dann als verfrüht
oder als allzu optimistisch erscheinen, wenn wir von einer physikalischen
Theorie letztlich doch erwarten, daß sie ein mathematisch wohl begründe-
tes und konsistentes Modell eines Teils der Erscheinungen darstellt, oder
sich doch durch klar beschreibbare Vereinfachungen oder Näherungen von
einem derartigen Modell ableiten läßt.

Noch heute, mehr als fünfzig Jahre nach Diracs bahnbrechender Arbeit
[1] über das quantisierte elektromagnetische Feld in Wechselwirkung mit
Materie, kann die Quantenfeldtheorie nicht den Anspruch erheben, eine
mathematische Theorie zu sein, jedoch sind die Bemühungen um ein tieferes
Verständnis der mathematischen Probleme quantisierter Wellenfelder viel-
fältig und alt. In diesem Vortrag will ich versuchen, einige Ergebnisse
der vergangenen zehn Jahre zusammenzustellen und kurz zu beschreiben.

Drei einander ergänzende Hauptströmungen kann man unschwer unterscheiden:

Im Rahmen der *Formalen Störungstheorie* wurde das Problem der Renormierung
sehr eingehend analysiert und kann heute als gelöst betrachtet werden.
Als besonderer Durchbruch der siebziger Jahre ist die Renormierung nichtabelscher Eichtheorien zu betrachten. Zusammenfassende Darstellungen von
Methoden und Resultaten samt vollständiger Referenzlisten findet man in
[2,3].

In der *Axiomatischen Quantenfeldtheorie* versucht man, die allgemeinen
Strukturen, die jeder relativistischen Theorie quantisierter Felder zugrunde liegen sollten, herauszuschälen, diese als erfüllt zu postulieren
und daraus möglichst detaillierte Folgerungen zu ziehen, siehe [4,5,6].

In der axiomatischen QFT wird die Frage offen gelassen, ob die postu-
lierten Prinzipien verträglich seien mit der Existenz einer nichttrivi-

alen Streumatrix. Durch Konstruktion expliziter Modelle versucht man
diese Frage zu beantworten in der *Konstruktiven Quantenfeldtheorie*.
Außerdem wird hier auch die detaillierte Struktur der Modelle analysiert,
wobei die physikalisch relevanten Größen wie z.B. Spektren von Energie-
Impuls Operator von besonderem Interesse sind. Anhand der Modelle kann
auch die Rolle der formalen (da meist divergenten) Störungsentwicklung
klargestellt werden. Für Übersichtsartikel und Referenzen siehe z.B.
[7,8,9]. Ich werde nichts über das störungstheoretische Renormierungs-
problem sagen. Von den vielen Resultaten der axiomatischen QFT will ich
nur die herausgreifen, welche mit der euklidischen Formulierung im Zusammenhang stehen. Unser Hauptaugenmerk werden wir auf Methoden und Resultate der konstruktiven Quantenfeldtheorie richten, aber auch hier
werden wir uns mit einer Auswahl begnügen müssen.

Ich werde mich bemühen, den Aufwand an mathematischem Formalismus möglichst klein zu halten und nur die grundlegenden Ideen und Methoden zu
skizzieren, selbst auf das Risiko hin, dadurch die *Grundabsicht* zu ver-
wischen: mittels strenger Mathematik den großen Problemen der Elementar-
teilchenphysik auf den Grund zu gehen. Außerdem werden meine Ausführungen
kaum eine Ahnung geben von dem Reichtum und der Vielfalt der angewandten
mathematischen Methoden, noch von der Komplexität und Schwierigkeit man-
cher Beweise und schon gar nicht von der stimulierenden Wirkung, die
viele der beschriebenen Untersuchungen auf die Mathematik gehabt haben.

II.   Axiomatische Quantenfeldtheorie

In diesem Kapitel will ich versuchen, eine kurze Einführung in die euklidische Formulierung der Quantenfeldtheorie und eine Auswahl von sogenannten Rekonstruktionstheoremen zu geben. Diese Resultate der axiomatischen Quantenfeldtheorie liegen heute praktisch jeder Konstruktion
von Modellen zu Grunde.

Wightman Axiome

Wir beginnen mit einer Aufzählung der grundlegenden Postulate (Axiome)
wie sie von Gårding und Wightman vorgeschlagen worden sind [4,5,6]. Der
Einfachheit halber beschränken wir uns auf ein neutrales Skalarfeld, obschon die hier zitierten Resultate (wie den nötigen aber offensichtli-
chen Modifikationen) auch für beliebig viele Felder mit beliebigem Spin
gelten.

H i l b e r t   R a u m : Der Raum der physikalischen Zustände ist ein
komplexer Hilbert Raum H .

F e l d o p e r a t o r : Auf H operiert ein Feldoperator ϕ , der
eine operatorwertige, temperierte Distribution ist; $\phi(f) = \int \phi(x) f(x) dx$

ist definiert auf einem dichten Bereich D (unabhängig von f) und für beliebige Ψ_1, Ψ_2 aus D gilt $<\Psi_1, \phi(f) \Psi_2> = <\phi(\bar{f})\Psi_1, \Psi_2> \cdot \phi(f) \subset D$.

R e l a t i v i s t i s c h e   K o v a r i a n z : Auf H gibt es eine unitäre Darstellung der eigentlichen inhomogenen Lorentzgruppe $(a, \Lambda) \longrightarrow U(a, \Lambda)$ $a\varepsilon R^4$, $1\varepsilon L_+^\uparrow$ derart, daß $U(a, \Lambda) \phi(x) U^{-1}(a, \Lambda) = \phi(\Lambda x + a)$. Der Bereich D ist invariant unter $U(a, \Lambda)$.

S p e k t r u m : Sei $U(a, 1) = e^{i(a^O H - \vec{a}\vec{P})}$ (d.h. $H, \vec{P}$ sind die infinitesimalen Erzeugenden der Translationen: Energie- und Impulsoperator). Dann soll Null ein (einfacher) Eigenwert von $(H, \vec{P})$ sein mit Eigenvektor Ω , dem *Vakuum*. Ω sei in D . Der Rest des Spektrums von $(H, \vec{P})$ ist enthalten im Vorwärtslichtkegel V_+ .

L o k a l i t ä t : $\phi(x) \phi(y) - \phi(y) \phi(x) = 0$ falls $x - y$ ein raumartiger Vektor ist.

V o l l s t ä n d i g k e i t : Als D kann man wählen die Menge aller endlichen Linearkombinationen von Vektoren der Form

$$\phi(f_1) \phi(f_2) \ldots \phi(f_n)\Omega \quad , \quad f_i \text{ und } n \text{ beliebig.}$$

Bemerkung zur Notation: Wir schreiben a, x, usw. für Vierervektoren und $a = (a^O, \vec{a})$, $x = (x^O, \vec{x})$ wenn wir Zeitkomponenten a^O, x^O und Raumkomponenten $\vec{a}, \vec{x}$ trennen wollen.

Hauptresultate

Es ist erstaunlich, daß schon aus diesen wenigen grundlegenden Annahmen eine ganze Reihe experimentell überprüfbarer Folgerungen gezogen werden können, insbesondere das TCP Theorem und der Zusammenhang von Spin und Statistik. Außerdem gelingt es (unter Zuhilfenahme einiger Zusatzannahmen, im Wesentlichen über das Spektrum von $(H, \vec{P})$) eine Teilcheninterpretation und eine Streutheorie herzuleiten. In diesem Zusammenhang lassen sich auch Dispersionsrelationen und Ungleichungen sowie asymptotische Beziehungen zwischen Streuamplituden beweisen. Siehe [5,6] und dort angegebene Referenzen.

Analytische Fortsetzung, Schwingerfunktionen

Eine relativistische Quantenfeldtheorie läßt sich auch beschreiben durch ihre Vakuumerwartungswerte, die sogenannten Wightman Distributionen:

$$W_n(x_1 \ldots x_n) = <\Omega, \phi(x_1) \ldots \phi(x_n) \Omega>$$

Ausgehend von den Wightman Axiomen läßt sich zeigen, daß die W_n folgende Eigenschaften haben:

$$
\text{(W)} \quad
\boxed{
\begin{array}{l}
\text{- Relativistische Kovarianz} \\
\text{- Positivität} \\
\text{- Spektrum Eigenschaften} \\
\text{- Lokalität} \\
\text{- Cluster Eigenschaften (falls } \Omega \text{ eindeutig)} \\
\text{- Regularität}
\end{array}
}
$$

Für eine präzise Formulierung siehe [4,5,6]. Wir begnügen uns mit einigen Erläuterungen.

- Die Positivität drückt aus, daß die W_n Skalarprodukte von Vektoren im Hilbertraum H sind.
- Die Spektrumseigenschaft schränkt den Träger der Fouriertransformierten der W_n ein.
- Die Lokalität besagt, daß $W_n(\ldots x_i, x_{i+1}, \ldots) = W_n(\ldots x_{i+1}, x_i \ldots)$ falls $x_i - x_{i+1}$, raumartig ist.
- Die Clustereigenschaft bedeutet, daß für raumartige Vektoren a
$$
\lim_{l \to \infty} W_{n+m}(x_1 \ldots x_n, y_1 + la, \ldots y_m + la) = W_n(x_1 \ldots x_n)\, W_m(y_1 \ldots y_m).
$$
Sie gilt *nur* falls Ω der einzige invariante Vektor ist.
- Die Regularität besagt, daß die W_n temperierte Distributionen sind.

In der Tat sind die Wightman Distributionen *Randwerte analytischer Funktionen* $W_n(z_1 \ldots z_n)$, welche man heuristisch erhält aus der Formel

$$
W_n(z_1 \ldots z_n) = \langle \Omega, \phi(x_1) e^{-\zeta_1 P} \phi(x_2)\, e^{-\zeta_2 P} \ldots e^{-\zeta_{n-1} P} \phi(x_n) \Omega \rangle
$$

wobei $z_{k+1} - z_k = x_{k+1} - x_k + i\zeta_k$, $\mathrm{Re}\,\zeta_k$ in V_+ ,

d.h. indem man in den Vakuumerwartungswerten zwischen je zwei Feldoperatoren einen Operator der Form $e^{-\zeta P}$, $\zeta P = \zeta^0 H - \vec{\zeta} \cdot \vec{P}$ schiebt. Da als ζ nur komplexe Vierervektoren mit $\mathrm{Re}\,\zeta$ im Vorwärtslichtkegel V_+ zugelassen werden, ist dank der *Spektrumseigenschaft* $e^{-\zeta P}$ immer ein beschränkter Operator mit Norm 1.

Im Analytizitätsgebiet der $W_n(z_1 \ldots z_n)$ enthalten sind die *Euklidischen Punkte*, definiert durch $z_k = (ix_k^0, \vec{x}_k)$, x_k *reelle* Vierervektoren, $k = 1, 2 \ldots n$, und $x_k \neq x_{k'}$ für alle $k \neq k'$.
Wir setzen (für Euklidische Punkte $(z_1 \ldots z_n)$): $S_n(x_1 \ldots x_n) = W_n(z_1 \ldots z_n)$.
Dies sind die *Schwingerfunktionen* oder Euklidischen Greensfunktionen.
Formal gilt auch für $x_1^0 < x_2^0 < \ldots < x_n^0$

$$
S_n(x_1 \ldots x_n) = \langle \Omega, \hat{\phi}(x_1) \ldots \hat{\phi}(x_n) \Omega \rangle , \quad \text{wobei}
$$
$$
\hat{\phi}(x) = e^{-x^0 H} \phi(o, \vec{x})\, e^{x^0 H}
$$
(Der Übergang $W_n \longrightarrow S_n$ oder $\phi \longrightarrow \hat{\phi}$ wird in der Literatur oft

"Wick Rotation" genannt).

Wieder ausgehend von den Wightman Axiomen kann man für die Schwingerfunktionen folgende Eigenschaften beweisen:

(S)

> - Euklidische Kovarianz
> - Positivität
>
> - Symmetrie
> - Cluster Eigenschaft (falls Ω eindeutig)
> - Regularität

Für eine präzise Formulierung siehe [7,10]. Man beachte folgendes:

- Euklidische Kovarianz bedeutet, daß $S_n(x_1 \ldots x_n)$ invariant (kovariant
  im Fall von Feldern mit Spin $\neq$ 0) ist unter $x_i \longrightarrow Rx_i + a$, wobei R
  für eine orthogonale Transformation des R^4 steht.
- Die Spektrumseigenschaft von (W) scheint in (S) kein Gegenstück zu
  haben. Sie ist jedoch eine Konsequenz von Regularität, Positivität und
  (Translations-) Invarianz von S_n.
- Symmetrie bedeutet, daß $S_n(x_1 \ldots x_n)$ eine symmetrische Funktion bzgl.
  $x_1 \ldots x_n$ ist. Diese Eigenschaft ersetzt die komplizierte Lokalitäts-
  eigenschaft in (W), siehe [5,Seite 85].
- Für die Regularitätseigenschaft gibt es verschiedene Versionen, siehe
  z.B. [10] und die Diskussion weiter unten.

Rekonstruktionstheoreme

Die Einführung und Diskussion der Wightman Distributionen und der Schwinger Funktionen wird gerechtfertigt durch die Tatsache, daß sowohl (W)
als auch (S) den Gåring-Wightman Axiomen äquivalent sind. Mit anderen
Worten, aus einem Satz von W_n (oder S_n) die (W) (oder (S)) erfüllen, lassen sich H, Ω, U(a,1) und $\phi(x)$ rekonstruieren, so daß alle
Axiome erfüllt sind. Siehe [4,5,6], [10,11] und auch [12,13,14]. Schema-
tisch dargestellt gilt also:

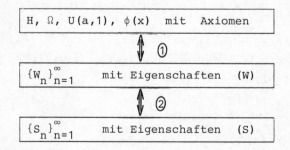

H, Ω, U(a,1), $\phi(x)$ mit Axiomen

$\Updownarrow$ ①

$\{W_n\}_{n=1}^{\infty}$ mit Eigenschaften (W)

$\Updownarrow$ ②

$\{S_n\}_{n=1}^{\infty}$ mit Eigenschaften (S)

Das Rekonstruktionstheorem (1) ist die Grundlage der axiomatischen QFT und soll hier nicht weiter diskutiert werden.

Der Zusammenhang zwischen Schwinger Funktionen und QFT spielt heute eine ausschlaggebende Rolle in der konstruktiven QFT, da Schwinger Funktionen mathematisch viel leichter zugänglich sind als Wightman Distributionen oder gar Feldoperatoren. Man überzeugt sich leicht davon, daß die Eigenschaften (S) eine viel einfachere Struktur haben als (W). Außerdem stellt es sich aber auch noch heraus, daß die Schwinger Funktionen von Bose QFT

- als Erwartungswerte von *stochastischen Prozessen* (Markoff Prozessen) interpretiert werden können,
- die *Momente eines Maßes* auf einen unendlich dimensionalen Raum (z.B. auf dem Schwartzschen Raum $S'(R^4)$) sind,
- mathematisch die Struktur von Korrelationsfunktionen von Modellen der *klassischen statistischen Mechanik haben.*

Diese Eigenschaften folgen zwar nicht aus den Axiomen, gelten jedoch für die üblicherweise untersuchten Modelle. Darum spielen in der konstruktiven QFT Ideen und Methoden aus der Theorie der stochastischen Prozesse, der Funktionalintegration und aus der statistischen Mechanik eine große Rolle und umgekehrt haben diese drei Gebiete manche Anregungen aus der QFT empfangen, siehe z.B. [11,15-19].

Die Regularitätsbedingung in (S)

Die Regularitätsbedingung für Schwinger Funktionen kann leicht variiert werden - insbesondere sind Verschärfungen möglich, welche die Rekonstruktion der relativistischen Theorie sehr erleichtern [12,14] oder aber auch zusätzliche interessante Strukturen für die rekonstruierte Theorie implizieren.

a) Die Erwartungswerte von zeitgeordneten Produkten

$$T\phi(x_1)\ldots\phi(x_n) = \phi(x_{i_1})\ldots\phi(x_{i_n})$$

$(i_1\ldots i_n)$ eine Permutation von $(1,\ldots n)$

so daß $x_{i_1}^o \leq x_{i_2}^o \leq \ldots x_{i_n}^o$

spielen eine wichtige Rolle in der LSZ Theorie [6,20,21], da aus ihnen mittels der Reduktionsformel die Matrixelemente des Streuoperators berechnet werden können. Obige Definition ist jedoch formal und es ist nicht klar, ob in jeder Wightman Theorie zeitgeordnete Produkte definiert werden können.

Wenn man jedoch von einem Satz von Schwinger Funktionen ausgeht, welche (S) erfüllen mit einer leicht modifizierten Regularitätseigenschaft, dann *existieren* in der rekonstruierten Wightman Theorie die zeitgeordneten Produkte und haben alle Eigenschaften, welche man auf Grund obiger formalen Definition erwarten würde [22]; siehe auch [23,24].

b) Neben der axiomatischen Formulierung der QFT nach Wightman, gibt es auch einen algebraischen Ansatz: die Haag-Kastler Axiome [25]. Hier sind die fundamentalen Objekte die lokalen Observablenalgebren. Wann können im Rahmen einer Wightman Theorie solche Algebren definiert werden, so daß sie alle Axiome von Haag-Kastler erfüllen? Wiederum ist eine hinreichende Bedingung gegeben durch eine verschärfte Regularitätsannahme für die Schwinger Funktionen der Theorie [26].

Diese beiden Beispiele haben illustriert, daß die Schwinger Funktionen in axiomatischen Problemen sehr nützliche Objekte sein können. Im nächsten Kapitel wollen wir zeigen, daß auch in der konstruktiven QFT der Zugang über die Schwinger Funktionen (oder deren erzeugendes Funktional) sehr einfach ist - auf Grund der bisherigen Erfahrungen ist man beinahe versucht zu sagen: der einzig vernünftige!

III. Konstruktive Quantenfeldtheorie

Konstruktion von Modellen

Ein Modell für die Wightman Axiome kann man im Prinzip finden, indem man Schwinger Funktionen konstruiert und die Eigenschaften (S) kontrolliert. Der Ausgangspunkt einer solchen Konstruktion ist ein *Euklidisches freies Feld* $\Phi(x)$, das definiert ist durch $\Phi(x)\,\Phi(y) = \Phi(y)\,\Phi(x)$ für *alle* x, y und

$$<\Omega,\ \Phi(x)\,\Phi(y)\ \Omega> = \frac{1}{(2\pi)^2} \int \frac{e^{ip(x-y)}}{p^2+m_o^2}\ d^4p = (-\Delta+m_o^2)^{-1}\ (x,y)$$

Äquivalent kann man auch sagen, $\Phi(x)$ ist ein Gauss'scher Prozess mit Mittelwert 0 und Kovarianz $(-\Delta+m_o^2)^{-1}$, [7, 11]. Damit werden Methoden und Techniken der Funktionalintegration und der Theorie der stochastischen Prozesse verfügbar für die konstruktive QFT.

Die Schwinger Funktionen einer freien Feldtheorie sind dann gegeben durch

$$S^{freies\ Feld}(x_1 \ldots x_n) = <\Omega,\ \Phi(x_1) \ldots \Phi(x_n)\ \Omega>$$

Sie haben natürlich alle Eigenschaften (S) , aber sie führen zu einer trivialen Streumatrix S = 1 !

Der Lagrange Formalismus legt es nun nahe, für eine *Wechselwirkende Theorie* mit Wechselwirkungsdichte $\lambda L_I(\phi)$ den *formalen* Ansatz zu machen

$$S_n(x_1 \ldots x_n) \;"="\; \frac{<\Omega, \; \Phi(x_1) \ldots \Phi(x_n) e^{-\lambda \int L_I(\Phi(y)) dy}{}_\Omega>}{<\Omega, e^{-\lambda \int L_I(\Phi(y)) dy}{}_\Omega>} \tag{*}$$

Wir bemerken, daß eine formale Potenzreihenentwicklung dieses Ansatzes nach Potenzen von λ zu der Euklidischen Version der Gell-Man Low-Formel und zu den (Euklidischen) Feynman Diagrammen führt.

Vom formalen Ansatz gelange man zu einer mathematischen Konstruktion in zwei Schritten:

1. Schritt: Regularisieren

Der formale Ausdruck $e^{-\lambda \int L_I(\Phi(y)) dy}$ wird ersetzt durch die mathematisch wohldefinierte Funktion $e^{-\lambda \int_\Lambda L_I(\Phi_\kappa(y)) dy}$. Hier bedeutet Λ ein *endliches Volumen* in R^4 und κ steht für einen *ultraviolett cutoff* (z.B. $\Phi_\kappa(x) = \int \Phi(y) \, \chi_\kappa(x-y) dy$, χ_κ ist genügend oft differenzierbar und strebt gegen $\delta(x)$ für $\kappa \to \infty$. Nun definiert man gemäß der Formel (*) regularisierte Schwingerfunktionen $S_n^{\Lambda, \kappa}$ und zeigt, daß sie außer der Euklidischen Kovarianz alle Eigenschaften (S) haben. Positivität folgt leicht bei geeigneter Wahl der Regularisierung und Symmetrie ist automatisch erfüllt, wohingegen die Regularität und die Clustereigenschaft harte Arbeit erfordern, da man die Abschätzungen gleichmäßig in Λ und in κ haben muß für den zweiten Schritt.

2. Schritt: Grenzwert studieren

Man zeigt, daß $S_n^{\Lambda, \kappa}$ für $\kappa \to \infty$ und $\Lambda \to R^4$ einen Grenzwert hat und definiert $S_n(x_1 \ldots x_n) = \lim\limits_{\Lambda \to R^4} \lim\limits_{\kappa \to \infty} S_n^{\Lambda, \kappa}(x_1 \ldots x_n)$. Die so definierten Schwinger Funktionen werden alle Eigenschaften (S) haben, wenn der Grenzwert nicht davon abhängt, wie die Folge der Volumina Λ gewählt wurde.

Zusammenhang mit Statistischer Mechanik

Als spezielles Beispiel einer Regularisierung nennen wir die *Gitterapproximation* [15,16]: der ultraviolett cutoff wird eingeführt, indem man Raum-Zeit durch ein kubisches Gitter (mit Gitterkonstanten κ^{-1}) ersetzt. Dann wird $S_n^{\Lambda, \kappa}(x_1 \ldots x_n) =$

$$\frac{1}{Z} \int \Phi_{i_1} \ldots \Phi_{i_n} \; e^{-\frac{1}{2} \sum\limits_{\Lambda, n.n.} (\Phi_i - \Phi_{i'})^2 - \frac{m_o^2}{2} \sum\limits_\Lambda \Phi_i^2 - \lambda \sum\limits_\Lambda L_I(\Phi_i)} \; \prod\limits_\Lambda d\Phi_i \; .$$

Hier ist $\Phi_i = \Phi(x_i)$ die "Spinvariable" im Gitterpunkt x_i, Z die Normierung analog zu (*), $\sum$ bedeutet Summation über alle Gitterpunkte x_i im Volumen $\overset{\Lambda}{}$ und $n.n.$ bedeutet, daß nur über nächste Nachbarn x_i und $x_{i'}$ summiert wird. Obige Formel definiert aber auch eine Korrelationsfunktion eines Gittermodells der *klassischen statistischen Mechanik* mit Zweikörperwechselwirkung zwischen nächsten Nachbarn. Bosequantenfeld-

theorie und klassische statistische Mechanik werden also durch mathematisch äquivalente Formalismen beschrieben. Dieser Zusammenhang erweist sich als äußerst fruchtbar. Als Beispiel erwähnen wir die Korre-
lationsungleichungen, die auf diesem Weg in die konstruktive QFT eingeführt und sofort zu einem wichtigen Hilfsmittel wurden, siehe z.B. [16].

Ü b u n g : Man wähle L_I so, daß das Ising Modell als Grenzfall einer
Quantenfeldtheorie dargestellt werden kann [27].

Natürlich treten bei der Durchführung des oben beschriebenen Programmes
alle die wohlbekannten Renormierungsprobleme auf. In 4 Raum-Zeit Dimensionen haben sie denn auch bisher die Konstruktion eines nichttrivialen
Modelles verhindert. Die Divergenzen (der Feynman Integrale z.B.) werden
milder, wenn man Modelle in 2 oder 3 Raum-Zeit Dimensionen studiert;
hier ist der Durchbruch gelungen, wie wir im Folgenden sehen werden.

Resultate

(d = Dimension von Raum-Zeit) Am vollständigsten analysiert sind Modelle
mit d = 2 und $L_I(\Phi) = P(\Phi)$, ein nach unten beschränktes Polynom: die sogenannten $P(\phi)_2$ *Modelle*. Die meisten Resultate sind auch bewiesen worden
für

$$
\begin{array}{llll}
d = 2 & L = \sin\varepsilon\phi & \text{(Sinus-Gordon Modell)} \\
d = 2 & L = e^{\alpha\Phi} & \text{(Høegh-Krohn Modell)} \\
d = 2 & L = \overline{\Psi}\Psi\Phi & \text{(Yukawa Modell; } \Psi: \text{ Fermi Feld)} \\
d = 3 & L = \phi^4 & ((\phi^4)_3\text{-Modell})
\end{array}
$$

Sei im Folgenden $P(\phi) = \phi^{2n} + \sum_{k=0}^{n-1} a_k \phi^{2k}$ für irgend ein $n = 1,2,\ldots$i
d = 2.

1) Für alle $L_I(\Phi) = \lambda P(\Phi)$ existieren die Schwinger Funktionen und erfüllen (S), eventuell ohne Cluster Eigenschaft. [11,12,15]

2) Sei $L_I(\Phi) = \lambda P(\Phi)$ und λ/m_o^2 genügend klein. Dann:

   a) Die Cluster Eigenschaft gilt; das Vakuum ist eindeutig [28].

   b) Das Spektrum von $(H,\vec{P})$ sieht entweder aus wie in Fig. 1 oder wie
      in Fig. 2.

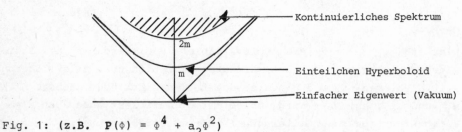

Fig. 1: (z.B. $P(\Phi) = \phi^4 + a_2\phi^2$)

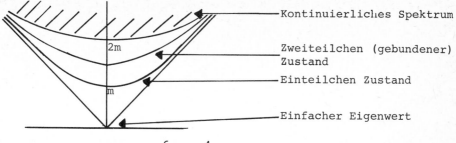

Fig. 2: (z.B. $P(\phi) = \phi^6 - \phi^4$)

Im Fall von Fig. 2 kann man die Masse des gebundenen Zustandes berechnen als [29]

$$m_G = 2m \ (1 - \frac{9}{8} \ (\frac{\lambda}{m^2})^2 + O(\lambda^3))$$

c) Zeitgeordnete Produkte existieren und die Streumatrix S ist nicht die Identität, d.h. die Modelle beschreiben nichttriviale Streuung. [23,24,30]

d) Die Schwinger Funktionen, die Wightman Distributionen, die Streu-matrixelemente und die Masse m des beschriebenen Teilchens haben asymptotische Entwicklungen nach Potenzen von λ, gegeben durch die übliche Feynman Diagramm Entwicklung [31,32].

e) Das rekonstruierte relativistische Feld $\phi(x)$ erfüllt die Feld-gleichung

$$(\Box + m^2) \ \phi = - \lambda \overset{..}{:} Q'(\phi) \overset{..}{:}$$

(Hier bedeutet $\overset{..}{:} \overset{..}{:}$ Wickordnung bezüglich des physikalischen Vakuums Ω, Q' ist die erste Ableitung eines Polynoms Q, welches in höch-ster Ordnung mit P übereinstimmt. Die Beziehung zwischen P und Q ist ein-eindeutig), [33].

3) Sei speziell $P(\phi)$ von vierter Ordnung. Dann sind die Taylorreihen in λ von den Schwinger Funktionen, der Masse m und der Feldstär-kenrenormierung Z nicht nur asymptotische Reihen, sondern *Borel summierbar*. Das heißt, daß diese Größen durch ihre Taylorkoeffizienten eindeutig bestimmt sind, obschon die Taylorreihen divergieren. Mit anderen Worten: die Störungstheorie bestimmt die Modelle vollständig und eindeutig [34]. Man beachte jedoch, daß dieser Schluß nicht aus dem Studium der Störungsreihe *allein* gezogen werden konnte.

4) Sei $\lambda L_I(\phi) = (\phi^2 - \sigma^2) \ / \ \sigma^2 + \mu\phi$, $m_o = 1$. Dann

a) Für $\mu \neq 0$ ist das Vakuum eindeutig (L_I hat ein eindeutiges Minimum).

b) Für $\mu = 0$ (L_I hat dann zwei Minima) und σ genügend groß, wird die Symmetrie $\Phi \to -\Phi$ gebrochen und es gibt zwei verschiedene Vakua (Phasen). Für jede der Phasen separat gelten wieder alle Eigenschaften (S). [18]

B e m e r k u n g : In zwei Dimensionen tritt das Phänomen der Goldstone Bosonen nicht auf [35]. Jedoch kann man zeigen für $d = 3$ und $\vec{\Phi} = (\Phi_1, \ldots \Phi_N)$, $\lambda L_I = (\vec{\Phi}^2 - \sigma^2)/\sigma^2$ + Renormierungsterme, daß die $O(N)$ - Symmetrie gebrochen werden kann und somit masselose Goldstone Bosonen auftreten [36].

5) a) Es gibt $P(\phi)_2$ Modelle für welche Phasenübergänge ohne spontane Symmetriebrechnung stattfinden [36]; außerdem gibt es Modelle mit mehr als zwei koexistierenden Phasen [37,38].

b) Für das pseudoskalare Yukawa-Modell und $d = 2$ gibt es auch spontaneSymmetriebrechung und einen Phasenübergang.

Es ist zu betonen, daß obige Liste sehr unvollständig ist und daß auch die zitierten Resultate jeweils nicht in der größtmöglichen Allgemeinheit dargestellt wurden. Ich hoffe jedoch einen Eindruck von der Art der studierten - und gelösten - Probleme vermittelt zu haben.

IV. Offene Probleme

Unter den offenen Problemen haben die folgenden eine gute Chance, bald gelöst zu werden:

a) Asymptotische Vollständigkeit für $P(\phi)_2$,

b) Borelsummierbarkeit in der Mehrphasenregion von $P(\phi)_2$ oder Yukawa$_2$ Modellen,

c) Axiomatik und Konstruktion von nichtabelschen Eichtheorien.

Für die folgenden Probleme ist eine Prognose bedeutend schwieriger:

d) Konstruktion eines nichttrivialen Modelles in $d = 4$,

e) Besseres mathematisches Verständnis von Infrarotproblemen und dem Verhalten in der Nähe kritischer Punkte.

Oder kurz zusammengefaßt:

Wie bringt man c und ℏ unter einem Hut?

Referenzen

[1] P. A. M. Dirac, Proc. Royal Soc. of London, Series A, <u>114</u>, 243 (1927)

[2] K. Hepp, in: Statistical Mechanics and Quantum Field Theory, C. DeWitt, R. Stora (editors), Gordon and Breach, New York (1971)

[3]   A. Wightman, G. Velo, (editors), Renormalization Theory (Nato
        Advanced Study Institute No. 23), Reidel Pub. 1976

[4]   R. F. Streater, A. S. Wightman, PCT, Spin and Statistics, Ben-
        jamin (1964)

[5]   R. Jost, The general Theory of Quantized Fields, AMS Publ. (1965)

[6]   N. N. Bogolubov, A. A. Logunov, I. T. Todorov, Axiomatic Quantum
        Field Theory, Benjamin (1975)

[7]   G. Velo, A. S. Wightman (editors), Constructive Quantum Field
        Theory, Springer Lecture Notes in Physics, Vol. 25, (1973)

[8]   M. Levy, P. Mitter (editors), New Developments in Quantum Field
        Theory and Statistical Mechanics, Plenum Press (1977)

[9]   L. Streit (editor), Quantum Dynamics: Models and Mathematics,
        Springer Verlag, Wien-New York (1976)

[10]   K. Osterwalder, R. Schrader, Commun. Math. Phys. $\underline{31}$, 83-112 (1973)
        and $\underline{42}$, 281-305 (1975)

[11]   E. Nelson, J. Funct. Anal. $\underline{12}$, 97-112 (1973)

[12]   J. Fröhlich, Helv. Phys. Acta $\underline{47}$, 265 (1974), Ann. de l'Institut
        H. Poincaré $\underline{21}$, 271-317 (1974), Ann. Phys. $\underline{97}$, 1-54 (1976)

[13]   G. Hegerfeldt, Commun. Math. Phys. $\underline{35}$, 115 (1974)

[14]   J. Glimm, A. Jaffe, in [8]

[15]   F. Guerra, L. Rosen, B. Simon, Ann. Math $\underline{101}$, 111-259 (1975)

[16]   B. Simon, The $P(\Phi)_2$ Euclidean (Quantum) Field Theory, Princeton
        University Press (1973)

[17]   J. Fröhlich, B. Simon, T. Spencer, Commun. Math. Phys. $\underline{50}$ 79-95
        (1976)

[18]   J. Glimm, A. Jaffe, T. Spencer, Commun. Math. Phys. $\underline{45}$, 203-216
        (1975), Ann. Phys. $\underline{101}$, 610-630 und 631-669 (1976)

[19]   S. Albeverio, R. Høegh-Krohn, Commun. Math. Phys. $\underline{68}$, 95-128
        (1979)

[20]   H. Lehmann, K. Symanzik, W. Zimmermann, Il Nuovo Cimento $\underline{1}$, 205
        (1955) and $\underline{6}$, 319 (1957)

[21]   K. Hepp, Commun. Math. Phys. $\underline{1}$, 95 (1965)

[22]   J.-P. Eckmann, H. Epstein, Commun. Math. Phys. $\underline{64}$, 95-130 (1979)

[23]   K. Osterwalder, R. Seneor, Helv. Phys. Acta $\underline{49}$, 525-535 (1976)

[24]   K. Osterwalder, in [9]

[25]   R. Haag, D. Kastler, J. Math. Phys. $\underline{5}$, 848 (1964)

[26]   W. Driessler, J. Fröhlich, Ann. Inst. Henri Poincaré, Sectiona,
        $\underline{27}$, 221-236 (1977)

[27]   J. Rosen, The Ising Model Limit of Φ^4 Lattice Fields,(Rocke-
       feller Univ. Preprint)

[28]   J. Glimm, A. Jaffe, T. Spencer, in [7]

[29]   J. Dimock, J.-P. Eckmann, Commun. Math. Phys. 51, 41-54 (1976)

[30]   J.-P. Eckmann, H. Epstein, J. Fröhlich, Ann. Inst. Poincaré 25,
                                          1-34, (1976)
[31]   J. Dimock, in [9]

[32]   J.-P. Eckmann, H. Epstein, Borelsummability of the Mass and the
       S-Matrix in ϕ^4 Models, Commun. Math. Phys., to appear

[33]   R. Schrader, Fortschritte der Physik 22 (1974) 611-631

[34]   J.-P. Eckmann, J. Magneh, R. Seneor, Commun. Math. Phys. 39,
       251-271 (1975)

[35]   S. Coleman, Commun. Math. Phys. 31, 259-264 (1973)

[36]   J. Fröhlich, Acta Phys. Austr. Suppl. 15, 133-269 (1976)

[37]   K. Gawedzki, Commun. Math. Phys. 59, 117-142 (1978)

[38]   S. Summers, Harvard Thesis, 1979

METHODS OF DIFFERENTIAL GEOMETRY IN GAUGE THEORIES AND GRAVITATIONAL THEORIES

Julius Wess, Technische Universität, Karlsruhe

It has become more and more customary to consider Einstein's theory of gravitation as a gauge theory. Einstein's theory, however, is a geometrical theory and it seems therefore natural to try to apply geometrical methods to gauge theories as well. This has proven to be successful in connection with supersymmetric gauge theories and with supergravity.

It is the purpose of this lecture to present the essential geometrical concepts which allow from a common point of view a formulation of gauge theories, of Einstein's gravitational theory, of supergauge theories and of supergravity as well. Im am going to list these basic concepts [1,2] , without being too precise in their definition.

1. **Manifolds**, this is usually something which locally looks like R^m , and which allows us to speak about differentiability.

2. **Tensor fields**, a linear space which admits a local basis, $\sigma^i(x)$ and which serves as a representation space of a Lie group:

$$\sigma'^i(x) = \sigma^e(x) \ G_e{}^i(x) \ , \tag{1}$$

$$G(x) = e^{i\Lambda(x)} \ , \qquad \Lambda(x) = \sum_r \lambda_r(x) T^r$$

A special tensor field is the tangent space of a manifold and its dual, the cotangent space. If $e_a(x)$ is a basis of the tangent space and $e^a(x)$ a basis of the cotangent space we can use the tensor product

$$e_{a_1} \otimes e_{a_2} \ \dots \otimes e_{a_r} \otimes e^{b_1} \otimes \dots \otimes e^{b_s}$$

as a basis of the r-times contravariant and s-times covariant tensor space.

3. **Forms**, a p-form being an antisymmetric, p-times covariant tensor field. We define the exterior product

$$e^{a1} \wedge e^{a2} = - e^{a2} \wedge e^{a1} \tag{2}$$

and we write a p-form as follows:

$$w = e^{a1} \wedge e^{a2} \wedge \ldots \wedge e^{a_p} w_{a_p \ldots a_1}(x)$$

The exterior derivation is a mapping of a p-form into a p+1-form with the defining properties:

$$d(w_1 + w_2) = dw_1 + dw_2 \tag{3}$$

$$d(w_1 \wedge w_2) = w_1 \wedge dw_2 + (-)^q dw_1 \wedge w_2 \tag{4}$$

$$d \cdot d = 0 \tag{5}$$

w_2 being a q form.

Exterior derivatives of local coordinate lines, dx^m form a natural basis in the cotangent space. This basis is related to the one above via the Viel-(Vier-) Bein field $e_m{}^a(x)$:

$$e^a = dx^m e_m{}^a(x) . \tag{6}$$

The Viel-Bein field is supposed to have an inverse:

$$\tilde{e}_b{}^m e_m{}^a = \delta_b^a .$$

4. Connection. The exterior derivative maps a p-form into a p+1 form but it does not map a tensor field into a tensor field. If $\sigma^i(x)$ transforms like a tensor field, $\sigma' = \sigma G$, we find for its exterior derivative an inhomogeneous transformation law:

$$d\sigma' = \sigma dG + d\sigma G. \tag{7}$$

In order to be able to define a derivative, which maps a p-form into a p+1-form and a tensor field into a tensor field we have to introduce the connection form. This is a one-form with the transformation properties:

$$\phi' = G^{-1} \phi G + G^{-1} d G . \tag{8}$$

The connection form can be chosen to be Lie-algebra valued:

$$\phi = dx^m \phi_{mr} T^r , \tag{9}$$

T^r are the generators of the group. It is easy to verify that the co-variant derivative:

$$D\sigma = d\sigma - \sigma\phi = dx^m D_m \sigma \qquad (10)$$

transforms like a tensor field:

$$(D\sigma)' = (D\sigma) \ G \ . \qquad (11)$$

5.   **Structure Equation**. It is possible to construct a tensor field in terms of the connection and its derivatives:

$$F = d\phi - \phi\phi \ . \qquad (12)$$

F is a 2-form, $F = \frac{1}{2} dx^m \wedge dx^n F_{nm}$, it is Lie-algebra valued if ϕ is, and F is a tensor field:

$$F' = G^{-1} F G \ . \qquad (13)$$

The covariant derivatives of tensor fields and the tensor field F are in general all the independent tensorial quantities which can be used to formulate covariant differential equations (field equations). If we try to construct new tensorial quantities through the repeated use of the exterior derivative we obtain identities instead. This is due to the fact that $d \cdot d = 0$.   These identities are called:

6.   **Bianchi Identities**.

From   (10)   follows:

$$d \ D \ \sigma \ = D \ \sigma \ \phi - \sigma \ F \ , \qquad (14)$$

and from   (12):

$$d \ F = \phi \ F - F \ \phi \ . \qquad (15)$$

Let me now explicate these various concepts within the respective theory:

I.   **Gauge Theories**:

The manifold is R^4,   the tensor fields are 0-forms, *i.e.* fields which transform under a compact Lie group. The connection is Lie algebra valued:

$$\phi = dx^m A_m(x) = dx^m A_{mr}(x) T^r \quad , \tag{16}$$

the component fields $A_{mr}(x)$ are called Yang Mills potentials. The tensor field F is called Yang Mills field:

$$F = \frac{1}{2} dx^m \wedge dx^n F_{nm} \tag{17}$$

$$F_{nm} = \partial_n A_m - \partial_m A_n + [A_n, A_m] \quad .$$

The covariant derivative (10) is the covariant Yang Mills derivative:

$$D_m \sigma^i = \partial_m \sigma^i - \sigma^j T^r{}_j{}^i A_{mr}(x)$$

and the Bianchi identity (14) expresses the fact that covariant derivatives do not commute, but their commutator yields the Yang Mills field:

$$(D_m D_n - D_n D_m)\sigma = - \sigma F_{mn} \tag{18}$$

The Bianchi identity (15) becomes the well known cyclic condition:

$$D_e F_{mn} + D_m F_{ne} + D_n F_{em} = 0 \quad . \tag{19}$$

• II.   Gravitational Theory:

The manifold is locally isomorphic to R^4. The basic tensor fields are the cotangent vector fields. They are 1-forms.

$$e^a = dx^m e_m{}^a(x) \quad . \tag{20}$$

The structure group is the Lorentz group:

$$e'^a = e^b L_b{}^a \tag{21}$$

The transformation acts on tangent space indices only. Tangent space indices can be raised and lowered by the Lorentz metric η^{ab}. Space indices (dz^m) can be related to tangent space indices via the Vierbein. The connection is Lie algebra valued, this means:

$$\phi_{mab} = - \phi_{mba} \tag{22}$$

The covariant derivative of the cotangent vector field is a 2-form and it is called torsion form:

$$de^a - e^b \phi_b{}^a = \Omega^a = \frac{1}{2} dx^m \wedge dx^n \Omega_{nm}{}^a \quad .$$

(23)

The tensor field F is the Riemannian curvature tensor:

$$F_a{}^b = \frac{1}{2} dx^m \wedge dx^n R_{n\,m\,a}{}^b \quad .$$

(24)

F is a 2-form and Lie-algebra valued:

$$R_{mnab} = -R_{nmab} = -R_{mnba} \quad .$$

(25)

The Bianchi identities yield the following relations beween curvature tensor and torsion:

$$\sum_{cycl.\,edc} \{ D_e \Omega_{dc} - R_{edc}{}^a + \Omega_{ed}{}^f \Omega_{fc}{}^a \} = 0$$

(26)

$$\sum_{cycl.\,edc} \{ D_e F_{dca}{}^b + T_{ed}{}^f R_{fca}{}^b \} = 0 \quad .$$

The indices of the curvature tensor and of the torsion have all been converted to tangent space indices with the help of the Vierbein field.

Supersymmetric Theories:

The underlying manifold is the superspace, introduced by Salam [3] and Strathdee. We denote its elements by:

$$z^M \sim (x^m, \theta^\mu \; \bar{\theta}_{\dot\mu}) \quad .$$

(27)

Here we have used two-component spinor notations. The parameters $\theta^\mu, \bar{\theta}_{\dot\mu}$ are anticommuting quantities (elements of a Grassmann-algebra). The commuting parameters are elements of R^4 .

$$z^M z^N = (-)^{nm} z^N z^M \quad ,$$

(28)

n is a function of the index N , it is 0 if $N = n$, vectorial, it is 1 if $N = \mu$ or $\dot\mu$, spinorial.

Tensor fields are superfields, these are functions of superspace and can be understood in terms of a power-series expansion in the anticommuting variables. This will always be a polynomial of finite order.

$$\sigma^i(z) = s^i(x) + \theta^\mu \chi^i_\mu(x) + \ldots \qquad + \theta\theta\bar{\theta}\bar{\theta}t^i(x) \quad . \tag{29}$$

Structure group elements will be functions of z as well.

$$\sigma'(z) = \sigma(z) \, G(z) \quad . \tag{30}$$

Tangent space and cotangent space can be introduced for superspace as well, we denote the respective basis by

$$E^A(z) \qquad \text{and} \qquad E_A(z) \quad .$$

The notions of forms can be generalized by defining the exterior product with a sign factor:

$$E^{A1} \wedge E^{A2} = -(-)^{a_1 a_2} E^{A2} \wedge E^{A1} \tag{31}$$

A p-form:

$$W = E^{A1} \wedge \ldots \wedge E^{Ap} \, W_{A_p \ldots A_1}(Z) \tag{32}$$

The coefficient function $W_{A_p \ldots A_1}(Z)$ will have mixed symmetry properties, in this case p can be arbitrary large.

The exterior derivative can be defined analogously:

$$d(w_1 + w_2) = d\,w_1 + dw_2$$

$$d(w_1 \wedge w_2) = w_1 \wedge dw_2 + (-)^q \, dw_1 \wedge w_2 \tag{33}$$

$$d \cdot d = 0 \quad .$$

The super-Vielbein relates the basis E^A in the contangent space to the natural basis dz^M:

$$dz^M E_M(z)^A = E^A$$

$$\tilde{E}_B{}^M E_M{}^A = \delta^A_B \tag{34}$$

Connection form, covariant derivative, curvature and Bianchi identities are completely analogous:

$$\phi = dz^M \phi_M$$

(35)

$$\phi' = G^{-1} \phi G + G^{-1} dG$$

$$D\sigma = d\sigma - \sigma\phi$$

(36)

$$d\phi - \phi\phi = F = \frac{1}{2} dz^M \wedge dz^N R_{NM} \ .$$

(37)

$$d(D\sigma) - (D\sigma)\phi = -\sigma F$$

(38)

$$dF = \phi F - F\phi$$

III. Supersymmetric Yang Mills Theories:

Manifold and tensor fields are as discussed above. Structure group is a compact Lie-group. It is convenient to choose a basis in the cotangent space which is adjusted to supersymmetry transformations:

$$e^A = dz^M e_M{}^A$$

$$e_M{}^A = \begin{pmatrix} \delta_m^a & 0 & 0 \\ -i\sigma_{\mu\dot\mu}^a \bar\theta^{\dot\mu} & \delta_\mu^\alpha & 0 \\ -i\theta^\beta \sigma_{\beta\dot\gamma}^a \epsilon^{\dot\gamma\dot\mu} & 0 & \delta_{\dot\alpha}^{\dot\mu} \end{pmatrix}$$

(39)

In this basis, the coordinate lines become curves which are generated by supersymmetry transformations:

$$\{Q_\alpha, \bar Q_{\dot\alpha}\} = 2i \ \sigma_{\alpha\dot\alpha}^m \partial_m$$

(40)

With help of the inverse Viel-Bein it is possible to define a suitable basis in the tangent space:

$$D_A = \tilde e_A{}^M \frac{\partial}{\partial z^M}$$

(41)

These derivatives have the property that they commute or anticommute with the generators of supersymmetry:

$$\{D_\alpha, Q_\beta\} = 0, \qquad [D_a, Q_\alpha] = 0$$

(42)

In the case of extended supersymmetry:

$$\{Q^i_\alpha, \bar{Q}_{\dot\alpha j}\} = 2i\delta^i_j \cdot \sigma^m_{\alpha\dot\alpha} P_m \tag{43}$$

all the various quantities are defined analogously, *e.g.*:

$$z^M \sim (x^m, \theta^\mu_i, \bar{\theta}^j_\mu) \tag{44}$$

IV. Supergravity:

The manifold is locally the superspace. The basic tensor fields are the cotangent vector fields:

$$E^A = dz^M E_M{}^A(z) \tag{45}$$

The structure group is the Lorentz-group again, the transformation acts on tangent space indices in a reducible form:

$$E'^A = E^B L_B{}^A , \tag{46}$$

The generators of the Lorentz transformation have the Lie-algebra structure [4] :

$$\Lambda_{ab} = - \Lambda_{ba} ,$$

$$\Lambda_{\alpha\beta} = \Lambda_{\beta\alpha}, \quad \Lambda_{\dot\alpha\dot\beta} = \Lambda_{\dot\beta\dot\alpha} , \tag{47}$$

$$\sigma^a_{\alpha\dot\alpha} \sigma^b_{\beta\dot\beta} \Lambda_{ab} = - 2\epsilon_{\alpha\beta} \Lambda_{\dot\alpha\dot\beta} + 2\epsilon_{\dot\alpha\dot\beta} \Lambda_{\alpha\beta}$$

Other components of the generator, like $\Lambda_{a\alpha}$ etc. are zero.

The connection form

$$dz^M \phi_{MA}{}^B$$

and the curvature form

$$\frac{1}{2} dz^M \wedge dz^N R_{NMA}{}^B$$

are Lie-algebra valued, they have the same structure as the generators Λ . These tensors, therefore decompose into a set of irreducible tensor components. They are, again, related via the Bianchi indentities.

Within the scheme, developed so far, we have constructed a set of quantities in terms of which it is possible to describe the dynamics of the respective physical system. The usual procedure is to postulate a Lagrangian and to derive the equations of motion from it. This Lagrangian should satisfy a certain set of postulates. It should be invariant under the structure group. We have developed the tensor calculus essentially with the aim to list all the possible invariants. Furthermore, the differential equations derived from the Lagrangian should be not higher than of second order, they should be local and give raise to a causal propagation. Ghosts - negative contributions to the energy - should not appear as physical degrees of freedom, they should rather be gauge degrees of freedom and should not contribute to any physical process.

To my knowledge, there is nothing in the framework of differential geometry which gives a hint to how such a Lagrangian should be constructed. It is not even clear for what type of geometries such a Lagrangian exists. For the examples listed above, such a Lagrangian exists, but it is of a quite different structure each time. The requirements mentioned above, however, make the Lagrangian quite unique. Let me list the various theories:

I.   Yang Mills theories:

$$L = - \frac{1}{4} \, \mathrm{Tr} \, F_{mn} F^{mn} \, .$$
(48)

There are no constraints to be solved.

II.   Einstein's theory of gravitation:

The torsion is set equal to zero. This is a constraint equation and allows to express the connection in terms of the Vier-Bein.

$$\Omega = 0$$

$$\phi_{mn,l} = -\frac{1}{2} \, (e_l{}^a \partial_n e_{ma} - e_n{}^a \partial_m e_{la}$$

$$+ \, e_m{}^a \partial_l e_{na} - e_m{}^a \partial_n e_{la}$$

$$+ \, e_n{}^a \partial_l e_{ma} - e_l{}^a \partial_m e_{na})$$
(49)

The metric tensor is given through:

$$\sigma_{mn} = e_m{}^a e_{na}$$
(50)

and the Lagrangian is:

$$L = \frac{1}{2k} \, R_{mn\,ab} \; e^{\,ma} \, e^{\,nb} \, \det e \qquad (51)$$

The Bianchi identities reduce to the well-known set of equations:

$$\underset{cycl.\,edc}{\Sigma} \; R_{edc}{}^{a} = 0$$

$$(52)$$

$$\underset{cycl.\,edc}{\Sigma} \; D_{e} \, R_{dca}{}^{b} = 0 \, .$$

III.   Supersymmetric Yang Mills theories:

The Lagrangian is known for supersymmetry [2], (N = 1) and for N = 2 in the case of extended supersymmetry [5].

   N = 1:

The constraint equations are

$$F_{\alpha\beta} = F_{\dot\alpha\dot\beta} = F_{\alpha\dot\beta} = 0 \, . \qquad (53)$$

The Bianchi indentities, subject to the condition (53) can be solved and they yield:

$$F_{a}{}^{\alpha} = -\frac{1}{8} \, \bar\sigma_{a}{}^{\dot\beta\alpha} \overline{W}_{\dot\beta} \, , \qquad F_{a}{}^{\dot\alpha} = -\frac{1}{8} \, \bar\sigma_{a}{}^{\dot\alpha\alpha} W_{\alpha}$$

$$F^{ab} = -\frac{i}{64} \, \{ (\sigma^{a}\bar\sigma^{b})_{\alpha}{}^{\beta} (D_{\beta} W^{\alpha} + D^{\alpha} W_{\beta})$$

$$(54)$$

$$-(\bar\sigma^{a}\sigma^{b})^{\dot\alpha}{}_{\dot\beta} (D_{\dot\alpha} \overline{W}^{\dot\beta} + D^{\dot\beta} \overline{W}_{\dot\alpha}) \}$$

$$D_{\dot\alpha} W_{\beta} = 0$$

The Langrangian can be written in terms of the chiral superfields W_{α}:

$$L \sim T_{r} (W^{\alpha} W_{\alpha} + \overline{W}_{\dot\alpha} \, \overline{W}^{\dot\alpha}) \qquad (55)$$

In the case of supersymmetric theories, the Langrangian density has to be integrated over a superspace volume.

   N = 2:

The constraints are different:

$$F^{\cdot}_{\alpha i, \; \beta}^{\;\; j} = 0 \; , \tag{56}$$

$$F^{ij}_{\alpha\beta} + F^{ij}_{\beta\alpha} = 0 \; , \quad F^{\cdot}_{\alpha i', \; \beta j} + F^{\cdot}_{\beta i', \; \alpha j} = 0$$

The Bianchi identities yield:

$$F^{ij}_{\alpha\beta} = \varepsilon_{\alpha\beta} \; g^{ij} W \; , \quad F^{\cdot}_{\alpha i', \; \beta j} = \varepsilon^{\cdot\cdot}_{\alpha\beta} \; g_{ij} \; \bar{W}$$

$$F^{\;\; j}_{\alpha\beta} = \frac{i}{4} \; \sigma_{a\beta\rho}^{\cdot} D^{\rho j} W$$

$$F^{\cdot\beta}_{aj} = - \frac{i}{4} \; \bar{\sigma}^{a\beta\rho} D_{\rho j} \bar{W} \tag{57}$$

$$F_{ab} = - \frac{1}{16} \; \{ (\bar{\sigma}_{ab})^{\cdot\alpha}_{\;\;\beta} D^{\cdot}_{\alpha i} D^{\dot\beta i} W - (\sigma_{ab})_{\alpha}^{\;\;\beta} D^{\alpha i} D_{\beta i} \bar{W} \}$$

$$D^i_\alpha \; W = 0$$

The Lagrangian was found to be proportional to:

$$L \sim \text{Tr} \; (WW + \bar{W}\bar{W}) \; . \tag{58}$$

IV. Supergravity:

It is only for the case $N = 1$ that the Lagrangian is known [6] .
It is:

$$L = \det E \tag{59}$$

To get the right equations of motion, however, the following constraint equations have to be solved first:

$$\Omega_{\alpha\beta} \; \underline{\gamma} = 0 \quad ,$$

$$\Omega^{\;\; c}_{\alpha\beta} = \Omega^{\;\; c}_{\beta\alpha} = 2i\sigma^{c\cdot}_{\alpha\beta} \; , \quad \Omega^{\;\; c}_{\alpha\beta} = \Omega^{\cdot\cdot c}_{\alpha\beta} = 0 \quad ,$$

$$\Omega_{\underline{\alpha} \; b}^{\quad c} = \Omega_{a\underline{\beta}}^{\quad c} = 0 \tag{60}$$

$$\Omega_{ab}^{\quad c} = 0$$

Underlined indices might be dotted or undotted.

To solve these constraint equations it is again of great advantage to use the Bianchi identities [7] .

The dynamics of all the systems, which have been discussed here, can be formulated with quantities which have a well-defined meaning in differential geometry. The question might be asked if also the Lagrangian, though different in each case, could have a well-defined meaning in differential geometry.

For what type of geometries does a Lagrangian exist, how can it be found?

References

[1]   W. Thirring, Gauge Theories of Gravitation, Facts and Prospects of Gauge Theories, P. Urban, Springer Verlag 1978.

[2]   J. Wess, Supersymmetry - Supergravity, Lecture Notes in Physics, 77 Springer Verlag.

[3]   A. Salam and J. Strathdee, Nucl. Phys. 76B (1974) 477

[4]   J. Wess and B. Zumino, Phys. Lett. 66B (1977) 361

[5]   R. Grimm, M. Sohnius and J. Wess, Nucl. Phys. B133 (1978) 275

[6]   J. Wess and B. Zumino, Phys. Lett. 74B (1978) 51

[7]   R. Grimm, J. Wess and B. Zumino, Nucl. Phys. B152 (1979) 255

QUANTENFELDTHEORIE UND TOPOLOGIE

Bert Schroer, Freie Universität Berlin

1.Einleitende Bemerkungen

In diesem Vortrag soll versucht werden, anhand zweier exemplarischer
Beispiele zwei neue Entwicklungen in der Quantenfeldtheorie vorzuführen
in denen mathematische Begriffe der Topologie eine Rolle spielen.

Ich darf zunächst daran erinnern, daß die Grundlagen der Quantenfeldtheorie gleich nach der Entdeckung der Quantenmechanik, also bereits in den
zwanziger Jahren entwickelt wurden. Es handelt sich um eine physikalische
Theorie, die Einstein zumindest in ihren Anfängen gekannt hat; der Quantisierung des Maxwell-Feldes lag die Photonenhypothese Einsteins von
1905 zugrunde. Aus philosophischen Gründen stand Einstein den quanten-
theoretischen Prinzipien mit wachsendem Lebensalter zunehmend kritischer
gegenüber. Es ist zu vermuten, daß die neueren topologischen und diffe-
rentialgeometrischen Methoden,wie sie in den letzten Jahren in der
Quantenfeldtheorie zur Anwendung kamen,wegen ihres geometrischen Gehalts
der Einsteinschen Ideenwelt näher gestanden hätten. Viele dieser Methoden
und Begriffe sind erst innerhalb der letzten zwanzig Jahre entwickelt
worden. Um neuere Entwicklungen der Quantenfeldtheorie richtig einordnen
zu können ist es zweckmäßig, ihre älteren Entwicklungsetappen ins Gedächtnis zurückzurufen. Die erste hinsichtlich ihrer Grundgleichungen
völlig ausgearbeitete Quantenfeldtheorie war die Quantenelektrodynamik,
die Lagrange-Feldtheorie der Wechselwirkungen zwischen Elektronen,
Positronen und Photonen. Innerhalb kurzer Zeit wurden die Übergangsampli-
tuden für zahlreiche Prozesse in unterster Ordnung der Störungsentwicklung berechnet. In dieser Pionierphase wurden konzeptionelle Aspekte
hintenangestellt. In vielen Rechnungen wurden die Fourierkoeffizienten
von wechselwirkenden Feldern d.h. die Feldquanten fälschlicherweise mit
physikalischen Teilchen identifiziert, was mitunter zu einer heillosen
Verwirrung von Observablen und nichtobservablen Größen führte. Ein für
die begriffliche Klärung wichtiger Begriff, die "S-Matrix" (Streuoperator)
wurde zwar bereits Anfang der 4Oer Jahre eingeführt, es brauchte jedoch
weitere 10 Jahre, um den Zusammenhang zwischen Quantenfeldtheorie und
der S-Matrix besser zu verstehen. Damit war es dann möglich zu zeigen,
daß die in höheren Ordnungen der Störungstheorie auftretenden Divergenzen

in nichtobservable Teile der Theorie absorbiert werden können und daß insbesondere die Observable S-Matrix bzw. physikalische Wirkungsquerschnitte sich eindeutig berechnen lassen. Diese sogenannte Renormierungstheorie ist vor wenigen Jahren durch die Einbeziehung aller renormierbarer Wechselwirkungen (insbesondere die Lagrange Theorien nichtabelscher Eichtheorien) zu einem gewissen Abschluß gelangt. Sie wurde ein zentraler Bestandteil der modernen Quantenfeldtheorie.

Mitte der 50er Jahre wurde der Zusammenhang zwischen den asymptotischen Teilchenzuständen der S-Matrix und den wechselwirkenden Heisenbergfeldern in einer von der (zumindestens im Falle der starken Wechselwirkungen fragwürdigen) Störungstheorie unabhängigen Weise untersucht. Dieses vertiefte mathematische Studium der Quantenfeldtheorie, auch "Axiomatik" genannt, hat zwar zu keinen direkten neuen Voraussagen geführt, war jedoch für eine begriffliche Klärung sehr wichtig. Z.B. wurde damals klar, daß wenige lokale Felder zu einem reichhaltigen Spektrum von Teilchenzuständen Anlaß geben können (Problem der gebundenen Zustände). Außerdem wurden gewisse, aus den allgemeinen Prinzipien der Einstein-Kausalität und des Teilchenspektrums folgende,Summenregeln, die sog. "Dispersionsrelationen", mit dem Experiment als konsistent nachgewiesen. Aus dieser Zeit stammt auch der einzige in der Entwicklung der Quantenfeldtheorie aufgetretene "Glaubenskrieg": die Abspaltung einer reinen S-Matrix Methode (d.h. ohne Quantenfelder). Dieser Weg erwies sich mangels dynamischer Substanz als eine Sackgasse. Auch die "axiomatische Methode" teilte dasselbe Schicksal, wenigstens dort wo sie nicht zu ihren Wurzeln, der Lagrange'schen Feldtheorie, zurückfand.

Obwohl von der Feldtheorie losgelöste phänomenologische Modelle mitunter erstaunliche Erfolge aufwiesen, so entstand der eigentliche Gewinn an Erkenntnis meistens durch die Konfrontation derartiger Modelle mit der Quantenfeldtheorie. Dabei zeigte es sich, daß diejenigen Elemente des phänomenologischen Modells,die sich mit feldtheoretischen Vorstellungen als konsistent herausstellten, in den meisten Fällen als die längerfristig erfolgreichen erwiesen.

Es ist nicht meine Aufgabe (ich verweise den Leser auf andere das Thema Feldtheorie behandelnden Symposiumsvorträge), die Entwicklung zu schildern, die schließlich Ende der 60er bzw. Anfang der 70er Jahre zur Eichtheorie der elektro-schwachen Wechselwirkung bzw. zur "Quantenchromodynamik" (QCD) der starken Wechselwirkung führten. Eine nichtabelsche Eichtheorie wirft eine Reihe von neuen begrifflichen und mathematischen Problemen

auf. Diese Probleme treten einem in besonders augenfälliger Weise in der
Quantenchromodynamik entgegen (in dem elektro-schwachen Modell von Salam
und Weinberg sind einige dieser Probleme durch den eingebauten Higgs-
Mechanismus "kaschiert").Das Hauptproblem ist das Verständnis des physikalischen Teilchenspektrums in einem Modell,in welchem aufgrund von Infra-
roteigenschaften die fundamentalen Felder sehr weit von den physikalischen Teilchen entfernt sind.Ein Teilaspekt dieses Problems ist das
"Quark-Confinement" d.h. die Frage, warum die so nützlichen in der
Lagrange Funktion auftretenden Quark-Felder nicht als Teilchen in Erscheinung treten. Ein vermutlich eng damit zusammenhängendes Problem
ist die Frage der Vakuumstruktur nichtabelscher Eichtheorien. Bei den
Versuchen, diese Fragen zu verstehen, haben mathematische Begriffsbildungen der Topologie (Homotopie, Kohomologie, deformationsinvariante
Eigenschaften elliptischer Differentialoperatoren) eine zunehmende Rolle
gespielt. In quasiklassischen Approximationen treten diese neuen Strukturen zunächst als "Solitonen"[1]und "Instantonen" in Erscheinung[2].
"Solitonen" sind eng mit dem Teilchenspektrum verknüpft während "Instantonen" mit der Vakuumstruktur zusammenhängen. Der von quasiklassischen
Approximationen unabhängige Teil der "Solitonphysik" läßt sich mit Hilfe
der aus der statistischen Mechanik (Ising-Modell) stammenden Kramers-
Wannier Dualität und der "dualen" Operatoren (Ordnungs- und Unordnungs-
felder) in adäquater Weise in die Quantenfeldtheorie einbeziehen.Die
quantenfeldtheoretische Auswirkung von Instantonen ist der in der Funktionalintegration auftretende θ-Winkel. In der QCD ist der θ-Winkel
mit der topologischen Bedeutung der Axial-Anomalie verknüpft. Wir werden
jeweils die Dualitätsalgebra und die Relation zwischen Anomalie-Gleichung
und θ-Vakua an einem illustrativen leichtverständlichen Beispiel erläutern und dann einige allgemeinere Bemerkungen daran anknüpfen. Dabei
soll hier nicht einmal der Versuch unternommen werden, mathematische
Begriffsbildungen wie Homotopie und Kohomologie systematisch darzustellen.
Der tiefere Grund für die Relevanz topologischer Methoden in der gegenwärtigen Quantenfeldtheorie erscheint mir folgender : das Hauptanliegen
topologischer Methoden innerhalb der Mathematik ist das strukturelle
Verständnis von (endlich- und-unendlich dimensionalen)Mannigfaltigkeiten.
Dabei ist die in der Kohomologie- und Homotopie-Theorie sichtbare"Groß-
struktur" am weitesten ausgearbeitet. Auf der anderen Seite sind Quantenfeldtheoretische Modelle wie z.B. die Quantenchromodynamik trotz ihrer
einfachen Grundgleichungen von beträchtlicher konzeptioneller und mathematischer Komplexheit. Es erscheint deshalb realistisch am gegenwärtigen
Entwicklungszustand der QCD sich von der Anwendung dieser Methoden ein
qualitatives strukturelles Verständnis zu erhoffen.

2.Topologische Solitonen und "Dualität"

Auf "topologische Solitonen" wurde man zuerst in der klassischen Feldtheorie aufmerksam.[1] Die einfachsten Beispiele findet man in zwei-dimen-
sionalen Modellen. Wir erinnern an zwei Beispiele:

a) A^4 Theorie mit dem Energiefunktional

$$E(A) = \frac{1}{2} \int_{-\infty}^{\infty} dx \left[(\frac{\partial A}{\partial t})^2 + (\frac{\partial A}{\partial x})^2 \right] + \int_{-\infty}^{\infty} V(A) dx \tag{1}$$

wobei das Potential $V(A)$

$$V(A) = - \frac{1}{2} \mu^2 A^2 + \frac{1}{4} \lambda A^4 \tag{2}$$

zwei nichttriviale Minima besetzt:

$$A_o = \pm \frac{\mu}{\sqrt{\lambda}} \tag{3}$$

Diese Minima sind absolute Minima von $E(A)$ d.h. klassische Vakua.
Klassische Konfigurationen endlicher Gesamtenergie müssen asymptotisch
für $x \to \pm\infty$ gegen eines dieser Minima streben:

$$A(x,t) \xrightarrow[x \to \pm\infty]{} A_o \tag{4}$$

Stetige interpolierende Funktionen mit dieser Eigenschaft bilden zwei
inequivalente Klassen

$$A \xrightarrow[x \to +\infty]{} \frac{\mu}{\sqrt{\lambda}} \quad \text{und} \quad A \xrightarrow[x \to \pm\infty]{} \mp \frac{\mu}{\sqrt{\lambda}} \tag{5}$$

die beiden anderen Möglichkeiten die Werte $\pm \frac{\mu}{\sqrt{\lambda}}$ auf die unendlich fernen
Punkte zu verteilen sind wegen der Symmetrie $A \to -A$ zu (5) physikalisch
äquivalent; ein Sachverhalt der insbesondere in der quantisierten Theorie
eine wesentliche Bedeutung erhält. Die Existenz zweier nichtäquivalenter
nichtdeformierbarer Klassen drückt sich mathematisch in der Homotopie-
Sprache folgendermaßen aus:

$$\pi_o (Z_2) = Z_2 \tag{6}$$

Z_2 sind die beiden asymptotischen Feldwerte (die Vakuummannigfaltigkeit)
und die nullte Homotopie-Gruppe entspricht den beiden unendlich fernen
Punkten.

Die nichttriviale Klasse wird "Kinksektor" genannt und in ihr gibt es ein
nichtriviales statisches Minimum von $E(A)$ (topologisches Soliton):

$$A_{kink} (x) = - \frac{\mu}{\sqrt{\lambda}} \tan h \frac{\mu}{\sqrt{2}} \tag{7}$$

welches nach Anwendungen von Poincaré-Transformationen eine 3-parametrige
Schaar von Minima erzeugt.

b) Sinus-Gordon Theorie

Hier wählen wir für V(A) die periodische Funktion

$$V(A) = 1 - \cos A \tag{8}$$

mit klassischen Vakua: $A_o = 2\pi n$

Die Vakuummannigfaltigkeit ist isomorph zu Z und deshalb haben die Deformationsklassen

$$\pi_0 \text{(Vakua)} = \pi_0(Z) = Z \tag{9}$$

eine additive Struktur, d.h. die Deformationsklassen lassen sich durch eine additive Ladung charakterisieren. Die statischen Lösungen

$$A_{sol} = \mp 4 \text{ arc tg } e^x \tag{10}$$

entsprechen Solitonen und Antisolitonen mit Ladungen ± 1 und auch für höhere Ladungssektoren lassen sich (zeitabhängige!) Lösungen angeben.

Die Verallgemeinerung auf höhere Dimensionen führt wegen des Derrick-Theorems[1] mehr oder weniger zwangsläufig[2] zu Eichtheorien mit skalaren Feldern. Da zu einer Zeit die unendlich fernen Punkte (d.h. die Richtungen) eine d-2 parametrige Mannigfaltigkeit bilden (d= Raumzeitdimension), handelt es sich bei höher-dimensionalen topologischen Solitonen um das Studium von:

$$\pi_{d-2} \text{ (Vakua)} \tag{11}$$

Den meisten Modellen liegt eine Symmetriegruppe zugrunde (im vorhergehenden Fall Z_2 bzw. Z) die durch die Vakuumlösungen auf eine Untergruppe H "heruntergebrochen" wird; die Vakuumfeldmannigfaltigkeit ist dann ein homogener Raum[1][2] G/H. Statische nichttriviale Lösungen in der Klasse $\pi_1(G/H)$ heißen Vortices während man bei $\pi_2(G/H)$ von Monopolen redet. Die Konstruktion expliziter analytischer Lösungen ist oft sehr schwierig und nur in gewissen Grenzfällen möglich, die rechnerischen Schwierigkeiten der Konstruktion von Multivortex- oder Multimonopol-Lösungen waren bisher unüberwindbar. Die klassischen Lösungen werden als semiklassische Approximation einer Quantenfeldtheorie behandelt, die den Vortices oder Monopolen entsprechendem quantentheoretischem Gebilde sind Teilchen, jedoch besitzen sie wegen ihrer starken Kopplung an elektromagnetische Felder eine Reihe von exotischen Eigenschaften: sie sind in einem viel stärkerem Maße "Infrateilchen" (d.h. treten nur zusammen mit einer langreichenden Photonwolke auf) als dies bei Elektronen der Fall ist.

An dieser Stelle ist eine Warnung am Platz: semiklassische Einsichten haben mit feldtheoretischen Strukturen im allgemeinen wenig zu tun. Die Ausnahme bilden spezielle Systeme, für die auf Grund von (in der Feldtheorie unendlich vielen) Erhaltungssätzen das semiklassische Spektrum exakt ist (Beispiel: das Wasserstoffatom, die Sinus-Gordon Theorie).

Um die Quantenfeldtheorie von Kinks zu verstehen, ist man deshalb in letzter Zeit einen anderen Weg gegangen: den der Untersuchung der Dualitätsstruktur[3)4)]. Wir studieren im folgenden diesen Begriff anhand eines speziellen Modells.

Diskussion eines Z_2 topologischen Kink-Modells mit Hilfe der dualen Algebra.

Wir betrachten ein zweidimensionales selbstkonjugiertes Diracfeld d.h. einen Majoranaspinor.

$$\psi(x) = \frac{1}{\sqrt{2\pi}} \int \left(e^{-ipv} u(p)\, a(p) + e^{ipx} v(p)\, a^+(p) \right) \frac{dp}{2p_0} \tag{12}$$

$$p = m\,(ch\theta, sh\theta) \tag{13}$$

In der γ_5-diagonalen Realisierung der γ-Matrizen haben die Spinoren die Gestalt:

$$u = \frac{\sqrt{m}}{2} \begin{pmatrix} -e^{-\theta/2} \\ e^{\theta/2} \end{pmatrix} \quad , \quad v = \frac{\sqrt{m}}{2} \begin{pmatrix} e^{-\theta/2} \\ e^{\theta/2} \end{pmatrix} \tag{14}$$

d.h.: $\quad \psi_1^+ = -\psi_1 \, , \, \psi_2^+ = \psi_2 \tag{15}$

Eine Bogoljubov-Valatin Rotation zur Zeit t=o

$$\psi_1 \longrightarrow \cos\vartheta(x)\,\psi_1(x) + \sin\vartheta(x)\,\psi_2(x) = \psi_1'$$
$$\psi_2 \longrightarrow \cos\vartheta(x)\,\psi_2(x) + \sin\vartheta(x)\,\psi_1(x) = \psi_2' \tag{16}$$

kann als kanonische Transformation formal durch einen Operator U implementiert werden:

$$U(\vartheta) = \frac{1}{\langle\ \rangle} \exp i \int_{-\infty}^{\infty} (\vartheta(x'): \psi_1\,\psi_2\,) \, dx' \tag{17}$$

Die Division durch den Vakuumerwartungswert ergibt die zweckmäßige Normierung: $\langle U \rangle = 1$. Dieser Operator führt nur dann aus dem physikalischen Hilbertraum nicht heraus (d.h. erzeugt keine Zustände unendlicher Energie) falls er im asymptotischen Bereich mit der Hamiltondichte kommutiert:

$$\mathcal{H}(x) \, U(\vartheta) \underset{x \to \pm\infty}{=} U(\vartheta) \, \mathcal{H}(x) \tag{18}$$

Dies entspricht der endlichen Energiebedingung im vorher diskutierten klassischen Fall. Der Unterschied der asymptotischen Werte von ϑ muß also ein ganzes Vielfaches von π betragen. Da wir o.B.d.A. bei $x=+\infty$ zu Null normieren können, bedeutet dies:

$$n\pi \xleftarrow{\; x'=-\infty \;} \vartheta(x') \xrightarrow{\; x'=+\infty \;} 0 \tag{19}$$

Der neben dem Vakuumsektor vorhandene einzige nichttriviale Kink-Sektor entspricht $n=1$, d.h. die multiplikative Kink-Struktur korrespondiert zum klassischen A^4 Modell.

Es ist nun zweckmäßig zum lokalen Limes

$$\vartheta(x') = \pi \, \Theta(x-x') \tag{20}$$

überzugehen. In diesem Fall wird aus der Bogoljubov-Valatin Transformation eine lokale und damit Lorentzinvariante Vertauschungsrelation:

$$U(x) \, \psi(y) = \begin{cases} \psi(y) \, U(x) & x < y \\[2mm] -\psi(y) \, U(x) & x > y \end{cases} \tag{21}$$

wobei $\quad U(x) \equiv U(\vartheta) \tag{22}$

mit ϑ wie in (20)

Ein Formalismus für die systematische Lösung von Vertauschungsrelationen des Typs (21) mit freien Feldern wurde von Sato et.al.[6] ausgearbeitet. Dasjenige U welches sich durch einen lokalen Grenzwert aus einer Bogoljubov-Valatin Drehung um π ergibt, hat die Form:[6][5]

$$U(x) = \; : \exp - \frac{1}{2\pi} \int d\Theta_p \, d\Theta_q \; \{ \operatorname{cth} \frac{\Theta_p - \Theta_q - i\varepsilon}{2} \; a^+(\Theta_p) a(\Theta_q) e^{i(p-q)x}$$

$$+ (\operatorname{th} \frac{\Theta_p - \Theta_q}{2} \; a^+(\Theta_p) a^+(\Theta_q) e^{i(p+q)x} + h.konj.) \} \tag{23}$$

Vermittels einer kurzen Abstandsentwicklung von U mit ψ kann man sich ein weiteres lokales skalares Feld V(x) verschaffen:

$$V(x) = \text{führender Term in } \lim_{y \to x} U(x) \cdot \psi(y) \tag{24}$$

$$= : \int (e^{-ipx} \, e^{\frac{i\pi}{4}} \, a(\Theta) + h.c) \, d\Theta \; U(x) :$$

Der führende Term in der Entwicklung für kurze Abstände ist für beide Komponenten von ψ (bis auf einen numerischen Vorfaktor) der gleiche. Aus (24) ergibt sich die Vertauschungsrelation mit ψ :

$$V(x) \; \psi(y) = \begin{cases} -\psi(y) \; V(x) & x < y \\ \psi(y) \; V(x) & x > y \end{cases} \tag{25}$$

Die skalaren Bosonfelder U und V erfüllen die Vertauschungsrelation:

$$U(x) \; V(y) = \begin{cases} V(y) \; U(x) & x < y \\ -V(y) \; U(x) & x > y \end{cases} \tag{26}$$

Diese algebraische Relation heißt Z_2-Dualitätsrelation[7]. Die Z_N duale Algebra eines zweidimensionalen Modells ist durch folgende Vertauschungsrelation definiert:

$$U(x) \; V(y) = \begin{cases} V(y) \; U(x) & x < y \\ e^{\frac{2\pi i}{N}} \; V(y) U(x) & x > y \end{cases} \tag{27}$$

Es wäre interessant, eine explizite feldtheoretische Realisierung dieser Z_N Algebra für $N > 2$ zu konstruieren.

Wegen der Lorentzinvarianz der Felder können in allen Vertauschungsrelationen (21),(25),(26),(27) die Ungleichungen $x \lessgtr y$ auch gelesen werden als:

   x links bzw. rechts raumartig relativ zu y.

Für das Folgende ist es zweckmäßig, für Operatoren U und V mit den Vertauschungsrelationen (21),(22),(26) als eine generische Bezeichnungsweise μ und σ einzuführen. Dann stellen die konkreten Operatoren U und V in (23)(24) eine Realisierung dieser Vertauschungsrelationen dar. Es entsteht die Frage, ob es noch weitere Realisierungen gibt. Offenbar führt die Vertauschung $U \leftrightarrow V$ zu einer Änderung von Vorzeichen und ist deshalb keine Realisierung. Falls man jedoch zusätzlich konstante Fermion Operatoren c und c^+ mit

$$\{c, \psi\} = o \quad , \quad \{c^+, \psi\} = o$$

einführt erhält man als eine weitere Realisierung[5]

$$\mu(x) = i(c+c^+)V(x) \tag{28}$$
$$\sigma(x) = (c+c^+)U(x)$$

Diese zweite Realisierung lebt zwar in einem größeren Hilbertraum, denn zusätzlich zum Fockraum des freien Feldes gibt es jetzt ein entartetes Vakuum (Eigenzustände zu $c+c^+$):

$$|0_{\pm}\rangle = \frac{1}{2}(|o\rangle \pm c^+|o\rangle) \tag{29}$$

jedoch ist dieser neue Freiheitsgrad ein "spurionischer" d.h. nicht dynamischer. Wegen der Eindeutigkeit von U in (21) als Grenzfall einer Bogoljubov-Valatin Drehung ist offenbar der Trick der Einführung von "Spurionen" die einzige Möglichkeit zu inäquivalenten Realisierungen zu kommen. Die Einführung von weiteren Spurionen bringt keine neue Realisirung. Die beiden Realisierungen entsprechen zwei Phasen,

Phase I : $\langle\mu\rangle = 1$, $\langle\sigma\rangle = o$ (30a)

Phase II : $\langle\sigma\rangle_{\pm} = \pm 1$, $\langle\mu\rangle_{\pm} = o$ (30b)

In der Phase I beschreibt der Operator σ einen "Kink" des μ-Feldes, während in der Phase II die Situation sich umkehrt. Z.B. gilt in Phase I in einem normierten Kinkzustand: $|\sigma(f)\rangle = \int\sigma(x)|o\rangle f(x)d^2x$

$$\langle\sigma(f)|\ \mu(x)\ |\sigma(f)\rangle = F(x) \xrightarrow[x\to\mp\infty]{} \pm 1 \tag{31}$$

Eine weitergehende Untersuchung [8][9] zeigt, daß der exponentielle Abfall der Zweipunktkorrelation:

$$\langle\ \sigma(x)\ \sigma(y)\ \rangle\ _{|x-y|\to\infty}\ e^{-c|x-y|} \tag{32}$$

quasiklassisch verstanden werden kann als eine Kondensation von μ entlang eines dünnen Schlauches, der x und y miteinander verbindet.[3][8]

Beim Übergang von Phase I nach Phase II findet hinsichtlich der reinen μ und σ Korrelationen nur eine Vertauschung der Korrelationsfunktionen statt (Selbstdualität), die Anwesenheit der Spurionen macht sich in den gemischten Funktionen bemerkbar.

Unsere bisherige Betrachtung war lediglich eine mathematische Illustration von Quantenkinks und der Z_2-Dualitätsalgebra. Eine weitere Überlegung für deren Details wir den Leser auf die Literatur [9][5] verweisen zeigt, daß μ und σ gerade die Skalenlimits des Unordnungs- bzw. Ordnungsparameters des zweidimensionalen Ising-Modells sind. Die euklidischen

Fortsetzungen der Korrelationsfunktionen sind mit den von Mc Coy, Tracy und Wu[10] berechneten identisch. Dabei entsprechen die Phasen I und II den Skalenlimites $T \rightarrow T_c \pm o$. Die Dualitäsalgebra als Vertauschungsrelationen zwischen Ordnungs- und Unordnungsparameter wurden für höhere dimensionale Modelle insbesondere für Eichtheorien von 't Hooft[3] eingeführt und untersucht. Sie spielen in der Diskussion des "Quark-Confinement" Problems eine wichtige Rolle.

Die hier diskutierte Ising-Feldtheorie ist ein Spezialfall einer allgemeinen Klasse von Z_N-Feldtheorien deren Spektren und S-Matrix exakt bekannt sind[11]. Das Problem der Rekonstruktion von μ und σ aus der bekannten S-Matrix , das sogenannte feldtheoretische "bootstrap" Programm ist für diese Modelle außer für den relativ trivialen Fall der Ising S-Matrix[12] bisher nicht durchgeführt. Es ist überraschend, daß die nach der trivialen Feldtheorie des freien Feldes einfachste Klasse von nichttrivialen Feldtheorien bereits eine reichhaltige topologische "Kink"-Struktur besitzen, wie sie in der Z_N Dualität zum Ausdruck kommt.

3. Topologische Instantons und euklidische Funktionalintecrale

Ein zweites exemplarisches Beispiel welches repräsentativ für eine neuere Entwicklung in der Feldtheorie steht, ist die QED_2, die zuerst von Schwinger untersucht wurde[13]. Es ist bekannt, daß die Berechnung von Korrelationsfunktionen mit Hilfe der euklidischen Funktionalintegration über Fermionfelder die Kenntnis zweier wichtiger Größen erfordert, nämlich die A_μ abhängigen Green's Funktionen der Dirac Gleichung.

$$i\gamma_\mu \; D_\mu \; G(x,y;A_\mu) = -\delta(x-y) \tag{33}$$

und die Determinante des Dirac-Operators oder

$$\Gamma = - \ln \frac{\det i \not{D}}{\det i \not{\partial}} \tag{34}$$

Falls die A_μ-Konfiguration zu einer nichttrivialen Topologie gehören, d.h. eine nichtverschwindende Windungszahl besitzen:

$$n = \frac{e}{4\pi} \int \varepsilon_{\mu\nu} F_{\mu\nu} = \frac{e}{2\pi} \oint A_\mu dx_\mu \tag{35}$$

erfordert die Ausrechnung von G und Γ ein genaueres mathematisches Verständnis[14]. Man geht zuerst von R^2 zum Einpunkt-kompaktifizierten Raum R_c^2 mit der Metrik

$$ds^2 = (\frac{2R^2}{R^2+x^2})^2 \; \Sigma dx_i^2 \tag{36}$$

über. Der Dirac Operator hat konforme Transformationseigenschaften:

$i\not{D} \to i\not{D}_c$

$$i\not{D}_c = (\frac{R^2+x^2}{2R^2})^{+3/2} i\not{D} (\frac{2R^2}{R^2+x^2})^{-1/2} \qquad (37)$$

Das diskrete selbstadjungierte Eigenwertproblem

$$i\not{D}_c u_i = \lambda_i u_i \qquad (38)$$

mit dem inneren Produkt $(u,w) = \int u^+ \sqrt{\det g}\, w\, d^2x$ kann auch geschrieben werden als:

$$i\not{D} u_i = \lambda_i \frac{2R^2}{R^2+x^2} u_i \qquad (39)$$

$$(u,w) = \int u^+ \frac{2R^2}{R^2+x^2} w\, d^2x \qquad (40)$$

Die λ_i und u haben die Dimension $\frac{1}{2}$ in Masseneinheiten, somit hat das innere Produkt (40) die Dimension $-\frac{1}{2}$. Ein orthonormales System von u_i enthält also einen willkürlichen Längenparameter $\frac{1}{\mu}$:

$$(u_i u_k) = \delta_{ik} \frac{1}{\mu} \quad , \quad \lambda_i \neq \lambda_k$$

Dieser Formalismus [15] legt folgende Definition von G und Γ nahe [16]:

$$G(x,y;A) = -\mu \sum_{\lambda_i \neq 0} \frac{u_i(x)\, u_i^+(y)}{\lambda_i} \qquad (41)$$

$$\Gamma = -\frac{1}{2}\ln \frac{\det i \not{D}_c}{\det i \not{\partial}_c} = \lim_{s \to 0} \{(\zeta_A'(s,R) - \zeta_0'(s,R) + \ln \mu [\zeta_A(s,R) - \zeta_0(s,R)]\}$$

$$\text{mit:} \quad \zeta_A(s,R) = \sum_{\lambda_i > 0} \frac{1}{(\lambda_i^2)^s}$$

wobei der Index A=o die ζ-Funktion des freien kompaktifizierten Dirac-operators bedeutet. Der ln μ abhängige Term ergibt sich dadurch, daß wir Γ mit Hilfe der dimensionslosen Eigenwerte definieren:

$$\Gamma = \lim_{s \to 0} \frac{d}{ds} \sum_{\lambda_i > 0} (\frac{\mu^2}{\lambda_i^2})^s \Big|_o^A \qquad (42)$$

Es gibt nun in der QED$_2$ einen eindeutigen Zusammenhang zwischen Nullwerten in der euklidischen Dirac-Gleichung (39) und der Windungszahl n (35). Falls die Windungszahl verschwindet, bedeutet der R$_c^2$ Formalismus keine Abänderung zum üblichen euklidischen Formalismus: alle "klassischen" Größen wie Green Funktionen der euklidischen Spinor-Felder sind (wegen der konformen Invarianz) identisch mit den üblichen euklidischen Funktionen z.B. (41) ist nur eine diskrete Darstellung von (33). Für die "Einloop" Größen wie Determinanten, induzierte Ströme etc. bedeuten die R$_c^2$ Methode eine (natürliche) Regularisierung. Es ist eine typische Eigenschaft von Γ's zweidimensionaler Eichtheorien, daß sie bei verschwindender Windungszahl R-unabhängig werden, z.B. in der QED$_2$:

$$\Gamma = \frac{e^2}{4\pi} \int F_{\mu\nu}(x) \; D(x-y) \; F_{\mu\nu}(y) \; d^2x \; d^2y \tag{43}$$

$$D(\xi) = -\frac{1}{4\pi} \ln \mu^2 \xi^2$$

Dies hängt mit der fehlenden Anomalie des Energie-Impuls-Tensors zusammen [15]. In QCD$_4$ ist Γ R abhängig und der Limes R→∞ muß zusammen mit der Wellenfunktions- und Kopplungskonstanten-Renormierung nach der Funktionalintegration über die Eichfelder A$_\mu$ vorgenommen werden.

Die Variation von Γ nach A$_\mu$ führt zum induzierten Strom:

$$- ej_\mu = \frac{\delta\Gamma}{\delta A_\mu(x)} \tag{44}$$

Die Benutzung des DeWitt-Seley Formalismus [17] des zu i$\not{D}_c$ gehörigen Wärmeleitungsproblems

$$\left(-\frac{\partial}{\partial\tau} + (i\not{D}_c)^2\right) h(x,y;\tau) = o \; , \; h(x,y;o) = \frac{1}{\sqrt{\det g}} \delta(x-y) \tag{45}$$

gestattet eine rigorose Herleitung der Anomalie-Gleichung des Axial-Stromes:

$$- \tilde{\partial}_\mu j_\mu = \partial_\mu j_{\mu 5} = \frac{e^2}{2\pi} \varepsilon_{\mu\nu} F_{\mu\nu} + 2 \frac{2R^2}{R^2+x^2} \sum_i u_i^+(o) \; \gamma_5 u_i(o) \tag{46}$$

wobei als letzter Term die orthonormierten Nullmoden von i$\not{D}_c$ auftreten. Diese Funktional-Differentialgleichung für Γ hat die Lösung [15]:

$$\Gamma = \frac{e^2}{4\pi} \int F_{\mu\nu}(x) \; D(x-x') F_{\mu\nu}(x') d^2x \; d^2x' + \text{tr} \ln N[A] - n \ln \mu R$$

wobei N[A] eine aus den unnormierten Nullmoden gebildete endliche Matrix ist. Die Nullmoden und die für n>o modifizierten Green's Funktionen (41) hängen von R ab, aber bei der Berechnung der QED_2 Korrelationsfunktionen heben sich die R abhängigen Terme heraus. Man erhält eine quadratische induzierte Wirkung [15], deren Funktionalintegration die Vakuumerwartungswerte im sogenannten Θ-Vakuum ergibt. Das feldtheoretische Vakuum ist als Konsequenz des Vorhandenseins nichttrivialer topologischer Konfigurationen (die Minima der induzierten Wirkung heißen induzierte "Instantonen") entartet, und diese Entartung wird (analog zum Falle des Josephson-Winkels in der Supraleitungstheorie) durch einen Winkel Θ parametrisiert [18][19][20]. Die Einführung eines Fermion Massenterms würde zur Aufspaltung der verschiedenen Θ-Vakua führen. In der 4-dimensionalen Quantenchromodynamik gibt es sogar elementare Instantonen [21] i.e. klassische Minima der reinen Eichtheorie ohne Fermionen. Hier ist jedoch bisher eine Ausintegration der Fermionen und eine Bestimmung der Minima der induzierten Wirkung in Anwesenheit von Quarks noch nicht durchgeführt. Die hier anhand der QED_2 geschilderte Methode läßt sich auf QCD_2 übertragen und gestattet eine Bestimmung von Γ und G. [22] Die Eigenschaften der induzierten Wirkung, die zum "Colour-Screening" und zum "Quark-Confinement" führen, sind bisher nur in abelschen zweidimensionalen Theorien verstanden [23][24].

Ich hoffe, durch die Diskussion dieses Beispiels auf die Relevanz topologischer Gesichtspunkte in der Funktionalintegration hingewiesen zu haben.

Referenzen und Fußnoten

1. S.Coleman, Lectures at the E.Majorana Summer School, Erice,Sicily 75.
2. L.J.Boya, J.F.Carinena and J.Mateos, Fortschritte der Physik **26**,175. (1978). In diesem Übersichtsartikel kann der Leser die wichtigsten Referenzen über topologische Methoden finden.
3. G.'t Hooft, Nucl.Phys. **B138**(1978),1.
4. S.Mandelstam, Phys.Rev. **D19**(1975)3026.
5. B.Schroer and T.T.Truong, Z_2-Duality Algebra in D=2 Quantum Field Theory, FU preprint (1979) to be published in Nucl.Phys.B.
6. M.Sato, T.Miwa and M.Jimbo, Proc.of the Japan Academy, **53A**(1977)6.
7. Der Dualitätsbegriff in der statistischen Mechanik geht auf Kramers und Wannier zurück. Der "Unordnungsparameter" d.h. das zum Ordnungsfeld duale Feld wurde zuerst von Kadanoff eingeführt. L.P.Kadanoff und H.Ceva, Phys.Rev.B3 (1971)3918. Die duale Algebra ist zuerst von 't Hooft in systematischer Weise untersucht worden, s.Ref.3).

8. E.Fradkin und L.Susskind, Phys.Rev. D17(1978) 2637.

9. M.Sato, T.Miwa und M.Jimbo, Field Theory of the 2-dimensional Ising Model in the scaling limit,RIMS preprint 207 (1976) unpublished.

10. B.M.Mc Coy, C.A.Tracy und T.T.Wu, Phys.Rev.Letters 38 (1977)793.

11. R.Koberle und J.A.Swieca,"Z_N Field Theories, Universidade de Sao Paulo, Sao Carlos, preprint 1979.

12. B.Berg, M.Karowski und P.Weisz, FUB-HEP 78/16, to be published in Phys.Rev.D.

13. J.Schwinger, Phys.Rev.128,2425 (1962).

14. B.Schroer, Schladming Lectures 1978, Acta Physica Austriaca, Suppl.XIX,155-202 (1978).

15. M.Hortaçsu, K.D.Rothe und B.Schroer, Generalized QED_2 and Functional Determinants, FU Berlin preprint, March 1979.

16. S.Hawking, Commun.Math.Phys.55 (1977)133.

17. The Computations based on DeWitts formalism were performed in N.K.Nielsen, Nordita preprint 1978, unpublished.

18. C.Callan, R.Dashen und D.Gross, Phys.Lett.D13(1976)3398.

19. R.Jackiw und C.Rebbi, Phys.Rev.D13 (1976).

20. G.'t Hooft , Phys.Rev.D12 (1976)3432.

21. A.A.Belavin, A.M.Polyakov, A.S.Schwarz und Yu.S.Tynpkin, Phys.Lett. 59B (1975)85.

22. N.K.Nielsen, K.D.Rothe und B.Schroer, "Fermionic Green's Function and Functional Determinant in QCD_2",FUB/HEP 5/May 79.

23. L.V.Belvedere, K.D.Rothe, B.Schroer and J.A.Swieca,Nucl.Phys.B 153 (1979) 112.

24. H.Rothe, K.D.Rothe and J.A.Swieca, "Screening versus Confinement", PUC preprint 24/78 to appear in Phys.Rev. D.

ARE FORCES BETWEEN LEPTONS AND QUARKS PHENOMENA OF A GAUGE FIELD THEORY?

Peter Minkowski, Institute for Theoretical Physics, University of Bern,
    Switzerland

> Die Wissenschaft erzieht den Menschen zum
> wunschlosen Streben nach Wahrheit und Objek-
> tivität, sie lehrt den Menschen, Tatsachen
> anzuerkennen, sich wundern und bewundern zu
> können.
>
>   Lise Meitner

I shall tentatively defend the answer to the question raised in the title: yes.

 To this end we shall pursue the intricate relationship

  local gauge invariance - universality of interaction

considering common aspects of gravity, electrodynamics, chromodynamics and weak interactions (based on the gauge group $SU2_L xU1$) in the light of quantitative tests and qualitative differences.

1) Gravity $\longleftrightarrow$ gauge symmetry:

    Coordinate transformations

$$x^\mu \longrightarrow y^\mu = y^\mu(x)$$

   (4-dimensional space-time)

The universal local source of gravity is the energy-momentum density according to the Einstein equations

$$R_{\mu\nu} - \frac{1}{2} g_{\mu\nu} R = - 8\pi G_N \theta_{\mu\nu} (c^{-4}) \qquad (1)^{(FN1)}$$

The nonrelativistic limit corresponds to

$$(\theta_{\mu\nu} - \frac{1}{2} g_{\mu\nu} \theta^\alpha{}_\alpha) \underset{\mu=\nu=0}{\longrightarrow} \frac{1}{2} c^2 \rho_m \qquad (2)$$

$$c^2 R_{00} \longrightarrow \text{div } \vec{g} \; ; \quad g^i = - c^2 \Gamma^i_{00} = - (\text{grad}\Phi)_i$$

In eq. (2) ρ_m denotes the mass density, $\vec{g}$ the acceleration by gravity and Φ the gravitational (Newton) potential.

The action associated with eq. (1) is

$$S = \int d^4 x \left\{ \frac{1}{16\pi G_N} c^3 \sqrt{|g|} \; [\underset{(R)}{G}] + \sqrt{|g|} \; c^{-1} L_m \; \binom{\text{matter}}{\text{fields}} \right\}$$

$$G = g^{\nu\beta} [\Gamma^\kappa_{\nu\beta} \Gamma^\mu_{\kappa\mu} - \Gamma^\kappa_{\nu\mu} \Gamma^\mu_{\beta\kappa}] \tag{3}$$

$$L = L [\Gamma^\rho_{\sigma\tau} \longleftrightarrow \partial_\rho g_{\sigma\tau}; \; g_{\mu\nu}; \; \partial_\alpha \psi, \; \psi]$$

$\psi = \{\psi_1, \ldots, \psi_N\}$ denotes the set of matter fields including fermions in particular. This implies that L explicitly depends on the vierbein fields $h^\mu(a)$ and their first derivatives

$$g^{\mu\nu} = \underset{a,a'}{\Sigma} \; h^\mu(a') \; h^\nu(a) \; \eta_{aa'}$$

$$h^\mu(a) = h^a(\mu) \; ; \; \eta_{ab} = \begin{pmatrix} 1 & & & \\ & -1 & & \\ & & -1 & \\ & & & -1 \end{pmatrix} \tag{4}$$

Since L in eq. (3) does not explicitly depend on space-time nor upon orientation in space-time there exist 10 locally conserved quantities:

- energy momentum density (τ^μ_ν)
- angular momentum density $(M^\mu_{\alpha\beta})$

$$\tau^\mu_\nu = \left\{ \begin{array}{l} \partial_\nu h^{\alpha(\beta)} L, \; \partial_\mu h^{\alpha(\beta)} \\ + \partial_\nu \psi L, \; \partial_\mu \psi \end{array} \right\} - \delta^\mu_\nu \cdot L$$

$$M^\mu_{\alpha\beta} = \begin{bmatrix} \eta_{\alpha\alpha'} \; x^{\alpha'} \tau^\mu_\beta - \eta_{\beta\beta'} \; x^{\beta'} \tau^\mu_\alpha & \begin{array}{l} (= \text{orbital angular momentum} \\ \quad \text{density}) \end{array} \\[2mm] \Omega_{\alpha\beta}(h) \; {}^{\sigma\tau} L, \; \partial_\mu h^\sigma(\tau) & (= \text{spin density}) \\[2mm] + \frac{1}{1} S_{\alpha\beta}(\psi) \; L, \; \partial_\mu \psi & \end{bmatrix} \tag{5}^{\text{FN2}}$$

The densities in eq. (5) are affine tensors. Nevertheless we shall envisage the local conservation laws

$$\partial_\mu \tau^\mu_\nu = 0 \; ; \quad \partial_\mu M^\mu_{\alpha\beta} = 0 \tag{6}$$

first, and the globally conserved momentum and angular momentum

$$P_\nu = \int_{t=\text{const.}} d^3 x \; \tau^o_\nu \; ; \quad M_{\alpha\beta} = \int_{t=\text{const.}} d^3 x \; M^o_{\alpha\beta} \tag{7}$$

second.[FN3] In view of the extended observation of the binary pulsar PSR 1913+16[6] which shows a decrease of the revolution period compatible with the quadrupole nature of gravitational waves[1,3,4,7] we look

forward to the experimental traceback of the energy-momentum transport
by these waves, at least in principle.[8]

It is typical of a nonabelian gauge structure that gravitons carry themselves the quantities which are also sources of the associated field-ener-
gy and angular momentum (S = 2).

Let us assume (for now) that gravity is *the* gauge theory setting the
scene for charge like local gauge transformations to have a physical
meaning.

<u>The universality aspect - signal of gauge invariance</u>

The exchange of gravitions between a field generating heavy mass e.g.
the sun and a test particle (Fig. 1)

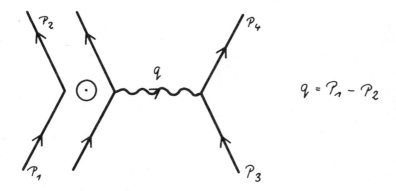

$$q = P_1 - P_2$$

Fig. 1: <u>Graviton exchange diagram</u>

shows the universal nature of the graviton-matter coupling

$$T \sim (P_{3,4\mu} \; P_{3,4\nu}) \; (P_{1,2\alpha} \; P_{1,2\beta})$$

$$P_{1,2} = \frac{1}{2}(P_1 + P_2) \; ; \quad P_{3,4} = \frac{1}{2}(P_3 + P_4)$$

If the test particle moves nonrelativistically the above can be tran-
scribed to the Newtonian gravitational field

$$\vec{g} = - \frac{G_N \; m_\odot}{r^2} \; \frac{\vec{x}}{r} = - \text{grad } \Phi \tag{8}$$

Universality here means equivalence of inertial and gravitational mass

$$m \; \ddot{\vec{x}} = m \; \vec{g} \tag{9}$$

$ \;\; \hookrightarrow E/c^2 \longrightarrow$ independent of other properties of the test particle

The pair $(\vec{g}, \Phi)$ leads the way to a parallel with electromagnetism

$$(\vec{g}, \Phi) \longleftrightarrow (\vec{E}, \phi)$$

The parallel levels for the general (nonstatic case) are

$$\Phi \longrightarrow g_{\mu\nu} \longleftarrow \text{potentials} \longrightarrow A_\mu \longleftarrow \phi$$

$$\vec{g} \longrightarrow \Gamma^\rho_{\sigma\tau} \longleftarrow \text{forces} \longrightarrow F_{\mu\nu} \longleftarrow \vec{E}$$

$$R_{\mu\nu} - \frac{1}{2} g_{\mu\nu} R = -8\pi G_N c^{-4} \theta_{\mu\nu} \longleftarrow \text{field equations} \longrightarrow \partial_\mu F^{\nu\mu} = j^\nu \text{ charges}$$

Going back to the static situation we consider the energy density

$$e_g = \frac{1}{8\pi G_N} \vec{g}^2 \quad \text{FN4)} \qquad\qquad e_{cl} = \frac{1}{2} \vec{E}^2$$

critical
distance

$$a_{kr} = \frac{1}{2} R_{Schwarzschild} \qquad\qquad R^{cl.}_{electron}$$

$$\int_{r \geq a_{kr}} d^3 x \, e_g = \frac{1}{2} m_\odot c^2 \qquad\qquad \int_{r > R^{cl}_e} d^3 x \, e_{el} = \frac{1}{2} m_e c^2$$

$$\rho_m = m_\odot \delta^3(\vec{x}) \qquad\qquad \rho^e_{cl} = e\, \delta^3(\vec{x})$$

$$a_{kr} = G_N m_\odot c^{-2} \qquad\qquad R^{cl}_e = \alpha \hbar c \frac{1}{m_e c^2}$$

The critical distances a_{kr}, R^{cl}_e are related to gravitational[9] and electromagnetic collapse respectively.

$$R^{cl}_e \qquad\qquad = \alpha \lambda e = \alpha^2 a_{Bohr}$$

$$2.8 \times 10^{-13} \text{cm} \quad 3.9 \times 10^{-11} \text{cm} \quad 0.53 \times 10^{-8} \text{cm}$$

$$\alpha^{-1} \cong 137.0360$$

The electromagnetic collapse stops because

$$a_{Bohr} \gg R^{cl}_e \qquad\qquad\qquad (10)$$

Atomic stability merely reflects the unexplained fact that

$$\alpha << 1$$

The relations corresponding to eq. (9) for gravity are

$$G_N \, m_\odot \, c^{-2} = a_{kr} = \alpha_g \frac{\hbar}{mc} = (\alpha_g)^2 \, a_B \quad (g)$$

$$\alpha_g = (\hbar c)^{-1} \, G_N \, m_\odot m = \frac{m_\odot m}{m_{Pl}^2} \tag{11}$$

$$m_{Pl} = (\frac{G_N}{\hbar c})^{-1/2} = \begin{cases} 1.22.10^{19} \text{ GeV}/c^2 \\ 2.2.10^{-5} \text{ g} \end{cases}$$

α_g can be $>> 1$ since $m_\odot >> m_{Pl}$. This implies that quantum effects will not immediately stop the gravitational collapse provided the imploding mass is appropriately chosen. Stable versus unstable storage of gravitational energy during the collapse thus becomes a prime question.

Atomic stability does not prevent the mass distribution of the electron from containing relativistic velocities. The Thomson cross section

$$\sigma_{k \, \xrightarrow{\gamma} \, o}^{\gamma e} = \frac{8\pi}{3} (R_e^{cl})^2 \tag{12}$$

determines the size of the electron to be of the order of R_e^{cl}. Thus if its mass is thought to be concentrated on the equator with radius R_e^{cl} (relative to the spin axis) one obtains

$$\frac{\hbar}{2} \approx R_e^{cl} P_\phi = \alpha \hbar c \frac{1}{m_e c^2} \frac{m_e v}{\sqrt{1 - \frac{v^2}{c^2}}}$$

$$\xrightarrow{\qquad} \frac{v}{c} \approx (1 + 4\alpha^2)^{-1/2} \tag{13}[FN5]$$

There exist definite indications of the asymptotically free nature of gravity at short distances due to the positivity of energy density.[10] The total energy density generated by the sun and its gravitational field is shown in Fig. 2. It gives rise to an effective (position dependent) mass seen e.g. by mercury on its orbit

$$m_\odot^{eff} (r) = \int d^3 \, \rho_e(x) \frac{1}{c^2} \tag{14}$$
$$|x| \le r$$

$$\lim_{r \to \infty} m_\odot^{eff} (r) = m_\odot \approx 1.99.10^{33} g$$

The effect on the mercury perihelion advance

$$\Delta\phi/_{\text{revolution}} = \frac{6\pi\ a_{kr}\odot}{[a(1-\epsilon^2)]\ _{\text{mercury}}}$$ (15)

transformed to the century can be interpreted as giving rise to a reduction of about 7 (arc) seconds.

$$\delta\phi \cong (50-7)"/_{\text{century}}$$ (16)[FN6]

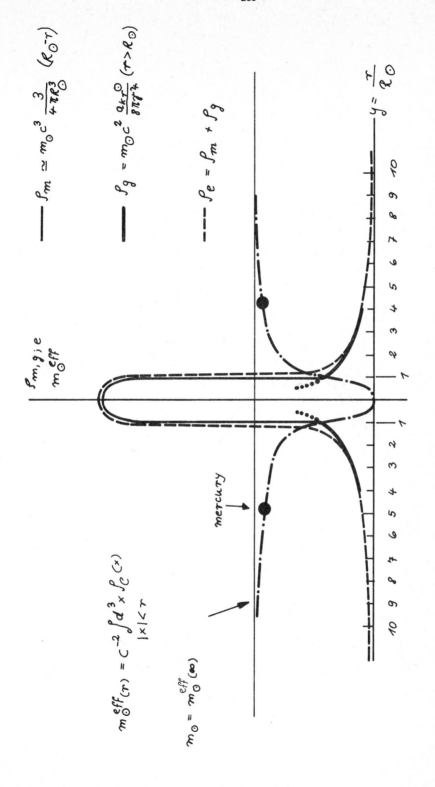

Fig. 2: <u>Total energy density and effective mass of the sun in its own gravitational field</u>

2)   Electrodynamics ⟵————⟶   gauge symmetry:
                                gauge transformation associated with
                                local conservation of electric charge

$$A_\mu(x) \longrightarrow A_\mu(x) + \partial_\mu \chi(x)$$

photon field

$$e(x) \longrightarrow \exp[-ie\chi(x)].e(x)$$

electron (charge carrier) field

Let us tune to the dynamics of the hydrogen atom. The stability problem
reappears in the form of the renormalization procedure, i.e. in the
question of the domain of applicability of renormalized perturbation
theory. The photon exchange diagram (Fig. 3) exhibits the universal nature of the photon-charge carrier coupling.

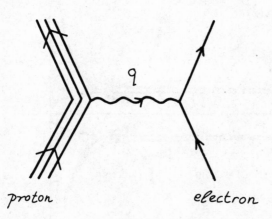

Fig. 3: <u>Photon exchange diagram</u>

$$T \propto (\gamma^\mu Q)_{proton} \ (\gamma^\nu Q)_{electron}$$

$$(17)$$

$$Q_p = |e| \ ; \ Q_p + Q_e = 0$$

The renormalization constants of the QED Lagrangean (electron part) are

$$L = Z_3 \ (-\frac{1}{4} F_{\mu\nu} F^{\mu\nu}) - \frac{Z_3}{2\eta_{bare}} \ (\partial_\mu A^\mu)^2 \qquad +$$

⌊→ Fermi gauges

$$- Z_1 \cdot e(\bar{e}\gamma^{\mu}e) \; A_{\mu}$$

$$Z_2 \; (\bar{e}i\gamma^{\mu} \tfrac{1}{2}\overleftrightarrow{\partial}_{\mu} \; e) \quad + \quad \text{coupling of}$$
$$\gamma \text{ to proton}$$

$$- \frac{Z_2}{Z_m} \; m_e \; (\bar{e}e) \tag{18}$$

Vaccum polarization generates in analogy with the situation in eq. (14) an effective position dependent charge of the proton (taken for definiteness as generating the electromagnetic field) which is experienced by the electron in its orbit. This effect and the level splittings associated with LS-coupling, the Lamb shift and the hyperfine interaction of the electron magnetic moment (orbital and intrinsic) with the proton magnetic moment are shown in Fig. 4 [FN7] [12].

We shall envisage three tests of QED:

- Lamb shift
- hyperfine splitting of the ground state of muonium (μ^+e^-) $1S_{1/2}$
- anomalous magnetic moments of electron and muon.

The agreement between theory and experiment to about one part per million calls for a serious effort to understand the remainder in the corresponding asymptotic series in α.

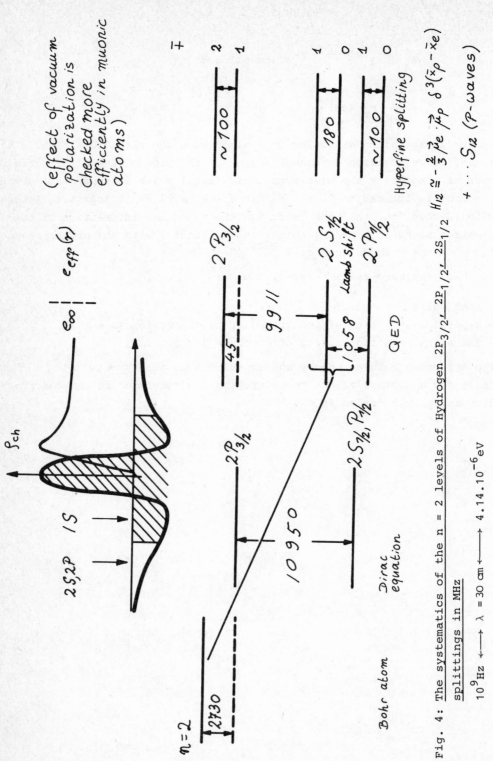

Fig. 4: The systematics of the n = 2 levels of Hydrogen 2P₃/₂, 2P₁/₂, 2S₁/₂ splittings in MHz

10^9 Hz ⟷ λ = 30 cm ⟶ 4.14·10^{-6} eV

Lamb shift[13,14]

theory	Mohr	1057.864(14)		'defines' theoretical
	Erickson	1057.916(10)	(MHz)	uncertainty
experiment	Lundeen Pipkin	1057.893(29)		
	Andrews Newton	1057.862(20)		

μ^+e^- hyperfine splitting (e^+e^-)[15,16]

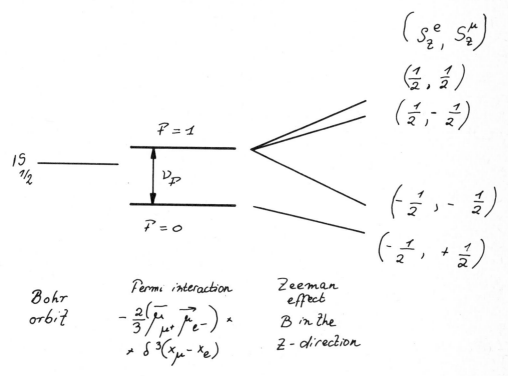

Fig. 5: <u>Zeeman effect recombination of the spin triplet and singlet lines</u>

The (total) magnetic moments of electron and muon are

$$\vec{\mu}_{\mu^+} = \frac{|e|\hbar}{2m_\mu c}(1 + a_\mu)\,\vec{\sigma}_\mu \qquad\qquad \vec{\sigma} = 2\cdot\vec{s}$$

$$\vec{\mu}_{e^-} = -\frac{|e|\hbar}{2m_e c}(1 + a_e)\,\vec{\sigma}_e \qquad\qquad a_l = \left(\frac{g-2}{2}\right)_l \qquad\qquad (19)$$

In terms of $a_{\mu,e}$ the hyperfine splitting (without external magnetic field) is given by

$$\nu_F = \frac{c^2}{h} \frac{8}{3} \left(\frac{\frac{m_e \alpha}{m_e}}{1+\frac{m_e}{m_\mu}}\right)^3 \frac{\alpha}{m_e m_\mu} \left\{ \begin{array}{l} (1+a_\mu)(1+a_e)+ \\ +O[\alpha\frac{m_e}{m_\mu}\log\frac{m_\mu}{m_e},\alpha^2] \end{array} \right\}$$

(20)

$$\cong 4,464,186 \frac{(1+a_\mu)(1+a_e)}{(1+\frac{\alpha}{2\pi})^2} \quad \text{kHz}$$

The main correction involves the terms proportional to $\log\frac{m_\mu}{m_e} \cong 5.3$ in eq. (20). The comparision of theory and experiment yields

$$\begin{array}{ll} \nu_{th} = & 4,463,297.9 \\ & \pm 7.2 \\ \nu_{exp} = & 4,463,302.35 \\ & \pm .52 \end{array} \quad \text{kHz}$$

(21)

Anomalous magnetic moment of electron and muon[17,18]

The structure of the dependence of $a_{e,\mu}$ in eq. (19) on α and on mass ratios (within QED restricted to electron and muon) is

$$a_e = A(\alpha) + B(\frac{m_e}{m_\mu}, \alpha)$$

$$a_\mu = A(\alpha) + B(\frac{m_\mu}{m_e}, \alpha)$$

(22)

$$A = \sum_{n=1}^{N} A_n(\frac{\alpha}{\pi})^n + R_N(A); \quad B = \sum_{n=1}^{N} B_n(\frac{\alpha}{\pi})^n + R_N(B)$$

$$A_1 = \frac{1}{2} \quad \text{(Schwinger)}$$

$\frac{\alpha}{\pi}$: effective (short distance) expansion parameter at least up to $N = 3$. For the electron we compare the following contributions

$a_e \cdot 10^{12}$

experiment	1 159 652 410	(200)
QED (e,μ)	1 159 652 473	(260)
hadronic from $e^+e^- \longrightarrow$ hadrons	2	
weak	$5 \cdot 10^{-2}$	
gravitational (Gastmans, Berends)[19]	$1 \cdot 10^{-33} \rightarrow O[(\frac{m_e}{m_{pl}})^2 \cong 1.75 \cdot 10^{-45}$	

$$\rightarrow a^e_{graviton} = \frac{7}{4\pi}(\frac{m_e}{m_{pl}})^2$$

and correspondingly for the muon

$a_\mu \cdot 10^{12}$

experiment	1 165 922 000	(9 000)
theory (total)	1 165 920 700	(9 800)
includes:		
hadronic	66 700	(9 400)
weak	2 100	($\sim$ 400)

Assuming the correctness of QED one can use the data to determine α and compare it with the precise measurements involving the AC Josephson-effect[20]

	$(\alpha)^{-1}$		Error ppm
experiment (ref.20)	137.035 987	(29)	0.21
from $\frac{g_e-2}{2}$ and QED	137.035 981	(29)	0.21

The effective (electric) charge distribution-rescaling:[21]

The rescaling equation for a one particle irreducible Greens function (in QED) is of the form

$$\{\mu\frac{\partial}{\partial\mu} + \beta(e)\frac{\partial}{\partial e} - \delta(e)\, m_e\frac{\partial}{\partial m_e}\}\, \Gamma =$$

$$= \{2\gamma_V\, \eta\cdot\frac{\partial}{\partial\eta} + \Sigma\gamma_\alpha(e,\eta)\}\, \Gamma \qquad (23)$$

field
type
(γ,e)

The universal functions β,δ controling the rescaling of charge and mass respectively need a specification of the renormalization procedure (although they are gauge invariant) which respects the sought universality i.e. is in particular independent of the electron mass or of the mass parameter m_e in eq. (23).[FN8]

Dimensional regularization does satisfy this criterion:

$\varepsilon = \frac{4-n}{2}$ ⠀⠀⠀⠀⠀⠀⠀⠀⠀⠀ n: variable dimension of space time

$$(e_\varepsilon)^2 = e^2\, (\mu^2\, \frac{e^\gamma}{4\pi})^\varepsilon \qquad (24)$$

$\gamma = 0.5772156649$: ⠀⠀⠀ Euler's constant

e: renormalized charge (in four dimensions)

$$a = \frac{e^2}{16\pi^2} = \frac{\alpha}{4\pi}$$

$$\frac{\beta(e)}{e} = -2 \operatorname*{Res}_{\varepsilon=0} \log (Z_3)^{1/2} = - b(a)$$

$$b(a) = \sum_{n=1}^{N} b_n a^n + R_N(b) \qquad (25)$$

$$\delta(a) = 2 \operatorname*{Res}_{\varepsilon=0} \log Z_m = \sum_{n=1}^{N} \delta_n a^n + R_n(\delta)$$

The anomalous dimension functions γ_e, γ_V associated with electron and photon wavefunction renormalization are in general gauge dependent quantities. For Fermi gauges in QED we have

$$\gamma_V = -2 \operatorname*{Res}_{\varepsilon=0} \log \frac{Z_3^{1/2} Z_2}{Z_1} = - b(a) \qquad (26)$$

$$Z_2 = {}'Z_1$$

as a consequence of the Ward identity. γ_e does depend on the gauge

$$\gamma_e(a,\eta) = -2 \operatorname*{Res}_{\varepsilon=0} \log (Z_2)^{1/2}$$

$$Z_2 = {}'Z_2 (a, \eta, \varepsilon) \qquad (27)$$

$\eta = (Z_3)^{-1} \eta_{bare}$ denotes the Fermi gauge parameter.

The (known) coefficients in the power expansion of β, δ with respect to a are all rational (dimensionless) numbers

$$b_1 = -\frac{4}{3} \qquad b_2 = -4 \qquad b_3 = +2\left(\frac{11}{3}\right)^2 \qquad (28)^{23}$$

$$\delta_1 = 6 \quad \longleftrightarrow \quad \delta m_e = m_e \frac{3\alpha}{4\pi} \log \frac{\Lambda^2}{m_e^2}$$

The so defined mass m_e is not the physical mass of the electron,[FN9] rather we have

$$m_e^{ph} = m_e \ F(e, \frac{m_e}{\mu}) \qquad (29)$$

and for

$$\mu = m_e^{ph} , \ e_\mu = e_{ph}$$

$$m_e^{ph} = m_e \ \Phi(e_{ph}) \qquad (30)$$

F, Φ in Eq. (29) and (30) are quite nontrivial functions of their arguments.

The effective charge distribution can be determined from the photon vacuum polarization function in the limit $m_e \longrightarrow 0$.

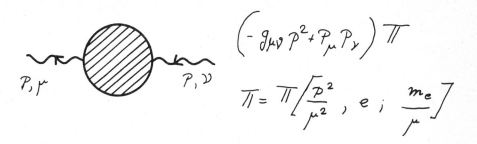

$$\left(- g_{\mu\nu} p^2 + p_\mu p_\nu\right) \Pi$$

$$\Pi = \Pi\left[\frac{p^2}{\mu^2}, e; \frac{m_e}{\mu}\right]$$

Fig. 6: <u>Vacuum polarization diagram</u>

$$\Pi \cong - \frac{e^2}{12\pi^2} \log \left(\frac{p_E^2}{\mu^2}\right) \; ; \quad p_E^2 = \vec{p}^2 - p_o^2$$

$$(d_{\mu\nu}(p))_{transverse} = {}'[- g_{\mu\nu} + \frac{p_\mu p_\nu}{p^2}]\frac{1}{p2} \quad \frac{1}{1 - \frac{e^2}{12\pi^2} \log \frac{p_E^2}{\mu^2}} \qquad (31)$$

$$(1+\Pi)(p_E^2) \longleftrightarrow \varepsilon(p_E^2) \qquad \qquad \underbrace{1 + \Pi}$$

ε: dielectric constant

Setting

$$p_E^2 = \langle\frac{1}{\vec{z}^2}\rangle = \frac{1}{r^2} \; ; \quad r\mu \ll 1$$

we obtain the effective charge distribution as shown in Fig. 4 and Fig. 7 below

$$e_{eff}(r) = \frac{e_\infty}{\varepsilon(r)} \cong (1 + \frac{e_\infty^2}{12\pi^2} \log \left(\frac{1}{r^2\mu^2}\right)) \cdot e_\infty$$

$$r\mu \ll 1 \qquad \qquad \downarrow \qquad \qquad \ll 1$$

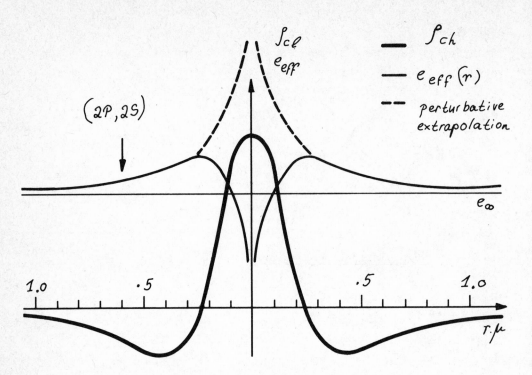

Fig. 7: <u>Charge density and effective charge for the proton, showing the instability at short distances of the perturbation expansion in α</u>

The breakdown of perturbation expansions in a (in pure QED) occurs for

$$\exp\left(\frac{3\pi}{2\alpha}\right) \approx 10^{280} = \frac{P_{kr}}{m_e} \tag{32}$$

Of course the weak interactions interfere with QED at much lower energies.

For $p_E \gg m_e$, e_{eff} satisfies the rescaling equation

$$e_{eff} \longrightarrow \bar{e} \ ; \ \frac{d}{dt}\,\bar{e} = \beta(\bar{e}) = -\,\bar{e}\,b\,(\bar{a}) \tag{33}$$

$$2t = \log\left(\frac{p_E^2}{\mu^2}\right)$$

3)   Chromodynamics   ⟵————————⟶   Gauge symmetry:[24]
     as substrate of   local gauge transformation of
     strong interactions   eight color charges forming
                                                      the group $SU3_c$.

We can (tentatively) discern three generations of tricolored quarks and leptons

$$\begin{pmatrix} u_1 & u_2 & u_3 & \nu_e \\ d_1 & d_2 & d_3 & e^- \end{pmatrix} \qquad \begin{pmatrix} c_1 & c_2 & c_3 & \nu_\mu \\ s_1 & s_2 & s_3 & \mu^- \end{pmatrix} \qquad \begin{pmatrix} t_1 & t_2 & t_3 & \nu_\tau \\ b_1 & b_2 & b_3 & \tau \end{pmatrix}$$

$$\qquad\qquad I \qquad\qquad\qquad\qquad II \qquad\qquad\qquad\qquad III$$

1,2,3: red, green, blue[25]

The generic notation for quark fields shall be

$$q_s^c \quad \begin{array}{l} c \quad \text{color} \quad (1,2,3) \\ s \quad \text{flavor} \quad (u,d,s,c,b,t,\ldots) \end{array}$$

Despite perfect (practically perfect) confinement of quarks inside color-less hadrons[26] the (current quark) masses are well defined.[27]

$$m_u \cong 4.5 \text{ MeV} \qquad m_d \cong 7.5 \text{ MeV} \qquad m_s \cong 150 \text{ MeV} \tag{35}$$
$$3 \qquad : \qquad 5 \qquad : \qquad 100$$

The error on the mass ratios in eq. (35) is about 15%, due to higher order chiral symmetry breakings which are being studied at present.[27,28] The absolute values of the three light quark masses depend on the non-perturbative determination of ratios of matrix elements

$$\langle h,p | \bar{q}\gamma_\mu q | h,p \rangle / \langle h,p | \bar{q}.q | h,p \rangle = P_\mu F_h(p_h^2) \tag{36}$$

which are inherently uncertain. Nevertheless an order of magnitude esti-mate yields

$$m_s \cong m_{Y^*}(1385) - m_\Delta(1232) \cong 150 \text{ MeV} \tag{37}$$

The dividing line between light and heavy flavors is drawn by the re-normalization group invariant mass μ^* governing physical phenomena in the chiral limit(s) corresponding to

$$m_u \longrightarrow 0 \; ; \; (\text{or } m_u, \, m_d \longrightarrow 0 \; ; \; \text{or } m_u, \, m_d, \, m_s \longrightarrow 0, \, \ldots)$$

The evaluation of μ^* from deep inelastic lepton hadron scattering on which we will focus next, yield for the time being the following limits

$$300 \text{ MeV} \leq \mu^* \leq 700 \text{ MeV} \tag{38}$$

The masses of the heavy flavors are less certain

$$m_c \cong 1.25 - 1.5 \text{ GeV} \; ; \; m_t = ?^{[29]} \; ; \; m_b \cong 4.1 - 4.5 \text{ GeV} \tag{39}$$

The universal nature of color charge is (as in Fig. 1,3) apparent in the gluon exchange diagram (Fig. 8) in which a heavy antiquark $\bar{Q}$ can be thought to generate the gluon field (configuration) in which a light quark q evolves.

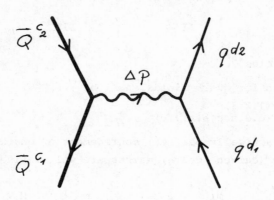

Fig. 8: <u>Gluon exchange diagram</u>

$$T \propto (g\gamma^\mu)_{\bar{Q}} \cdot (g\gamma^\mu)_q$$

$$g^{\bar{Q}} = g \; (\frac{\chi^A}{2})_{c_1 c_2} \quad ; \quad g^q = g \; (\frac{\chi^B}{2})_{d_2 d_1}$$

(40)

g: color coupling constant.

The gluon (gauge) fields are:

potentials V_μ^a, a = 1,..., 8   ; $B_\mu = i g \frac{\chi^a}{2} V_\mu^a$

field strengths $V_{\mu\nu}^a = \partial_\nu V_\mu^a - \partial_\mu V_\nu^a - g f_{abc} V_\nu^b V_\mu^c$

(41)

$$B_{\mu\nu} = \partial_\nu B_\mu - \partial_\mu B_\nu + [B_\nu, B_\mu]$$

The quark and gluon terms in the QCD Lagrangean are of the form (eq.(18))

$$z_3 \; [-\frac{1}{4} (\partial_\nu V_\mu^a - \partial_\mu V_\nu^a) (\partial^\nu V^{\mu a} - \partial^\mu V^{\nu a})]$$

$$L = + z_1 g \; (\partial_\nu V_\mu^a) f^{abc} V^{\nu b} V^{\mu c}$$

$$z_4 g^2 \; [-\frac{1}{4} f^{abc} f^{ab'c'} V_\nu^b V_\mu^c V^{\nu b'} V^{\mu c'}] \qquad +$$

(42)

$$Z_2 \, \bar{q} \, {}^c_s \, i \, \partial \, {}^\mu \, \overleftarrow{\frac{1}{2} \partial_\mu} \, q \, {}^c_s$$

$$- \frac{Z_2}{Z_m} \, \bar{q} \, {}^c_s \, [m_{st} \, \frac{1+\gamma_5}{2} + m^*_{ts} \, \frac{1-\gamma_5}{2}] \, q \, {}^c_t$$

$$- \hat{z}_1 g \, \bar{q} \, {}^c_s \, \gamma^\mu (\frac{\lambda^a}{2})_{cc'} \, q \, {}^{c'}_s \, v^a_\mu$$

Note that $z_1 g = z_3^{3/2} \, g_{bare}$ and $\hat{z}_1 = z_2 \, z_3^{1/2} \, g_{bare}$.

The renormalization procedure relies in the nonabelian case on an extended set of Ward identities.[30,31]

Rescaling - effective charge and effective mass:

The rescaling equations (for one particle irreducible Greens function) show the same structure as in QED (eq. (23))

$$\kappa = \frac{g^2}{4\pi} \, , \quad k = \frac{g^2}{16\pi^2} = \frac{\kappa}{4\pi}$$

$$\{ \mu \, \frac{\partial}{\partial \mu} - 2b(k) \, k \, \frac{\partial}{\partial k} - \delta(k) \, \sum_{s,t} \, [m_{st} \, \frac{\partial}{\partial m_{st}} + m^*_{st} \, \frac{\partial}{\partial m^*_{st}}] \} \Gamma =$$

$$= (2\gamma_V \, \eta \, \frac{\partial}{\partial \eta} + \sum_{\substack{field \\ type}} \gamma_\alpha \, (g, \eta)) \, \Gamma \tag{43}$$

The universal rescaling functions for charge and mass are

$$\frac{\beta(g)}{g} = - b(k) = - 2 \, \underset{\varepsilon=0}{\mathrm{Res}} \, \log \, (\frac{(z_3)^{3/2}}{z_1})$$

$$\delta(k) = 2 \, \underset{\varepsilon=0}{\mathrm{Res}} \, \log z_m \, ; \quad g^2_\varepsilon = g^2 (\frac{\mu^2 e^\gamma}{4\pi})^\varepsilon \tag{44}$$

and the anomalous dimension functions

$$\gamma_V = - 2 \, \underset{\varepsilon=0}{\mathrm{Res}} \, \log z_3^{1/2}$$

$$\tag{45}$$

$$\gamma_q = - 2 \, \mathrm{Res} \, \log z_2^{1/2}$$

As in QED (eq. (28)) the (known) coefficients in the Taylor expansion of b, δ with respect to k are rational numbers

$$b(k) = \sum_{n=1}^{N} b_n k^n + R_N(b)$$

$$\delta(k) = \sum_{n=1}^{N} \delta_n k^n + R_N(\delta)$$

$$b_1 = 11 - \frac{2}{3} n_{fl} \quad , \quad b_2 = 102 - \frac{38}{3} n_{fl}, \quad \cdots$$

$$\delta_1 = 8 \tag{46}$$

There are two (universal i.e. flavor independent) polarization functions

$$\text{charge:} \quad g^2 \longrightarrow g^2_{eff} (p^2) = \frac{g^2(p_o{}^2)}{\varepsilon(\frac{p^2}{p_o{}^2})} = \bar{g}^2 \tag{47}$$

$$\text{mass:} \quad m \longrightarrow m_{eff} (p^2) = \bar{m}$$

$\bar{g}^2$, $\bar{m}$ are determined by b, δ

$$\tau = \log\left(\frac{\mu^2}{\mu^{*2}}\right) .$$

$$\frac{d}{d\tau} \bar{k} = - \bar{k} \cdot b(\bar{k}) \quad ; \quad \bar{k}(\mu = \mu^o) = k_o \tag{48}$$

$$\frac{d}{d\tau} \log \bar{m} = - \frac{1}{2} \delta(\bar{k}) \quad ; \quad \bar{m}(\mu = \mu_o) = m_o$$

$$\downarrow$$

$$\frac{d}{d\bar{k}} \log \bar{m} = \frac{\delta}{2 \, b \cdot \bar{k}}$$

Eq. (48) implies

$$\frac{\bar{m}}{m_o} = \exp\left[- \int_{\bar{k}}^{k_o} \frac{\delta(k')}{2 \, b(k')} \frac{dk'}{k'} \right] \tag{49}$$

For $b_1 > 0$ ($n_{fl} \le 16$) chromodynamics exhibits ultraviolet stability (asymptotic freedom) $\longleftrightarrow$ infrared instability of renormalized perturbation expansions in k. Then we have asymptotically

$$\mu \to \infty , \quad \bar{k} \to 0$$

$$\frac{\bar{m}}{m_o} \sim \text{const} (\bar{k})^{-\frac{\delta_1}{2b_1}} \quad (\underset{n_{fl}=4}{\longrightarrow} \frac{12}{25} \, , \, \underset{n_{fl}=6}{\frac{4}{7}})$$

$$(b_1\bar{k})^{-1} \sim \log \frac{\mu^2}{\mu^{*2}} + \frac{b_2}{b_1^2} \log\log (\frac{\mu^2}{\mu^{*2}}) +$$

$$+ 0 \left[(\log \frac{\mu^2}{\mu^{*2}})^{-1} \right] \tag{50}$$

Eq. (50) defines μ^* i.e. the quantitative way short distance phenomena retain the characteristic hadronic scales. A plot of $\bar{k} = \frac{\bar{g}^2}{4\pi}$ versus $\log \frac{(\mu^2)}{(\mu^{*2})}$ is shown in Fig. 9.

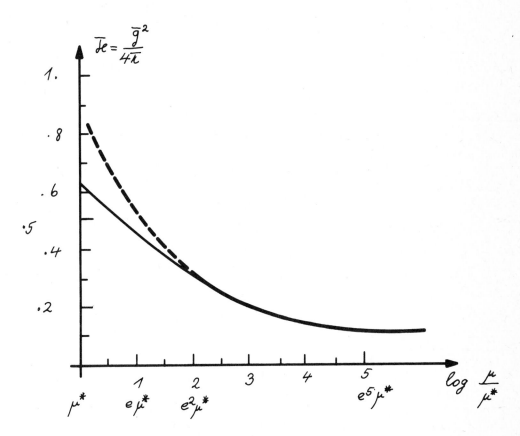

Fig. 9: <u>The effective coupling constant</u> $\bar{k} = \frac{\bar{g}^2}{4\pi}$ <u>as a function of the relative mass scale</u> $\frac{\mu}{\mu^*}$

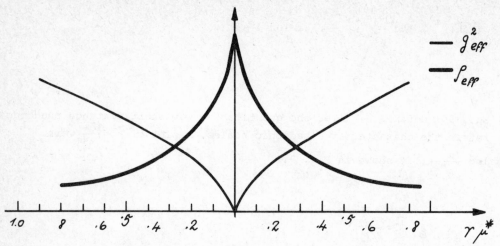

Fig. 10: <u>Color charge density and effective charge showing stability</u>
<u>at short distances</u>

The color charge density and effective charge in Fig. 10 show an analogous behaviour to gravity (Fig. 2).

Since

$$\bar{m} \sim (m_{r=0}) \cdot [\log(\frac{\mu^2}{\mu^{*2}})]^{-\frac{\delta_1}{2b_1}} \tag{51}$$

the curve for effective mass would look similar at short range to the
one for effective charge. We have $\frac{\delta_1}{2b_1} = \frac{12}{15}$ if $n_{fl} = 4$ and $\frac{4}{7}$ for $n_{fl} = 6$.

<u>Detailed tests for structure functions of deep inelastic lepton-nucleon</u>
<u>scattering</u>

We consider electromagnetic or weak quanta emitted by inelastically
scattered leptons (electrons, muons, neutrinos) which upon absorption
by a nucleon, probe its structure (Fig. 11).

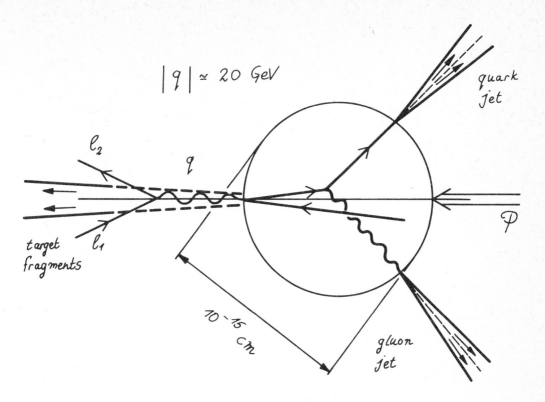

Fig. 11: <u>3 jet event, no jet in the current fragmentation region,</u>
<u>target fragments seen</u>

The event structure in Fig. 11 has every resemblence with the stochastic interpretation of the Feynman path integral. For $Q^2 = - q^2$ large enough a perturbation expansion in powers of $\kappa(Q^2)$ is meaningful.

The most frequent event signature is not the one shown in Fig. 11, rather it is longitudinally oriented along the target-current axis for E_1, Q^2 large enough (Fig. 12) in the frame where $q^o = 0$.

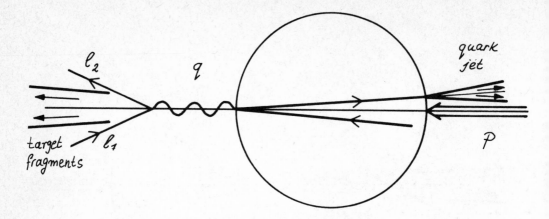

Fig. 12: <u>2 jet event, one each yielding target and current fragments</u>

The cross sections for neutrino (antineutrino) nucleon scattering are given by

$$\nu = p \cdot q \quad , \quad x = \frac{Q^2}{2\nu} \quad ; \quad y = (\frac{E_h}{E_\nu}) \, \text{lab.} \quad = \frac{\nu}{l_1 \cdot p}$$

$$\frac{d\sigma^{\nu N \to \mu^- X}}{dx \, dy} \cong \frac{G_F^2 \, m_N E_\nu}{\pi} \quad x \, \{q(x, Q^2) \quad + \quad (1-y)^2 \bar{q} \, (x, Q^2)\}$$

$$\frac{d\sigma^{\bar{\nu} N \to \mu^+ X}}{d \, x \, dy} \cong \frac{G_F^2 \, m_N E_{\bar{\nu}}}{\pi} \quad x \, \{\bar{q}(x, Q^2) \quad + \quad (1-y)^2 q\}$$

$$d\sigma^{eN \to eX} \propto \frac{5}{18} \quad x \quad (q + \bar{q})$$

$$\qquad \qquad \quad \llcorner_{\to \, <e_q^2>_{u,d}}$$

(52)

In eq. (52) $q = (u_p + d_p) \, (x, Q^2)$ denotes the number of u and d quarks per unit interval in x in the proton, the bar referring correspondingly to antiquarks.

The total neutrino (charged current) cross section is proportional to the neutrino (lab) energy up to the highest energies measured $E_\nu \lesssim 200$ GeV

$$\sigma^{\nu N \to \mu^- X} = 0.7 \cdot 10^{-38} \text{cm}^2 \, (\frac{E_\nu^{\text{lab}}}{1 \, \text{GeV}}) \qquad (53)[33]$$

The momentum fraction of the (target) nucleon carried by quarks is approximately given by

$$\varepsilon_q = \int\limits_o^1 dx \quad x(q + \bar{q}) \cong \frac{18}{5} \int dx \, F_2^{[e(p+n) \, 1/2]} \qquad (54)[34]$$
$$5 \le Q^2 \le 15 \, (\text{GeV})^2$$

From QCD it follows

$$\lim_{Q^2 \to \infty} \varepsilon_q(Q^2) = <e_q^2>^{-1} \lim_{Q^2 \to \infty} \int dx \ F_2^{[e(p+n)1/2]}$$

$$= \frac{3n_{fl}}{3n_{fl}+16}$$

(55)

This is equal to 0.36, o.45, and o.53 for $n_{fl} = 3,4,6$ respectively.

Neglecting terms of order m_{lepton}/E there are three structure functions determining the current correlation function(s)

$$\bar{W}_{\mu\nu}^{\pm e.m.}(q,p,...) = \frac{1}{4\pi} <Np| \int d^4 x e^{+iqx}$$

$$\cdot (j_\mu^{\pm e.m.}(x) \ j_\nu^{\mp e.m.}(o)) | Np>$$

$$W_{\mu\nu} = [-\delta_{\mu\nu} + \frac{q_\mu q_\nu}{q^2}] \ F_1 + (P_\mu - q_\mu \frac{\nu}{q^2}) \ (P_\nu - q_\nu \frac{\nu}{q^2}) \cdot \frac{1}{\nu} F_2 +$$

$$+ \frac{i}{2} \varepsilon_{\mu\nu\alpha\beta} \ p^\alpha q^\beta \frac{1}{\nu} F_3$$

(56)

The quark distribution functions in eq. (52), (54) are given in terms of the structure functions $F_{1,2,3}$

$$F_T = 2 \ x \ F_1 \ , \quad F_L = F_2 - 2 \ x \ F_1$$

$$xq + (1-y)^2 \ x\bar{q} \ \to \ \frac{1}{2} \ x(2F_1 + F_3) + \frac{1}{2}(1-y)^2 \ (2F_1 - F_3)$$

$$+ \ (1-y)F_L$$

(57)

$$\left. \begin{array}{l} 2F_1^\nu = 2F_1^{\bar\nu} \cong q + \bar{q} \\ F_3^\nu = F_3^{\bar\nu} \cong q - \bar{q} \end{array} \right] \begin{array}{c} \text{for isoscalar} \\ \text{targets} \end{array} \qquad F_L^{\nu,\bar\nu} \cong 0$$

<u>Theoretical results ($\to F_3$ here)</u>[351]

We focus in the following on the moments of F_3 as determined from neutrino (antineutrino) nucleon scattering

$$M_n^3(Q^2) = \int_0^1 dx \ x^{n-1} F_3(x, Q^2) \ ; \ n = 1,2,...$$

(58)

$$F_3^{\nu p} \cong (d - \bar{u})_{proton}; \ F_3^{\nu n} \cong (u - \bar{d})_{proton}$$

In the limit $Q^2 \to \infty$ the Gross-Llewellyn sum rule becomes exact

$$\int dx \, (F_3{}^{\nu p} + F_3{}^{\nu n}) = (M_1{}^3)_p + (M_1{}^3)_n \underset{Q^2 \to \infty}{\to} 3 \tag{59}$$

The Bjorken scaling Ansatz is

$$(M_n{}^3)_{target} = (C_n)_{target} \cdot 1 \tag{60}$$

where the constants C_n are characteristics of the target wave function (its quark part), expectation values of the local quark bilinears

$$O_n^{st} = \bar{q}_s^c \, \gamma_{\mu_1} \, \overset{\cdot}{(\gamma_5)} \, (\overset{\leftrightarrow}{D}_{\mu_2} \ldots \overset{\leftrightarrow}{D}_{\mu_n})^{cc'} q_t^{c'}$$

$$D_\mu = \partial_\mu + B_\mu$$

$$<Np|O_n{}^{(jj)}|Np> = P_{\mu_1} \ldots P_{\mu_n} \, (C_n{}^{(jj)})_N + (g_{\mu_1\mu_2} P_{\mu_3} \ldots P_{\mu_n}) \ldots \tag{61}$$

For the radiative (gluon) corrections to eq. (60) the anomalous dimension functions of the operators O_n^{st} (nonsinglet (NS) part here) play an important role

$$(O_n^{st})_{bare} = Z_n^{NS} O_n^{st}$$

$$\gamma_n = -2 \, \text{Res} \log Z_n^{NS} = \gamma_n^1 \cdot k + \gamma_n^2 \, k^2 + \ldots \tag{62}^{35_2}$$

$$\gamma_n^1 = \frac{8}{3} \{1 - \frac{2}{n(n+1)} + 4 \sum_{j=2}^{n} \frac{1}{j}\}$$

γ_n^2 : given in the Appendix.

An appropriate summation of O_n^{st} yields the (renormalized) bilocal operators

$$\Sigma(O_n^{st})_{\mu_1 \ldots \mu_n} z^{\mu_2} \ldots z^{\mu_n} \frac{(-1)^{n-1}}{(n-1)!} =$$

$$= \bar{q}^{\,c}(z) \, \gamma_{\mu_1} \, (\gamma_5) \, (P \exp - \int_o^z d\zeta^n B_\mu)^{cc'} q^{c'}(o) \tag{63}$$

Perturbative QCD corrections to the scaling behaviour in eq. (60) can be expanded in terms of the running coupling constant[35] (eq. (50))

$$(M_n{}^{(3)})_{target} \to (C_n)_{target} [1 + \bar{k} \, (B_n + P_n)] \tag{64}$$

$$x \, (b_1 \, \bar{k})^{\gamma_n^1/2b_1} \, (1 + O \, (\bar{k}))$$

$$\bar{k} = \frac{\kappa(Q^2)}{4\pi} \underset{Q^2 \to \infty}{\sim} [b_1 \log \frac{Q^2}{\mu^{*2}} + \frac{b_2}{b_1} \log \log \frac{Q^2}{\mu^{*2}}]^{-1} (1 + 0(\log^{-1}\frac{Q^2}{\mu^{*2}}))$$

All the dependence on the target wave function remains within the constants C_n (eq. (60))

$$B_n = \frac{4}{3} \left\{ \begin{array}{l} 3 \sum_{j=1}^{n} \frac{1}{j} - 4 \sum_{j=1}^{n} \frac{1}{j^2} - \\ - \frac{2}{n(n+1)} \sum_{j=1}^{n} \frac{1}{j} + 4 \sum_{j=1}^{n} \frac{1}{j} \sum_{k=1}^{j} \frac{1}{k} + \\ + \frac{3}{n} + \frac{4}{n+1} + \frac{2}{n^2} - \frac{4n+2}{n(n+1)} - 9 \end{array} \right\}$$

(65)

$$P_n = \frac{\gamma_n^2}{2b_1} - \frac{b_2 \, \gamma_n^1}{2b_1^2}$$

The constants B_n, P_n, γ_n^1, γ_n^2, b_1, b_2 are independent of the target structure (wave function) as is the behaviour of $\bar{k}$ on Q^2, and μ^*; the constants are all rational numbers. [FN10]

We turn to the tests of the first order parametrization of the scale breaking effects (M_n^3). The first six moments of F_3 have been extracted from the CERN neutrino experiment involving the collaboration operating the British European Bubble Chamber (BEBC) and the CERN-Dortmund-Heidelberg-Saclay collaboration.[36]

The first order approximation to the scale breaking is given by

$$Q^2 \frac{d}{dQ^2} \log M_n^3 \cong -\frac{1}{2} (\gamma_n^1)^{NS} \bar{k}(Q^2) =$$
$$-\frac{1}{8\pi} (\gamma_n^1)^{NS} \bar{\kappa}(Q^2)$$

(66) [FN11]

$F_3(x,Q^2)_{isoscalar}$ is an even function of x in the extended (crossed) region

$$-1 \leq x \leq 1, \qquad Q^2 > 0$$

The natural moments are the even ones ($n = 1,3,5,...$) for which the relation to the matrix elements of the local operators O_n holds. Nevertheless one may consider as an interpolation the unnatural (odd) moments as well ($n = 2,4,6,...$).

The anomalous dimension coefficients $(\gamma_n^1)^{NS}$, separated into natural and unnatural ones, are

$$\gamma_1^1 = 0 \qquad\qquad \gamma_2^1 = 2.67 = \frac{8}{3}$$

$$\gamma_3^1 = 4.17 = \frac{25}{6} \qquad\qquad \gamma_4^1 = 5.23 = \frac{157}{30} \qquad\qquad (67)$$

$$\gamma_5^1 = 6.07 = \frac{91}{15} \qquad\qquad \gamma_6^1 = 6.75 = \frac{709}{105}$$

We note the ratios

$$\frac{\gamma_5^1}{\gamma_3^1} = 1.456 \qquad\qquad \frac{\gamma_6^1}{\gamma_4^1} = 1.290 \qquad\qquad (68)$$

governing the linear relations

$$\frac{d \log M_5^3}{d \log M_3^3} \cong \frac{\gamma_5}{\gamma_3} \qquad\qquad \frac{d \log M_6^3}{d \log M_4^3} \cong \frac{\gamma_6}{\gamma_4} \qquad\qquad (69)$$

Eq. (66), (68) and (69) are compared with the corresponding experimental results in Figs. 13 - 15.

A rough analysis of M_3^3 between $Q^2 = 20 \text{ GeV}^2$ and 100 GeV^2 yields

$$\frac{d \log M_3^3}{d \log Q^2} \cong \frac{\Delta \log M_3^3}{\Delta \log Q^2} \cong 0.086 \cong \frac{4.17}{8\pi} \kappa(Q = 4.5 \text{ GeV}) \qquad\qquad (70)$$

$$\rightarrow \kappa(Q = 4.5 \text{ GeV}) \cong 0.5$$

Tracing this value of $\bar{\kappa}$ in Fig. 9 we see that $Q = 4.5$ GeV is on the edge of the applicability of eq. (50) for $\bar{\kappa}$.

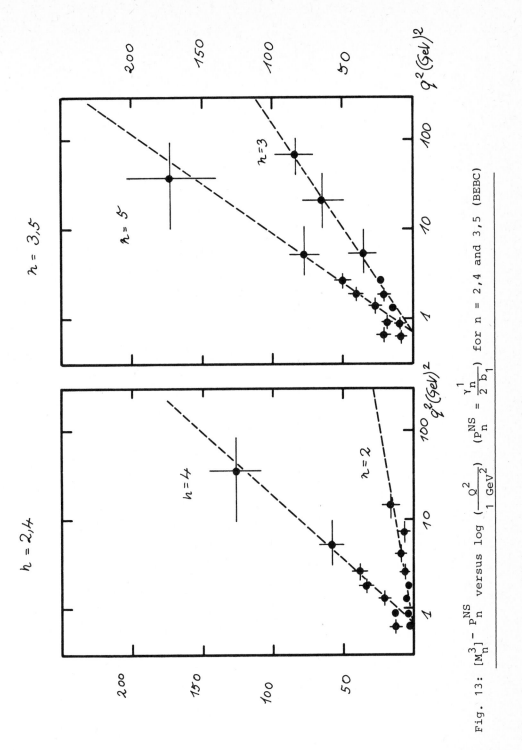

Fig. 13: $[M_n^3] - P_n^{NS}$ versus $\log\left(\frac{Q^2}{1\ \text{GeV}^2}\right)$ $\left(P_n^{NS} = \frac{\gamma_n^1}{2\ b_1}\right)$ for n = 2,4 and 3,5 (BEBC)

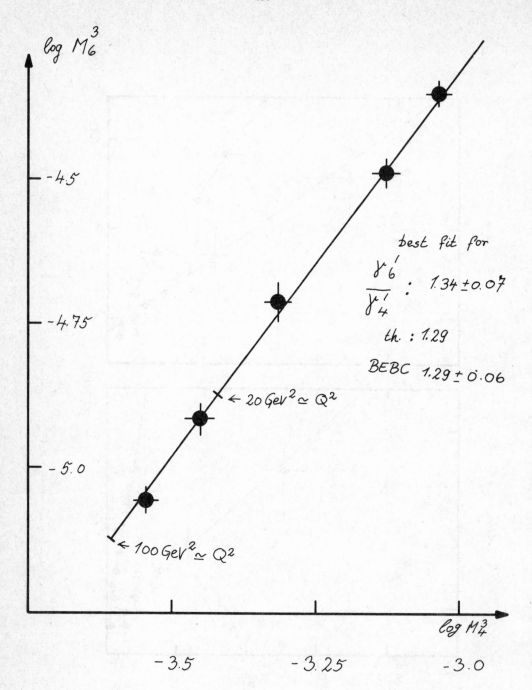

Fig. 14: log M_4^3 versus log M_6^3 (CDHS)

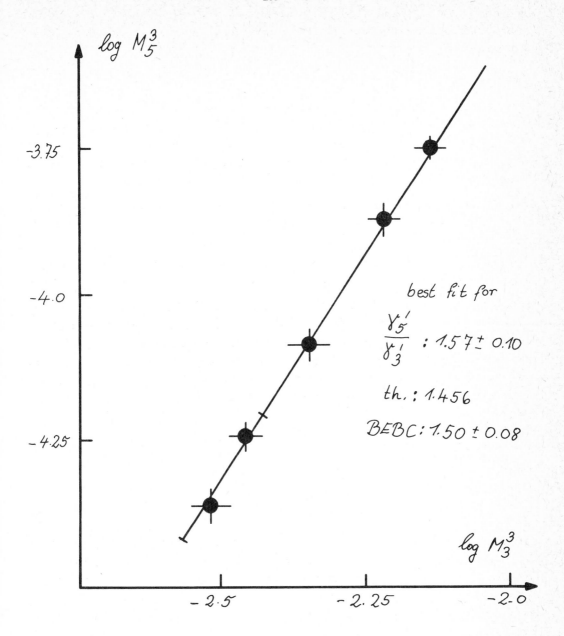

Fig. 15: <u>log M$_5^3$ versus log M$_3^3$ (CDHS)</u>

The two loop (second order) corrections do become important when the (n dependent) interval (FN11)

$$(Q_1)^2 \leq Q^2 \leq (Q_2(n))^2$$

is transgressed. In order to make this statement precise a physically meaningful choice of $Q_1^2(n)$ is necessary. For Q^2 small enough there are powerlike corrections of perturbative (quark masses) and more important nonperturbative nature which are completely unaccounted for by the parametrizations of eq. (66) (one loop) and eq. (65) (two loops).

An illustration is Fig. 16 where large scale breaking effects appear in M_4^3 for $Q^2 < 3$ GeV^2 depending on whether elastic $\nu p \to \mu^- n$ events are included or not.[38]

These scale breaking effects should not be mistaken for the asymptotic ones (for large Q^2) in eq. (65) and (66).

Clearly further refined tests over as wide a Q^2 interval as possible are desirable. Before an adequate precision of the small scale breakings in the large structure functions F_1, F_3 can be obtained it is mandatory to know the small structure function F_L within appropriate errors. So far it has not been possible to exploit the full wealth of the two loop calculations[35] from the experimental side.

The nature of the long range color force

We turn from short distances

$$r \ll (\mu^*)^{-1} \cong 0.4.10^{-13} \text{cm}$$

to characteristic and large distances

$$r \geq (\mu^*)^{-1}$$

Extending the effective charge (Fig. 9) to the effective (distance dependent) quark mass, both amenable to perturbation expansions in $\bar{k}$ at short distances, the qualitative features of these functions are shown in Fig. 17.

In trying to understand the long range limit of the color force from the observed spectrum of nonstrange mesons (Fig. 18) one is struck by the (surprising) simplicity.

The following comments shall illustrate these findings[26]

Die Bahnen sind Ellipsen...

die Periheldrehung ist 0 trotz v/c = 0 (1)... .

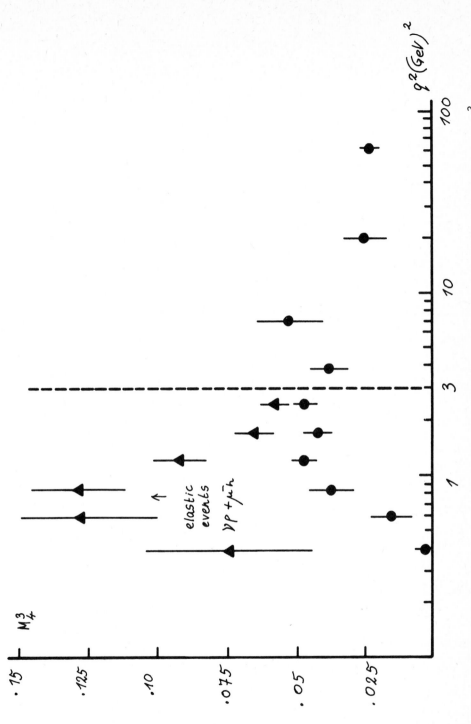

Fig. 16: <u>Influence of elastic $\nu p \rightarrow \mu^- n$ scattering on the scale breaking effects in M_4^3.</u>

Fig. 17: $\underline{m_{eff}, g^2_{eff}/4\pi}$ as functions of distance r. Also shown is a
characterization of bound systems in terms of mean velocity
in orbit v/c and an imaginary changed relative value of
Planck's constant $\vec{H}/\hbar$

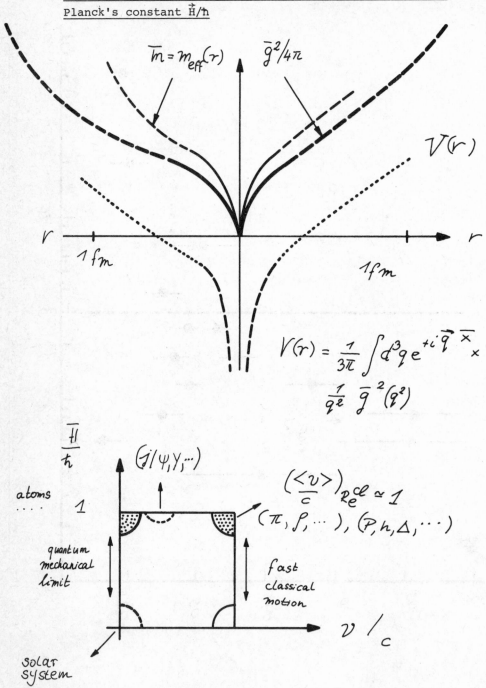

$$\bar{m} = m_{eff}(r) \qquad \bar{g}^2/4\pi$$

$$V(r)$$

$$1\,fm \qquad 1\,fm$$

$$V(r) = \frac{1}{3\pi}\int d^3q\, e^{+i\vec{q}\cdot\vec{x}} \times$$
$$\frac{1}{q^2}\,\bar{g}^2(q^2)$$

$$\frac{\bar{H}}{\hbar}$$

$$(J/\psi, Y, \cdots)$$

atoms

$$1$$

$$\left(\frac{\langle v \rangle}{c}\right)^{cl}_{Re} \simeq 1$$

$$(\pi, \rho, \cdots), (P, n, \Delta, \cdots)$$

quantum
mechanical
limit

fast
classical
motion

$$v/c$$

solar
system

$$E^2 = \left(\frac{M^2}{4}\right)\substack{\text{bound} \\ \text{system}} = \vec{p}^2 + m_{eff}^2(r) - \text{const...}$$

$$m_{eff}^2(r) \xrightarrow[r \to \infty]{} \lambda^2 r^2 \quad , \quad \lambda \cong \frac{m_\rho^2}{4} \cong \frac{1}{8\alpha'} \qquad (71)$$

α': slope of Regge trajectories

- the Baryons show a corresponding simplicity,

$$\vec{L} \cdot \vec{S} \ (\vec{L}_{ij} \ \vec{S}_{ij})\text{-couplings are} \sim 0.^{39}$$

- the hyperfine interaction which for a potential $\sim \frac{1}{r}$ is given by

$$- \frac{8}{3} \vec{\mu}_1 \vec{\mu}_2 \ (\frac{\vec{\sigma}_1 \vec{\sigma}_2}{4}) \ \delta^3(\vec{x} - \vec{y})$$

appears within $m_{eff}^2 \ (S_1, S_2)$ as an r-independent term

$$\lambda \, (\frac{\vec{\sigma}_1 \vec{\sigma}_2}{4})$$

$\mu_{1,2}$: magnetic moments of e.g. electron and proton in hydrogen.

- the intense quark-gluon emission and absorption processes which characterize the random walk[40] generate a Gaussian wave function apart from short distance corrections

- the gluon radiation damping

$$\Gamma_{\rho \, (\omega)} \cong 150 \ \text{MeV} \quad \Gamma_{\eta' \to \eta \pi \pi} \leq 300 \ \text{keV}$$

is quite small.

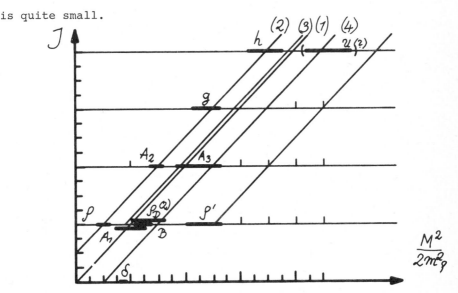

Fig. 18: Spin versus (mass)^2 in units of $2 \ m_\rho^2$ of the nonstrange mesons forming the pattern of the (approximate) trajectories 1(π,B,A$_3$,...), 2(ρ,A$_2$,g,h,...), 3(A$_1$,...), 4(δ,ρ_D,...)

4) The weak interactions $\longleftrightarrow$ Gauge symmetry:[41] local gauge trasformations based on the gauge group $SU2_L \times U1$ (including as an appropriately mixed gauge boson the photon $\rightarrow$ 2)).

We shall only sketch an outline in the present perspective.[42] The effective current current Hamiltonian (for energies much smaller than m_W) involving the three generations of eq. (34) is

$$H_W = \frac{4G_F}{\sqrt{2}} [j_\mu^+ \, j^{\mu -} + j_\mu^{(n)} \, j^{\mu \, (n)}]$$

$$j_\mu^+ = (\bar{u}, \, \bar{c}, \, \bar{t})_L \, \gamma_\mu \, (\theta_{ik}^{(h)}) \begin{pmatrix} d \\ s \\ b \end{pmatrix}_L + \ldots$$

$$(\bar{\nu}_e, \, \bar{\gamma}_\mu, \, \bar{\nu}_\tau)_L \, \gamma_\mu (\theta_{ik}^{(1)}) \begin{pmatrix} e^- \\ \mu^- \\ \tau \end{pmatrix}_L + \ldots$$

(72)

$$G_F = (1.0262 \pm 0.0002) \times 10^{-5} \, m_o^{-2} : \text{Fermi constant}$$

The hadronic mixing matrix $\theta^{(h)}$ governing the charged currents $j_\mu^\pm$ can (for three generations) be reduced to the form[43]

$$\theta^{(h)} = \begin{pmatrix} c_1 & s_1 c_3 & s_1 s_3 \\ -s_1 c_2 & c_1 c_2 c_3 - e^{i\delta} s_2 s_3 & c_1 c_2 s_3 + e^{i\delta} s_2 c_3 \\ s_1 s_2 & -c_1 s_2 c_3 - c_2 s_3 e^{i\delta} & -c_1 s_2 s_3 + c_2 c_3 e^{i\delta} \end{pmatrix}$$

(73)

$$(s_1, \, c_1) = (\sin\theta_c, \, \cos\theta c) \longrightarrow \text{Cabibbo angle}$$

$$\theta_c \cong 13.0^o \text{ from } S(\bar{S})\text{-decays}$$

$$(s_2, \, c_2) = (\sin\theta_2, \, \cos\theta_2) \longrightarrow \text{charmed particle}$$
$$ \text{decays}$$
$$ \text{b-decays}$$
$$ \ldots\ldots$$

$$\delta \longleftrightarrow \text{phase parameter governing CP-violation}$$

Universality is reflected through the relations

$$G_F = G_\mu = (\cos\theta_c)^{-1} G_\beta \tag{74}{}^{44}$$

The neutral current is independent of $\theta^{(h)}$:

$$j_\mu^{(n)} = j_\mu(I_3^W) - \sin^2\theta_W \, j_\mu^{e.m.}$$

$$I_3^W = \left\{ \begin{array}{l} \frac{1}{2} \ (u,c,t,\nu_c,\nu_\mu,\nu_\tau)_L \\[2ex] -\frac{1}{2} \ (d,s,b,e^-,\mu^-,\tau^-)_L \\[2ex] 0 \ \text{otherwise} \end{array} \right. \tag{75}$$

The weak mixing angle θ_W yields the mass eigenstates (Z,A) of the electrically neutral gauge bosons in terms of the $SU2_L$ eigenstates W^3 (member of an $SU2_L$ triplet) and B ($SU2_L$ singlet):

$$Z_\mu = \cos\theta_W \ W_\mu^3 - \sin\theta_W \ B_\mu$$

$$A_\mu^{(\gamma)} = \sin\theta_W \ W_\mu^3 + \cos\theta_W \ B_\mu \tag{76}$$

The coupling constants of $SU2_L$ and $U1$ are related to the unit electric charge

$$SU2_L \ (W^\pm, W^3): \ g_{SU2} = \frac{e}{\sin\theta_W} \qquad (> e)$$

$$g' = \frac{e}{\cos\theta_W}$$

$$t_g^2 \theta_W = \left(\frac{g'}{g_{SU2}}\right)^2 \longrightarrow = \frac{3}{5} \ \text{in the symmetry limit of a unifying gauge group in a chain like e.g.:}$$

$$E_6 > SO10 > SU5 \longrightarrow SU2_L \times U1 \times SU3_c \tag{77}$$

The masses of the W and Z gauge bosons follow a pattern similar to eq. (77)

$$m_o = (\frac{\pi\alpha}{\sqrt{2}\ G_F})^{1/2} = 37.291 \text{ GeV}$$

$$m_W^{\pm} = \frac{m_o}{\sin\theta_W} = 77.8 \pm 3.2 \text{ GeV} \tag{78}$$

$$m_Z = \frac{m_W}{\cos\theta_W} = 88.7 \pm 2.6 \text{ GeV}$$

for $\sin^2\theta_W = 0.23 \pm 0.03$[42]

We may ask whether one generation of fermions corresponds to a (physically) irreducible representation of an as yet unspecified gauge group, unifying the fragments

$$\text{gravity x } SU3_c \text{ x } SU2_L \text{ x } U1$$

of commuting gauges.

E.g. the left handed fermion fields

$$\begin{pmatrix} u_1\ u_2\ u_r\ \nu_e & \bar{u}_1\ \bar{u}_2\ \bar{u}_3\ N_e \\ d_1\ d_2\ d_3\ e^- & \bar{d}_1\ \bar{d}_2\ \bar{d}_3\ e^+ \end{pmatrix} \tag{79}$$

$$\underbrace{\qquad\qquad}_{SU2_L \text{ doublets}} \underbrace{\qquad\qquad}_{SU2_L \text{ singlets}}$$

can form the 16 dimensional spinor representation of SO10. As a consequence of the completeness of the fermion representation - no matter what the gauge group may be - we have the relations

$$g_{SU3_c} = g_{SU2_L}$$

$$\sin^2\theta_W = \frac{Sp\ (I_3^W)^2}{Sp\ (Q^2)} = \frac{3}{8} = 0.375$$

$$\neq 0.23 \pm 0.03$$

The large discrepancy between the symmetry calue and the experimental value of $\sin\theta_W$ leaves the alternative of either no way or a long way to go (in energy) until the sought symmetry limit is reached.

We shall conclude exhibiting the rescaling behaviour of the (inverse square) effective coupling constants corresponding to the groups U1, $SU2_L$, $SU3_c$ respectively.[45]

$$\frac{4\pi}{(g')^2} \cdot \frac{3}{5} = \frac{3}{5} \frac{\cos^2\theta_W \ (\bar{\mu})}{\alpha} \qquad \longleftrightarrow \quad U1$$

$$\frac{4\pi}{(g_{SU2_L})^2} = \frac{\sin^2\theta_W \ (\bar{\mu})}{\alpha} \qquad \longleftrightarrow \quad SU2_L \qquad\qquad (80)$$

$$\frac{1}{\kappa} \qquad\qquad\qquad \longleftrightarrow \quad SU3_C$$

as a function of $\log_{10} \ (\frac{\bar{\mu}}{100 \ GeV})$ in Fig. 19.

Despite large uncertainties the four types of gauge theories (gravity, electrodynamics, chromodynamics, weak interactions) do show the tendency to become comparable in strength in the region of energies between 10^{15} and 10^{21} GeV.

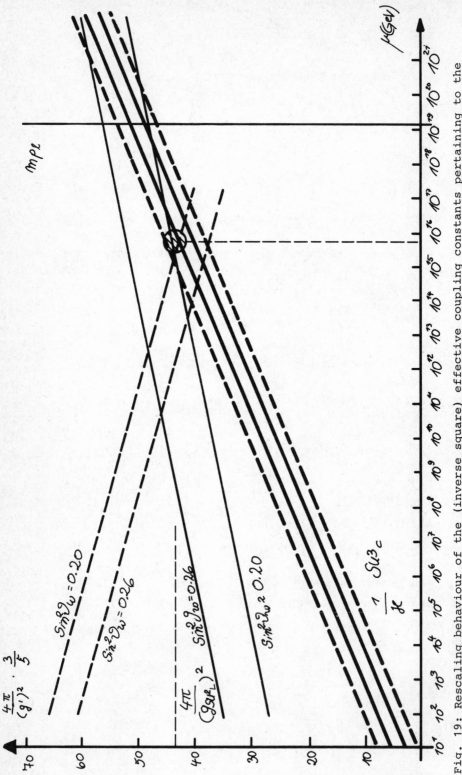

Fig. 19: Rescaling behaviour of the (inverse square) effective coupling constants pertaining to the gauge groups U1, SU2$_L$ and SU3$_c$.

Footnotes

FN1)   We use the following conventions related to the metric tensor
$g_{\mu\nu}$, the Christoffel symbol $\Gamma^{\mu}_{\alpha\beta}$ and the curvature tensor
$R^{\mu}{}_{\nu;\alpha\beta}$ 1)

$$ds^2 = g_{\mu\nu}\, dx^{\mu}\, dx^{\nu} = \left\{ \begin{array}{l} > 0 \ \text{time like} \\ < 0 \ \text{space like} \end{array} \right\} \text{distances}$$

$$\Gamma^{\mu}_{\alpha\beta} = g^{\mu\nu}\Gamma_{\nu;\,\alpha\beta}; \quad \Gamma_{\nu;\,\alpha\beta} = \tfrac{1}{2}\,(\partial_{\alpha}g_{\nu\beta} + \partial_{\beta}g_{\nu\alpha} - \partial_{\nu}g_{\alpha\beta})$$

$$R^{\mu}_{\nu;\,\alpha\beta} = \partial_{\beta}\Gamma^{\mu}_{\nu\alpha} - \partial_{\alpha}\Gamma^{\mu}_{\nu\beta} +$$

$$+ \Gamma^{\mu}_{\beta\kappa}\Gamma^{\kappa}_{\nu\alpha} - \Gamma^{\mu}_{\alpha\kappa}\Gamma^{\kappa}_{\nu\beta}$$

$$R_{\nu\beta} = R^{\mu}{}_{\nu;\,\mu\beta} = \partial_{\beta}\Gamma^{\mu}_{\nu\mu} - \partial_{\mu}\Gamma^{\mu}_{\nu\beta} +$$

$$\Gamma^{\mu}_{\beta\kappa}\Gamma^{\kappa}_{\nu\mu} - \Gamma^{\mu}_{\mu\kappa}\Gamma^{\kappa}_{\nu\beta}$$

$$R = g^{\nu\beta}R_{\nu\beta} = g^{\nu\beta}\,\partial_{\beta}\,\Gamma^{\mu}_{\nu\mu} - g^{\nu\beta}\,\partial_{\mu}\Gamma^{\mu}_{\nu\beta} +$$

$$+ g^{\nu\beta}(\Gamma^{\mu}_{\beta\kappa}\Gamma^{\kappa}_{\nu\mu} - \Gamma^{\mu}_{\mu\kappa}\Gamma^{\kappa}_{\nu\beta})$$

$$\sqrt{|g|} \quad G = R\sqrt{|g|} \quad \text{(modulo a total divergence)}$$

$$G = g^{\nu\beta}\,[\Gamma^{\kappa}_{\nu\beta}\Gamma^{\mu}_{\kappa\mu} - \Gamma^{\kappa}_{\nu\mu}\Gamma^{\mu}_{\beta\kappa}]$$

FN2)   Finite Lorentz transformations are parametrized and represented in
the following way

$$\Lambda\,(\omega) = \exp \tfrac{1}{2}\,\omega^{\alpha\beta}\Omega_{\alpha\beta} \longleftrightarrow \exp\,(\tfrac{1}{2i}\omega^{\alpha\beta}S_{\alpha\beta})$$

The matrics

$$\Omega_{\alpha\beta},\ \tfrac{1}{i}\,S_{\alpha\beta}$$

satisfy the same commutation rules as the operator

$$\tfrac{1}{i}\,M^{(orbital)}_{\alpha\beta} = x_{\alpha}\,\partial_{x\beta} - x_{\beta}\,\partial_{x\alpha};\quad x_{\alpha} = \eta_{\alpha\alpha'}\,x^{\alpha'}$$

$$[\Omega_{\mu\nu}, \Omega_{\sigma\tau}] = \eta_{\mu\tau}\,\Omega_{\nu\sigma} + \eta_{\nu\sigma}\,\Omega_{\mu\tau}$$

$$- \eta_{\mu\sigma}\,\dot{\Omega}_{\nu\tau} - \eta_{\nu\tau}\,\dot{\Omega}_{\mu\sigma}$$

Within the Dirac-Clifford algebra of γ matrices S is given by

$$S_{\mu\nu} = \frac{i}{4}\,[\gamma_\mu\,\gamma_\nu]; \quad \{\gamma_\mu\,\gamma_\nu\} = 2\eta_{\mu\nu}$$

FN3) The discussion of the localization of gravitational energy goes back to Einstein[2] in particular with respect to the emission of gravitational radiation[3] by a binary system of stars leading to a decrease of the revolution period according to

$$-\frac{dE}{dt} = \frac{G_N}{10c^5}\,\{\,\sum_{i,k}(\overset{\cdots}{D}_{ik})^2 - \frac{1}{3}\,(\,\overset{3}{\underset{i}{\sum}}\,\overset{\cdots}{D}_{ii})^2\,\}$$

$$D_{ik} \cong \frac{m_1 m_2}{m_1 + m_2}\,z_i z_k\,(t) = \int d^3\xi \rho_m(\xi)\,\xi_i \xi_k$$

$$\vec{z} = (\vec{x}_1 - \vec{x}_2)_{\text{center of mass}}$$

A valid resolution of the problems related to the canonical energy momentum density τ^μ_ν has been given by Möller.[4] The reason for the difficulties is the intertwining of

$$\tau^\mu_\nu$$

with an associated quantity derived from the angular momentum density

$$K^\mu_\nu = \eta^{\alpha\beta}\,\partial_\alpha\,M^\mu_{\beta\nu}$$

FN4) The gravitational energy stored in $1\ m^3$ on the surface of the earth is

$$\int_{1m^3} d^3x\,e_g = (6.672)^{-1}(9.806)^2\,\frac{1}{8\pi}\,10^{18}\ erg =$$
$$= (5.7 \cdot 10^4)\ Mw\cdot sec$$

This corresponds to 190 seconds running time of a 300 Mw nuclear power plant with an active core of $1\ m^3$.

FN5) The total energy is

$$m_e c^2 \quad (\text{not } m_e c^2 \, \frac{1}{\sqrt{1-v^2/c^2}} \;)$$

due to the binding forces.

FN6) The question of asymptotic freedom in gravity was subject of repeated discussions in seminars at the University of Bern, a detailed resulting treatment can be found in ref. (11).

FN7) Much of the material on tests of QED I have adapted from a colloquium on this subject by B. Lautrup (Bern, 1979).

FN8) B. Schroer informed me that this was well known from studying critical behaviour in statistical mechanics, see e.g. ref. (22).

FN9) A similar mass parameter appears in constructive field theories as discussed by K. Osterwalder at this symposium.

FN10) It was noted by Bardeen, Buras, Duke and Muta[35] that the nonrational factor $\log (e^{\gamma} \frac{1}{4\pi})$ can be absorbed within their parametrization of $\bar{k} = \bar{k}(\frac{\mu^2}{\Lambda^2})$ into a rescaling of $\Lambda \to \bar{\Lambda}$. It is their $\bar{\Lambda} \cong 500$ MeV which corresponds to our parameter μ^*.

FN11) The applicability of this approximation to a (n dependent) limited Q^2 interval

$$(Q_1)^2 \leq Q^2 \leq (Q_2(n))^2$$

such that

$$Q_2^2(n) / (Q_1)^2 << \exp(\frac{8\pi}{\gamma_n^1} \frac{1}{\bar{k}(Q^2)})$$

has been stressed in ref. (37). For $n = 3$ and $(Q_1)^2 = 20$ Gev^2 the above condition reads

$$\frac{|Q_2(3)|}{|Q_1|} << \exp(3.0 \frac{1}{\kappa}) \cong 329$$

$$\text{for } \kappa(Q_1^2) = 0.52$$

i.e. for

$$|Q_1| = 4.5 \text{ GeV}, \quad |Q_2(3)| << 1470 \text{ GeV}.$$

Appendix: The analytic form of γ_n^2 [35]

$$\gamma_n^2 = -\frac{2}{9} \left\{ \begin{array}{l} 16\, S_1(n)\, \dfrac{2\,n+1}{n^2(n+1)^2} + \\[2ex] 16\,[2\,S_1(n) - \dfrac{1}{n(n+1)}]\,[S_2(n) - S_2^1(\tfrac{n}{2})] \\[2ex] + 64\,\tilde{S}(n) + 24\,S_2(n) - 3 - 8\,S_3^1(\tfrac{n}{2}) \\[2ex] -8\,\dfrac{3n^2+n^2-1}{n^3(n+1)^3} - 16\,(-1)^n\,\dfrac{2n^2+2n+1}{n^3(n+1)^3} \end{array} \right\}$$

$$+4\left\{ \begin{array}{l} S_1(n)\,[\,\dfrac{536}{9} + 8\,\dfrac{2n+1}{n^2(n+1)^2}\,] \\[2ex] -16\,S_1(n)\,S_2(n) + S_2(n)\,[\,-\dfrac{52}{3} + \dfrac{8}{n(n+1)}\,] \\[2ex] -\dfrac{43}{6} - 4\,\dfrac{151n^4+263n^3+97n^2+3n+9}{9n^3(n+1)^3} \end{array} \right\}$$

$$+\frac{2}{3}n_{fl}\left\{ \begin{array}{l} -\dfrac{160}{9}\,S_1(n) + \dfrac{32}{3}\,S_2(n) + \dfrac{4}{3} \\[2ex] + 16\,\dfrac{11n^2+5n-3}{9n^2(n+1)^2} \end{array} \right\}$$

$$S_i(n) = \sum_{j=1}^{n} (j)^{-i} , \quad \tilde{S}(n) = \sum_{j=1}^{n} (-1)^j\,\frac{S_1(j)}{j^2}$$

$$S_i^1(\tfrac{n}{2}) = \left\{ \begin{array}{ll} S_i(\tfrac{n}{2}) & n \text{ even} \\[1ex] S_i(\tfrac{n-1}{2}) & n \text{ odd} \end{array} \right\}$$

Acknowledgements

I should like to thank the members of the theory group in Bern for many controversial and enlightening discussions and likewise B. Lautrup, A. Buras, D. A. Ross, H. D. Politzer, R. M. Barnett, M. Gell-Mann, R. P. Feynman, P. Ramond, S. Brodsky, H. Quinn, J. D. Bjorken, W. A. Bardeen, J. Rosner, C. Quigg, B. Schroer, R. Schrader, R. Seiler, J. M. Combes, E. Seiler, K. Osterwalder, H. Lehmann, E. C. Nelson and R. Jost.

References

1. see eg.
   W. Pauli, Relativitätstheorie, Editore Boringhieri, Torino, 1963

   B. S. DeWitt, Dynamical Theory of Groups and Fields, Gordon and Breach New York, 1964; Phys. Reports 19C (1975) 295.

   S. Weinberg, Gravitation and Cosmology, John Wiley and Sons, New York, 1972.

2. A. Einstein, Berliner Ber. (1915) 778.

   see in particular the criticism by E. Schrödinger, Phys. Z. 19 (1918) 4.

3. A. Einstein, Über Gravitationswellen, Berliner Ber. (1918) 154.

4. C. Möller, K. Danske. Vidensk. Selsk. mat.-fys. Skr. 34 (1964) 66; Annals of Phys. 12 (1961) 118; K. Danske Vidensk. Selsk. mat.-fys. Skr. 1 (1961) and concerning the emission of gravitational waves: Phys. Letters 3 (1963) 329.

   see also the contributions of C. Möller and J. Plebanski in Proceedings of the Conference on Theory of Gravitation, Warsaw 1964, p. 31 and 45.

5. see e.g.
   R. Jost, The General Theory of Quantized Fields, American Mathematical Society, Providence, Rhode Island, 1965, p. 26.

6. J. H. Taylor, L. A. Fowler and P. M. McCulloch, Measurement of General Relativistic Effects in the Binary Pulsar 1913+16, Nature 277 (1979) 437.

7. R. Epstein, Astrophys. J. 216 (1977) 92

8. J. Weber, General Relativity and Gravitational Waves, Interscience Publishers Inc., New York, 1961.

9. see R. Penrose, contribution to this symposium.

10. E. S. Fradkin and G. A. Vilkoviski, On Renormalization of Quantum Field Theory in Curved Space Time, University of Bern preprint 1976.

11. H. Bebie and H. Leutwyler, Lecture Notes on Mechanics and Relativity, University of Bern 1979.

12. see e.g.
    G. P. Lepage, Two Body Bound States in Quantum Electrodynamics, SLAC preprint Slac-212 (1978)

    W. E. Caswell, G. P. Lepage and J. Saperstein, Phys. Rev. Letters 38 (1977) 488

13. for the theoretical treatment see
    B. Lautrup, A. Petermann and E. de Rafael, Phys. Reports 3C (1972) 193
    P. J. Mohr, Phys. Rev. Letters 34 (1975) 1050
    G. W. Erickson, Phys. Rev. Letters 27 (1971) 780

14. for the experiments see

    S. R. Landeen and F. M. Pipkin, Phys. Rev. Letters $\underline{34}$ (1975) 1368

    D. A. Andrews and G. Newton, Phys. Rev. Letters 37 (1976) 1254

15. for the theoretical treatment see

    B. Lautrup et al., op. cit. ref. 13.

    T. Fulton, D. A. Owen and W. Repko, Phys. Rev. Letters 26 (1971) 61

16. for the experiments see

    D. E. Casperson, T. W. Crane, A. B. Denison, P. O. Egan, V. W. Hughes, F. G. Mariam, H. Orth, H. W. Reist, P. A. Souder, R. D. Stambaugh, P. A. Thompson and G. zu Putlitz, Phys. Rev. Letters 38 (1977) 956

17. for theoretical contributions see

    P. Cvitanovic and T. Kinoshita, Phys. Rev. $\underline{D10}$ (1974) 4007

    R. Barbieri and E. Remiddi, Nucl. Phys. $\underline{B90}$ (1975) 233

    J. Calmet and A. Petermann, Phys. Letters $\underline{47B}$ (1973) 369; $\underline{58B}$ (1975) 449

18. for a review of the g-2 experiments see

    F. H. Combley and F. Picasso, Phys. Reports $\underline{14C}$ (1974) 1

    F. H. Combley in Proceedings of the 1975 Int. Symposium on Lepton and Photon Interactions at High Energies, Stanford, 1975, p. 913

19. F. Berends and R. Gastmans, Phys. Letters $\underline{55B}$ (1975) 311

20. on the new determination of α from the AC Josephson effect see

    P. T. Olsen and E. R. Williams in Atomic Masses and Fundamental Constants 5, J. H. Sanders and A. H. Wapstra editors, Plenum, New York, 1976, p. 538

21. E. C. G. Stückelberg and A. Petermann, Helv. Phys. Acta $\underline{26}$ (1953) 499

    M. Gell-Mann and F. Low, Phys. Rev. $\underline{95}$ (1954) 1300

    C. G. Callan, Phys. Rev. $\underline{D2}$ (1970) 1541

    K. Symanzik, Comm. Math. Phys. $\underline{18}$ (1970) 227

    for the universal rescaling of mass see

    G. t'Hooft, Nucl. Phys. $\underline{B61}$ (1973) 455

    S. Weinberg, Phys. Rev. $\underline{D8}$ (1973) 3497

22. L. P. Kadanoff, Rev. Mod. Physics $\underline{39}$ (1967) 395

    F. Jegerlehner and B. Schroer, Acta Physica Austriaca, Suppl. XI (1973) 389

23. E. de Rafael and J. L. Rosner, Annals of Phys. $\underline{82}$ (1974) 369

24. Y. Nambu in Preludes in Theoretical Physics, Amsterdam, 1966

    (Early idea in the right direction)

H. Fritzsch, M. Gell-Mann and H. Leutwyler, Phys. Letters <u>47B</u> (1973) 365

S. Weinberg, Phys. Rev. Letters <u>31</u> (1973) 494

the property of asymptotic freedom in nonabelian gauge theories was demonstrated explicitly by

H. D. Politzer, Phys. Rev. Letters <u>30</u> (1973) 1346

D. Gross and F. Wilczek, Phys. Rev. Letters <u>30</u> (1973) 1343

and implicity by

I. Khriplovich, Soviet Journal of Nucl. Phys. <u>10</u> (1970) 235

G. t'Hooft (1972) unpublished

25.   for the theory of color perception by the human eye see e.g.

H. v. Helmholtz, Handbuch der Physiologischen Optik II, Leopold Voss Verlag, Leipzig, 1911, p. 101

H. J. Mey, Farbentheorie, zum 100. Jahrestag von Maxwells Dreifarbendemonstration, Neue Züricher Zeitung, Mittwoch, 17.5.61 (Mittagsausgabe, Blatt 9, Nr. 1845 (71))

26.   for a recent discussion see e.g.

P. Minkowski, Asymptotic Freedom - Infrared Instability, University of Bern preprint, 1979

H. Leutwyler and J. Stern, Annals of Phys. <u>112</u> (1978) 94; Nucl. Phys. <u>B133</u> (1978) 115; Phys. Letters <u>73B</u> (1978) 75

27.   M. Gell-Mann, R. Oakes and B. Renner, Phys. Rev. <u>175</u> (1968) 2195

S. L. Glashow and S. Weinberg, Phys. Rev. Letters <u>20</u> (1968) 224

H. Leutwyler, Nucl. Phys. <u>76B</u> (1974) 431

J. Gasser and H. Leutwyler, Nucl. Phys. <u>94B</u> (1975) 269

S. Weinberg, On the Problem of Mass, in a Festschrift for I. I. Rabi, New York Academy of Sciences, New York, 1977

P. Minkowski and A. Zepeda, Hadron Masses and Current Algebra Masses, University of Bern preprint, 1979

28.   for a review see e.g.

H. Pagels, Phys. Reports <u>16C</u> (1975) 219

29.   H. Fritzsch, Phys. Letters <u>73B</u> (1978) 457; CERN preprint TH-2640, 1979

30.   renormalizability of nonabelian gauge theories was proven by

G. t'Hooft, Nucl. Phys. <u>B33</u> (1971) 173; <u>B35</u> (1971) 167

31.   A. A. Slavnov, University of Kiev preprints ITP 71-83E (1971); Theor. and Math. Phys. <u>10</u> (1972) 153

C. Becchi, A. Rouet and R. Stora, Annals of Phys. <u>98</u> (1976) 287

32.   W. E. Caswell, Phys. Rev. Letters <u>33</u> (1974) 244

D. R. T. Jones, Nucl. Phys. <u>B75</u> (1975) 531

33.   B. C. Barish, Experimental Aspects of High Energy Neutrino Physics, Phys. Reports <u>39C</u> (1978) 279

34. see e.g.

R. E. Taylor in Proceedings of the 1975 Int. Symposium on Lepton and Photon Interactions at High Energies, Stanford (1975) p. 679

35)$_1$ H. Georgi and H. D. Politzer, Phys. Rev. $\underline{D9}$ (1974) 416

D. Gross and F. Wilczek, Phys. Rev. $\underline{D8}$ (1973) 3633; $\underline{D9}$ (1974) 980

H. Politzer, Phys. Reports $\underline{14C}$ (1974) 129

35)$_2$ E. G. Floratos, D. A. Ross and C. T. Sachrajda, Nucl. Phys. $\underline{B129}$ (1977) 66 and CERN preprints TH-2566, TH-2570 (1978)

A. J. Buras, E. G. Floratos, D. A. Ross and C. T. Sachrajda, Nucl. Phys. $\underline{B131}$ (1977) 308

W. A. Bardeen, A. J. Buras, D. W. Duke and T. Muta, Phys. Rev. $\underline{D18}$ (1978) 3998

see also

O. Nachtmann, Nucl. Phys. $\underline{B63}$ (1973) 237

36. for the CDHS collaboration see the contributions of

K. Kleinknecht and K. Tittel in Proceedings of the 1977 Int. Symposium on Lepton and Photon Interactions at High Energies, Hamburg, 1977, p. 271 and 309

J. G. H. de Groot et al., University of Dortmund preprint, 1978, 12/78

for the Aachen-Bonn-CERN-London-Oxford-Saclay collaboration (BEBC) see

P. C. Bosetti et al., University of Oxford preprint, 1978, Report No. 16/78

37. H. Fritzsch and P. Minkowski, Nucl. Phys. $\underline{B76}$ (1974) 365; Phys. Letters $\underline{49B}$ (1974) 462

38. for a critical discussion of the significance of the QCD parametrization of scale breaking effects in moments of deep inelastic structure functions as compared to the existing data see

L. Abbott and M. M. Barnett, SLAC preprint, 1979

39. see e.g. the discussion by

N. Isgur and G. Karl, Phys. Letters $\underline{72B}$ (1977) 109; $\underline{74B}$ (1978) 353; Phys. Rev. $\underline{D18}$ (1978) 4187; Phys. Rev. Letters $\underline{41}$ (1978) 1269

40. E. Nelson, contribution to this symposium

41. S. L. Glashow, Nucl. Phys. $\underline{22}$ (1961) 579

A. Salam and J. C. Ward, Phys. Letters $\underline{13}$ (1964) 168

S. Weinberg, Phys. Rev. Letters $\underline{19}$ (1967) 1264

A. Salam in Elementary Particle Theory, ed. by N. Svartholm, Stockholm, 1968

42. for a detailed discussion see

J. Iliopoulos, contribution to this symposium

H. Fritzsch, Flavourdynamics, Contribution to the XIX Int. Conf. on High Energy Physics, Tokyo, 1978

43. M. Kobayashi and K. Maskawa, Progr. Theor. Phys. $\underline{49}$ (1973) 652

44. A. Sirlin, Nucl. Phys. $\underline{B71}$ (1974) 29 and New York University preprint, 1978

D. H. Wilkinson, Nature $\underline{257}$ (1975) 189

45. H. Georgi, H. R. Quinn and S. Weinberg, Phys. Rev. Letters $\underline{33}$ (1974) 451

DER DUALISMUS VON FELD UND MATERIE IN DER ALLGEMEINEN RELATIVITÄTSTHEORIE

Peter Mittelstaedt, Universität Köln

1. Einleitung und Problemstellung

In der allgemeinen Relativitätstheorie werden die Beziehungen zwischen dem raum-zeitlichen Führungsfeld, der Materie und dem von dieser Materie erzeugten Gravitationsfeld behandelt. Abgesehen von verschiedenen Verallgemeinerungen dieser Theorie, die sich auf ihre Anwendung in der Kosmologie, in der Nähe von Singularitäten oder im Bereich quantenmechanischer Phänomene beziehen, darf sie als eine richtige Theorie angesehen werden, die im Rahmen der bisher erreichten experimentellen Genauigkeit in Übereinstimmung mit allen Beobachtungs-Daten steht. Wir wollen daher im folgenden von der Voraussetzung ausgehen, daß die allgemeine Relativitätstheorie Einsteins das Raum-Zeit-Kontinuum, die Materie und die Gravitation in einer Weise verbindet, die der empirischen Realität entspricht.

Unter diesen Voraussetzungen kann man rückblickend die Frage stellen, welche wissenschaftlichen Ziele Einstein mit der Aufstellung einer "Allgemeinen Relativitätstheorie" verfolgt hat und ob diese Zielsetzungen in der tatsächlich geschaffenen und inzwischen anerkannten Theorie verwirklicht worden sind. Wir wollen diese Frage hier aber nicht in der soeben formulierten Allgemeinheit behandeln, sondern sie auf das Problem einschränken, ob das raum-zeitliche Führungsfeld und die Materie als selbständige oder voneinander abhängige Entitäten zu betrachten sind. Entsprechend diesen beiden Möglichkeiten werden wir von einer dualistischen bze. monistischen Theorie des Raum-Zeit-Feldes und der Materie sprechen.

Die möglicherweise vorhandene Abhängigkeit dieser beiden Entitäten kann wenigstens unter zwei voneinander getrennten Gesichtspunkten behandelt werden: Erstens *ontologisch* in dem Sinne, daß einer dieser Entitäten eine unmittelbare Realität zukommt, während die andere nur mittelbar, durch Hypstasierung, aus der ersten hervorgeht. Zweitens *kausal* in dem Sinne, daß das eine der beiden Phänomene die Ursache des anderen ist, oder jeden falls eine kausale Beziehung zwischen beiden besteht. Wir werden im folgenden diese beiden Aspekte in Betracht ziehen, wenn wir der Frage nachgehen, ob die Allgemeine Relativitätstheorie eine monistische oder eine dualistische Theorie von Raum, Zeit und Materie ist.

2. Historische Vorbemerkungen

Den Rückblick auf die historischen Vorstufen der Allgemeinen Relativitätstheorie möchte ich beginnen lassen bei Isaak Newton, weil hier die Grund-
postulate bereits in einer mit der heutigen Physik vergleichbaren Form
ausgesprochen werden. In den *Philosophiae naturalis principia mathematica* von
1687 hat Newton seine Konzeption des Raum-Zeit-Kontinuums in einfacher
und unmißverständlicher Form ausgedrückt: "Der absolute Raum bleibt vermöge seiner Natur und ohne Beziehung auf einen äußeren Gegenstand stets
gleich und unbeweglich" und "Die absolute, wahre und mathematische Zeit
verfließt an sich und vermöge ihrer Natur gleichförmig und ohne Beziehung
auf irgend einen äußeren Gegenstand". Raum und Zeit werden hier als für
sich bestehende Substanzen aufgefaßt, die den allgemeinen Rahmen darstellen, in dem Materie und deren Bewegung beschrieben wird. In diesem Rahmen
wird das Trägheitsgesetz und die anderen Axiome der Mechanik formuliert,
und in diesem Rahmen werden die Bewegungen der Himmelskörper dargestellt,
ohne daß der Rahmen selbst mit einem empirischen Objekt indentifiziert
wird. Eine solche Identifikation mit "dem Mittelpunkt des Weltensystems"
bzw. der "Bewegung der Himmelskörper" wird aber, wenn auch nur ungenau,
für möglich gehalten.

Diese ohne Zweifel dualistische Konzeption von Raum, Zeit und Materie ist
allerdings schon zu Newtons Zeit nicht unwidersprochen geblieben. In seinem Briefwechsel mit Clarke hat Leibniz (1715 - 1716) darauf hingewiesen,
daß der absolute Raum bei der Beschreibung der Bewegung von Körpern gar
nicht auftritt, sondern daß immer nur die Bewegung eines Körpers relativ
zu einem anderen Körper von Bedeutung ist. Der Raum ist keine an sich bestehende Substanz, sondern "Raum" ist vielmehr der Inbegriff aller Stellen gleichzeitig existierender Dinge. - Ein ähnlicher Gedanke findet sich
bei Berkeley, der in seiner Schrift *De motu* (1721) bemerkt, daß der Raum
Newtons gar nicht verifiziert werden kann und daß das, woran wir den Zustand der Ruhe von dem der Bewegung unterscheiden, nicht der Raum, sondern der Fixstern-Himmel sei. - Wir wollen diesen von Leibniz und Berkeley
geäußerten Gedanken als *kinematische Relativität* bezeichnen. Falls keine anderen Beobachtungen hinzukommen, führt diese kinematische Relativität auf
eine monistische Konzeption von Raum-Zeit und Materie, in der primär die
bewegte Materie gegeben ist, Raum und Zeit hingegen nur sekundär als Ord-
nungsgesichtspunkte hinzukommen.

Newton hatte jedoch in seinen Prinzipien bereits vorsorglich ein Argument formuliert, das es ermöglichte, derartige Einwendungen entgegenzutreten, und das durch ein ebenso einfaches wie geniales Gedankenexperiment unterstützt wurde. Wegen seiner großen Bedeutung für die folgenden Überlegungen sei dieses "Eimer-Experiment" noch einmal kurz skizziert: Ein mit Wasser gefüllter Eimer wird am Seil aufgehängt und das Seil anschließend durch Drehen des Eimers verdrillt. Wenn man den an dem verdrillten Seil hängenden Eimer plötzlich losläßt, so wird er sich sehr schnell in der entgegengesetzten Richtung drehen. Dabei kann man beobachten, daß sich das Wasser zunächst mitbewegt und daß erst allmählich die Drehbewegung des Eimers auf das Wasser übertragen wird. In der ersten Phase des Versuches, in der das Wasser nicht rotiert, bleibt die Wasseroberfläche eben, in der zweiten Phase, in der das Wasser mit dem Eimer rotiert, ist die Oberfläche jedoch gewölbt. - Newton schließt aus diesem Experiment, daß das Wasser bei einer Bewegung relativ zum Eimer keine Veränderung erfährt, daß aber bei einer Bewegung relativ zum absoluten Raum Trägheitskräfte auftreten, die die Oberfläche krümmen. Der absolute Raum ist daher direkt beobachtbar.

Newton hat nur zum Nachweis des absoluten Raumes ein Experiment angegeben. Man kann jedoch leicht ein analoges "zweites Eimerexperiment" zum Nachweis der absoluten Zeit angeben, indem man etwa einen freien Rotator als Uhr für die absolute Zeit auffaßt und diesen vergleicht mit einem durch Reibung gebremsten Rotator: Derjenige Rotator, der sich gegenüber der absoluten Zeit "bewegt", wird Trägheitskräfte erfahren, die zu einer Deformation führen können, während der mit der absoluten Zeit synchron bewegte Rotator keine Deformation erfährt. Durch diese Trägheitskräfte wird daher die "Bewegung" gegenüber der absoluten Zeit, und damit auch diese selbst, beobachtbar.

Formal liegen, worauf schon Hermann Weyl hingewiesen hat, der unterschiedlichen Raum-Zeit-Struktur bei Newton und Leibniz verschiedene Invarianzgruppen zu Grunde. Während die Newtonsche Raum-Zeit - abgesehen von Dilatationen - gegenüber der 7-parametrigen *elementaren Gruppe* E invariant ist, ist die Leibniz'sche Raum-Zeit gegenüber der viel umfassenderen *kinematischen Gruppe* K invariant, die, wenn man auch die Absolutheit der Zeit aufgeben würde, von 7 willkürlichen Funktionen abhängt. Vergleicht man diese beiden Konzeptionen mit den hier maßgebenden Eimer-Experimenten, so wird deutlich, daß die Newtonsche Raum-Zeit zu viel, die Leibniz'sche Raum-Zeit hingegen zu wenig Struktur hat. Durch die beiden erwähnten Experimente wird vielmehr die 10-parametrige Galilei-Gruppe G als die für die Physik maßgebende Invarianzgruppe ausgezeichnet. Während die Existenz der Trägheitskräfte gegen die Leibniz-Struktur spricht, spricht die dynamische Äquivalenz aller Inertialsysteme gegen die Newton'sche Raum-Zeit.

Auf dem Hintergrund dieser Situation wird die Bedeutung der Argumentation von Ernst Mach deutlich. Mach hat darauf hingewiesen, daß durch das Eimer-Experiment Newtons und die dabei auftretenden Trägheitskräfte gar nicht eine Bewegung gegenüber dem absoluten Raum nachgewiesen wird. Das Experiment zeige vielmehr, daß bei einer Bewegung des Wassers gegenüber dem Eimer *keine* Trägheitskräfte auftreten, während bei einer Bewegung des Wassers gegenüber dem Fixsternhimmel - ähnlich Berkeley - Trägheitskräfte zu beobachten sind. Ob der Fixsternhimmel die Ursache der Trägheitskräfte ist, oder ob es sich hier vielmehr um eine zufällige Koinzidenz des Trägheitskompass mit dem Sternenkompass handelt, wird von Mach nicht diskutiert. Wesentlich für ihn ist, daß offensichtlich der absolute Raum nicht nur kinematisch unbeobachtbar ist, - was schon Leibniz wußte - sondern daß auch dynamisch, durch Trägheitskräfte, der absolute Raum in keinem Experiment auftritt.

Mach hat versucht, diesen auf den Raum bezogenen Gedankengang, der heute als Mach'sches Prinzip bezeichnet wird - auch auf die absolute Zeit zu übertragen. Von einer Uhr kann man immer nur sagen, ob sie relativ zu einer anderen Uhr richtig geht, nicht aber ob sie die absolute Zeit anzeigt. Auch in dem oben erwähnten "zweiten Eimerexperiment" mit einem freien und einem gedämpften Rotator wird man nur sagen können, daß in demjenigen Rotator, der gegenüber einer geeigneten kosmischen Zeitskala - Mach spricht von der Entropie des Weltalls - gedämpft ist, Trägheitskräfte auftreten, sonst aber nicht. Es ist natürlich wiederum völlig offen, ob die als kosmische Uhr verwendete Entropie des Weltalls die Ursache dieser Trägheit ist, oder ob es sich um eine zufällige Koinzidenz von Weltzeit und Trägheitszeit handelt. Durch diesen Gedankengang, den man als ein auf die Zeit bezogenes "zweites Mach'sches Prinzip" bezeichnen könnte, wird aber deutlich, daß nicht nur kinematisch sondern auch dynamisch die absolute Zeit in keinem Experiment auftreten wird. Durch diese Überlegung hat Mach im Prinzip einen Weg gewiesen, wie die aus der Kinematik bereits bekannte Relativität des Raumes und der Zeitskala auch auf die Dynamik übertragen werden könnte. Durch eine vollständige Elimination der Begriffe des absoluten Raumes und der absoluten Zeit aus den physikalischen Theorien würde dann deutlich, daß die maßgebliche, der Physik zugrunde liegende Invarianz-Gruppe die schon erwähnte kinematische Gruppe K ist, - die von 7 willkürlichen Funktionen abhängt. Damit wäre die stärkste mögliche Gegenposition zu der dualistischen, auf Newton zurückgehenden Konzeption von Raum, Zeit und Materie erreicht. - Die Mach'schen Ideen waren aber zunächst nur ein sehr allgemeines Programm, denn die klassische Newton'sche Mechanik entspricht als Theorie nicht den Mach'schen Prinzipien, sondern ist invariant gegenüber der 10-parametrigen Galilei-Gruppe G.

3. Die allgemeine Relativitätstheorie Einsteins

Einstein war durch seine Untersuchungen zur speziellen Relativitätstheorie zu dem Ergebnis gelangt, daß nicht die Galilei-Gruppe, sondern die 10-parametrige Lorentz-Gruppe L die der Physik zugrunde liegende Invarianzgruppe sei. Von Einstein selbst und anderen Physikern seiner Zeit wurden infolgedessen die wesentlichen Theorien der damaligen Physik durch entsprechende Lorentz-invariante Theorien ersetzt. Einstein war sich allerdings darüber im klaren, daß auch eine Lorentz-invariant formulierte Physik - ähnlich wie die alten galilei-invarianten Theorien - nur die Relativität in Bezug auf die Klasse der Inertialsysteme zum Ausdruck bringt und den absoluten Raum und die absolute Zeit jedenfalls noch in dem Maße enthält, in dem diese Größen durch das Auftreten von Trägheitskräften erkennbar sind. Tatsächlich ist die spezielle Relativitätstheorie in gleichem Maße der Mach'schen Kritik ausgesetzt wie die klassische Newtonsche Mechanik.

Um eine allgemeine Relativitätstheorie aufstellen zu können, in der alle, auch die gegen den absoluten Raum beschleunigten Bezugssysteme gleichberechtigt sind, griff Einstein daher auf die Mach'schen Ideen zurück, ergänzte und konkretisierte sie aber wesentlich. Während Mach nur die möglicherweise zufällige Koinzidenz des Trägheitskompass mit dem Sternenkompass konstatiert hatte, ergänzte Einstein diese Feststellung durch die These, daß zwischen beiden Phänomenen ein kausaler Zusammenhang insofern besteht, als das eine, nämlich der Sternenhimmel, die Ursache des anderen sei. Auf Grund der damals durch genaue Beobachtungen erhärteten Erkenntnis der Äquivalenz von träger und schwerer Masse vermutete Einstein weiter, daß der Kausalzusammenhang durch die *Gravitationswechselwirkung* zwischen den Fixsternen und den beschleunigten Körpern hergestellt würde. *Die lokal an beschleunigten Körpern auftretenden Trägheitskräfte wären dann erklärbar durch die Gravitationswechselwirkung der globalen Massenverteilung im Kosmos.* Die empirisch gesicherte Äquivalenz von schwerer und träger Masse wäre dann eine Folge dieses Prinzips, daß wir in der konkreten hier angegebenen Formulierung als *Mach-Einstein-Prinzip* bezeichnen wollen.

Man hätte, was Einstein nicht getan hat, auch das auf die Zeit bezogene "Zweite Mach'sche Prinzip" mit Hilfe der Gravitationswechselwirkung konkretisieren können. Die Metrik der Zeit, wie sie etwa an einem freien Rotator festgestellt werden kann, wäre dann erklärbar durch die Gravitationswechselwirkung, die von globalen Bewegungen der kosmischen Massen induziert wird, etwa von einer Expansion oder Rotation der Galaxien. Daß ein derartiges, zweites Mach-Einstein-Prinzip weder von Einstein selbst, noch in der späteren Literatur diskutiert worden ist, liegt wohl vor allem daran, daß der empirische Gehalt eines solchen Prinzips noch schwerer zu fassen ist als der des ersten Mach-Einstein-Prinzips.

Es ist bereits auf Grund dieser Bemerkungen deutlich, daß bei der Formulierung einer "allgemeinen" Relativitätstheorie die Theorie der Gravitation eine zentrale Rolle spielen wird. Es soll hier nicht der lange und wenig systematische Weg nachgezeichnet werden, auf dem Einstein schließlich zu seiner metrischen Theorie der Gravitation gelangt ist. Neben dem Mach-Einstein-Prinzip waren dabei noch einige andere Gesichtspunkte wie das Äquivalenzprinzip und das Prinzip der minimalen Kopplung von Bedeutung, ohne daß diese Postulate als axiomatische Voraussetzungen der Theorie angesehen werden dürfen. Sie haben vielmehr die Funktion heuristischer Gesichtspunkte besessen, die Einstein die Aufstellung seiner "Allgemeinen Relativitätstheorie" ermöglichten.

In dieser Theorie, deren Einzelheiten hier nicht beschrieben werden müssen, wird der Zusammenhang zwischen der Materie als Quelle eines Gravitationsfeldes und dem metrischen Raum-Zeit Kontinuum hergestellt, das die Bewegung der Materie bestimmt und für das Weyl die suggestive Bezeichnung Führungsfeld eingeführt hat. Dieser Zusammenhang besteht einerseits in der Einstein'schen Feldgleichung

$$R_{\mu\nu} - \frac{1}{2} R \, g_{\mu\nu} = -\kappa \, T_{\mu\nu}$$

andererseits in der Bewegungsgleichung

$$m \frac{d^2 x^\mu}{ds^2} + m \Gamma^\mu_{\alpha\beta} \frac{dx^\alpha}{ds} \frac{dx^\beta}{ds} = F^\mu$$

eines Massenpunktes unter dem Einfluß einer nichtgravitativen Kraft F^μ. Die Feldgleichung verbindet die geometrischen Feldgrößen $g_{\mu\nu}$, R und $R_{\mu\nu}$ eines Riemann'schen Raumes mit dem Energie-Impuls-Tensor $T_{\mu\nu}$ der Materie, während die Bewegungsgleichung zeigt, daß die kräftefreien Bahnen mit den Autoparallelen des affinen Zusammenhangs $\Gamma^\mu_{\alpha\beta}$ dieses Riemann'schen Raumes übereinstimmen.

Von dieser Theorie, die anscheinend in der gewünschten Weise die Materie mit dem Gravitationsfeld und dieses mit dem Führungsfeld verbindet, sollte man zunächst erwarten, daß in ihr das Mach-Einstein-Prinzip enthalten ist und daß somit der absolute Raum und die absolute Zeit vollständig eliminiert sind. Die Theorie sollte daher invariant gegenüber einer entsprechenden kinematischen Gruppe sein, worauf die allgemeine Kovarianz der Feld- und -Bewegungsgleichungen hinzudeuten schien. Auch Einstein teilte, jedenfalls anfangs, diese Erwartung und bezeichnete daher die von ihm geschaffene Theorie als "Allgemeine Relativitätstheorie".

Diese Erwartung hat sich jedoch nicht erfüllt. Für den Fall, daß global keine Materie existiert, also $T_{\mu\nu}$ überall verschwindet, sollte nach dem Mach-Einstein-Prinzip auch kein durch $g_{\mu\nu}$ induziertes Führungsfeld existieren, durch das bei nichtgeodätischen Bewegungen Trägheitskräfte hervorgerufen werden. Einstein vermutete daher, daß bei überall verschwindendem $T_{\mu\nu}$ auch der metrische Tensor $g_{\mu\nu}$ verschwinden muß. Tatsächlich jedoch besitzt die Einsteinsche Feldgleichung auch im materiefreien Fall eine große Anzahl von Vakuumlösungen, deren bekannteste der pseudo-euklidische Minkowski-Raum ist. Allgemein verschwindet im Vakuum zufolge der Feldgleichung nur der zu $R_{\mu\nu}$ und R proportionale Anteil des Riemannschen Krümmungstensors $R_{\mu\nu\kappa\lambda}$, nicht aber der verbleibende Weyl'sche Konform Tensor $C_{\mu\nu\kappa\lambda}$, - der somit als der durch die Einsteinschen Gleichungen nicht bestimmte Riemannsche Krümmungstensor des Vakuums aufgefaßt werden kann.

Die verschiedenen möglichen Vakuumlösungen lassen sich daher auch in übersichtlicher Weise durch die Eigenschaften der entsprechenden Weyl-Tensoren klassifizieren. Sie sind nicht durch die Feldgleichungen bestimmt und können daher im Rahmen der Einsteinschen Theorie frei gewählt werden. Diese Freiheit überträgt sich insofern auch auf Lösungen mit nichtverschwindendem Materietensor, als auch in diesem Fall das metrische Feld nicht schon allein aus den Feldgleichungen bestimmbar ist, sondern erst durch die Vorgabe von frei wählbaren Randbedingungen oder Symmetrieeigenschaften eindeutig wird. Es ist also in jedem Fall ein von der Materie unabhängiger Anteil des Führungsfeldes vorhanden, der sich nicht auf die Gravitationswirkung entfernter Massen zurückführen läßt.

Aus diesen Ergebnissen wird deutlich, daß die von Einstein tatsächlich geschaffene Theorie keine "allgemeine Relativitätstheorie" ist, in der im Sinne der Mach'schen Ideen nur der Materie eine unmittelbare Realität zukommt, während Raum und Zeit nur mittelbar als Ordnungsschemata dieser Materie existieren. Die Einsteinsche Theorie ist vielmehr eine *dualistische* Theorie, in der sowohl die Materie als auch das Raum-Zeit Kontinuum als eigenständige Entitäten vorkommen und miteinander in Verbindung gebracht werden. Im Gegensatz zur Newtonschen Theorie ist diese Verbindung in der Einsteinschen Theorie aber viel enger: Die Trägheitswirkung der materiefreien Raum-Zeit und die Gravitationswirkung der Materie sind im Einsteinschen Führungsfeld in unlösbarer Weise miteinander verbunden und können weder experimentell noch theoretisch voneinander getrennt werden.

Diese unlösbare Einheit beruht auf der empirisch gesicherten Äquivalenz von träger und schwerer Masse, die auch der wichtigste Anhaltspunkt für das Mach-Einstein-Prinzip gewesen war. Man wird daher fragen, ob die in

diesem Prinzip zum Ausdruck gebrachte Hypothese, daß nämlich die bei der
Bewegung von Massen auftretende Trägheit auf Gravitation zurückführbar
ist, nicht doch in irgendeiner Form in der Einsteinschen Theorie enthalten ist. Tatsächlich ist das wenigstens teilweise der Fall: Eine im Führungsfeld gleichförmig rotierende materielle Kugelschale erzeugt im Mittelpunkt der Kugel Gravitationskräfte, die sich qualitativ nur wenig von
den aus dem Newton'schen Eimerexperiment bekannten Trägheitskräften unterscheiden, quantitativ aber nur einen Teil der Trägheitskräfte ausmachen,
die durch Rotation gegenüber dem Führungsfeld induziert würden. Das Mach-
Einstein Prinzip ist daher in der Allgemeinen Relativitätstheorie nur in
dem Sinne verwirklicht, daß man sich die in dem rotierenden Eimer auftretenden Trägheitskräfte zusammengesetzt denken kann aus einem Beitrag
des materiefreien Führungsfeldes und einem Beitrag, der von der Relativ-
Bewegung gegenüber den kosmischen Massen herrührt.

Ähnlich verhält es sich mit dem auf die Zeit bezogenen, oben erwähnten
zweiten Mach-Einstein Prinzip. Denkt man sich die metrische Zeit an einem
Ort durch einen freien Rotator realisiert, so werden in einem gegenüber
diesem Körper gedämpften Rotator Trägheitskräfte auftreten, die ihren Ursprung etwa in der verzögerten Rotation gegenüber dem Fixsternhimmel haben
sollten. Tatsächlich induziert eine beschleunigt rotierende Kugelschale
im Mittelpunkt der Kugel Gravitationskräfte, die qualitativ den erwähnten
zeitlichen Trägheitskräften entsprechen. Quantitativ stellen sie aber
wiederum nur einen Teil der Kräfte dar, die bei beschleunigter Rotation
gegenüber dem Führungsfeld auftreten würden. Man kann sich daher auch die
bei einer "Bewegung" gegenüber der absoluten Zeit auftretenden Trägheitskräfte allenfalls zusammengesetzt denken aus einem Beitrag des materiefreien Führungsfeldes und einem Beitrag, der von der "Bewegung" gegenüber
einer kosmischen Zeitskala herrührt.

Die Allgemeine Relativitätstheorie Einsteins ist daher keine "allgemeine"
Relativitätstheorie, in der im Sinne des Mach-Einstein Prinzips der absolute Raum und die absolute Zeit eliminiert worden wären. Es handelt sich
bei dieser Theorie vielmehr um eine relativistische Fortentwicklung der
Newtonschen Raum-Zeit Theorie, wobei die Gravitation wegen ihrer Ununterscheidbarkeit von der Trägheit mit in das raum-zeitliche Führungsfeld
einbezogen ist. Das Führungsfeld ist somit zwar über die Gravitation von
der Materie beeinflußt, es existiert aber unabhängig von dieser als eine
eigenständige, physikalische Realität.

4.   Die Theorie der Gravitation im Vakuum

Die bisherigen Überlegungen sollten deutlich gemacht haben, daß das Mach-
Einstein Prinzip in der *Allgemeinen Relativitätstheorie* nicht - oder allenfalls

in rudimentärer Form - enthalten ist, und daß es sich bei dieser Theorie im wesentlichen um eine Theorie der Gravitation, nicht aber um eine allgemeine Relativitätstheorie handelt. Dann sollte aber das Mach-Einstein Prinzip auch als heuristischer Gesichtspunkt bei der Formulierung der Theorie entbehrlich sein. Zwar ist dieses Prinzip, wie wir gesehen haben, keine direkte Voraussetzung der Allgemeinen Relativitätstheorie, es spielt aber als heuristischer Gesichtspunkt doch insofern eine Rolle, als die Begriffsbildungen von dem Prinzip deutlich beeinflußt sind. So verzichtet etwa die Allgemeine Relativitätstheorie wegen der erhofften Zurückführbarkeit der Trägheit auf Gravitation auch in der theoretischen Begriffsbildung darauf, Trägheitsfeld und Gravitationsfeld überhaupt zu unterscheiden, oder es wird wegen der vermeintlichen allgemeinen Relativität ein besonderer Wert auf die kovariante Formulierung gelegt.

Die Tatsache, daß das Mach-Einstein Prinzip in der Allgemeinen Relativitätstheorie nicht realisiert und daher als Voraussetzung in jeder Form entbehrlich ist, wird noch ergänzt durch die Feststellung, daß dieses Prinzip aus heutiger Sicht auch als Programm überholt sein dürfte. Während nämlich Mach und Einstein nur mit der vergleichsweise einfachen Aufgabe konfrontiert waren, die einzige damals lokal meßbare Eigenschaft des Vakuums, nämlich das Trägheitsfeld, durch Rückgriff auf die Gravitation der globalen Massenverteilung zu erklären, müßte die heutige Physik zahlreiche weitere Funktionen des Vakuums erklären, um die materiefreie Raum-Zeit aus der Physik eliminieren zu können. Aus heutiger Sicht ist das Vakuum nicht nur der Ursprung der Trägheitskräfte, sondern zugleich ein Kontinuum virtueller Elementarteilchen und damit der Ursprung von Erzeugungs- und Vernichtungsprozessen reeller Materie (und Antimaterie). Wegen der großen Fülle dieser Phänomene, und da hier anders als bei der Trägheit ein empirischer Hinweis wie die Äquivalenz von träger und schwerer Masse fehlt, - so ist der Versuch, die quantenphysikalischen Eigenschaften des Vakuums auf den Einfluß der globalen Massenverteilung zurückzuführen, auch als Programm wenig überzeugend.

Bei dieser Sachlage erscheint es sinnvoll, sich noch einmal in die Situation vor Aufstellung der Allgemeinen Relativitätstheorie zurückzuversetzen. Im Rahmen der speziellen Relativitätstheorie ist das Raum-Zeit-Kontinuum durch den gegenüber der 10-parametrigen Lorentz-Gruppe invarianten Minkowski-Raum gegeben. Ohne auf die Allgemeine Relativitätstheorie Bezug zunehmen, sind auf dieser Basis und in einer unabhängigen Entwicklungslinie zahlreiche Lorentz-invariante Feldtheorien für die verschiedenen Elementarteilchen aufgestellt worden, deren bekannteste die Maxwell'sche Theorie des elektromagnetischen Feldes bzw. des Photons ist. Allgemein lassen sich die stabilen Elementarteilchen durch die irreduziblen

Darstellungen [m,s] der Lorentz-Gruppe, d.h. durch die Parameter der
Masse m und des Spins s, charakterisieren. Es ist daher naheliegend,
auch die Theorie der Gravitation als eine Lorentz-invariante Feldtheorie
zu formulieren und dadurch die Gravitationstheorie in eine Gestalt zu
bringen, die frei ist von den in der Allgemeinen Relativitätstheorie enthaltenen und durch das Mach-Einstein Prinzip geprägten Begriffsbildung.

Eine Lorentz-invariante Theorie der Gravitation wurde zuerst von Nordström 1912 konzipiert. Wegen der unrichtigen Behandlung der Lichtausbrei-
tung in dieser Theorie, und wegen der Suggestion, die von dem erfolgrei-
cheren Einsteinschen Programm ausging, wurde diese Forschungsrichtung
aber zunächst nicht weiter verfolgt. Erst einige Jahrzehnte später wurde
unter dem Eindruck der Quantenfeldtheorie das Problem einer Lorentz-invarianten Gravitationstheorie wieder aufgegriffen. Ausgehend vom Minkowski'schen Raum-Zeit-Kontinuum konnten Gupta, Thirring u.a. zeigen, daß
das Gravitationsfeld sich als ein masseloses, symmetrisches Tensorfeld im
Sinne einer Lorentz-invarianten Feldtheorie auffassen läßt, das minimal
an den Energie-Impuls-Tensor der Materie gekoppelt ist. Die in der skalaren Nordström-Theorie aufgetretenen Mängel ließen sich in der formal
reichhaltigeren Tensor-Theorie ohne Schwierigkeiten beheben. Darüber
hinaus zeigte sich, daß sich die Bahnen von Lichtstrahlen und Massenpunkten im Gravitationsfeld als Geodäten eines Riemannschen Raumes auffassen lassen, der somit erst sekundär durch die Ausmessung der Raum-
Zeit mit Hilfe von Lichtstrahlen und Massenpunkten gegeben ist.

Die Feld- und Bewegungsgleichungen dieser Lorentz-invarianten Gravita-
tionstheorie sind lokal äquivalent zu den entsprechenden Gleichungen der
Einsteinschen Theorie. Das Gravitationsfeld unterscheidet sich daher
nicht grundsätzlich von anderen, in der heutigen Lorentz-invarianten
Feldtheorie behandelten Feldern. Die einzige zunächst nicht auffällige
Besonderheit dieses Feldes besteht in der Kopplung an den Energie-Impuls
Tensor der Materie als Quelle dès Feldes. Das hat zunächst die hochgradi-
ge Nichtlinearität der Feldgleichungen zur Folge. Weiterhin stimmen die
Bewegungsgleichungen von Lichtstrahlen und frei fallenden Massenpunkten
mit den Geodätengleichungen in einem Riemannschen Raum überein. Dieser
Riemann-Raum erweist sich daher als die mit Lichtstrahlen und Massenpunkten observable Raum-Zeit-Struktur, während der der Theorie zugrundelie-
gende Minkowski-Raum die Rolle eines unbeobachtbaren Darstellungsraumes
spielt. Es ist daher sicher zweckmäßig, den unbeobachtbaren Minkowski-
Raum aus der Formulierung der Theorie zu eliminieren, und diese direkt
auf die observale Riemann'sche Raum-Zeit zu beziehen. Auf diese Weise erhält man die Allgemeine Relativitätstheorie.

Gegen diese Interpretation, die die Gravitationstheorie vollständig in den Rahmen der dualistischen Konzeption der Feldtheorie einordnet, sind sowohl physikalische als auch wissenschaftstheoretische Einwände vorgebracht worden. Physikalisch ist zunächst festzuhalten, daß die Einstein'schen Feldgleichungen in Abwesenheit von Materie eine große Klasse von Vakuum-Lösungen zulassen, während die Lorentz-invariante Gravitationstheorie nur den Minkowski-Raum als Vakuum-Lösung zu besitzen scheint. Hier ist aber zu bemerken, daß die Wahl des Minkowski-Raumes als Ausgangsmetrik im Grunde willkürlich ist, und daß sinngemäß auch eine andere Vakuum-Lösung hätte verwendet werden können. Da die Konstruktion der observablen Metrik aus der vorausgesetzten Vakuum-Metrik ein rein lokales Verfahren ist, ist nur die Lorentz-Invarianz der Feldgleichungen, nicht aber die der Hintergrunds-Metrik von Bedeutung. Wodurch andererseits die Vakuum-Raum-Zeit bestimmt ist, ist weder in der Lorentz-invarianten Gravitationstheorie noch im Rahmen der Allgemeinen Relativitätstheorie bekannt. In beiden Theorien muß die Vakuum-Raum-Zeit als ein kontingenter Erfahrungsinhalt hinzugenommen werden.

Aus wissenschaftstheoretischer Sicht kann und ist der Einwand erhoben worden, daß der Versuch, die Riemann'sche Struktur des Raum-Zeit-Kontinuums auf der Basis des Minkowski-Raumes durch Gravitationswirkungen zu erklären, ein Rückfall in die Phase vor der Entdeckung der Riemann'schen Raum-Zeit-Struktur darstellt, - ähnlich einer Erklärung von Quanteneffekten durch verborgene Parameter im Rahmen der klassischen Physik. Tatsächlich besteht hier eine gewisse Analogie insofern, als in der Lorentz-invarianten Gravitationstheorie der Minkowskiraum eine unbeobachtbare Struktur darstellt, die sich auch rechnerisch nicht aus den Beobachtungsdaten rekonstruieren läßt. Das Festhalten an der Riemannschen Raum-Zeit allein wegen des angeblichen wissenschaftlichen Fortschritts ist jedoch nichts anderes als eine *Dogmatisierung des allgemein-relativistischen Paradigmas*, die übersieht, daß das *feldtheoretische Paradigma* eine empirische gleichwertige Alternative darstellt, die theoretisch einige bemerkenswerte Vorteile besitzt.

Zwar könnte es zunächst so erscheinen, als sei die Reinterpretation der Allgemeinen Relativitätstheorie als eine Lorentz-invariante Feldtheorie nichts anderes als die schon von Carnap (1922) und Reichenbach (1924) erörterte Einführung von *universellen Kräften*. Das wäre in der Tat reiner Konventionalismus. Im vorliegenden Fall ist die Lorentz-invariante Feldgleichung des Tensorfeldes jedoch nicht ad hoc zum Zwecke der Rekonstruktion der Einsteinschen Theorie aufgestellt worden, sondern diese ergibt sich nahezu zwangsläufig im Kontext der Klassifikation von Feldern durch die irreduziblen Darstellungen der Lorentz-Gruppe. Daß die von einem

masselosen, symmetrischen Tensorfeld mit minimaler Kopplung an den Energie-Impuls Tensor ausgeübten Kräfte dann universell sind, ist richtig,
darf aber nicht als ein Mangel der Theorie angesehen werden. Man wird
vielmehr sagen dürfen, daß die Lorentz-invariante Gravitationstheorie
insofern eine größere Erklärungsfähigkeit besitzt als die Einstein'sche
Theorie, als sie nicht nur die Abweichungen von der geodätischen Bewegung durch Kräfte erklärt - sondern auch den Riemannschen Charakter der
Metrik selbst auf Gravitationskräfte zurückführt und damit erklärbar
macht.

Unabhängig von derartigen möglichen Vorteilen zeigt aber eine als Feldtheorie in einer Vakuum-Raum-Zeit formulierte Gravitationstheorie, daß
die Allgemeine Relativitätstheorie nicht nur das Mach-Einstein Prinzip
nicht enthält, sondern daß darüber hinaus auch alle in der Begriffsbildung enthaltenen Einflüsse der Mach'schen Ideen aus der Formulierung der
Theorie eliminiert werden können. Die Allgemeine Relativitätstheorie läßt
sich auf diese Weise als eine *manifest dualistische Theorie* darstellen, in
der das leere Raum-Zeit-Kontinuum und die Materie als eigenständige Entitäten vorkommen. Daß diese beiden Entitäten sich in mannigfacher Weise
kausal beeinflussen, steht außer Zweifel: Die Materie als Quelle der Gravitation erzeugt ein Gravitationsfeld und beeinflußt damit das Führungsfeld. Die Vakuum-Raum-Zeit induziert andererseits in bewegter Materie
Trägheitskräfte und ist, unter geeigneten Voraussetzungen, zugleich der
Ursprung der Erzeugung von reeller Materie (und Antimaterie) durch quan-
tenmechanische Prozesse. - Worauf es im Rahmen der hier diskutierten
Problemstellung ankommt, ist, daß die materiefreie Raum-Zeit als eigenständige physikalische Realität aufgefaßt werden kann, die unabhängig von
- und möglicherweise auch vor - aller Materie existiert. In Anlehnung an
Newton könnte man daher auch heute noch sagen: "Die absolute Raum-Zeit
existiert an sich und ohne Beziehung auf einen äußeren Gegenstand".

MATHEMATICAL THEORIES AND PHILOSOPHICAL INSIGHTS IN COSMOLOGY

Roberto Torretti, Universidad de Puerto Rico, USA

Philosophers of science often regard physics as stemming from the interplay of theories and facts. By a *theory* we are to understand here the interpreted theory of a mathematical structure, *i.e.* a theory-*cum*-model;
by a *fact*, a description of observations couched in such language that
it can be seen to agree or to disagree with a theorem of the relevant
theory. A classical statement of this view is contained in Max Born's
lecture *Experiment and Theory in Physics*,[1] delivered in 1943. In it
Born harshly criticized the British cosmologists, A. S. Eddington and
E. A. Milne, for resorting to philosophical principles as basic premises
in the construction of physical theories. In a letter to Born of September 7, 1944, Albert Einstein made the following comment on this lecture
and on his friend's negative stance towards philosophical speculation in
physics:

> *Ich habe mit viel Interesse Deinen Vortrag gegen die Hegelei*
> *gelesen, welche bei uns Theoretikern das Don Quijote'sche*
> *Element ausmacht oder soll ich sagen, den Verführer? Wo dies*
> *Übel oder Laster aber gründlich fehlt, ist der hoffnungslose*
> *Philister auf dem Plan.*[2]

Contrasting his own methodology of physics and philosophy of nature with
that propounded by Born, Einstein added:

> *Ich [glaube] an volle Gesetzlichkeit in einer Welt von etwas*
> *objektiv Seiendem, das ich auf wild spekulativem Wege zu*
> *erhaschen suche.*[3]

Einstein's devotion to the 'Quixotic element' in science, his firm per-
suasion that facts alone cannot suffice as a source of guidance for the
formulation and acceptance of theories is shown also in several other
texts, one of which I shall have occasion to quote later.

General views concerning the nature of things, whether obtained through
insight or by wild speculation, appear to be indispensable for choosing
a physically viable theory from the luxuriant jungle of conceivable mathematical structures, infinitely many of which can be made to fit, within the allowable margin of imprecision, any given set of facts. Indeed,
without some such views the physicist would be hard put to describe his
observations in terms agreeable with a mathematical theory. The interplay between mathematical theories and what I propose to call philosophical insights - or, if you wish, philosophical hunches - is therefore
no less essential for the growth of physics than the interplay between

theories and facts. Fortunately, philosophy of science has begun of late to direct to the former some of the attention it had paid to the latter.[4] Though the said interplay can be seen in all the fundamental fields of physical inquiry, it has been quite prominent in 20th century cosmology, partly because of the very nature of its subject-matter, but partly perhaps too because of the kind of factual evidence on which it must rely. I feel therefore that it might be worth while to illustrate with examples from cosmology a few basic questions regarding the interplay between philosophical insights and mathematical theories.

1. The first question that I wish to consider is a fairly obvious one: If we take a look at facts, we see that theories may be said to be *corroborated* or *refuted* by them; what can philosophical insights do for theories? In principle, they can of course determine them, by supplying their axioms. Descartes' cosmological programme (in his *Discourse on Method* , part V) should probably be understood as an attempt in this direction. And in this century, E. A. Milne quite deliberately set out to derive the entire system of the laws of physics from a few insights concerning the structure of the universe. However, Milne's cosmology is afflicted by the following difficulty, which is bound to arise in all enterprises of this kind: in the course of actually building a mathematical theory of nature that is generally adequate to facts but rests altogether on a priori principles, Milne has had to supplement his original allegedly insightful and at any rate philosophically plausible postulates with additional assumptions, that are more or less clearly *ad hoc* and for which he could provide at best a far-fetched justification. From the consideration that "if a rational understanding of the universe is possible, it ought to be possible to set up a consistent system of time-keeping throughout the universe",[5] Milne derives the conception of a family of "fundamental observers", each of which can make with his clock and theodolite a set of observations indistinguishable from those made by any other fundamental observer. Having *proved* that the coordinates employed by three collinear fundamental observers transform into one another by Lorentz transformations, he *postulates* that this holds good also for any set of fundamental observers in three-space.[6] Further assumptions are: that the universe has zero net angular momentum,[7] that the acceleration of a free particle is a unique function of its position, velocity and epoch,[8] etc. The willingness of Milne and the like of him to enlarge the *a priori* basis of their deductions as it might be required for the advancement of their theories has contributed not a little to the current distrust of speculation in natural philosophy.

A somewhat different approach is exemplified by Bondi and Gold's version

of the Steady State cosmology.[9] Here the theory is derived from a single
philosophical insight, embodied in the Perfect Cosmological Principle,
plus one or two undeniable facts, such as the darkness of the sky at
night or the frequency-shift of radiation received from distant sources.
The Perfect Cosmological Principle says that the universe is homogeous
in space *and* time, so that it looks more or less the same from any spa-
tiotemporal vantage point. It is justified by a typically philosophical
argument, reminiscent of Reichenbach's vindication of the straight rule
of induction.[10] Unless the looks of nature are always and everywhere
approximately the same we would not be entitled to extrapolate to the
entire breadth and duration of the universe the results of our exact
physical measurements, which are performed within a pitifully small region of it. Thus, *either* the Perfect Cosmological Principle is true, and
that disposes of the theories that contradict it, *or* the Perfect Cosmological Principle is false, but then the particular theories that may be
developed in opposition to it have scarcely any chance of being true instead. The Perfect Cosmological Principle, together with factual evidence
that the larger clusters of matter are receding from one another, implies
that most famous or infamous tenet of the Steady State Theory: the Continuous Creation of Matter. It is indeed ironic that a principle introduced to ensure the universal validity of our terrestrial physics should
thus lead to the negation of a deeply entrenched and well corroborated
local law. Before the discovery of the relic microwave background radiation made it unfashionable, the Steady State Theory was strongly resisted in the name of matter conservation. Thus, Mario Bunge forcefully
argued in 1962 that the acceptance of continuous creation would spell
disaster for scientifc thought and would make it indistinguishable from
magic.[11] Others, I must recall, were charmed by the dialectical piquancy
of a theory in which the sheer repetition of *creatio ex nihilo* is the
means of securing that *nihil sub sole novum*.[12]

Einstein's General Theory of Relativity bears witness to a more subtle
mode of action of insight on theory. Philosophical vision manifestly
guides the formulation of the theory, but is not incorporated into it
as an axiom or a set of axioms. Moreover, the theory, once formulated,
makes the philosophical vision more precise and can, in a sense, even be
said to modify it. Answering to E. Kretschmann's remark that general covariance is a trivial requirement that any phyical theory is capable of
fulfilling, Einstein declared in 1918 that General Relativity rests on
three principles: the Principle of Equivalence, the Principle of General
Covariance, and Mach's Principle.[13] If, following Gerald Holton,[14] we
classify the diverse ingredients or strands of a piece of scientific

thought into the empirical or factual, the analytic or logico-mathematical, and what Holton calls the thematic element, of which our "philosophical insight" is a proper or an improper part, it is apparent that
the Principle of General Covariance is a logico-mathematical requirement, while the Principle of Equivalence is a bold but reasonable extra-
polation from facts. The philosophical component of General Relativity
must therefore lie mainly with Mach's Principle. Now Mach's Principle
is not an axiom of the theory. Indeed, in the particular version proposed by Einstein in 1918, it is not even a theorem of it - though in a
modified version it has been said to state the boundary conditions prescribed for the field equations.[15]   But the insight which Mach's Principle is supposed to express certainly guided Einstein during the long
strenuous search that led to the formulation of General Relativity.[16]
In traditional philosophical language that insight can be stated thus:
Absolute space and absolute time - as well as absolute spacetime -, without matter, are physically inviable mathematical contraptions. This
thesis is implicit in the writings of the greatest pre-Newtonian natural
philosophers, Aristotele and Descartes, and was explicity defended by
Leibniz against Newton's spokesman, Samuel Clarke.[17]   It was also shared
by Kant, but this philosopher, believing that the conceptual framework
of Newtonian dynamics was a prerequisite of natural science, felt com-
pelled to conclude that nature itself, regarded as a scientific object,
was no less unreal, or "transcendentally ideal", than absolute space and
time had to be. The said insight also inspired Ernst Mach's celebrated
criticism of Newton's scholion on space and time.[18]   Einstein traces the
origin of his Mach principle to this Machian criticism.[19]   Since Newtonian
time had been superseded already in the Special Theory of Relativity, the
Mach Principle of General Relativity must be directed against Newtonian
space. Now the absoluteness of space is manifested, according to Newton,
through the phenomena of absolute acceleration.[20]   It might seem paradoxical that Newton should resort with such assurance to the concept of
an absolute acceleration, while providing no criteria and having no use
for the presumably more fundamental concept of absolute motion.

But this paradox was dispelled by Minkowski's interpretation of Special
Relativity and the subsequent reformulation of Newtonian theory in four
dimensions by Cartan and others.[21]   All that is needed to make sense of
an absolute acceleration without having to conceive an absolute motion
is that spacetime be endowed with a linear connection, plus whatever
structure is employed for singling out worldlines, *i.e.* spacetime curves
that are nowhere tangent to an hypersurface of simultaneity. The linear
connection then determines an intangible yet perfectly rigid network of

geodesic worldlines, which any material particle must follow if let loose at an arbitrary spacetime point. Absolutely accelerated motion is simply motion which deviates from this natural network, *i.e.* motion along a non-geodesic worldline. This reconstruction of the concept of absolute acceleration clearly brings out the effective physical meaning of Newton's talk of absolute space. In Newtonian and in special relativistic dynamics, spacetime, of itself, constrains force-free matter to follow a special kind of cosmic track. On the other hand, the presence of matter makes no difference in the structure of spacetime. According to Einstein, such lopsidedness in the mutual relationship between two physical entities is utterly at variance with everything we know of nature.[22] Such was the insight that Einstein sought to embody in the Mach Principle of 1918. This can be paraphrased as follows: The linear connection, or rather, the spacetime metric on which it depends in a relativistic theory, is fully determined by the distribution of matter. Now, if Special Relativity holds locally, the distribution of matter should be represented by the stress-energy tensor, a symmetric tensor field of order two, whose covariant divergence vanishes. Mach's Principle implies then that the components of the metric relative to a given spacetime chart must be obtained by integration of a system of differential equations relating the components of the stress-energy tensor with those of a tensor field constructed from the metric itself and its derivatives with respect to that chart. If Newton's gravitational theory is valid as a first approximation, the required equations must depend on the second derivatives of the metric. If we assume for simplicity's sake that no higher order derivatives are involved, the stated conditions uniquely determine Einstein's field equations (up to two arbitrary constants).[23] However, unless one understands the "distribution of matter" in a Pickwickian sense, the field equations do not agree with the above version of Mach's Principle, since they also have solutions if the stress-energy tensor is identically zero.[24] Later in life Einstein rejected the 1918 Mach Principle. On February 2, 1954, he wrote to Felix Pirani:

> *One shouldn't talk any longer of Mach's principle, in my opinion. It arose at a time when one thought that 'ponderable bodies' were the only physical reality and that in a theory all elements that are not fully determined by them should be conscientiously avoided. I am quite aware of the fact that for a long time, I, too, was influenced by this fixed idea.*[25]

Nevertheless, he unflinchingly held to the original insight that spacetime could not be allowed to act on matter without the latter acting on spacetime in its turn. On May 12, 1952, he wrote to Max Born, with regard to the apparent failure of one of the three classic effects of

General Relativity:

> *Wenn überhaupt keine Linienablenkung, keine Perihelbewegung und keine*
> *Linien-Verschiebung bekannt wäre, wären die Gravitationsgleichungen*
> *doch überzeugend, weil sie das Inertialsystem vermeiden (dies Gespenst,*
> *das auf alles wirkt, auf das aber die Dinge nicht zurückwirken). Es ist*
> *eigentlich merkwürdig, daß die Menschen meist taub sind gegenüber die*
> *stärksten Argumenten, während sie stets dazu neigen, Messgenauigkeiten*
> *zu überschätzen.*[26]

2. The second question that I wish to consider is, in a way, the inverse
of the first. What can theories do for insight? What can mathematical
thought contribute to the philosophical understanding of nature? Let us
look once more at the interplay between theories and facts. Not only do
theories, in connection with their underlying insights, provide the
framework for the scientific description of phenomena; they have often
predicted the existence of entirely unexpected things, like radio waves
or antimatter. Can theories do something analogous for insight? Can pure
mathematical thought give birth to a new view of nature? I am not sure
I know of a case in which this has happened. Francis M. Cornford once
argued that the idea of an absolute space was begotten by Greek geometry.[27] But even if he were right, one might still object that this was
not a contribution to insight but to obfuscation. Anyhow, it is evident
that mathematics has repeatedly contributed to make existing philosophical views "clear and distinct" and has thereby helped to show in what
sense they are tenable. This has occurred in many fields of philosophical inquiry. The importance of mathematics in contemporary logic and
foundational research is familiar to everyone. Less well known, but very
promising is the use of mathematics in ontology, as in Professor Bunge's
pioneering treatise.[28]

In the special domain of our present workshop, few ideas have been more
helpful and, I dare say, more fertile, than the mathematical concept of
a differentiable manifold, first introduced by Bernhard Riemann for the
philosophical purpose of elucidating the essence of physical space,[29]
and subsequently perfected by Christoffel, Ricci, Levi-Città, Weyl,
Cartan and others. As an example of what a mathematical theory can do
for our understanding of nature, I shall now examine a philosophically
significant application of the manifold concept in cosmology. Relativistic cosmology conceives the universe as a four-dimensional real
Hausdorff manifold endowed with a Lorentz metric. Such a manifold is
called a *spacetime*. The physical properties of the universe are represented by diverse scalar, vector, tensor or spinor fields on spacetime,
i.e. by smooth mappings of the manifold into several fibre bundles that
naturally go with it. (Indeed, the Lorentz metric is one such mapping,

into the bundle of symmetric covariant tensors of order two.) This conceptual framework has paved the way for a totally new approach to one of the oldest and most intractable problems of cosmology, the problem of the spatial extent and the temporal beginning of the universe. When Immanuel Kant, notwithstanding his beautiful early contributions to the development of a Newtonian world-view,[30] pronounced cosmology a pseudoscience, rooted in a fatal illusion of human reason, he counted this problem as the first of the four that forced him to draw this conclusion.[31] Kant maintained that the temporal succession of physical phenomena must have a beginning, because otherwise any current event would mark the end of an eternal process, which he deemed absurd. He also argued, less plausibly, that the universe must be finite in extent, because an infinite totality of simultaneously given things involves a completed yet infinite synthesis, which again, to his mind, was absurd. On the other hand, a universe finite in extent and developing from a definite starting-point raises the specter of an empty space lying beyond its confines and an empty time running before its beginning. Since Kant believed that empty space and time outside the world are no less absurd than a completed infinity or a terminating eternity, he declared cosmology impossible and judged the universe to be a mere idea, which is useful for regulating our scientific inquiries, but which generates inextricable contradictions as soon as it is held to be a genuine physical object, the proper subject-matter of a science of its own. By conceiving the universe as a differentiable manifold relativistic cosmology can deal at once with both horns of the Kantian dilemma. There is no difficulty in conceiving an infinite spacetime on which the matter fields take non zero values right out to infinity. (No difficulty, that is, if one has no qualms about the classical, Cantorean, conception of continua.) However, if one feels prompted by experience to reject the actual infinity of the universe, one can always abide by the other horn, for it is possible to have a spacetime both finite in extent and of bounded duration, without conjuring up an unthinkable void beyond it. We are all familiar with Riemann's conception of a finite yet unlimited world. Einstein's revival of it, in his "Kosmologische Betrachtungen" of 1917, may rightly be said to mark the beginning of modern cosmology.[32] We may regard a spacetime as spatially finite if every point of it lies on some compact spacelike hypersurface which cuts the spacetime into two components. (This characterization must be refined if closed timelike curves are permitted.) Within the theory of differentiable manifolds such a spacetime is not harder to conceive than, say, a cylindrical surface, through each of whose points passes a circle which cuts the

surface in two. On the other hand, the concept of a universe of bounded
duration is not without difficulties and deserves more attention. Let
us remark first that a spacetime can have many different shapes, provided that each point in it has a neighbourhood that can be charted onto
an open subset of R^4. Thus, a spacetime M can fulfil the following condition: for every point P in M and for every future-directed timelike curve c through P, parametrized by proper time, the domain of c
is bounded below. If, for simplicity's sake, we ignore the existence of
causal links over null curves – these can be dealt with by resorting to
a so-called generalized affine parameter, but it would be untimely to
go into such refinements now –,[33] it is evident that, if we happened
to live in a spacetime such as M, current events would not mark the
conclusion of any eternal process. M is therefore immune to Kant's
strictures against an eternal past. But M is no preceded by an empty
time. In order to see this more clearly, let us say that a point P in
M is a *starting-point* of a future-directed timelike curve $c: I \rightarrow M$,
if for every neighbourhood U of P there is a real number $t_U \in I$,
such that $c(t)$ lies in U for every $t \in I$ with $t \leqslant t_U$. We shall say that
such a curve is *past-inextendible* if no point in M is a starting-point
of it. Now the condition prescribed for M does not mean that every future-directed timelike curve in M has a starting-point, but rather that
every *past-inextendible* such curve in M has a domain of definition
that is bounded below. That such past-inextendible curves exist in M
can be seen by considering a field V of unit timelike vectors on M.
(Such a field exists if M, as we may sensibly assume, is time-orientable.) Let c be the maximal integral curve of V through a given point
P. The domain of c is bounded below. Let its greatest lower bound be
0. (This can be secured, if necessary, by regraduating the parameter.)
If c were defined at 0, $c(0)$ would be its starting-point. But then $c(0)$
would have an open neighbourhood U in M, V would be defined on all
U, and the maximal integral curve of V through $c(0)$ would have values
in the intersection of U with the past of $c(0)$. Hence c, contrary to
our assumption, would not be a maximal integral curve of V through
$c(0)$. Therefore c is not defined at 0. Moreover, since c is maximal,
it cannot be extended to a curve $\hat{c}$, defined at 0 and agreeing with c
on the positive side of 0. Consequently, there is no point in M that
is a starting-point of c.

The condition prescribed for our spacetime M obtains in the current
favoured Big Bang world-models. Let W be a Big Bang spacetime and let
c be any future-directed past-inextendible timelike geodesic in W,
parametrized by proper time. Suppose that c is complete, *i.e.* that it

has a value for every real number. Then there must be a real number t
such that $c(t)$ is the Big Bang, *i.e.* a point containing all the energy
in the universe. At such a point the stress-energy tensor, the curvature,
etc. would not be smooth. Since this is impossible by definition, no
such point can belong to W. Consequently, c is not complete and t is
the greatest lower bound of its domain. The Big Bang is often described
as the beginning of the world, or, at least, as the beginning of its
present dispensation. We see that this description is misleading. Big
Bang universes have their duration bounded below, but do not have a
proper beginning. There is in them no *first* instant that might tempt us
to ask what went on before it. For every given instant along a causal
line there is another instant preceding it, at which something was already going on. However, each event in the past of any given one lies
at a finite temporal distance from it. In this sense, in a Big Bang world
we can touch, so to speak, with the tip of our fingers -- provided we
manage to stretch them back in time some fifteen billion years -- the
radical contingency of nature. But an eternal world would be no less
contingent, as Leibniz noted in his tract *De rerum originatione radica-
li*.[34]

At this point, it may be instructive to compare the Big Bang with the
singularity found inside a black hole. Take a Schwarzschild field,
which is the simplest case in which a black hole can arise, and a Friedmann universe, which is the simplest kind of Big Bang universe. A
Schwarzschild field is a symmetric and static solution of Einsteins'
equations for empty spacetime. By Birkhoff's theorem, the solution holds
even if the requirement of staticity is relaxed, provided perfect symmetry is preserved at all times. The Schwarzschild solution involves a
constant of integration m, which is usually and very naturally interpreted as the mass of the material source of the field.[35] If all the
matter in the source lies within a sphere of radius less than $2Gm/c^2$
-- where G is the gravitational constant and c is the speed of light *in
vacuo* -- it must collapse to a space point at the axis of symmetry of
the field: the black hole singularity. At such a point the matter fields,
the curvature, the metric, etc. cannot be smooth, so we are led to conclude, once more, that there is no such point in our manifold. A Schwarzschild field must therefore be regarded as a spacetime punctured, or
rather scratched, at its axis of symmetry. However, most physicists will
not allow the mass m simply to vanish into the singularity. Of course,
one can strictly conceive of the Schwarzschild field about the singularity as the field left behind by a collapsing, *i.e.* literally self-an-
nihilating mass. The field is fully determined by the boundary conditions.

The idea that it must break down as soon as its source is no longer there to support it bespeaks an inability to think spatiotemporally and field-theoretically. But there is something philosophically disquieting about a chunk of matter thus fading away into nothingness. Hence the common feeling that black holes constitute a genuine paradox in General Relativity and can only be understood, if at all, from a post-relativistic, probably quantumtheoretical perspective. On the other hand, a Friedmann universe is derived from some simplifying assumptions concerning the current global distribution of matter. In such a universe, as one goes backward in time -- say as one moves towards the past along any parametric line of the Robertson-Walker time coordinate -- the density of matter indefinitely increases in one's neighbourhood. Indeed, if ϱ denotes the density component of the stress-energy tensor (in terms, say, of the Robertson-Walker coordinates) and c is a parametric line of the time coordinate, parametrized by Robertson-Walker time, the composite function $\varrho \cdot c$ grows beyond all bounds as the argument approaches a definite real number t_0. Consequently ϱ would be infinite at the spacetime point $c(t_0)$ if c were defined at t_0, that is to say, if t_0 were in the range of the Robertson-Walker time coordinate. But ϱ cannot become infinite at any point of spacetime, and hence we must conclude that $c(t_0)$ does not exist and that t_0 is not in the range of the time coordinate. However, this should not be objectionable even to those who would reject matter annihilation inside black holes, for the source of the Friedmann field is coeval with the field itself, so that the excision of the Big Bang singularity is not paradoxical in the way in which the excision of black hole singularities might be said to be. This is not to deny that a post-relativistic theory explaining black holes will probably throw a light on the Big Bang too and might eventually do away with it. Be that as it may, the preceding has made clear, I hope, that by means of Riemann's idea of a manifold one can conceive without paradox an unbounded yet finite universe, which has an age though it never was born.

3. I can barely touch on a third question that must be asked with regard to insights and theories. To clarify it I resort once again to the analogy of the interplay between theories and facts. There is no doubt that, even if there are no incorrigible observation statements and all phenomena can be described in different ways within diverse conceptual frameworks, there is something final about facts, by virtue of which they provide a touchstone for theories. Thus, now that all those stones have been brought back from the moon, one would not wish to maintain with Aristotle that it is made of incorruptible, pellucid, imponderable ether.

Is there a comparable finality about insights? Is there any way of arriving at definitive philosophical insights that set fixed boundaries to admissible theories? Some of the greatest philosophers have thought they had found a method for this and have devoted considerable efforts to its pursuit. In the last two centuries the most influential of them has doubtless been Kant, who made the following simple yet alluring discovery: if one can gain an insight into the prerequisites of scientific knowledge one will thereby obtain an insight into its subject-matter, for, as he put it, "die Bedingungen der Möglichkeit der Erfahrung überhaupt sind zugleich Bedingungen der Möglichkeit der Gegenstände der Erfahrung."[36] He believed he could claim, on these grounds, that Euclidean geometry and Newtonian chronometry, the continuity of qualitative change, mass conservation, causal determinism and the equality of action and reaction between distant bodies, were permanent features of physical science. As is well known, not one of them survived the advent of Relativity and the Quantum. But Kant's failure in the application of his method does not invalidate the method itself, and arguments in Kantian style are still used for vindicating basic scientific assumptions. I have already mentioned Bondi and Gold's justification of the Perfect Cosmological Principle. The following example is taken from Hawking and Ellis' treatise, *The Large Scale Structure of Space-Time*. Discussing the possibility that a relativistic spacetime might contain closed timelike curves - *i.e.* timelike curves homeomorphic to a circle - the authors write:

> However the existence of such curves would seem to lead to the possibility of logical paradoxes: for, one could imagine that with a suitable rocketship one could travel round such a curve and, arriving back before one's departure, one could prevent oneself from setting out in the first place. Of course there is a contradiction only if one assumes a simple notion of free will; but this is not something which can be dropped lightly since the whole of our philosophy of science is based on the assumption that one is free to perform any experiment.[37]

Hawking and Ellis allude to H. Schmidt's "Model of an oscillating cosmos that rejuvinates during contraction",[38] which contains closed timelike curves and in which the concept of free will is modified, but conclude that

> One would be much more ready to believe that space-time satisfies what we shall call the chronology condition: namely, that there are no closed timelike curves.[39]

The chronology condition is one of the hypotheses of an important theorem on the existence of singularities in spacetime, discovered by Hawking and Penrose in 1970.[40] Now, though it is certainly encouraging to hear two distinguished scientists argue, in a strictly scientific argument, from

the premise that freedom is a prerequisite of science, one must also bear
in mind that, as a matter of fact, one is not free to perform any experiment anywhere at any time. Indeed, physics must constantly extrapolate
from our terrestrial laboratories to spacetime regions where free experi-
mentation is physically impossible. Consequently, Hawking and Ellis' argument must imply either that science is illusory or that its possibility requires only that *some* timelike curves be open, and that they include
the worldlines of the materials out of which scientists and scientific
instruments are made. (By the way, the compactness of *some* timelike
curves is not more inimical to freedom than the existence of a Cauchy
surface in spacetime, which Hawking and Ellis do not seem to be willing
to exclude *a priori*, though they readily admit that "there does not seem
to be any physically compelling reason for believing that the universe
admits" one.)[41] The Kantian method of philosophical inquiry into the
prerequisites of science can also be said to inspire the current German
school of protophysics, but a discussion of their endeavours would be
somewhat out of place here, for not only do they show little sympathy
with all that Relativity philosophically stands for, but they tend to
dismiss cosmological theories as metaphysical fancies.[42]

A completely different approach to the subject of philosophical insight
was initiated by Edmund Husserl about 1900, giving rise to the so-called
Phenomenological Movement. Husserl maintained that we can grasp the features that make up the "essence" *(Wesen)* of phenomena, quite independently of whether the entities seemingly manifested by those phenomena actually exist or not. Such "grasp of essences" *(Wesenschau)* can be at-
tained by freely varying in imagination the phenomena actually experienced, the "essential" traits being precisely those that remain invariant under free variation. Husserl's approach exerted a considerable influence on the German and Central European *Geisteswissenschaften* after
the First World War and is now in vogue in American psychology, but its
effect on the philosophy of physics has been negligible. Nevertheless,
it is undeniable that a flair for such constant features of phenomena,
for what Husserl called "den invarianten allgemeinen Stil, in dem [unsere
empirisch anschauliche Welt] im Strömen der totalen Erfahrung verharrt",[43]
has always been a critically important ingredient of scientific genius.
Scientists who, in Max Born's words, "haben gar kein Gefühl für die innere Wahrscheinlichkeit einer Theorie",[44] have seldom made decisive
contributions to natural philosophy. And evidently such a feeling for
the intrinsic likelihood of theories cannot grow only out of a thorough
acquaintance with available mathematical devices and experimental results. Let me give one example of what I take to be Husserl's meaning.

It is usually assumed that the spacetime manifold of a physically viable
relativistic world-model must be time-orientable. Mathematically spoken:
the manifold must admit an everywhere defined smooth field of timelike
vectors. I should say that this assumption does not rest on the observation of particular facts, which could never authorize such a sweeping
generalization, but on an intuitive grasp of the apparent necessity that
every event E be surrounded by a neighbourhood inside which the events
to the future of E can be neatly discerned from the events to the past
of E (I note in passing that the time-orientability of spacetime must
not be confused with the so-called time asymmetry of some physical phenomena; the latter, as illustrated, $i.e.$, by heat flow, cannot even be
described unless time-orientation is taken for granted, for it consists
in the fact that certain types of events, A, B, C,... always happen so
that C is to the future of B when A is to its past, and never
take place in the inverse order.) However, we ought to be wary of rely-
ing too freely on our grasp of essences. I dare say that all of us would
readily assume that each point-event in the universe has a sizeable
neighbourhood that is contractible to it, if we were not acquainted with
physical speculations involving multiply connected spacetimes.[45] Even
if these speculations do not look promising, they are not absurd, and
we may not simply discard them as physically impossible just because
they defy our sense of plausibility. On such matters, the early reception and eventual acceptance of Einstein's Relativity have taught us a
lesson that is not easily forgotten: a theory can contradict some of
our most cherished beliefs concerning the nature of things and yet turn
out to agree with our genuine intuitions much better than the theory it
supersedes.

Summing up: neither Kant's nor Husserl's method can provide foolproof
philosophical insights, though of course, if one is sure that he has
lighted on a true prerequisity of scientific knowledge, or that he has
effectively grasped an invariant feature of the world, he will certainly
hold his discovery binding for science. The safest test of insights appears to be, not their self-evidence, which can be deceptive, but the
possibility of articulating them in a mathematical theory that yields
a satisfactory framework for the understanding of facts. In the end, as
it often happens philosophy, our discussion has not revealed us anything
that we did not know beforehand, but has only clarified what we had
always known: while facts depend on insightful theories for their coherent formulation, insights depend on theoretically conceived facts for
the corroboration of their adequacy. Mathematical theorizing, at their
prompting, weaves the web that holds insights and facts together.

Notes

1. M. Born, *Experiment and Theory in Physics*, Cambridge: at the University Press, 1943.

2. A. Einstein, H. und M. Born, *Briefwechsel 1916-1955*, München: Nymphenburger Verlagshandlung, 1969, p. 203. I thank the Estate of Albert Einstein for allowing me to reproduce the above and the remaining passages of Einstein's writings quoted in this paper.

3. Einstein-Born Briefwechsel, l.c., p. 204.

4. See, for example, N. Maxwell, "The rationality of scientific discovery", *Phi. Sci.*, 41 (1974) 123, 247; J. Agassi, *Science in Flux*, Dordrecht: Reidel, 1975; L.Laudan, *Progress and its Problems*, Berkeley: University of California Press, 1977.

5. E. A. Milne, *Modern Cosmology and the Christian Idea of God*, Oxford: Clarendon Press, 1952, p. 50.

6. E. A. Milne, *Kinematic Relativity*, Oxford: Clarendon Press, 1948, p. 39.

7. E. A. Milne, *Modern Cosmology and the Christian Idea of God*, p. 61, compare, however, *Kinematic Relativity*, p. 93.

8. E. A. Milne, *Kinematic Relativity*, p. 64.

9. H. Bondi and T. Gold "The Steady-State Theory of the Expanding Universe", *Mon. Not. R. Astron. Soc.*, Cambridge: at the University Press, 1960, Chapter XII.

10. H. Reichenbach, *The Theory of Probability*, Berkeley: University of California Press, 1949, paragraph 91.

11. M. Bunge, "Cosmology and Magic", *The Monist* 47 (1962), 116.

12. Compare the sympathetic stance towards the Steady State Theory taken in J. Merleau-Ponty's *Cosmologie du XXe Siecle*, Paris: Gallimard, 1965.

13. A. Einstein, "Prinzipielles zur allgemeinen Relativitätstheorie", *Annalen der Physik*, 55 (1918) 241. Cf. E. Kretschmann, "Über den physikalischen Sinn der Relativitätspostulate", *Annalen der Physik*, 53 (1917) 575.

14. G. Holton, *Thematic Origins of Scientific Thought*, Cambridge, Mass.: Harvard University Press, 1973, Introduction and Chapter 1; *The Scientific Imagination*, London: Cambridge University Press, 1978, Chapter I.

15. J. A. Wheeler, "Mach's Principle as Boundary Condition for Einstein Equations", in Chiu and Hoffmann, eds, *Gravitation and Relativity*, New York: W. A. Benjamin, 1964, pp. 303 - 349.

16. Compare A. Einstein, "Über das Relativitätsprinzip und die aus dem-
    selben gezogenen Folgerungen", *Jahrb. Radioakt. u. Elektronik*,
    4 (1907), p. 454; "Zum gegenwärtigen Stande des Gravitations-
    problems", *Phys. Ztschr.* 14 (1913), pp. 1254 f., 1260 f.,
    1265; A. Einstein and M. Grossmann, *Entwurf einer verallgemei-
    nerten Relativitätstheorie und einer Theorie der Gravitation*,
    Leipzig: Teubner, 1913, p. 6. See also Mach's obituary, by
    Einstein in *Phys. Ztschr.*, 17 (1916) 101.

17. *The Leibniz-Clarke Correspondence*, edited ... by H. G. Alexander,
    Manchester: Manchester University Press, 1956.

18. E. Mach, *Die Mechanik*, Darmstadt: Wissenschaftliche Buchgesell-
    schaft, 1963, pp. 216 ff.

19. A. Einstein, *Annalen der Physik*, 55 (1918), p. 243, footnote.

20. I. Newton, *Philosophiae Naturalis Principia Mathematica*, Cambridge,
    Mass.: Harvard University Press, 1972, vol. I, p. 51.

21. E. Cartan, "Sur les variétés à connexion affine et la théorie de
    la relativité généralisée", *Ann. Ec. Norm. Sup.* 40 (1923) 325;
    41 (1924) 1.

22. "It is contrary to the mode of thinking in science to conceive of a
    thing (the space-time continuum) which acts itself, but which
    cannot be acted upon". A. Einstein, *The Meaning of Relativity*,
    Princeton: Princeton University Press, 1956, pp. 55 f.

23. See, *e.g.*, S. Weinberg, *Gravitation and Cosmology* New York: Wiley,
    1972, pp. 133-135, 151-155; D. Lovelock, "The Einstein tensor
    and its generalizations", *J. math. physics*, 12 (1971) 498.

24. A. Einstein, "Prinzipielles zur allgemeinen Relativitätstheorie",
    Annalen der Physik, 55 (1918), p. 243; W. de Sitter, "On Ein-
    stein's Theory of Gravitation and its astronomical consequences",
    Monthly Not. R. Astr. Soc., 78 (1917) 3.

25. Quoted in G. Holton, "Mach, Einstein, and the Search for Reality",
    Boston Studies in the Philosophy of Science, VI, p. 194, n. 24.

26. Einstein-Born *Briefwechsel* (quoted in note 2), p. 258.

27. F. M. Cornford, "The Invention of Space", in *Essays in Honour of
    Gilbert Murray*, London: Allen & Unwin, 1936, pp. 215-235.

28. M. Bunge, *Treatise on Basic Philosophy*, vol. 3, Ontology I, Dord-
    recht: Reidel 1977.

29. B. Riemann, "Über die Hypothesen, welche der Geometrie zugrunde
    liegen", *Abh. K. Ges. d. Wiss. zu Göttingen*, Bd. 13 (1867).

30. I. Kant, *Allgemeine Naturgeschichte und Theorie des Himmels*,
    Königsberg/Leipzig: J. F. Petersen, 1755.

31. I. Kant, *Kritik der reinen Vernunft*, Riga: J. F. Hartknoch, 1781,
    pp. 424 ff., 517 ff., etc.

32. A. Einstein, "Kosmologische Betrachtungen zur Allgemeinen Relativi-
    tätstheorie" *Preuss. Ak. d. Wiss., Sitzber.* 1917, Pt. 1, pp.
    142-152. Cf. Riemann, "Über die Hypothesen...", p. 22.

33. B. G. Schmidt, "A new definition of singular points in General Relativity", *J. Gen. Rel. and Grav.* 1 (1971) 269. Cf. S. Hawking and G. F. R. Ellis, *The large scale structure of space-time*, Cambridge: at the University Press, 1973, p. 259.

34. *Die philosophischen Schriften von G. W. Leibniz*, hrg. von C. J. Gerhardt, Hildesheim: Olms 1965, vol. 7, p. 302.

35. The reason for this interpretation is clearly explained by Max Jammer in his *Concepts of Mass in classical and modern physics*, Cambridge Mass.: Harvard University Press, 1961, pp. 205 f.

36. Kant, *Kritik der reinen Vernunft*, (1781), p. 158.

37. S. W. Hawking and G. F. R. Ellis, *The large scale structure of space-time*, Cambridge: at the University Press, 1973, p. 189; my italics.

38. H. Schmidt in *J. Math. Phys.*, 7 (1966) 494.

39. Hawking and Ellis, l.c., p. 189.

40. Hawking and Ellis, l.c., p. 266, cf. S. W. Hawking and R. Penrose, "The singularities of gravitational collapse and cosmology", *Proc. Roy. Soc. London*, A 314 (1970) 529.

41. Hawking and Ellis, l.c., p. 206.

42. See, *e.g.*, P. Lorenzen, "Die allgemeine Relativitätstheorie als eine Revision der Newtonschen Gravitationstheorie" *Philos. Naturalis*, 17 (1978), p. 9. On protophysics see G. Böhme (ed.), *Protophysik* Frankfurt: Suhrkamp, 1976.

43. E. Husserl, *Die Krisis der europäischen Wissenschaften und die transzendentale Phänomenologie*, Haag: Nijhoff, 1954, p. 29.

44. Born an Einstein, 25. 8. 1923; *Briefwechsel*, l.c., p. 117.

45. J. A. Wheeler, *Geometrodynamics*, New York: Academic Press, 1962. Cf. R. W. Fuller and J. A. Wheeler, "Causality and multiply connected space-time", *Phys. Rev.*, 128 (1962) 919-929.

DIE PHILOSOPHISCHE RELEVANZ DER KOSMOLOGIE

Bernulf Kanitscheider, Justus Liebig-Universität, Giessen

I.   Analytische und synthetische Philosophie

Die analytische Philosophie, die in der Tradition des englischen Empirismus von Russell über Carnap bis Popper und Nagel denkt, schließt sich in ihren Überlegungen sicher viel enger an die Einzelwissenschaft an als die klassische Philosophie. Sie lehnt jedoch eine selbständige Naturphilosophie im traditionellen Sinne ab. Am deutlichsten wird diese Einstellung bei Wittgenstein.

> *Die richtige Methode der Philosophie wäre eigentlich die, nichts zu sagen, als was sich sagen läßt, also Sätze der Naturwissenschaft, also etwas, was mit Philosophie nichts zu tun hat und dann immer, wenn ein anderer etwas Metaphysisches sagen wollte, ihm nachzuweisen, daß er gewissen Zeichen in seinen Sätzen keine Bedeutung gegeben hat. Diese Methode wäre für den anderen unbefriedigend - er hätte nicht das Gefühl, daß wir ihn Philosophie lehrten - aber sie wäre die einzig streng richtige.*[1]

Aus diesem Zitat kann man folgende Grundhaltung herausfiltern. Befaßt sich ein Philosoph mit Naturwissenschaft, so hat er logische und semantische Analyse der *Sprache* der Naturwissenschaft durchzuführen. Es mag sein, daß er sich privat für die Aussagen und Gehalte interessiert, ex officio jedoch muß er sich mit der linguistischen Seite der Wissenschaft befassen. Ein anderer Gebrauch, etwa der *Ergebnisse* der Physik, für kognitive, philosophische Zwecke ist nicht erlaubt. Es gibt vor allem zwei Argumente, die im Geiste Wittgensteins gegen die Synthese einzelwissenschaftlicher Ergebnisse vorgebracht werden.

1.   Verknüpfungen von Ergebnissen aus verschiedenen Theorien einer Wissenschaft oder Verbindungen von Resultaten verschiedener Wissenschaften können in bezug auf ihren Wahrheitswert nie von der Philosophie beurteilt werden. Nur der Einzelwissenschaftler selbst kann mit Beobachtungen und Experimenten Tests und damit auch echte Validierungen durchführen.

2.   Jeder Satz liegt auf der gleichen sprachlichen Ebene wie ein Satz der Wissenschaft oder er ist Unsinn. Als Einwand gegen die erste Behauptung kann man folgendes vorbringen:

Ein synthetischer Gebrauch von Wissenschaft bedarf keiner neuen faktischen Aussagen und daher muß auch der Philosoph keine eigenständigen empirischen Tests durchführen. Die synthetische Verwendung materialer Aussagen bedingt demgemäß keine eigenverantwortliche empirische Entscheidung. Wohl aber kann der Philosoph in der Lage sein, ein neues Begriffsschema zu entwerfen, in dem schon vorhandene Ergebnisse verbunden werden. Es können Probleme hohen Allgemeinheitsgrades, die traditionell der Philosophie angehören und die eine starke begriffliche Komponente besitzen, behandelt werden. Die Synopsis von Einzelergebnissen geht mithin nicht über den spezifischen Aussagegehalt der Theorien einer Wissenschaft hinaus; das Kriterium hierfür kann darin gesehen werden, daß keine neuen mathematischen Strukturen zur Beschreibung und keine neuen Objekte, wie Felder, Teilchen, Kräfte oder dgl., eingeführt werden. Dies bedingt freilich nicht, daß dem Philosophen nur ein triviales Zusammenstellen von diskreten Forschungsergebnissen übrig bleibt. Das Zusammensehen vorher getrennter Elemente einzelner Theorien kann durchaus zu einem neuen ganzheitlichen Verständnis führen.

Am klarsten wird dies vielleicht, wenn man bedenkt, daß die Einzelwissenschaften die Welt nach epistemischen Kategorien zerlegt sehen. Die Realität zerfällt de facto aber nicht in einen physischen, chemischen, biotischen, psychischen und sozialen Bereich, sondern dieser Einteilung liegt eine arbeitsteilige Kompartimentalisierung zugrunde, die durch den Komplexitätsgrad der einzelnen Wirklichkeitsschichten hervorgerufen wird. Daß der Blick des Spezialisten für synthetische Zusammenhänge infolge der notwendigen Abstraktion verstellt sein kann, sieht man etwa am Problem des "Zeitpfeils". Analysiert der Psychologie die Gerichtetheit des menschlichen Zeiterlebnisses, konstatiert der Physiker in der Thermodynamik eine Anisotropie der Prozesse, bemerkt der Biologe die Einsinnigkeit der biologischen Evolution und berichtet zuletzt der Kosmologe von einer Asymmetrie der Expansion des Universums, so bedarf es eines zusätzlichen Problembewußtseins, um die Frage nach einem Zusammenhang dieser Richtungsauszeichnungen zu stellen. Wenn es möglich ist, eine hierarchische Ordnung in den bereichsspezifischen zeitlichen Asymmetrien zu entdecken, so braucht man eine synthetische Hypothese, die einerseits interdisziplinären Charakter besitzt und andererseits zeigt, daß naturphilosophische Probleme mehr erfordern als eine enzyklopädische Popularisierung einzelwissenschaftlicher Ergebnisse.

Gegen das zweite Argument kann man einwenden, daß die Dichotomie wissenschaftliche Tatsache-Unsinn später von Wittgenstein aufgrund eigener Argumente aufgegeben worden ist. In seiner Spätphilosophie verwendet er ganz andere Kriterien für die Bedeutungshaftigkeit sprachlicher Ausdrücke.[2] Heute hat sich die Überzeugung verstärkt, daß man durchaus rational verständlich, wenn auch metaphysisch über große naturale Zusammenhänge sprechen kann. Der Umschwung in der Auffassung hat sich zum großen Teil im Lager der analytischen Philosophie selbst vollzogen.[3] So wurde etwa von Quine gezeigt, daß man, um das Zusammenpassen verschiedener begrifflicher

Entwürfe beurteilen zu können, von den einzelwissenschaftlichen Theorien
und ihren faktischen Bezügen zurücktreten muß, daß man das durchführen
muß, was er einen "semantic ascent" genannt hat. Quine kommt zu diesem
Begriff auf dem Wege einer linguistischen Analyse der Existenzvorausset-
zungen von Theorien. Der semantische Aufstieg ist notwendig, um eine Be-
urteilungsplattform zu gewinnen, von der aus eine Übereinstimmung in der
Frage erzielt werden kann, über welche Objekte bestimmte Theorien reden.
Das Verfahren ist aber nicht auf ontologische Probleme beschränkt, es betrifft ganz allgemein die begriffliche Konfrontation verschiedener Gedan-
kengebäude. "The strategy is one of ascending to a common part of two
fundamentally disparate conceptual schemes, the better to discuss the
disparate foundations".[4]

Systematisch hat Smart 1968 den Begriff der synthetischen Philosophie wieder in die Diskussion eingeführt.[5] Unter anderem Titel ist diese Inten-
tion auch von Suppes aufgenommen worden, der eine ähnliche Bestrebung mit
dem Ausdruck "deskriptive Metaphysik" bezeichnet. "The tenets of a properly formulated descriptive metaphysics that is also properly supported by
the weight of scientific and commonsense evidence should form a broad conceptual framework for thinking about natural phenomena and human experience".[6]

Es fragt sich, wie man diesen Begriff sauber von der gewöhnlich verstan-
denen analytischen Philosophie trennen kann. Auf dem intensionalen Wege
einer expliziten Begriffsbestimmung wäre dies sicher recht mühsam. Einfacher läßt es sich durch eine extensionale Durchmusterung von Problemen
bewerkstelligen, indem man die in einschlägigen Zeitschriften und Mono-
graphien behandelten Themen durchgeht. Zwei Klassen von Problemen lassen
sich deutlich unterscheiden. Erstens finden Diskussionen *über* die Wissenschaft statt: z.B. wird die logische Struktur eines gültigen Argumentes,
das Prüfen von Hypothesen, das Bilden von Begriffen, die Konstruktion von
Theorien untersucht. Kennzeichen dieser kognitiven Ziele ist, daß von den
Unterschieden der spezifischen Wissenschaft abgesehen wird. Es sind Re-
flexionen des Wissenschaftlers auf jene strukturalen Terme, die *nicht*spe-
zifisch für einen Objektbereich sind. Man könnte diese Tätigkeit auch die
Suche nach *methodologischen Invarianten* nennen.[7] Bei der zweiten Klasse von
Problemen handelt es sich um Fragestellungen, bei denen die Ergebnisse
der Wissenschaft *verwendet* werden. Es sind also philosophische Überlegungen
mittels der Wissenschaft. Hier werden die Unterschiede der verschiedenen
Objektklassen berücksichtigt. Es gehen die materialen Komponenten wesentlich in die Diskussion ein. Der faktische Zusammenhang, der dabei verwendet wird, hat nicht den Charakter eines Beispiels, das beliebig ersetzt
werden kann, sondern ist konstitutiv für das Problem. Historisch gesehen

wurde die zweite Problemklasse von der Naturphilosophie behandelt. Metho-
dologisch betrachtet, verfuhr diese in der Vergangenheit allerdings meist
im Stil der rationalen Metaphysik. Nun hat sich aber *ein* Erbe des logischen Empirismus zweifellos durchgesetzt. Man.glaubt heute nicht mehr an
einen autonomen philosophischen Zugang zur Natur. Ein direktes Erfassen
naturaler Gesetzesstrukturen auf rationalem Wege ohne Erfahrungskontrolle
ist heute obsolet geworden. Wenn eine synthetische Philosophie heute erfolgreich arbeiten will, muß sie den Weg über die zu diesem Zeitpunkt
als fundamental angesehenen einzelwissenschaftlichen Theorien nehmen. Aus
ihnen müssen die zentralen ontologischen und epistemischen Komponenten
herausgefiltert werden. Vor 70 Jahren prägte Max Planck den Begriff des
physikalischen Weltbildes.[8] Damals war dieses auf zwei Pfeilern aufge-
baut: der Mechanik und dem Elektromagnetismus. Heute ist die Situation
viel unübersichtlicher geworden. Wenn man jedoch die Hoffnung auf ein ein-
heitliches Verständnis der Natur nicht aufgegeben hat, dann kann der Weg
nur über die fundamentalen Theorien der Physik führen.

Ein Mißverständnis sei am Ende dieser allgemeinen Überlegungen noch ausgeräumt: Analytische und synthetische Philosophie sind niemals Alterna-
tiven und es gibt keinen Grund, zwischen ihnen einen Konflikt oder eine
Konkurrenzsituation herbeizuführen. Synthetische Philosophie hat immer
auch eine begriffsklärerische Komponente, während andererseits eine logische Analyse der Wissenschaftssprache durchaus unabhängig von den faktischen Geltungen der wissenschaftlichen Aussagen durchgeführt werden kann.
Die Grenzgespräche zwischen den Wissenschaften erfordern sehr oft auch die
Elimination von Unsinn, sie sind aber, so wie ich sie hier verstehe, mehr
als die Aufklärung sprachlicher Mißverständnisse.

II.   Die Wissenschaft von der Welt im Großen

Zu den wichtigsten Untersuchungsthemen einer Naturphilosophie im eben eingeführten Sinne gehören neben den Problemen, die sich um die Struktur des
Raumes, den Aufbau der Materie, die Richtung der Zeit ranken, auch diejenigen, die mit dem großräumigen Aufbau der Welt zu tun haben, also dem Bereich der Kosmologie angehören. Die Frage nach dem großräumigen materialen und raumzeitlichen Aufbau der Welt wurde früher von einem philosophischen Fach, der metaphysica specialis beantwortet. Spätestens seit dem
Jahre 1917 ist die Kosmologie eine physikalische Disziplin geworden.[9]
Der Grund für dieses historische Grenzdatum ist darin zu sehen, daß mit
Einsteins Zylinderuniversum zum erstenmal ein von den kosmologischen Para-
doxa der klassischen Gravitationstheorie freies konsistentes Weltmodell
konstruiert wurde, welches quantitativ prüfbare Aussagen liefert. Auch
später, als Milne und McCrea die Möglichkeit von dynamischen Newtonschen
Weltmodellen entdeckten, änderte sich die Situation nicht grundlegend,

denn die klassischen Modelle sind nur unter Verwendung von zueinander beschleunigten Bezugssystemen brauchbar, die in der Newtonschen Begriffswelt keinen natürlichen Platz haben;[10] außerdem muß deren ursprüngliche theoretische Symmetrie insofern zerstört werden, als die Idee des absoluten Raumes aufgegeben, die der absoluten Zeit aber beibehalten werden muß.[11] Einsteins Zylinderuniversum machte quantitativ prüfbare Aussagen, die Testbarkeit kosmologischer Lösungen war damit gesichert und 1929 erfolgte auch die Widerlegung der Behauptung, daß unsere Welt durch das statische Zylinderuniversum darstellbar ist, durch die Entdeckung der kosmischen Dynamik.[12] Kommt man auf die metatheoretische Seite dieser Disziplin zu sprechen, so kann man zuerst einmal fragen, in welchem Sinne es in diesem Bereich eigentlich eigenständige Theorien gibt, die in besonderer Weise eine Megaphysik konstituieren? Sind kosmologische Theorien nicht einfach Anwendungen lokaler Theorien? Nun, von der *ontologischen* Seite betrachtet gibt es erst einmal keinen Grund dafür, daß der Aufbau der Welt im Großen unserem Mesokosmos wesentlich gleicht. *Methodisch* spricht eher einiges dagegen. Man hat schon mehrmals mit dem Prinzip von Fourier Schiffbruch erlitten, das dieser zuerst 1822 formuliert hatte,[13] daß nämlich physikalische Gesetze niemals einen inneren Bezug auf eine absolute Größenordnung enthalten.[14] Was sich schon um die Jahrhundertwende abzeichnete, war, daß die Gesetze des Mesokosmos, d.h. der mittleren makroskopischen Größenordnung, also die Gesetze der klassischen Mechanik und klassischen Elektrodynamik nicht auf die Mikrowelt übertragen werden konnten, sondern daß man eine neue Theorie brauchte, eben die Quantenmechanik. Das Wirkungsquantum h gibt danach an, wo die Übertragbarkeit endet und damit auch das Prinzip von Fourier gebrochen ist. Im astrophysikalischen Bereich gibt es eine dimensionslose Größe, die ebenfalls eine Grenze der Beschreibung ausdrückt. Das Verhältnis zwischen Schwarzschild-Radius R und Radius eines Körpers R gibt an, wann man zu einer relativistischen Formulierung übergehen muß. Für Hauptreihensterne ist $\frac{R}{R} \approx 10^{-6}$, für weiße Zwerge $\approx 10^{-4}$, aber für Neutronensterne und schwarze Löcher ist $\frac{R}{R} \approx 1$, was anzeigt, daß die sonst erfolgreiche lokale Theorie Newtons hier nicht mehr verwendbar ist.[15] In der Kosmologie gibt es für den mittleren Bereich der Laufzeit Modelle, die ein klassisches Newtonsches Analogon besitzen, aber in den frühen und späten Zeiten, wenn extreme Situationen vorliegen und hohe Gravitationsgezeitenkräfte wirken, dann wird auch hier die Einsteinsche Theorie Verwendung finden müssen.

Ein wesentlicher Unterschied ontologischer Art besteht zwischen dem astrophysikalischen und kosmologischen universe of discourse. Es gibt *viele* weiße Zwerge, Hauptreihensterne, Galaxien und Galaxienhaufen, aber es gibt nur *ein* Universum. Von den Gegnern der relativistischen Kosmologie wurde aus dieser Tatsache ein methodologisches Argument konstruiert, das die epistemische Sonderstellung der Kosmologie beinhaltet. Eine Theorie der Welt im ganzen darf keine Vielzahl von Lösungen enthalten, sie muß in dem Sinne streng sein, daß sie der Einzigkeit der Welt durch das Vorhandensein genau *eines* kosmologischen Modells Rechnung trägt. In der relativistischen Kosmologie wird die Einzigkeit des Bezugsobjektes nicht durch die Theorie selbst aufgefangen, sondern der Bestimmung der Randbedingungen überlassen. Wohl unterscheidet man lokale Lösungen, wie Schwarzschild-, Reissner-Nordstrøm- und Kerr-Metrik, von globalen Lösungen, wie Friedman-, Lemaître- und Einstein-de Sitter-Metrik. Der Unterschied zwischen diesen beiden Klassen ist aber nicht logisch streng bestimmt. Auch eine Welt, die ein einziges kollabiertes rotierendes Objekt enthält, das sich in einem asymptotisch flachen Raum befindet und dann durch eine Kerr-Metrik beschrieben würde, wäre in diesem hypothetischen Fall kosmologisch. Sieht man von der Einzigkeit des Objektes einmal ab, haben die astrophysikalischen und kosmologischen Theorien vieles gemeinsam: 1. Das Objekt der Untersuchung ist physisch unzugänglich. 2. Es kann in diesem Bereich nicht experimentiert werden. 3. Bis vor kurzem konnte Information nicht erzeugt werden; in jüngster Zeit ist man soweit gekommen, daß durch Raketenaufstiege und künstliche Satelliten die Information nicht mehr nur passiv abgewartet werden muß, sondern z.B. durch Röntgen-Detektoren wie Uhuru und Kopernikus der Informationsfluß künstlich verstärkt werden kann. 4. Trotzdem muß aber die lokal ankommende Information immer noch äußerst intensiv aufgeschlossen werden und sie enthält immer noch eine Reihe von theoretischen Annahmen, die nicht unmittelbar testbar sind. Bei der Eruierung der kosmischen Parameter H_o und q_o muß man z.B. mit Homogenitätsextrapolationen arbeiten, die die Eichung der bereichsabhängigen Entfernungsmaßstäbe ermöglichen.[16]

Möglicherweise besitzt der kosmische Bereich eine Besonderheit gegenüber den begrenzten astrophysikalischen Systemen. Es kann sein, daß es lokale Phänomene gibt, die Aufschluß über die Welt im Großen geben, daß man also, ohne es zu wissen, im Besitz kosmologischer Information ist. Dies hängt davon ab, ob es kausale Bindeglieder zwischen lokalen Tatsachen und der großräumigen Verteilung der Materie gibt. Zwei Kandidaten für eine solche Kopplung kennt man schon seit langem: das Olberssche Argument und das Mach-Prinzip. Sie wurden immer wieder als elektromagnetische und als mechanische Verbindung eines lokalen Phänomens mit einer globalen Struktur verstanden.[17] Eines ist jedoch klar, wenn eine solche Kopplung besteht, dann kann sie nur durch eine starke Theorie vermittelt werden. Ohne Hypothese über das Strahlungsverhalten der Sterne in der Vergangenheit und ohne einen Trägheitsübertragungsmechanismus ist weder die Dunkelheit des

mondlosen Nachthimmels noch das Zusammenfallen von Trägheits- und Fix-
sternkompaß wirklich kosmologische Information.

Dies zeigt schon einen klaren methodologischen Befund an. Nirgendwo trägt
die Theorie soviel Last wie in der Kosmologie. Die Beanspruchung des
"surplus meaning" ist in keinem Bereich derartig stark. Der riesige Be-
zugsbereich, die reference class, überschreitet den winzigen Testbereich,
die evidence class, gewaltig. Das Mißverhältnis ist so groß, daß von man-
chem Methodologen die Kosmologie auch nach ihrer quantitativen Fassung
eher als philosophische denn als eine physikalische Theorie geführt wird.
Ich glaube jedoch nicht, daß man diesen Schluß ziehen muß und daß man davon sprechen kann, daß die hypothetisch-deduktive Methode im kosmischen
Bereich versagt. Im Gegenteil zeigt die relativistische Kosmologie die
typischen Strukturmerkmale dieser Vorgangsweise. Aus den allgemeinen Feldgleichungen gewinnt man lokale und globale Lösungen. Der Übergang zu der
gegenwärtig relevanten Klasse von kosmologischen Lösungen erfolgt durch
das Postulat der Uniformität. Dadurch werden die homogenen und isotropen
Robertson-Walker-Friedman-Modelle ausgesondert. Aber noch ein weiteres
typisches Merkmal der hypothetisch-deduktiven Methode zeigt sich im kosmologischen Bereich. Die Theorie bezieht sich nicht direkt, sondern mittels eines *Modellobjektes* auf die Realität.[18] Im Modellobjekt liegen die
Idealisierungsannahmen verborgen. Die theoretische Physik muß in allen
Bereichen immer mit Vereinfachungen arbeiten. So verwendet etwa die klassische Partikelmechanik den Begriff des Massenpunktes. Sie idealisiert
bei der Behandlung himmelsmechanischer Probleme ganze Planeten als aus-
dehnungslose Punkte. In der gleichen Weise behandelt die Kosmologie die
Galaxien nicht in ihrem ungeheuer verwickelten realen Aufbau, sondern
idealisiert sie als Gas nicht wechselwirkender Teilchen. Die innere Struktur der Galaxien, die Tatsache, daß sie aus Sternen bestehen, ihre Haufen-
bildung werden ebenso vernachlässigt wie die Teilchenstruktur dieses Gases. Auf diese Weise gelangt man zum Modellobjekt der vollkommenen Flüssigkeit. Es ist nurmehr durch drei Eigenschaften gekennzeichnet: die
4-Geschwindigkeit u^α , die Dichte der Masseenergie ρ und der kinetische
Druck der Galaxien p. Alle anderen Eigenschaften, die sicherlich in der
Realität vorhanden sind, werden vernachlässigt, wie etwa Scherung, Aniso-
tropiedruck, Rotation und Zähigkeit. Im Gefolge dieser brutalen Verein-
fachung erhält man aber einen sehr übersichtlich zu handhabenden mathematischen Apparat. Das Modellobjekt besitzt einen einfachen Energiespan-
nungstensor $T_{\alpha\beta} = (\rho + p) u_\alpha u_\beta + \rho\, g_{\alpha\beta}$ und die Konsequenzen für die
Raum-Zeit-Struktur sind deren Homogenität und Isotropie

$$ds^2 = c^2 dt^2 - R^2(t) \frac{dr^2 + r^2(d\vartheta^2 + \sin^2\vartheta d\varphi^2)}{(1 + \frac{1}{4} kr^2)^2} .$$

Daß auch sie eine Idealisierung ist, ist leicht einzusehen. Die Feinstruktur der Raumzeit ist sicherlich weder homogen noch isotrop angesichts der lokalen Zusammenballungen von Materie. Wichtig für die Begründung der Behauptung, daß die physikalische Kosmologie nicht aus dem hypothetisch deduktiven Denken, wie es zuerst im logischen Empirismus erarbeitet worden ist, herausfällt, besteht darin, daß diese Idealisierung ausschließlich *vorläufigen* Charakter hat. Durch die Verwendung eines bestimmten Modellobjektes wird eine Klasse von physikalisch möglichen Welten ausgesondert. Es sind dies die Robertson-Walker-Friedman-Welten (RWF). Im allgemeinsten Fall der Theorie gibt es noch eine Konstante λ, die $\neq$ o sein kann. In der größten Zahl der Fälle arbeitet man mit einer Unterklasse der RWF-Welten, wo λ = o ist. Die Standardkosmologie, wie man sie heute versteht, verwendet nur den λ = o-Typ von Welten. Aber auch diese Auswahl ist empirisch revidierbar. Beatrice Tinsley hat Argumente dafür gebracht, daß die gegenwärtigen astrophysikalischen Daten nur in die weitere Klasse von $\lambda \neq$ o hineinpassen.[19] Natürlich können hier wieder nur diejenigen Modelle verwendet werden, die ein endliches Alter besitzen. Der Gebrauch eines Modellobjektes bedingt also keine apriorischen Vorentscheidungen. Im Prinzip kann die gesamte Klasse der RWF-Welten aus der Konkurrenz ausscheiden. Dann muß das Modellobjekt geändert werden. Es müssen neue Freiheitsgrade darin eingebaut werden, etwa Rotation und Scherung. Auch die Gesetzlichkeit verändert sich dann. Die einfachen Friedman-Gleichungen müssen durch die kompliziertere Raychaudhuri-Gleichung ersetzt werden. Die empirische Forschung hat dann die Freiheit, irgendein Element aus der weiteren Klasse der anisotropen und inhomogenen Welten auszuwählen. Gegenwärtig macht man von den anisotropen Modellen bereits Gebrauch. Sie spielen für die Frühzeit des Universums eine Rolle, denn es hat sich gezeigt, daß sie mit den uniformen Modellen kausal verbunden werden können, wenn man bestimmte Anisotropie-Dämpfungsmechanismen annimmt.

Es gibt noch einen Ansatzpunkt, der manchen Philosophen zur Befürwortung der These dient, daß die Kosmologie eine apriorische Wissenschaft ist oder doch zumindest starke apriorische Komponenten besitzt.[20] Die *Homogenität* der Materieverteilung und der Raumzeit ist lokal nicht überprüfbar. Logisch mit den Daten vereinbar ist es, daß die Umgebung unserer Galaxis atypisch für das gesamte Universum ist. Empirisch bestätigt ist nur die lokale Isotropie, also die isotrope Galaxienverteilung vom terrestrischen Standpunkt und daraus folgt formal nicht die Homogenität. Man braucht noch spezielle Zusatzannahmen, so etwa das *Kopernikanische Prinzip*, das in einer seiner vielen Formulierungen so lautet: Wir leben an keinem ausgezeichneten Ort des Universums. Mit der lokalen Isotropie und unter Annahme des Kopernikanischen Prinzips kann man dann logisch zur globalen Isotropie und von dieser zur Homogenität übergehen. Aber die lokale Isotropie ist mit der Negation des Kopernikanischen Prinzips vereinbar. Von dieser Tatsache hat erst kürzlich G.F.R. Ellis Gebrauch gemacht mit seinem sog. SSS-Universum.[21] Das ist eine sphärisch-symmetrische-statische Welt mit zwei ausgezeichneten Orten, unserem Aufenthaltsort und einer nackten Singularität, die alljene empirische Befunde erzeugt, die wir in der Standardkosmologie der Anfangssingularität zuordnen. Nach meiner Überzeugung bedeutet dies jedoch immer noch nicht, daß das Kopernikanische Prinzip als ein Stück Apriori-Metaphysik behandelt werden muß. In der normalen Verwendung bedeutet apriorische Gültigkeit die Unabhängigkeit von jeder *möglichen* Erfahrung.[22] Genau diese Situation liegt hier nicht vor. Es sind *technische* Gründe, wegen denen wir ausreichend entfernte Punkte im Universum nicht erreichen können, um das Kopernikanische Prinzip zu prüfen. Es ist also wichtig in diesem Zusammenhang, die technische, die faktische und die logische Unmöglichkeit zu unterscheiden. Wir brauchen uns nur zu überlegen, daß wir durchaus von fernen Kulturen Information erhalten können, daß etwa nach ihren Beobachtungen das Universum in jeder Richtung ganz und gar nicht gleich aussieht. Damit wäre das Kopernikanische Prinzip grundsätzlich falsifizierbar, denn es ist logisch und faktisch möglich, dagegensprechende Daten zu erhalten. John Wheeler führt sogar eine positive empirische Instanz an,[23] die zumindest die schwache Testbarkeit im Sinne einer Bestätigung bekräftigt, nämlich die Messung der Feinstrukturkonstante α für eine Gruppe von Galaxien, die $2 \cdot 10^9$ Lichtjahre entfernt sind. Bahcall und Schmidt haben 1967 diese Messung für Galaxien, die eine Rotverschiebung zwischen 0,17 und 0,26 aufweisen, durchgeführt mit dem Ergebnis, daß α mit dem lokal gemessenen Wert in der Größe von 3:1000 übereinstimmt.[24] Deshalb ist es viel sinnvoller, davon zu sprechen, daß das Kopernikanische Prinzip den Charakter einer vorläufigen festgesetzten *Randbedingung* besitzt, für die Einfachheitsgründe maßgebend sind; ebenso wie man solange wie möglich mit der λ = o-Unterklasse der RWF-Welten arbeitet, um die Komplexität der Modelle zu reduzieren, wobei jedoch die Reaktivierung von λ aus empirischen Gründen jederzeit möglich ist, gilt dies in gleicher Weise auch für die Bedingung der Homogenität und Isotropie. Auch hier sind schon Vorschläge gemacht worden, von dieser starken Symmetriebedingung abzuweichen, etwa um eine Lösung für das Homogenitätsparadoxon zu finden. Dieses besteht darin, daß Strahlung aus zwei Bereichen, die kausal unverbunden, d.h. durch Horizonte getrennt sind, identische Eigenschaften besitzt. Anisotrope Welten, wie sie der Kasner- und Mixmaster-Lösung entsprechen, wurden wieder aufgrund dieses empirischen Befundes verwendet, um vor der Friedman-Ära eine starke Durchmischung der Materie zu erzielen. Auch wenn der Mixmaster-Ansatz sich nicht als so erfolgreich erwies, wie man zuerst glaubte, ist das Methodische hier wesentlich, daß nämlich aufgrund empirischer Befunde von der Isotropiebedingung für frühe Zeiten des Universums abgewichen werden kann.

Ein letzter Grund, der gegen den apriorischen Status des Kopernikanischen Prinzips anzuführen wäre, ist die Vermutung, daß das Prinzip eine deduktive Konsequenz einer höherrangigen Theorie werden kann, nämlich einer Theorie der Galaxienentstehung. Eine solche Theorie würde dann an ganz anderen logischen Konsequenzen getestet, enthielte aber als untestbares Theorem einen Satz, der dem Kopernikanischen Prinzip äquivalent ist. In diesem Fall müßte die Validierung des Prinzips über die logische Kopplung mit der Theorie erfolgen.

Es gibt noch eine andere Überlegung mit dem Ziel, der Kosmologie den Status einer normalen physikalischen Disziplin zu rauben. Anders als in der lokalen Physik, so heißt es, kann es in der Kosmologie wegen der Einzigkeit des Bezugsobjekts keine Erklärungen geben.[25] Grundsätzlich sollte man hier die Lösung des *deskriptiven* Problems der des *explanativen* Problems trennen. Das deskriptive Problem betrifft den metrischen und topologischen Aufbau der Welt im Großen und die Bewegungsform, die Dynamik der Materie im Ganzen. Die Astrophysiker glauben heute grundsätzlich an die empirische Entscheidbarkeit des deskriptiven Problems. Das explanative Problem richtet sich auf die Frage, *warum* die Welt gerade durch ein bestimmtes Modell strukturiert ist und durch kein anderes. Selbstredend ist dies wesentlich anspruchsvoller. Die Forderung nach einem weitergehenden Erklärungsanspruch entspricht in starkem Maße Einsteins Erkenntnisziel, wie er es, allerdings in bezug auf die Feldtheorie, formuliert hat: "Wir wollen nicht nur wissen, *wie* die Natur ist (und *wie* ihre Vorgänge ablaufen), sondern wir wollen nach Möglichkeit das vielleicht utopisch und anmaßend erscheinende Ziel erreichen zu wissen, warum die Natur *so und nicht anders ist.*"[26]

Die Gegner der erklärungheischenden Fragestellung in der Kosmologie argumentieren folgendermaßen: Erklärungen setzen Gesetze voraus, jedes Gesetz muß über eine Vielzahl von Instanzen laufen. Die Universalität ist eine der wenigen wesentlichen Züge, auf die man sich bezüglich der Gesetzesartigkeit einer Aussage einigen wird. Jedes Gesetz hat die Aufgabe, die Gemeinsamkeit vieler Systeme und Prozesse herauszuheben. Nimmt man ein bekanntes lokales Gesetz, wie das des schiefen Wurfes, so ist die Parabelform der Bahn des Objektes dasjenige Element, was allen schiefen Würfen gemeinsam ist. Das Gesetz sagt danach aus, was bei den vielen akzidentellen Bestimmungen (Anfangsbedingungen) aller schiefen Würfe die bleibende Struktur ist. Gerade diese Überlegung läßt sich auf den Fall

der Kosmologie nicht übertragen. Anscheinend degeneriert der Gesetzes-
begriff hier, weil die nomologische Aussage sich nurmehr auf ein einziges Objekt bezieht. Es gibt in diesem Fall kein Kriterium mehr zur Tren-
nung von nomologischen und de facto-Eigenschaften. Offenbar haben Randbedingungen hier den gleichen Status wie Gesetze. Wenn das richtig ist,
gibt es auch keine Erklärungen mehr, denn das Gesetz ist das wesentliche
Element einer deduktiv-nomologischen Erklärung. Was läßt sich hiergegen
sagen? Nun, die logische Analyse stimmt erst einmal gar nicht mit dem
überein, was die theoretische Tätigkeit der Kosmologen tatsächlich umfaßt.
Schon 1927 beschloß Lemaître seine berühmte Arbeit über die Lichtausbrei-
tung in einem expandierenden Universum mit den Worten: "Il resterait à
se rendre compte de la cause de l'expansion de l'univers."[27] Er fragt
also nach einer kausalen Erklärung für eine typisch globale Eigenschaft
des Gesamtsystems. In der gegenwärtigen Kosmologie geht die explanative
Fragestellung von der Entdeckung der anisotropen und inhomogenen Modelle
aus. Man ist nicht mehr damit zufrieden, auf empirische Weise ein bestimmtes Element der Klasse der RWF-Welten zu erhalten und damit eine gute
Beschreibung der großräumigen Struktur zu liefern, sondern man will wissen, *warum* die physikalischen Gesetze die globale Uniformität gestatten,
ja, ob die Gesetze die Homogenität und Isotropie vielleicht sogar fordern:
"Accepting the agreement with observations we want to understand *why the
laws of physics should demand (rather than merely permit) a universe that is homogeneous
and isotropic to high accuracy on large scales.*"[28] Es wird damit die Frage nach
einem gesetzesartigen Zusammenhang gestellt, wonach ein Universum *notwendig*
zu späten Zeiten hochsymmetrisch wird. Zur Beantwortung der Frage muß man
wissen, wie sich ein Universum entwickelt, das zu frühen Zeiten stark
chaotisch gestartet wird. Chaos wird hier weder im Sinne einer völligen
Gesetzlosigkeit noch in der Bedeutung thermodynamischer Unordnung verstanden, sondern man meint damit eine Zufallsverteilung der Anfangsbedingungen
derart, daß ein weites Spektrum der Schwankungen in Dichte, Entropie und
lokaler Expansionsrate vorhanden ist. Der natürliche Glättungsprozeß, der
das Universum homogenisiert, sei es durch Mixmaster-Schwingungen, Neutri-
no-Viskosität oder Teilchenentstehung, steht dann nicht etwa mit dem 2.
Hauptsatz der Thermodynamik in Widerspruch, sondern die Anisotropiedämp-
fung ist gerade ein Fall von Entropieproduktion. Misner hat ein Beispiel
einer solchen Möglichkeit gegeben.[29] Das Kasner-Universum expandiert ani-
sotrop mit verschiedenen Raten entlang der x- und y-Achse, und entlang
der z-Achse zeigt es sogar Kontraktion. Ein Beobachter würde in einem
solchen Universum Rot- und Blau-Verschiebungen zugleich feststellen. Nun
ist die Kasner-Metrik eine Vakuumlösung, gehorcht also der Gleichung
$G_{\mu\nu} = 0$. Wenn man in eine solche Lösung etwas Materie hineingibt und

zwar als Flüssigkeit ohne Druck, dann dominiert zu frühen Zeiten (t→o)
die Krümmung der Raumzeit, sie beherrscht dort die Expansionsrate. Nach
einer charakteristischen Zeit gewinnt der Materieterm die Überhand und
die Metrik geht asymptotisch in das homogene und isotrope Friedman-
Modell mit k = o über. Dieser physikalische Zusammenhang hat nur eine
methodisch wichtige Konsequenz. Der *chaotische Frühzustand* kann gesetzes-
artig mit einem *symmetrischen Spätzustand* verbunden werden. Damit haben wir
die Möglichkeit, kosmologische Erklärungen eines bestimmten Typs zu formulieren, nämlich Antworten auf die Frage, *warum* das Universum gegenwärtig die raumzeitliche Struktur besitzt, die wir tatsächlich beobachten.
Es sind also *genetische Erklärungen*, die dynamische Evolutionsgesetze verwenden, welche nicht durch empirische Generalisationen zustande gekommen
sind, die hier in der Kosmologie auftauchen. Generalisationen sind natürlich nicht so einfach möglich, sie würden ja eine Pluralität von Welten voraussetzen und ein Weltensemble wird in der Standardkosmologie nicht
verwendet. Nur im Rahmen von solchen Ideen wie Dickes biologischer Selektion der Konstanten, Carters anthropischem Prinzip zur Erklärung der Zah-
lenkoinzidenzen und Wheelers Wiederingangsetzungsmodell des kosmischen
Kollapses, also dem sogenannten "reprocessing model and probabilistic
scattering in superspace", wird mit einer Vielzahl von Welten gearbeitet.
In dem ontologisch restriktiven Standardmodell gibt es dynamische Gesetze,
die innerweltliche Kopplungen von Zuständen beschreiben, welche jedoch
auch eine bestimmte Art von Erklärungen ermöglichen.

Was liefert nun das Wissen um solche Verbindungen? Generell, auch in der
lokalen Physik, ist man daran interessiert, möglichst viele randbedingungs-
artige de facto-Eigenschaften nomologisch zu reduzieren. Denn, je stärker
die gesetzesartigen Verknüpfungen zwischen den Eigenschaften des Modells
sind, um so größer ist der Kohärenzgrad des Modells. Die am weitesten geführte programmatische Forderung in Richtung der Reduktion von Kontingenz
stammt von Harrison.[30] Seine These, die man einen Hypermachismus nennen
könnte, ist eine Anwendung des bootstrap-Prinzipes von Chew. Chew behauptet, daß der Zustand eines Hadrons mit denen aller anderer Hadronen in
einer solchen Weise gekoppelt ist, daß keines für sich alleine vollständig
beschrieben werden kann. Und der kosmologische Hypermachismus besagt entsprechend, daß die Welt, so wie wir sie prima facie kontingent vorfinden,
in ihrer Existenzweise tatsächlich *einzig* ist, da die Eigenschaften jedes
Teilsystems mit denen aller anderen strikt nomologisch verbunden sind. Die
Verkettung der Untersysteme ist also so stark, daß das Modell nur eine
einzige Realisierung besitzen kann. Es ist kein Zweifel, daß die gegenwärtige Kosmologie eine solche Erweiterung der Beziehung, wie sie Mach einst
nur für die Trägheit formuliert hat, nicht erfüllt. Ich möchte sogar aus

einem anderen Begründungszusammenhang heraus die These vertreten, daß die
totale Reduktion von Kontingenz unmöglich ist. Dies kann man am besten verdeutlichen am Beispiel einer Theorie, die in ihrem Erklärungsanspruch viel
weiter geht als die Relativitätstheorie. Es ist der Ansatz von Schäfer und
Dehnen von 1977, der das Ziel besitzt, den *Ursprung der Materie* im Universum zu erklären.[31] Hier wird mit der Theorie des quantisierten Dirac-
Feldes in der gekrümmten Raumzeit die Teilchenentstehungsrate berechnet,
die vom expandierenden Universum verursacht wird. Dabei wird die Voraussetzung eines sphärischen Raumes gemacht, aber das Expansionsgesetz kann
beliebig sein. Die Theorie erreicht zwar ihr vorgegebenes Ziel nicht völlig. Es wird nicht die gesamte Materieentstehung erfaßt, denn zur Vermei-
dung von Divergenzen wird nur das *freie* quantisierte Dirac-Feld verwendet,
man berücksichtigt also nur die inkohärente Materie ohne Einbeziehung der
starken Wechselwirkung. Deshalb gilt die Theorie auch erst ab $t > 10^{-4}$ sec,
d.h. für eine Epoche, da die Hadronenära schon vorbei ist. Aber immerhin
gelingt es der Theorie doch, 90% der existierenden Materie zu erklären.
Was man hier jedoch methodisch sehen kann, ist für unser Problem sehr lehr-
reich: Die nomologische Kopplung zwischen den Grundelementen ist wesentlich stärker als in der Relativitätstheorie. Nicht nur wird die Teilchen-
zahl $N(t)$ durch die Feldgleichungen mit dem Verlauf der Skalenfunktion
$R(t)$ verbunden; die Dirac-Gleichung koppelt auch $R(t)$ mit $N(t)$, da in gekrümmten nichtstationären Räumen Teilchenerzeugung durch Vakuumfluktua-
tionen des quantisierten Materiefeldes auftritt. Das erkenntnistheoretische Ziel von Schäfer und Dehnen ist klar: "The properties of the universe should be deduced from provable natural laws alone",[32] d.h. rand-
bedingungsartige kontingente Elemente sollen auf starke gesetzesartige
Beziehungen zurückgeführt werden. Zweifellos ist in dieser Theorie der
Materieentstehung ein Schritt im Sinne des früher erwähnten Hypermachismus gelungen. Man sieht jedoch, auch die Reduktion von kontingenten Eigenschaften rekurriert auf etwas Gegebenes. Hier sind es die Eigenschaften
des physikalischen Vakuums, wie es durch die Quantenfeldtheorie beschrieben wird, welche unerklärt zurückbleiben. Selbst eine so abstrakte kosmologische Theorie fügt sich damit homogen in den Rahmen wohlbekannter er-
kenntnislogischer Zusammenhänge ein. Aus ihnen weiß man längst, daß Letzt-
begründungen unmöglich sind.[33] Weder die *Existenz* des Universums noch *alle*
seine *Eigenschaften* sind erklärbar. Wir können versuchen, das Universum durch
immer mehr logische Verknüpfungen zwischen den fundamentalen Eigenschaften
besser zu verstehen, aber den Gesamtaufbau auf einmal zu erklären, ohne
eine tiefere Realitätsschicht zu verwenden, ist aus der Logik der Erklärung heraus unmöglich.

III. Die Weltbildfunktion der Kosmologie und die Singularitäten der Raumzeit

Wir haben es am Beginn unserer Ausführungen expliziert: Zu den wesentlichen Elementen einer synthetischen Philosophie gehört die Auffassung, daß naturwissenschaftliche Problemlösungsansätze durchaus neues Licht auf alte philosophische Fragestellungen werfen können. Ja, man kann so weit gehen zu behaupten, daß naturphilosophische Überlegungen nur dem Grad der Allgemeinheit nach sich von naturwissenschaftlichen Gedankengängen unterscheiden und daß der Übergang zwischen beiden fließend ist. An keiner Stelle sieht man dies so deutlich wie am Rätsel der Singularitäten. Ursprung und Ende der Welt sind alte Themen der speziellen Metaphysik, die begrifflichen Hilfsmittel der Differentialgeometrie aber erlauben eine wesentlich genauere und schärfere Bestimmung der klassischen Problemlage. Nichts kennzeichnet diese Situation deutlicher als das Eindringen des Begriffes der Eschatologie in den physikalischen Wortschatz.[34] Wer andererseits heute Behauptungen über die ersten und letzten Dinge aufstellen will, wird an den Aussagen der relativistischen Kosmologie und Astrophysik nicht mehr vorbeikommen. Voraussagen über abnorme Zustände der Materie macht die Physik an zwei Stellen: im astrophysikalischen Bereich treten als Endzustand von kollabierten Sternen und Galaxien *lokale* Singularitäten auf und im kosmologischen Gebiet zeichnet sich die Existenz einer *Anfangssingularität* ab, deren symmetrisches Gegenstück, die Endsingularität, aber noch nicht gesichert ist.

Die mathematischen Physiker haben sich bemüht, Klarheit in den Begriff der Singularität zu bringen, der erst einmal ein umfassender Ausdruck für die vielen Weisen ist, in der eine Raumzeit-Mannigfaltigkeit pathologisch sein kann.[35] Als Explikat verwendet man meist den Begriff der geodätischen Unvollständigkeit (GU), d.h. aus dem diskreten Spektrum der Raumzeit-Pathologien wird die GU-Eigenschaft zur Schärfung des Begriffes der Singularität ausgesucht. Es gibt drei Weisen, in denen eine Raumzeit geodätisch unvollständig sein kann: zeitartig, raumartig und null, wobei diese drei Arten keineswegs in einer Mannigfaltigkeit vereint sein müssen. Die zeitartige GU hat eine physikalisch anschauliche Bedeutung. In einer solchen Raumzeit gibt es eine Geodäte, auf der die Geschichte eines sich frei bewegenden Beobachters oder Teilchens nicht über ein finites Intervall der Eigenzeit fortgesetzt werden kann. Dies bedeutet noch nicht, daß an diesem Punkt der Raumzeit unendlich hohe Krümmung herrscht oder daß dort irgendwelche gewaltsame Vorgänge ablaufen, jedoch zeigt es an, daß sicher der Beschreibungsrahmen der differenzierbaren Mannigfaltigkeit hier überschritten wird. Die Null-GU einer Mannigfaltigkeit kann man in analoger Weise für Teilchen mit $m_o = o$ formulieren; die raumartige GU

findet, da die Existenz von Tachyonen unbestätigt ist, in unserer Welt
vermutlich keine Anwendung. Mittels dieser Explikation ergibt sich dann
als *Minimumbedingung* für die Singularitätsfreiheit einer Raumzeit-Mannigfaltigkeit deren zeitartige und Null-geodätische Vollständigkeit (GV).[36]

Eigenartigerweise gibt es auch Raumzeiten, die GV aufweisen, aber zeitartige Kurven von begrenzter Beschleunigung und finiter Länge haben. Ein
Beobachter, der einer solchen Weltlinie mit einer Rakete entlangfährt,
wird nach einem endlichen Zeitintervall nicht mehr durch einen Punkt der
Mannigfaltigkeit dargestellt. Auch hier sagt die Mathematik nichts über
katastrophenähnliche Vorgänge an diesem Endpunkt voraus; das exakte
Schicksal des Objektes, seine mögliche Vernichtung, ist nicht Gegenstand
des mathematischen Modells. Das Existenzende ist eine spezielle *Interpretation* des Ereignisses, das dem frei fallenden, aber auch dem Raketenbeob-
achter widerfährt,[37] der man sich aber nicht anschließen muß. Der Grund
liegt darin, daß die GU nicht gleichbedeutend mit dem Vorhandensein einer
Krümmungssingularität ist, wo die Komponenten des Krümmungstensors divergieren. So gibt es etwa im Taub-NUT-Raum unvollständige Geodäten, und
dennoch bleiben alle Krümmungsskalare $R^{abcd}R_{abcd}$ endlich. In vielen Fällen werden Singularitäten mit unendlich großen Gravitations-Gezeiten-
Kräften verbunden sein. In diesem Fall ist die Deutung des Endpunktes der
Weltlinie als von katastrophaler Bedeutung für den Inhaber der Raumzeit-
Bahn sicher gerechtfertigt.

Nach dem Vorstehenden könnte man immer noch vertreten, daß solche Pathologien der Raumzeit eigentlich philosophisch irrelevant sind; wenn in ir-
gendwelchen abstrusen Raumzeiten, die Mathematiker konstruiert haben, sonderbare Verhältnisse herrschen, muß dies doch für die reale Welt keine Bedeutung besitzen. Die eigentliche naturphilosophische Relevanz der Singularitäten zeigte sich, als mathematische Sätze gefunden wurden, die diesen
Pathologien einen hohen ontologischen Dringlichkeitsgrad verliehen: Unter
sehr realistisch-plausiblen Bedingungen scheinen sie nämlich unvermeidlich zu sein. Das entscheidende Theorem wurde 1970 von Hawking und Penrose
bewiesen.[38] Danach besitzt eine Raumzeit *(M,g)* keine zeitartige und Null-
GV, wenn vier Voraussetzungen erfüllt sind: die *Energiebedingung*, wonach die
Energiedichte an jedem Punkt nicht-negativ ist, die *Allgemeinheitsbedingung*,
wonach *M* nicht zu speziell symmetrisch sein darf, die *Kausalitätsbedingung*,
wonach es in *M* keine geschlossenen zeitartigen Kurven gibt, und die *Krüm-
mungsbedingung*, der entsprechend *M* einen Punkt p enthält, dessen Ver-
gangenheitslichtkegel rekonvergiert.

An diesem Theorem fällt auf, daß es die Existenz der Singularität unter
sehr allgemeinen Voraussetzungen sichert. Keine von den Vieren kann als

artifiziell angesprochen werden, wenn man unsere reale Welt im Auge hat. Möglicherweise könnte man daran denken, die Kausalitätsverletzung als Ausweg aus dem Singularitätendilemma zu wählen. Allerdings steht man dann vor der Frage, welche Welt seltsamer ist, eine solche, die Singularitäten enthält oder eine solche, in der es geschlossene zeitartige zukunftsgerichtete Weltlinien gibt, die die Möglichkeit von Zeitreisen, Doppelgängertum und dgl. liefern. Zwar ist es nicht richtig, daß die Zulassung kausaler Schleifen in jedem Fall zu logischen Widersprüchen führt[39] - wenn dies gälte, dann hätte die Kausalitätsbedingung ja rein formalen Charakter - aber in diesem Zusammenhang ist entscheidend, daß es ein Singularitäten-Theorem gibt, das die Existenz solcher Raumzeit-Stellen voraussagt, *ohne* von der Kausalitätsvoraussetzung Gebrauch zu machen.[40] Damit besteht die Alternative, wie sie das obige Theorem ausdrückt, nicht generell und ein Ausweichen auf eine komplexere Kausalstruktur mit Einschluß von Wirkungsketten in die Vergangenheit wäre kein Ausweg aus dem Singularitätenproblem. Am ehesten wird man sich noch überlegen, ob nicht die Energiebedingung ($\rho \geqslant 0$, $\rho + \sum p_i \geqslant 0$, welche ausdrückt, daß die Gravitation immer anziehend wirkt) in extremen Situationen außer Kraft gesetzt werden kann; Vorschläge hierzu gibt es bereits;[41] sie haben aber noch keine allgemeine Anerkennung gefunden.

Nimmt man die Singularitäten als deduktive Konsequenzen einer bewährten Theorie der Gravitation ernst, so steht man vor drei philosophischen Folgeproblemen:

1. Als *methodologische* Konsequenz ergibt sich, daß die Relativitätstheorie physikalische Situationen voraussagt, in denen ihre eigene Gültigkeitsgrenze erreicht ist. Die Existenz von Kollapsaren, wie sie durch die Röntgendetektoren aufgewiesen wurden, muß als positiver Testfall angesehen werden. Anders ausgedrückt, das Vorhandensein einer überschweren Masse von $M > 10\ M_\odot$ im kalten, von thermonuklearen oder Rotationskräften nicht stabilisierten Zustand wäre eine Falsifikationsinstanz der Relativitätstheorie. Insofern ist der Nachweis von schwarzen Löchern als Erfolg der Relativitätstheorie zu verbuchen. Andererseits treten im Innern eines solchen Objektes Zustände auf, bzw. werden von der Theorie vorausgesagt, denen man entsprechend früherer Erfahrungen mit dem Auftreten von Singularitäten anderer Theorien keine Aussagekraft zuzubilligen gewillt ist.

2. Eine *epistemische* Folge ergibt sich angesichts des Horizontes, der eine echte Krümmungssingularität, wie sie durch den Gravitationskollaps erzeugt wird, umgibt. Ist ein solcher Horizont, eine kausale Einwegmembrane, schon von vornherein ein in der lokalen Physik unbekanntes wunderliches Gebilde, so löst jener Fall, wo er fehlt und die Singularität ihre

Wirkungen in das äußere Universum absenden kann, noch viel stärkere Probleme aus. Der Kollaps hochasymmetrischer Körper kann eine sogenannte *nackte Singularität* erzeugen, die dadurch gekennzeichnet ist, daß ein Beobachter sich diesem Objekt beliebig nähern und wieder entfernen kann. Wenn unsere Welt nackte Singularitäten enthielte, träte eine Situation ein, bei der der Zustand der Welt durch vergangene Ereignisse nur mehr unvollständig bestimmt ist, die asymptotische Zukunftsvorhersagbarkeit wäre nicht mehr gegeben.[42] Da die physikalischen Prozesse in der Singularität völlig unbekannt sind, wäre ein Universum, das auch nur eine Singularität ohne Horizont beherbergt, *epistemisch unbestimmt*, da die auslaufenden Wirkungen völlig unberechenbar wären. Der Horizont hat epistemologisch gesehen danach eine Doppelfunktion: Er verhüllt die Natur der Singularität für den äußeren Beobachter, er verhindert aber auch, daß - nach den gegenwärtig bekannten Gesetzen - unkontrollierte Wirkungen in den Außenraum gelangen. In der ersten Rolle erzeugt dies eine Erkenntnis-Begrenzung für die empirische Erforschung der Welt, in der zweiten wird damit aber die Vollständigkeit der Erkenntnis für das äußere flache Universum gesichert. Die Versuche von Penrose, seine kosmische Zensorhypothese streng zu beweisen, wonach es keine nackte Singularität gibt und auch der stark asymmetrische Kollaps immer zu einem schwarzen Loch führt, sind bis jetzt nicht geglückt.[43] So muß man in der gegenwärtigen Situation zwei Fälle ins Auge fassen: Wenn die kosmische Zensorhypothese gilt, ist das Universum nur an diskreten Stellen epistemisch unbestimmt, zumindest so lange, wie keine strenge Theorie der Quanten-Gravitation existiert, aber das gesamte äußere Universum kann mit den vorhandenen Gesetzen bewältigt werden. Wenn die kosmische Zensorhypothese falsch ist, dann erweitert sich die epistemische Unbestimmtheit auf den gesamten Teil der Raumzeit, der von der nackten Singularität beeinflußbar ist. In diesem Fall muß man annehmen, daß das Universum nicht mehr völlig nomologisch determiniert ist. Die Ursache bestimmter zukünftiger Ereignisse hängt in keiner durch bekannte Gesetze beschriebenen Weise von den vergangenen Ereignissen ab.

3. Am wenigsten geklärt scheint noch der *ontologische* Status von Singularitäten. Die bis in die 60er Jahre vertretene Auffassung, daß Singularitäten Artefakte von Koordinatensystemen sind, hat sich nicht halten lassen. Man hat gelernt, zwischen Koordinatensingularitäten und nicht wegtransformierbaren echten Krümmungssingularitäten zu unterscheiden. Die Standardauffassung besteht heute darin, die Singularität als Gültigkeitsgrenze der allgemeinen Relativitätstheorie anzunehmen, d.h. diesen Voraussagen keine ontologische Dignität zuzuschreiben. Uneinig ist man sich aber noch, wo die Grenze zu ziehen ist. Ein Dimensions- und Analogieargument zur Quantenelektrodynamik besagt, daß die Mannigfaltigkeitsstruktur bis

maximal zum Krümmungsradius $\sim 10^{-33}$ cm und Dichten von 10^{94}g/cm^3 verwendet werden kann. Hawking und Ellis vermuten den Bruch[44] in der Mannigfaltigkeitsstruktur irgendwo zwischen 10^{-15} und 10^{-33} cm. Aber auch eine solche Verletzung des normalen Beschreibungsrahmens für physikalische Prozesse hätte revolutionäre Folgen. Konstruiert man nämlich eine Fläche um die Region, wo die Krümmung $< 10^{-15}$ cm ist, zerfiele das Universum in zwei disjunkte Teile. Auf der einen Seite wäre das normale Mannigfaltigkeitsbild der Raumzeit verwendbar, auf der anderen jedoch müßte eine unbekannte Quantenbeschreibung der Raumzeit in Kraft treten. Bezüglich des Materieaustausches zwischen den beiden Regionen ergäbe sich sogar: "Matter crossing the surface could be thought of as entering or leaving the universe, and there would be no reason why that entering should balance that leaving".[45] Selbst dann also, wenn Singularitäten nur eine begriffliche Existenzweise zugebilligt wird, sind die Folgen für das physikalische Weltbild einschneidend. Dieser Eindruck verstärkt sich noch, wenn die Singularitäten als tatsächliches Element der physikalischen Realität begriffen werden.[46] Schließlich ist die Relativitätstheorie eine bisher überall erfolgreiche Theorie - auch die lang gesuchten Gravitationswellen scheinen nun zumindest indirekt nachgewiesen worden zu sein -,[47] warum soll man ihr mißtrauen an einer Stelle, wo es höchstens Vorurteile philosophischer Natur sind, welche die Ablehnung begründen? Ist es etwa in dem Fall, da die Anfangssingularität von bösartigstem Typ ist, wonach alle Materie in endlicher Eigenzeit unendliche Dichte erfährt, wirklich unmöglich, diese Voraussage der Relativitätstheorie ontologisch ernst zu nehmen? Misner versucht in einer Analogie zur Temperatur die begriffliche Situation von theologischen Assoziationen zu lösen und damit weltanschaulich zu entkrampfen. Sicherlich, Entstehung und Schöpfung von Materie ist sprachlich klar zu trennen, beide Begriffe erfüllen ganz unterschiedliche syntaktische und semantische Strukturen.[48] Überdies kommen ex nihilo Entstehungsvorgänge auch in Theorien wie der Steady State-Theorie vor, die von ihrer Zeitstruktur her gar nicht in den theologischen Rahmen paßt. Dennoch wird man an der Behauptung Misners zweifeln müssen, daß die Einführung eines absoluten Nullpunktes der Zeit nur eine Gewöhnungsfrage ist, die nach einiger Zeit genauso wenig wie der absolute Nullpunkt der Temperatur einige Grade unter der Zimmertemperatur noch jemanden aufregt. Der Temperatur-Nullpunkt setzt keine prinzipielle Erklärungsgrenze, im Gegenteil, die molekularkinetische Auffassung der Materie läßt verstehen, warum es ihn gibt und warum er durch kein physikalisches System exakt erreicht werden kann. Im Gegensatz dazu ist die Annahme eines echten Beginns des Universums bei $t = 0$ die Setzung eines Ereignisses, das selbst nicht mehr Gegenstand einer wissenschaftlichen Erklärung werden kann.

Zeit und Temperatur spielen eben durchaus keine epistemisch symmetrischen Rollen in der Physik. Es gibt danach mindestens ein Ereignis im Universum, das unerklärbar bleibt, und zwar ist dies nicht irgendein beliebiges, auf dessen Verstehen man in Ruhe verzichten könnte, sondern dasjenige, das kausaler Vorgänger aller anderen physikalischen Ereignisse ist, das nur Nachfolger hat, selber aber keinem anderen Ereignis nachfolgt. Angesichts dieser erkenntnislogischen Situation ist es forschungsstrategisch wesentlich sinnvoller, das genetische Prinzip von Lukrez zu bemühen, wonach es *keine* absoluten Entstehungs- und Vernichtungsvorgänge gibt und das Universum von $t = -\infty$ bis $t = +\infty$ reicht.[49] Der big bang und der big crunch - wenn es ihn entgegen den astrophysikalischen Befunden doch geben sollte[50] - sind nach diesem Vorschlag nicht *ontologische* Elemente der Natur, sondern *epistemische* Konsequenzen der Theorie.

Diese Hypothek auf eine zukünftige Theorie der Quantengravitation ist in diesem Sinne ein progressiveres Forschungsprogramm als die Absolutsetzung einer mathematischen Folgerung der vorhandenen Theorie.

Wie immer man diesen letzten Punkt beurteilen wird, es müßte doch deutlich geworden sein, wie tief heute die Ergebnisse der theoretischen Physik in die Lösungsansätze alter philosophischer Probleme eingreifen.

<u>Anmerkungen</u>

1. L. Wittgenstein, Tractatus Logico-philosophicus, Nr. 6.53

2. L. Wittgenstein, Philosophische Untersuchungen, Nr. 43

3. K. R. Popper, The nature of philosophical problems and their roots in science. In: Conjectures and Refutations, London 1963, S. 66-96

4. W. v. O. Quine, Word and Object, New York 1960, S. 272

5. J.J.C. Smart, Between Science and Philosophy, New York 1968, S. 11

6. P. Suppes, Probabilistic Metaphysics, Uppsala 1974, S. 23

7. Streng genommen müßte man darunter noch eine Hierarchie von strukturalen Termen mit eingeengter Reichweite unterscheiden; so wird der Regelkreis in *einigen*, aber nicht in *allen* Objektwissenschaften Verwendung finden.

8. M. Planck, Die Einheit des physikalischen Weltbildes. In: Wege zur physikalischen Erkenntnis, Leipzig 1933, S. 1 - 32

9. A. Einstein, Kosmologische Betrachtungen zur allgemeinen Relativitätstheorie. In: Sitz. Ber. der Preuß. Akad. der Wissensch., Berlin 1917, S. 142 - 152

10. D. W. Sciama, Kosmologie. In: P. C. Aichelburg und R. U. Sexl (Hrsg.), Albert Einstein, Sein Einfluß auf Physik, Philosophie und Politik, Braunschweig 1979, S. 223

11. W. Rindler, Essential Relativity, 2. Aufl. New York 1977, S. 223

12. E. P. Hubble, A relation between distance and radial velocity among extragalactic nebulae, Proc. Nat. Acad. Sci. U.S. 15 (1929), 169

13. J. B. J. Fourier, Théorie analytique de la chaleur, Paris 1822. Dt. Ausgabe: B. Weinstein, Analytische Theorie der Wärme, Berlin 1884, vgl. v.a. S. XIV

14. Dies besagt nicht, daß in ihnen nicht bestimmte Konstanten vorkommen dürfen, sondern, daß die Gültigkeit eines Gesetzes nicht durch eine absolute Größenordnung begrenzt ist.

15. H. und R. U. Sexl, Weiße Zwerge - schwarze Löcher, Hamburg 1975, S. 60

16. A. Sandage and G. A. Tamman, Steps towards the Hubble constant I-VI, Astrophys. Journ. 190-197 (1974-1975)

17. D. W. Sciama, The Unity of the Universe, London 1959, S. 84

18. Zu diesem Begriff vgl. M. Bunge, Concepts of Model. In: Method, Model and Matter, Dordrecht 1973, S. 91 - 113.

19. J. E. Gunn and B. M. Tinsley, An Accelerating Universe, Nature 257 (1975), S. 454-457

20. K. Hübner, Kritik der wissenschaftlichen Vernunft, München 1978, S. 265

21. G. F. R. Ellis, The expansion of the universe, Mon. Not. Roy. Astr. Soc. 184 (1978), S. 439-465

22. "Wird also ein Urteil in strenger Allgemeinheit gedacht, d.i. so, daß gar keine Ausnahme als möglich verstattet wird, so ist es nicht von der Erfahrung abgeleitet, sondern schlechterdings a priori gültig". I. Kant, Kritik der reinen Vernunft, hrsg. von R. Schmidt, Hamburg 1956, S. 40.

23. J. A. Wheeler, Conference summary. In: G. Shaviv and J. Rosen, General Relativity and Gravitation, GR7, Tel Aviv 1974, S. 299-344

24. J. N. Bahcall and M. Schmidt, Does the Fine-Structure Constant Vary with Cosmic Time? Phys. Rev. Lett. 19 (1967), S. 1294 - 1295

25. M. K. Munitz, The Logic of Cosmology. Brit. Journ. Phil. Sci. 13, 49 (1962)

26. A. Einstein, Über den gegenwärtigen Stand der Feldtheorie. Festschrift für A. Stodola, Zürich 1929, S. 126

27. G. Lemaître, Un univers homogène de masse constante et de rayon croissant, rendant compte de la vitesse radiale des nebuleuses extragalactiques, Ann. de la Soc. Scient. de Bruxelles, Vol. 47, A (1927), S. 49 - 59

28. C. W. Misner, K. S. Thorne and J. A. Wheeler, Gravitation. San Francisco 1973, S. 800

29. C. W. Misner, Mixmaster-Universe, Phys. Rev. Lett. $\underline{22}$,20 (1969) S. 1071 - 74

30. E. R. Harrison, Cosmological Principles II. The Physical Principles. Comments on Astrophysics and Space Physics $\underline{6}$ (1974), S. 29

31. G. Schäfer and H. Dehnen, On the Origin of Matter in the Universe. Astron. and Astrophys. $\underline{54}$ (1977), S. 823 - 836

32. G. Schäfer, H. Dehnen, a.a.O., S. 823

33. W. Stegmüller, Ergebnisse der analytischen Philosophie und Wissenschaftstheorie, Band I, 2. Aufl. 1973, S. 113

34. M. J. Rees, The Collapse of the Universe: An Eschatological Study. The Observatory $\underline{81}$ (1969), S. 193

35. R. Geroch, What is a Singularity in General Relativity? Ann. Phys. $\underline{48}$ (1968) S. 526 - 540

36. S. W. Hawking and G. F. R. Ellis, The Large-Scale Structure of Space-Time, Cambridge 1973, S. 258

37. C. W. Misner, K. S. Thorne and J. A. Wheeler, Gravitation, a.a.O., S. 934

38. S. W. Hawking, R. Penrose, The Singularities of Gravitational Collapse and Cosmology. Proc. Roy. Soc. London A $\underline{314}$ (1979), S. 529 - 548

39. B. Kanitscheider, Philosophie und moderne Physik. Darmstadt 1979, Kap. III

40. S. W. Hawking and G. F. R. Ellis, The Large-Scale Structure of Space-Time. a.a.O., S. 272, Theorem 4

41. H. Dehnen and H. Hönl, The Influences of Strong Interactions on the Early Stages of the Universe. Astrophys. and Space Science $\underline{33}$ (1975), S. 49 - 73

42. S. W. Hawking, Gravitationally collapsed objects of very low mass. Mon. Not. Roy. Astr. Soc. $\underline{152}$ (1971), S. 75 - 78

43. Vgl. dazu R. Wald, Space-Time and Gravity. Chicago 1977, S. 86.

44. S. W. Hawking and G. F. R. Ellis, a.a.O., S. 359

45. S. W. Hawking, G. F. R. Ellis, a.a.O., S. 363

46. Ch. W. Misner, Absolute Zero of Time, Phys. Rev. $\underline{186}$,5 (1969) S. 1328 - 1333

47. J. H. Taylor, L. A. Fowler and M. Mc Culloch, Measurements of general relativistic effects in the binary pulsar PSR 1913 + 16, Nature $\underline{277}$ (1979), S. 437 - 440

48. B. Kanitscheider, Philosophisch-historische Grundlagen der physikalischen Kosmologie, Stuttgart 1974, S. 186

49. M. Bunge, Causality. Cambridge (Mass.) 1959, S. 24

50. Vgl. dazu H. Elsässer: Entdeckung von Staubscheiben als Geburts-
     stätten der Sterne - Konsequenzen für die Geburt des Alls?
     Naturwissenschaftliche Rundschau 32,4 (1979), S. 142 - 143

DIE ROLLE DER MATHEMATIK IN EINER PHYSIKALISCHEN THEORIE

Günther Ludwig, Philipps - Universität, Marburg

Die Rolle der Mathematik in einer physikalischen Theorie ist nicht von
vornherein deutlich erkennbar, wenn man die Sprechweise der Physiker beobachtet. Diese Sprechweise scheint zunächst ein unentwirrbares Durcheinander von sprachlichen Formulierungen zu sein, die teils mathematischen
Charakter haben, teils aber gerade einen Bedeutungsinhalt zu beschreiben
versuchen. Zu alledem kommt noch eine weitere "Methode" des Redens hinzu,
bei der in telegrammartiger Sprechweise mathematisch komplizierte Sachverhalte nur sehr kurz "angedeutet" werden.

Geht man deshalb der Bedeutung der Mathematik in der Physik näher nach,
so erkennt man durchaus *mehrere* Seiten dieser Bedeutung. Da wird z.B. die
Mathematik benutzt, um konkrete Probleme, z.B. gerade auch Probleme der
Technik (d.h. der Anwendung der Physik) zu "berechnen". Da wird in mathematischer Form eine Begrifflichkeit entwickelt, um aus dieser Begrifflichkeit heraus, d.h. aus dieser mathematischen Vorstellungswelt heraus
Vorschriften und Normen für den Umgang des Menschen mit seiner natürlichen Umwelt zu entwickeln. Aber es gibt neben diesen immer auch etwas
technisch orientierten "Anwendungen der Mathematik" noch eine andere
Bedeutung der Mathematik in der Physik, nämlich als Erkenntnishilfsmit-
tel, was wir jetzt ein wenig genauer betrachten wollen. Diese Seite der
Anwendung von Mathematik in der Physik wollen wir als "physikalische
Theorie" bezeichnen.

Eine physikalische Theorie, kurz PT genannt, besteht in diesem Sinn
aus drei Teilen: einer mathematischen Theorie (kurz mit MT bezeichnet),
einer Anwendungsvorschrift (kurz mit (——) gekennzeichnet) und einem Bereich, den die Theorie beschreiben will und den wir den Wirklichkeitsbereich der Theorie (kurz mit W bezeichnet) nennen. In diesem Sinn ist
also PT = MT (——) W. Wir wollen versuchen, diesen Aufbau einer PT
in kurzer Form zu schildern; eine ausführliche Darstellung findet der
interessierte Leser in [1].

Zur Übersicht über unsere kurze Schilderung möge das folgende Diagramm
dienen, auf das wir im Text immer wieder hinweisen werden.

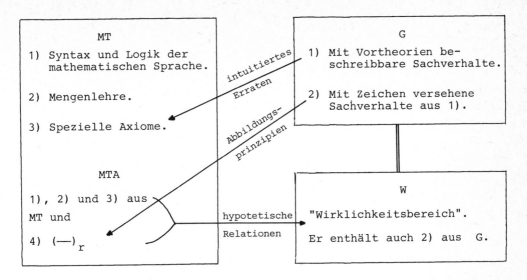

Wie schon zu Anfang betont, entspricht der jetzt zu schildernde Aufbau einer PT nicht dem "Zustand" physikalischer Theorien, wie man sie so in Lehrbüchern antrifft, sondern stellt einen gewissen Idealzustand dar, für dessen korrekte Formulierung sich die Physiker meistens keine Zeit nehmen, obwohl für sie dieser Idealzustand eigentlich immer mehr oder weniger bewußt als Leitbild dient.

Im Idealzustand einer PT erscheint eine mathematische Theorie MT zunächst als "etwas, was als mathematisches Gebilde *in sich* definiert ist, d.h. wohl definiert ohne jeden Bezug auf Anwendungen in der Physik. In diesem Sinn ist MT festgelegt durch die Syntax und Logik der mathematischen Sprache, durch die Axiome der Mengenlehre *und* durch weitere spezielle Axiome. Diese speziellen Axiome wählen die Mathematiker aus Mathematik-internen Gründen aus, d.h. (um es etwas überspitzt auszudrücken) aus Gründen, um "schöne" Sätze beweisen zu können. Im Rahmen einer PT werden die Axiome auf eine andere Art und Weise gefunden, die im obigen Diagramm mit dem Pfeil "intuitives Erraten" angedeutet ist. Wir werden am Schluß unserer Betrachtungen noch einmal auf dieses intuitive Erraten zurückkommen. Wie in der Mathematik üblich, setzen wir MT als in sich widerspruchsfrei voraus, obwohl wir heute natürlich wissen, daß eine solche Widerspruchsfreiheit nicht bewiesen werden kann, außer in relativer Form, d.h. relativ zu anderen als widerspruchsfrei vorausgesetzten Theorien.

Für eine physikalische Theorie ist von entscheidender Bedeutung die "Anwendung" von MT. Aber was heißt hier Anwendung? Sieht man sich das Vorgehen der Physiker an, so könnte man den Eindruck gewinnen, als ob die mathematischen Symbole selbst zu physikalischen Sachverhalten werden, weil die Physiker oft mathematische Symbole mit physikalischen Begriffen zu "bezeichnen" pflegen. Mathematische Symbole *direkt* mit physikalischen Sachverhalten *zu identifizieren*, wäre aber methodisch unsauber, da man offensichtlich in der Mathematik mit nur "gedachten" Dingen und Relationen und nicht mit realen Sachverhalten umgeht. Es kann also nur so sein, daß die Physiker mathematische Symbole als etwas wie "Zeichen" für physikalische Sachverhalte benutzen. Das bedeutet, daß zunächst unabhängig von MT so etwas wie physikalische Sachverhalte gegeben sein müssen. Diesen *vor* der Anwendung von MT gegebenen Bereich physikalischer Sachverhalte wollen wir als Grundbereich (kurz G) bezeichnen (siehe obiges Diagramm).

G wird auch manchmal der Anwendungsbereich der PT genannt. Eine begrifflich scharfe Abgrenzung von G im Rahmen von PT ist so gut wie nie möglich. Dies stört oft die Wissenschaftstheoretiker, ist aber typisch für die Physik. Diese müssen wir aber vor Augen behalten, wenn wir jetzt versuchen, G etwas näher zu schildern. Weiter unten werden wir auf das Abgrenzungsproblem noch einmal zurückkommen.

G enthält also die schon vor der betrachteten Theorie PT mögliche begriffliche Formulierung von festgestellten physikalischen Sachverhalten. Solche Sachverhalte können sowohl in der "freien" Natur gegeben sein als auch erst durch menschliche Handlungen zustandegekommen sein. Auf jeden Fall müssen es vorgekommene Sachverhalte und keine hypothetisch gedachten Sachverhalte sein. Die betrachteten Sachverhalte können entweder in der Alltagssprache beschreibbar und in diesem Sinn "unmittelbar" vorweisbar sein, oder aber sie sind mit Hilfe anderer Theorien (sogenannter Vortheorien, "vor" relativ zu der betrachteten PT) als Tatsachen aus den Wirklichkeitsbereichen dieser Vortheorien beschreibbar.

In einem nächsten Schritt muß man diese Sachverhalte aus G in die Sprache der MT aus PT übersetzen. Dieses Übersetzen geschieht mit Hilfe sogenannter Abbildungsprinzipien. Die sprachliche Formulierung dieser Abbildungsprinzipien ist ein gesondertes Problem. Es geschieht in zwei Schritten: Zunächst muß man bestimmte Sachverhalte aus G mit Zeichen versehen und zwar verschiedene Sachverhalte mit verschiedenen Zeichen. Liegt in G eine Formulierung mit Hilfe von Vortheorien vor, so sind schon Zeichen vorhanden, da es sich um Formulierungen in der Sprache der mathematischen Theorien aus den Vortheorien handelt; man hat dann nur darauf zu achten, daß nicht "versehentlich" Zeichen gleich sind. Liegen aber "unmittelbar" gegebene Sachverhalten vor, so muß man sozusagen neue Zeichen an bestimmten Sachverhalten anbringen. In einem zweiten Schritt sind dann die Feststellungen aus G mit Hilfe dieser

Zeichen in im Rahmen von MT sinnvoll formulierte Relationen zu übersetzen. Wir wollen dies hier nicht im Einzelnen erläutern (siehe [1]
§ 5), sondern nur an einem Beispiel verdeutlichen.

Zunächst ein simples Beispiel: MT sei definiert durch eine Menge M,
durch zwei einstellige Relationen s(x) und r(x) über M und durch
das einzige Axiom: $\forall x(r(x) \Rightarrow s(x))$. Der Grundbereich G sei "unmittelbar" gegeben und als begriffliche Beschreibung in G sei bekannt, was
ein Vogel ist, was ein Rabe ist und was "schwarz sein" bedeutet. Die vorgefundenen Vögel (die wir in diesem Falle nicht selber machen können) bezeichnen wir mit Buchstaben a_1, a_2, ... a_n als Zeichen wie mit einem
Vogelring. Anderen Sachverhalten wie z.B. Katzen und Hunden geben wir
erst gar keine Zeichen, da wir sie sowieso nicht bei der Anwendung unserer Theorie benutzen. Das Umschreiben der vorgefundenen Sachverhalte
in die Sprache von MT geschieht dann so, daß wir für: "Der Vogel mit
dem Zeichen a_i" in MT die Relation "$a_i \in M$" hinzufügen; für "Der Vogel
mit dem Zeichen a_k ist schwarz" fügen wir in MT hinzu: $s(a_k)$; für "der
Vogel a_l ist nicht schwarz" fügen wir "nicht $s(a_l)$" hinzu; ähnlich für
"der Vogel a_m ist ein Rabe" die Relation "$r(a_m)$".

Die wie an diesem Beispiel erläutert neu zu MT hinzugefügten Relationen, die Sachverhalte aus G beschrieben, haben wir in dem obigen
Diagramm kurz mit $(-)_r$ bezeichnet. MT zusammen mit den zusätzlichen
"Axiomen" $(-)_r$ bildet eine neue mathematische Theorie, die wir MTA
nennen wollen. MTA ist nie abgeschlossen, da immer neue Erfahrungen
hinzugefügt werden können. Für eine brauchbare Theorie erwarten wir,
daß MTA widerspruchsfrei ist, was man in der Physik meist so auszudrücken pflegt: Theorie und Experiment stimmen überein.

Das obige Beispiel mit den Vögeln ist sehr simpel, deshalb sei noch kurz
skizziert, wie etwa bei der Newtonschen Mechanik sich gravitierend an-
ziehender Massen vorgegangen werden kann. MT setzen wir als bekannt
voraus (siehe auch [1] § 7.3). Um diese Theorie z.B. mit der Bewegung
der Planeten, Monde und eventuell frei fliegender Raumschiffe zu vergleichen, muß G schon mit Hilfe von Vortheorien der raumzeitlichen
Vermessung festgelegt werden. Die gefundenen Orte zu verschiedenen Zeiten
sind auf ein Koordinatensystem umzurechnen, das nicht relativ zum Fixsternhimmel rotiert und in dem die "Sonne ruht". Das so vermessene Material aus G wird zunächst in der Weise mit Zeichen versehen, daß
jeder der Planeten, Monde, Raumschiffe ein Zeichen bekommt (im einfachsten Fall wird dabei oft eine Nummer benutzt). Dann fügt man die in G
vorliegenden Meßergebnisse in bekannter Weise in der Sprache von MT
zu MT hinzu. Dabei sind allerdings "Ungenauigkeiten" zu beachten, um

nicht Widersprüche zu erhalten. Wie dies geschieht, ist den Physikern geläufig; in [1] § 6 ist die Methode solcher "unscharfen Abbildungen" allgemeiner diskutiert.

Das letzte Beispiel zeigt, daß es keine großen Probleme dabei gibt, die in G schon in mathematischer Form vorliegenden Orts- und Zeitkoordinaten als Relationen in MT auszudrücken. Als begrifflich "neuartiger" Schritt bleibt allein, die Objekte "Erde", "Venus", usw. wie die einzelnen Raumschiffe mit Zeichen zu versehen, um so in der Sprache von MT formulierbare Relationen über "Massenpunkte" zu erhalten.

An beiden Beispielen kann man sich noch zwei typische Züge physikalischer Theorien verdeutlichen: erstens die Tatsache, daß "widersprüchliche Einzelfälle" nicht beachtet werden; und zweitens, daß die Logik beim Vergleich der Theorie mit Erfahrungen, d.h. beim Formulieren der Relationen aus $(—)_r$ nur eine sehr primitive Rolle spielt.

Quantoren wie z.B. "alle" treten *nur* in den Axiomen aus MT auf und *nicht* bei Aussagen über physikalische Sachverhalte. Die in $(—)_r$ aufzuschreibenden Relationen enthalten dagegen nur die logischen Verknüpfungen "und" und "nicht". Treten in $(—)_r$ Relationen auf, die zu einer widersprüchlichen Theorie MTA führen, so kann man sich fragen, ob man durch Weglassen einer oder nur "sehr weniger" Relationen aus der meist unübersehbaren Fülle von Relationen aus $(—)_r$ zu einer widerspruchsfreien MTA gelangen kann. Ist dies der Fall, so kann man einfach diese "wenigen "Sonderfälle" aus $(—)_r$ streichen. Z.B. möge in dem ersten Beispiel ein einziger Rabe festgestellt worden sein, der nicht schwarz ist. Dies ist für die Theorie kein Beinbruch. Die Relation $\forall x(r(x) \Rightarrow s(x))$ mit dem Quantor "alle" gilt nur in MT; man läßt dann für den Vogel a_α mit "$r(a_\alpha)$" die eine Relation "nicht $s(a_\alpha)$" einfach weg. Die Theorie bleibt weiter brauchbar. Diese "physikalische" Methode des Vergleichs von Theorie und Erfahrung ist deshalb sinnvoll, weil sowieso eine "all"-Aussage in MT nicht "physikalisch ernst" gemeint ist, d.h. eine Idealisierung in MT darstellt, die nur zum Ausdruck bringen soll, daß es nur "sehr selten", d.h. so gut wie fast nie vorkommt, nicht schwarze Raben vorzufinden.

Aber auch die Tatsache, daß in G oft nicht über Tatsachen exakt entschieden werden kann, sei es weil vorliegende Messungen mit Hilfe der Vortheorien nicht zu einem eindeutigen Resultat führen, oder sei es, weil die benutzten Begriffe keine klare Entscheidung ermöglichen (z.B. bei der Entscheidung ob ein Vogel schwarz sei oder nicht schwarz sei), stört das angegebene Verfahren nicht, da in $(—)_r$ eben *nur* diejenigen Tatsachen (in der Sprache von MT) aufzuschreiben sind, die eindeutig festgestellt werden konnten. Zweifelsfälle sind immer auszuscheiden.

Es ist nicht zu bestreiten, daß es in der Physik häufiger vorkommt, daß bei der Interpretation einer Theorie, d.h. bei dem Vergleich von MT mit Erfahrungen aus G das eben geschilderte Verfahren nicht eingehalten wird, z.B. weil man Arbeit sparen will. Natürlich hat dies manchmal die Konsequenz, daß Interpretationen unklar oder sogar falsch werden können. Die Allgemeine Relativitätstheorie ist dafür ein Beispiel: Légère Interpretationsverfahren benutzen ausgesuchte Koordinatensysteme mit der intuitiven Behauptung, daß in diesen die Interpretation "klar sei"; z.B. indem man einen "gekrümmten" Lichtstrahl zeichnet, der in der Nähe der Sonne vorbeigeht, so als ob man diese "krumme" Linie mit der Erfahrung vergleichen könnte. Tatsächlich muß man in der Allgemeinen Relativitätstheorie auf die lokalen Beobachtungssysteme wie z.B. eine Sternwarte zurückgehen. Diese lokalen Systeme sind aber schon im Rahmen der Vortheorien beschreibbar, wie z.B. der Aufbau und die Aufstellung eines Fernrohres durch Vortheorien erfaßbar ist. Die Interpretation hat dann zu erfolgen mit Hilfe der lokal feststellbaren Tatsachen, die in die Sprache der zur Allgemeinen Relativitätstheorie gehörigen MT ohne Interpretationsschwierigkeiten übersetzt werden können; z.B. durch Vergleich der Richtungen (!) des von Fixsternen einfallenden Lichtes mit und ohne Sonne.

Nach der Schilderung der Interpretation einer PT können wir nun noch einige Worte über das Problem der Abgrenzung des Grundbereiches sagen. Aus der Schilderung der Interpretation geht hervor, daß auf keinen Fall solche Sachverhalte zu G zu rechnen sind, bei denen die Abbildungsprinzipien versagen, d.h. bei denen nicht klar ist, wie man sie in die Form der Axiome (—)$_r$ zu übersetzen hätte. Das Hauptproblem entsteht aber dort, wo man zwar Tatsachen in Form von (—)$_r$ aufschreiben kann, diese aber zu Widersprüchen mit MT führen. Wollte man in solchen Fällen eine PT "wegwerfen", so müßte man alle physikalischen Theorien wegwerfen. Ein moderner Physiker wird überhaupt nicht erwarten, daß eine PT die Wirklichkeit "exakt" beschreibt, sondern vielmehr MT nur als ein *approximatives* Bild ansehen. Diese approximative Gültigkeit zeigt sich einmal darin, daß man beim Aufschreiben von (—)$_r$ *Ungenauigkeits*mengen berücksichtigen muß *und* daß man die Anwendung der Theorie auf einen Gültigkeitsbereich, d.h. auf einen *abgegrenzten* Grundbereich G beschränken muß.

Reichen die durch Vortheorien beschreibbaren Ungenauigkeiten aus, um für einen abgegrenzten Grundbereich G keine Widersprüche zu erhalten, so nennt man die Theorie eine in G im Rahmen der bisher erreichten

Meßgenauigkeiten bestätigte Theorie. Reichen die Meßungenauigkeiten nicht aus, so kann man *größere* Ungenauigkeitsmengen benutzen, um Widerspruchs-freiheit mit den Sachverhalten aus G zu erhalten. Diese Ungenauigkei-ten faßt man dann als ein Maß dafür auf, wie gut MT die Tatsachen aus G zu beschreiben gestattet. Eine solche endliche Güte von PT ist immer ein Ansporn, die Theorie PT zu verbessern, d.h. eine zu PT umfang-reichere PT' zu finden. Wie hierbei das Wort "umfangreicher" gemeint ist, können wir jetzt nicht genauer schildern (siehe dazu z.B. [1] § 8).

Es darf aber nicht die Tatsache bestritten werden, daß die Physiker auch ohne Kenntnis einer umfangreicheren Theorie allein auf Grund von Erfah-rungen im Umgang mit der Theorie PT in der Lage sind, Ungenauigkeiten und Grenzen von G zu beschreiben, ja sogar oft quantitativ abzuschät-zen, wie die Grenzen von G von den Ungenauigkeiten abhängen, d.h. wie mit Vergröberungen der Ungenauigkeiten die Grenzen von G erweitert wer-den können. Ja, ein sehr wesentlicher Teil der Bemühungen der Experimen-talphysik liegt gerade auf diesem Gebiet.

Dennoch ist es klar, daß vom theoretischen Standpunkt her die Abschätzung von Ungenauigkeiten und Grundbereich der alten Theorie PT erst dann be-grifflich deutlicher hervortritt, wenn eine umfangreichere Theorie PT' vorhanden ist.

Im historischen Entwicklungsprozeß der Physik ergeben sich oft tiefgrei-fende Probleme der Interpretation eines mathematischen Bildes MT , weil man sich erst allmählich herantasten muß, wie der Vergleich von MT mit schon "vor" dieser Theorie beschreibbaren Tatsachen zu erfolgen hat. Die Quantenmechanik ist ein typisches Beispiel hierfür. Obwohl schon *Bohr* im Prinzip den richtigen Weg erkannte, daß zur Interpretation der Quanten-mechanik die "klassische" Beschreibung der Meßapparate notwendig ist, wurde nachträglich wieder dieser Weg blockiert durch den Versuch, die Quantenmechanik als "allgemein gültig" hinzustellen. So kommt die Dis-kussion um die Interpretation der Quantenmechanik zu keinem endgültigen Abschluß. Wie es möglich ist, auch die Quantenmechanik auf eine Weise zu interpretieren, die sich auf einen durch Vortheorien beschreibbaren Grundbereich bezieht, ist in [2] dargestellt.

Noch klarer als oben dargestellt läßt sich die Interpretation einer PT herausarbeiten, wenn MT in Form einer "axiomatischen Basis" vor-liegt. Die Einführung einer axiomatischen Basis hat aber einen noch wesentlich tieferen Grund, der in dem obigen Diagramm durch das Käst-chen "Wirklichkeitsbereich" charakterisiert ist. Dazu kommt noch die Frage nach der "Möglichkeit" physikalischer Prozesse.

Wir wollen versuchen kurz zu erklären, was wir unter einer axiomatischen

Basis verstehen wollen (nähere Einzelheiten kann man in [1] § 7.3 finden). Dazu führen wir zunächst den Begriff äquivalenter Theorien ein.
Zwei Theorien PT_1 und PT_2 sollen äquivalent heißen, wenn

1. der Grundbereich G derselbe ist und damit die in G
vorliegenden Sachverhalte durch Abbildungsprinzipien
$(—)_1$ bzw. $(—)_2$ sowohl in MT_1 wie MT_2 ausdrückbar
sind; und wenn

2. "MT_1A widerspruchsfrei $\Longleftrightarrow MT_2A$ widerspruchsfrei"
für alle möglichen Anwendungen A gilt.

So sind z.B. folgende zwei Theorien äquivalent: Die Allgemeine Relativi-
tätstheorie mit MT_1 in der "alten" Koordinatenschreibweise und die
Allgemeine Relativitätstheorie mit MT_2 in einer koordinatenfreien
Schreibweise.

Als *axiomatische Basis* einer PT bezeichnet man eine solche äquivalente
Form, in der MT folgenden Bedingungen genügt:

Die in MT benutzten Mengen und Relationen zur Formulierung der Aussagen $(—)_r$ (die man deshalb kurz Bildmengen und Bildrelationen nennt),
sollen gerade die in MT eingeführten Basismengen (d.h. in MT nicht
aus anderen Mengen hergeleiteten Mengen) und die in MT eingeführten
Grundrelationen, d.h. die sogenannten Strukturterme in MT sein. Die
in MT mit Hilfe dieser Grundmengen und Grundrelationen formulierten
speziellen Axiome (siehe das Diagramm am Anfang) heißen dann *Naturgesetze*.

Liegt eine PT nicht in Form einer axiomatischen Basis vor, so brauchen
also die Grundmengen und Grundrelationen aus MT keine, zumindest keine
unmittelbare physikalische Bedeutung zu haben, da die Abbildung von G
auf die Relationen $(—)_r$ mit Hilfe abgeleiteter Mengen und abgeleiteter
Relationen erfolgt. So liegt z.B. die Quantenmechanik normalerweise nicht
in Form einer axiomatischen Basis vor, wenn man als MT die Theorie eines Hilbertraumes H benutzt. Es ist unklar, ob die Elemente von H
eine physikalische Bedeutung haben und was die Relationen wie Addition
von Vektoren, Multiplikation von Vektoren mit komplexen Zahlen und das
innere Produkt $\langle x,y \rangle$ physikalisch bedeuten könnten. Zumindest können
alle diese Größen nicht mit dem durch Vortheorien zur Quantenmechanik
beschreibbaren Grundbereich der Herstellungs- und Meßapparate der Mikrosysteme *direkt* in Beziehung gesetzt werden. Vielleicht lassen sie sich
als indirekte feststellbare Strukturen der Mikrosysteme interpretieren,
wenn man von einer axiomatischen Basis ausgeht (siehe z.B. [2]).

Aber was heißt "indirekt messen"? Mit Recht hat *Einstein*, einmal in einer

Unterhaltung mit Heisenberg betont, daß erst eine Theorie festlegt, was alles gemessen werden kann.

Indirekte Messungen werden in der Physik heutzutage in nicht mehr übersehbarer Fülle durchgeführt. Z.B. werden relativ zur Theorie der Elektrodynamik elektrische Ladungen, elektrische Ströme, elektrische und magnetische Felder indirekt gemessen über die direkte Messung z.B. von Kräften.

Gerade in einer axiomatischen Basis läßt es sich am saubersten definieren, was man mit indirekt messen meint, da ja die Grundmengen und Grundrelationen auf der Basis des Grundbereiches "direkt interpretierbar" sind. Deshalb ist auch jede in einer axiomatischen Basis MT ableitbare Struktur auf Grund ihrer Ableitung physikalisch interpretiert. Läßt sich dann ein abgeleiteter Term E eindeutig durch eine MTA (d.h. MT mit geeigneten experimentellen Daten A angereichert) festlegen, so kann man E als "indirekt gemessen" bezeichnen, gemessen mit Hilfe der in A aufgeschriebenen, allein schon in G feststellbaren Sachverhalte. Es kann aber sein, daß ganz verschiedene A_1, A_2, ... dasselbe E festlegen, d.h. daß es verschiedene "Meßmethoden" zur indirekten Messung von E geben kann (siehe genauer [10] §§ 10.4 und 10.5).

Unter der Menge des auch indirekt Gemessenen verstehen wir dann den Wirklichkeitsbereich W einer Theorie (siehe das Diagramm am Anfang). So kann ein sogenanntes "schwarzes Loch" zu einem "wirklichen" kosmischen Objekt werden auf der Basis der Allgemeinen Relativitätstheorie *und* von terrestrisch durchgeführten Messungen A . Ja, für den Physiker wird auf diese Weise sogar die Metrik "innerhalb" eines solchen schwarzen Loches indirekt meßbar, obwohl lokal, d.h. sozusagen von den Stellen innerhalb des Loches keine direkte Information mehr nach außen dringen kann; es genügt aber, daß durch Beobachtung A und Theorie MT die globale Situation *festgelegt* ist.

Aber nicht nur Wirklichkeitsaussagen, sondern auch modale Aussagen über mögliche Vorgänge können mit Hilfe einer PT mit axiomatischer Basis MT gemacht werden. Aber wie geschieht das, obwohl doch in MT keine modalen Aussagen auftreten?

Es ist unmöglich, dies hier genauer und systematisch zu schildern; deshalb sei der Leser auf [1] § 10.4 verwiesen. Es sei hier nur kurz auf Aussagen in und über MT hingewiesen, die man physikalisch als modale Aussagen über die Wirklichkeit interpretiert. In MTA können Aussagen der Form auftreten: "Es gibt Elemente ...". Solche Aussagen können unter gewissen Bedingungen zur Interpretation führen: Diesen Elementen entsprechende physikalische Sachverhalte sind "möglich". Oder über MT

und eine Relation R kann folgender Sachverhalt ausgesagt werden: R
kann ohne Widerspruch zu MTA hinzugefügt werden und auch "nicht R "
kann ohne Widerspruch zu MTA hinzugefügt werden. Auch solche Situationen können zu Möglichkeitsaussagen über physikalische Sachverhalte
führen.

Mit Hilfe des mathematischen Bildes MT in Form einer axiomatischen
Basis ist es in der hier nur angedeuteten Weise möglich, die üblicherweise in der Physik benutzten modalen Aussagen zu gewinnen. Dies ist
besonders wichtig für die Quantenmechanik. Mit Hilfe einer axiomatischen
Basis für die Quantenmechanik ist es so z.B. möglich, folgende modalen
Aussagen zu gewinnen (siehe hierzu [1] und [2]): Hat man ein Elektron
präpariert, d.h. hergestellt, so *ist es möglich*, den Ort oder den Impuls zu messen. *Es ist* aber *physikalisch ausgeschlossen*, daß Ort und
Impuls zusammen gemessen werden. Während *über die Möglichkeiten*, Ort
oder Impuls zu messen, *verfügt* werden kann, sind die verschiedenen *Ansprechmöglichkeiten* einer wirklich aufgestellten Ortsmeßapparatur nicht
mehr verfügbar, da diese Ansprechmöglichkeiten einer Wahrscheinlich-
keitsstruktur unterworfen sind (siehe auch [1] § 12).

Mit diesen letzten Bemerkungen sind wir bis an den Grenzen dessen vorge-
stoßen, was ein mathematisches Bild MT innerhalb einer PT methodisch
leisten kann, um das "physikalische Sprechen" auf eine festere Basis zu
stellen.

Zum Schluß müssen wir aber auf zwei wichtige Probleme hinweisen, die
nicht mit Hilfe von MT lösbar, ja deren Lösung oder Entscheidung schon
"vorher" gefunden sein muß.

Es ist einmal das Problem, das in dem am Anfang dargestellten Diagramm
mit dem Pfeil "intuitives Erraten" gekennzeichnet ist. Für die Entdeckung
neuer oder umfangreicherer Theorien gibt es keine Methode noch irgendwelche Normen; alles ist erlaubt. Ja, gerade etwas "Verrücktes" ist gesucht; denn jedes der schon vorhandenen Bilder MT enthält Idealisierun-
gen, die immer dort eingeführt werden, wo ein physikalisches Problem zunächst ungelöst bleiben mußte (siehe [1] § 9). Lösung eines solchen
Problems heißt daher immer Umstoß einer vorher gemachten, vielleicht
auch sehr vertraut gewordenen Idealisierung. In diesem Sinne kann die
Entwicklung der Physik schon anarchisch und revolutionär wirken. Tatsächlich aber zeigt sich *nachträglich* nur, daß die neue Theorie umfangreicher als eine alte Theorie ist, ja daß man den Anwendungsberich G
der alten Theorie erst so recht im Rahmen der neuen Theorie abschätzen
kann. So gesehen wiederum wird keine alte Theorie umgestoßen; vielmehr
läßt die neue Theorie "mehr" an Strukturen der Wirklichkeit sichtbar

werden als die alte Theorie.

Ein zweites Problem ist ebenfalls durch den formalen Aufbau PT = MT(——)W einer PT nicht lösbar, nämlich das Problem der Anerkennung einer Theorie. Und doch haben wir bei allen unseren Überlegungen immer vorausgesetzt, daß wir keine "falsche" Theorie betrachten. Wir haben immer so getan, als ob es klar war, daß wir durch MT etwas von der Wirklichkeit erkennen können. Alle von Wissenschaftstheoretikern vorgeschlagenen Kriterien für die Anerkennung einer Theorie haben sich als unzulänglich erwiesen. Als Physiker mußte man immer wieder erklären, daß kein Kriterium die persönliche Entscheidung der Physiker voll trifft, eine Theorie endgültig zu akzeptieren. Der oben angedeutete Entwicklungsprozeß, der eben doch als immer umfangreichere Erfassung von Wirklichkeitsstrukturen gedeutet werden *kann*, bestätigt die endgültige Anerkennung von Theorien, auch wenn der Entwicklungsprozeß selbst nicht immer geradlinig verläuft.

Aber nicht nur, daß wir eine PT als richtige Beschreibung von Wirklichkeitsstrukturen anerkennen, ist notwendige Voraussetzung für die auf den vorigen Seiten gebrachten Überlegungen; um modale Aussagen machen zu können, müssen wir PT als eine *abgeschlossene* Beschreibung des geeignet abgegrenzten Grundbereichs G voraussetzen, d.h. voraussetzen, daß keine Naturgesetze "vergessen" wurden. Solche Entscheidungen sind viel schwieriger zu fällen; und auch Physiker sind sich manchmal nicht einig in der Beurteilung, ob eine Theorie abgeschlossen sei. Die eigentliche Meinungsverschiedenheit zwischen *Bohr* und *Einstein* (nachdem in Diskussionen einige Bedenken gegen die Wiederspruchsfreiheit der Quantenmechanik ausgeräumt werden konnten) bestand darin, daß *Bohr* die Quantenmechanik als Theorie von Atomen und Molekülen für abgeschlossen hielt, während *Einstein* bis zum Ende seines Lebens davon überzeugt war, daß die Quantenmechanik nicht abgeschlossen ist, eben "unvollständig" ist, wie es *Einstein* formulierte. Das sogenannte *Einstein-Podolski-Rosen-Paradoxon* ist nicht widersprüchlich im Rahmen der Quantenmechanik (im Rahmen einer axiomatischen Basis für die Quantenmechanik läßt es sich durchsichtig beschreiben; siehe [2]), sondern soll eben im Sinne der Verfasser "zeigen", daß die Quantenmechanik nicht abgeschlossen sein könne. Vielleicht war aber tatsächlich der Zustand der Quantenmechanik vor der Aufstellung einer axiomatischen Basis zu schlecht, um in überzeugender Weise eine solche Diskussion um die Abgeschlossenheit der Quantenmechanik führen zu können.

References

[1] G. Ludwig, Die Grundstrukturen einer physikalischen Theorie,
     Berlin-New York, Springer 1978.

[2] G. Ludwig, Einführung in die Grundlagen der Theoretischen Physik,
     Kapitel XIII, Vieweg 1976.

  -, A Theoretical Description of Single Microsystems, in: W.C. Price,
     S.S. Chissick, (Eds.), The Uncertainty Principle and Foundations
     of Quantum Mechanics, John Wiley, 1977.

  -, Foundations of Quantum Mechanics, Berlin-New York: Springer
     (2.ed. of Grundlagen der Quantenmechanik; Translation in pre-
     paration).

  -, Axiomatische Basis der Quantenmechanik, Berlin-New York: Springer,
     (in preparation).

THE MATHEMATICAL ORIGINS OF GENERAL RELATIVITY AND OF UNIFIED FIELD THEORIES

Elie G. Zahar, London School of Economics, Great Britain

Abstract:

In this paper I discuss the heuristic role which mathematics plays in physical disco-.
very: first through the surplus structure which mathematics injects into physical principles which are given a mathematical formulation; secondly , through the realist interpretation of certain mathematical entities which appear at first sight to be devoid of
any physical meaning. I then try to account for this dual role of mathematics in terms
of a single philosophical principle, namely Meyerson's principle of identity. I finally
apply these considerations to the study of two important questions; the questions namely
of the continuity between STR and GTR (STR = Special Theory of Relativity, GTR = General
Theory of Relativity) and of the emergence both of General Relativity and of the Unified
Field Theories of Weyl, Eddington and Schrödinger-Einstein.

I   Role of Mathematics within Physics

In their preface to *Methods of Mathematical physics* Hilbert and Courant
express their regret that, towards the beginning of the XXth Century,
all fruitful interplay between mathematics and physics had almost stopped.
They write:

> *Since the seventeenth century physical intuition has served as a vital
> source for mathematical problems and methods. Recent trends and fashions
> have however weakened the connection between mathematics and physics;
> mathematicians, turning away from the roots of mathematics in intuition,
> have concentrated on refinement and emphasised the postulational side
> of mathematics, and at times have overlooked the unity of their science
> with physics and other fields. In many cases physicists have ceased to
> appreciate the attitudes of mathematicians. This rift is unquestionably
> a serious threat to science as a whole...*

It is well-known that empirical science has often exerted a beneficial
influence on the development of mathematics. Physics sets problems for
which urgent mathematical solutions are required; as a result certain
branches of pure mathematics receive a powerful impetus and develop
rapidly. For a long time mathematics and physics were regarded as identical disciplines; or rather people were not clear about the differences
between them. Euclidean geometry for example was taken to describe the
properties of real, albeit idealised, space. Its postulates, as distinct
from the axioms, were accepted because they supposedly expressed self-
evident truths about physical space. In Einstein's words,[1] geometry
constituted one of the oldest physical theories. In the preface to his
Principia Newton treats geometry as a branch of mechanics, *i.e.* as a
branch of physics: 'Therefore geometry is founded in mechanical practice
and is nothing but that part of universal mechanics which accurately
proposes and demonstrates the art of measuring.'

In the above examples physics played the dominant role in that it was
physics which gave rise to the mathematical problems and to a large
extent dictated the type of mathematics to be used. In his Theory of
Fluxions[2] Newton constructed the Calculus specifically for the study
of differentiable motions: the fluent variable was the time and the
fluxion the instantaneous velocity. So Analysis, *i.e.* the discipline
which dominated mathematical thinking for over two centuries owes it
origins to physics. Similarly, the study of differential equations and
the development of what is now known as Advanced Calculus were closely
connected with the articulation of the Newtonian Programme in the XVIII th
century. A similar process took place in the XIXth century when Faraday,
using 'line of force' as a new physical concept, enunciated laws of which
Maxwell later gave a mathematical formulation. This powerfully contributed to the development of vector algebra and vector analysis. These
examples thus illustrate the heuristic role of physics with regard to
mathematics; in each of the cases so far considered it was physics which
led to the discovery of an appropriate mathematical theory. It looks as
if, whenever mathematics and physics have interacted, the latter set the
pace. Has this always been the case? I would like to examine whether the
reverse process has ever taken place, *i.e.* whether the solution of a
mathematical problem has ever led to the discovery of a physical theory.
Prima facie, this process ought to be possible. There is no reason why
the mathematical rendering of some vague physical principle should not
unexpectedly capture more of the truth than the scientist was initially
aware of. This question concerns the heuristic role of mathematics vis-
à-vis physics. There are two important ways in which mathematics further
physical discovery:

I.1   Increase of content through translation into a mathematical lan-
       guage.

As mentioned above, the scientist may start from an intuitive physical
principle. Through being translated into some mathematical theory available at the time, the principle is generally modified; it may in particular acquire some surplus structure[3]   and thus become a stronger physical assumption. For example, Fresnel set out to give a mathematical
formulation to his conjecture that light is a wave process through the
ether. He instinctively resorted to the periodic function with which he
was most familiar, namely sin x . His original assumption that light
is a wave phenomenon is obviously weaker than his final hypothesis that
the wave is representable by the function $\sin(2\pi t/T)$.

It has often happened that certain physical hypotheses which require very specific mathematical theories for their formulation were proposed after, and sometimes immediately after, the mathematical theories were elaborated. This seemingly pre-established harmony between the mathematics and the subsequent physics has been a cause for puzzlement among historians and philosophers of science. One example of such a coincidence is the relationship between the mathematics of Hilbert space and the formulation of Quantum Mechanics. Another very good example is the development of Riemannian Geometry prior to the emergence of GTR. It was as if Gauss, Riemann, Ricci and Levi-Civita had Einstein in mind when constructing their non-Euclidean geometries. Einstein himself considered his discoveries as a continuation of Gauss's and Riemann's work. Yet Einstein was a physicist and Riemann a pure mathematician. In his *Geometry and Experience* Einstein strictly distinguished between geometry qua mathematical system on the one hand and geometry qua physical theory on the other. It is prima facie strange that mathematicians should anticipate the work of later physicists. The solution of this puzzle is two fold. On the one hand there is no pre-established harmony between mathematics and physics. The physicist forces, or impresses, his principles into an existing mathematical framework. The physical principles thus acquire the surplus structure of the mathematical system which is used. When Einstein decided he could not account for gravity in Special Relativistic terms, he turned to his friend Grossmann, a working mathematician steeped in the Riemannian tradition. It is small wonder that Grossmann used Riemannian geometry to cast the field equations for empty space in the form $R_{ij} = 0$. In this way Grossmann and Einstein geometrised gravitation. The harmony between mathematics and physics is thus forcibly established, not *pre*-established. On the other hand, however, Riemannian geometry had to be of such a nature as to lend itself to the use to which it was put by Einstein and Grossmann. In other words, this geometry had to have an empirical or a quasi-empirical character which allowed physics to be imbedded in it. That this was in fact the case is clearly indicated by the following quotation from Riemann's *Über die Hypothesen, welche der Geometrie zugrunde liegen*:

> *Es wird daraus hervorgehen, daß eine mehrfach ausgedehnte Größe verschiedener Maßverhältnisse fähig ist und daß der Raum also nur einen besonderen Fall einer dreifach ausgedehnten Größe bildet. Hiervon aber ist eine notwendige Folge, daß die Sätze der Geometrie sich nicht aus allgemeinen Größenbegriffen ableiten lassen, sondern daß diejenigen Eigenschaften, durch welche sich der Raum von anderen denkbaren dreifach ausgedehnten Größen unterscheidet, nur aus der Erfahrung entnommen werden kann ... Diese Tatsachen sind wie alle Tatsachen nicht notwendig, sondern nur von empirischer Gewissheit, sie sind Hypothesen; man kann also ihre Wahrscheinlichkeit, welche innerhalb der Grenzen der Beobachtung allerdings sehr groß ist, untersuchen...* [4] *(p. 272-272).*

Riemann's approach to geometry is quasi-empirical in that it is based
on the notion of measured distance between neighbouring points. Riemann
physicalised mathematical geometry, thus making it possible for Einstein
and Grossmann to geometrize gravitation. We should however remark that
empirical considerations alone, *i.e.* without idealising a priori assumptions, do not determine the structure of Riemannian Geometry. Riemann
himself realised that the differentiability condition for example is not
derivable from experience.

> *Ich werde diese Aufgabe nur unter gewissen Beschränkungen behandeln*
> *und beschränke mich erstlich auf solche Linien, in welchen die Ver-*
> *hältnisse zwischen den Größen dx - den zusammengehörigen Änderungen*
> *der Größen x - sich stetig ändern ... Die Frage über die Gültig-*
> *keit der Voraussetzungen der Geometrie im Unendlichenkleinen hängt*
> *zusammen mit der Frage nach dem inneren Grunde der Maßverhältnisse*
> *des Raumes. Bei dieser Frage, welche wohl noch zur Lehre vom Raume*
> *gerechnet werden darf, kommt die obige Bemerkung zur Anwendung, daß*
> *bei einer diskreten Mannigfaltigkeit das Prinzip der Maßverhältnisse*
> *schon in dem Begriffe dieser Mannigfaltigkeit enthalten ist, bei*
> *einer stetigen aber anders woher hinzukommen muß.*[5]

There is another assumption of whose a priori character Riemann seems
to be unaware, namely that space is determined by a metric, *i.e.* by one
fundamental symmetric tensor g_{mn} . That this is an a priori assumption
is shown by the fact that, after 1916, purely affine geometries were
developed which could accommodate the facts then known to Einstein.

Let me say that I am by no means committed to the Kantian notion of the
apriori. I use 'apriori' in the sense of 'not imposed by experience'.
This conception of the a priori is perfectly compatible with Mach's.
Mach[6] identified a priori knowledge with the instinctive, innate knowledge which may have arisen through the gradual adaption of the mind to
its surroundings, *i.e.* through natural selection. Thus, for Mach, a
priori knowledge is inheritable, *i.e.* genetically imbedded in the organ-
ism.

Going back to Riemann, we see that metricity and differentiability are
a priori assumptions which belong to the surplus structure of geometry
with respect to physics. Thus, forcing physics into the Riemannian mould
imposed more constraints than are warranted by observational results.
Does this imply that the later hypotheses of Weyl, Eddington and Schrödinger were more 'empirical' or more 'inductive', than GTR ? By no
means. Metricity was not replaced by laws induced from experience, but
by different *geometrical* assumptions. Weyl[7] links the presence of the
electromagnetic field to the change in *length* of a *parallely transported*
vector. Eddington proposes to explain gravitation and electromagnetism
by means of the *symmetric and anti-symmetric parts* of the Ricci *tensor*.[+8]

As is well known, the Ricci tensor is obtained by *contraction* from the curvature tensor B^m_{ijn} ; and B^m_{ijn} is obtained from the affinity Γ^m_{ij} by considering the variation of a vector taken round an infinitesimal loop. Eddington derives the symmetry in u and v of Γ^m_{uv} from a *commutativity* condition. Schrödinger[9] constructs his theory by applying a variational principle to a certain integral; where the integrand is what he and Eddington consider as the simplest *invariant density* built out of the curvature tensor, namely $\sqrt{-\det[R_{ij}]}$. I mention all these details in order to make the following point: length, parallel transport, contraction, commutativity, tensor densities and variations of integrals are *mathematical* notions not directly abstracted from experience. The physicist operates at the mathematical level, hoping that his operations mirror certain features of reality. However, he is not very clear as to how this mirroring takes place, so he lets himself be guided by the syntax, or by the symbolism, of some mathematical system.

Of course, these seemingly abstract unified field theories have novel empirical consequences. It is precisely because Weyl's hypothesis had some undesirable physical consequences that Einstein rejected it. This illustrates the point already made that physics is strengthened by mathematical surplus structure.

Mathematical speculation thus seems to have played a dominant role in the development of Einstein's programme. However, it can also be said that the problems which physics faced after 1916, *e.g.* the fact that GTR does not treat gravitation and electromagnetism on a par, gave rise to new types of non-Riemannian geometry. Can we not say with equal right that physics played a dominant heuristic role with regard to mathematics? In my opinion the relationship between mathematics and physics is best described in dialectical terms as a to and fro movement between two poles. One moves from physical principles to idealising mathematical assumptions; then back to some more physics; then forward to fresh mathematical innovations with ever increasing surplus structure. The so called harmony between physics and mathematics is not a miracle but the result of an arduous process of mutual adjustment.

I.2   Relalistic Interpretation of Mathematical Entities.

There is a second way in which mathematics can play a fundamental role in physical discovery. The usual method in theoretical physics is to give mathematical expression to some physical hypothesis, then use logico-mathematical techniques to draw consequences from the hypothesis. In doing this, the physicist may resort to a number of mathematical operations; these are sometimes in the nature of 'tricks' or 'gimmicks' which simplify the deductive process. Duhem[10] pointed out that it would be foolish to insist on giving a physical interpretation to *all* mathematical quantities and operations used in a scientific theory. Duhem is evidently right: adding lengths corresponds to juxtaposing rods; multiplying lengths may correspond to the construction of rectangular areas but multiplying t by $\sqrt{-1}$, through convenient, does not appear to be susceptible of any physical interpretation. However, through trying to find a *realist* interpretation of certain mathematical entities which seem at first sight to be devoid of any physical meaning, the scientist may be led to a physical conjecture.[11] This can occur in two different ways.

II.2a   The first and straightforward way is through an increase in the empirical content of a given theory. By being realistically interpreted, part of the mathematical scaffolding becomes physically meaningful, hence testable, at least in principle. The theory remains syntactically unchanged, but its observational content, *i.e.* its contact with 'observable' reality, is extended. Dirac's equation for the electron, his explanation of spin and especially his discovery of the positron are a case in point. Dirac proposed a relativistic equation which was found to possess negative energy solutions. Prima facie such solutions cannot be physically interpreted. By insisting on interpreting the negative solutions, Dirac predicted the existence of the positron: the absence of an electron charge - e and energy - E was interpreted as the presence of a positron, that is of a particle of charge + e and energy + E .

II.2b   The second way in which the realistic interpretation of mathematical entities leads to new discoveries is both more complex and more interesting than the previous one. The theory itself is altered in this process; *i.e.* the syntactical expression of the physical principles themselves does not escape unscathed. This usually occurs as follows. We start from some hypothesis H which we may want to modify for reasons we need not go into. We express H in an equivalent form[12] , H*(t) say, which brings out a certain mathematical entity t . t is then given a realist interpretation which subsumes it under a philosophical category, *e.g.* the category of substance. This category obeys some very general laws, *e.g.* conservation or symmetry laws. It is found that H*(t) violates these laws. A breakthrough is achieved when H*(t) is modified into H'(t) , where H'(t) is chosen in such a way that it conforms to the laws in question. Let us give an example: by using variational methods, Einstein re-expressed his field equations for free space, namely $R_{im} = 0$, in the form:

$$\frac{\partial}{\partial \chi^n} (g^{mp}\Gamma^n_{ip}) = - (t^m_i - \frac{1}{2} \delta^m_i t)$$

where $t \underset{\text{Def}}{=} t^m_m$ and the coordinates are chosen in such a way that $g = -1$ throughout the frame of reference. It turns out that the matrix t^m_i obeys a conservation law; *i.e.* that

1. $t^m_{i/m} = 0$, where $t^m_{i/m} \underset{\text{Def}}{=} \frac{\partial}{\partial \chi^m} (t^m_i) = $ ordinary divergence of t^m_i .

2. Thus: $(R_{mi} = 0) \Longleftrightarrow (\frac{\partial}{\partial \chi^n} (g^{mp}\Gamma^n_{ip}) = - (t^m_i - \frac{1}{2} \delta^m_i t))$.

For the case where non-gravitational energy is present, Einstein had ge-neralised $R_{im} = 0$ to:

3. $R_{im} = - kT_{im}$ where T_{im} is the (non-gravitational) stress-energy tensor.

Einstein realised that 2 is inadmissible, because the right-hand side of the equation, but not in general its left hand side, has vanishing divergence. It turns out that:

4. $(R_{im} = -k \, T_{im}) \Longleftrightarrow (\frac{\partial}{\partial \chi^m} (g^{np}\Gamma^m_{ip}) = (-(t^n_i - \frac{1}{2}\delta^n_i t) - kT^n_i))$.

In view of the conservation law 1 , Einstein interpreted t^m_i as a gravitational stress-energy matrix, despite the fact that t^m_i is frame-dependent. (t^m_i is not a tensor). He then noticed that the right member of 4. is not symmetric in T^m_i and t^m_i , that symmetry ought to ob-tain if both entities represent stress-energy. Einstein consequently modified the field equations into

5. $\frac{\partial}{\partial \chi^n} (g^{mp} \, \Gamma^n_{ip}) = - \left[(t^m_i + kT^m_i) - \frac{1}{2} \delta^m_i (t + T) \right]$

where the right-hand side is now symmetric in t^m_i and kT^m_i . It turns out that 5. is equivalent to:

6. $R_{im} - \frac{1}{2} g_{im} R = - kT_{im}$

and that $(t^m_i + kT^m_i)$ obeys a conservation law, namely:

7. $(t^m_i + kT^m_i)\big|_m = 0$. Note that 6. constitute the 'correct' field

equations.

Thus, against Duhem, I maintain that the attempt to interpret mathematical entities realistically proves fruitful even in situations where prima facie such an attempt seems bound to fail. (non-tensor character of t_i^m).

II   Meyerson's Principle of Identity

In this section I shall give a brief exposition of Meyerson's philosophy of science. I shall argue that this philosophy accounts *in terms of a single principle* both for the applicability of mathematics to physics and for the role which conservation laws and certain philosophical notions like that of substance play in the progress of science. This single Meyersonian principle allows of many formulations which, though distinct, bear to one another a sort of family resemblance. According to Meyerson,[13] all explanations, whether scientific or commonsensical, spring from one basic tendency of the human mind; namely the tendency to deny diversity and change; or to assert the existence of constants behind the fleeting appearances; or to explain the Many in terms of the One; or to subsume the flux of Becoming under the immutability of Being. The best formulation, in my opinion, is that the human mind inevitably tends to deny diversity and assert sameness or identity both in space and in time. For the human mind, only the undifferentiated One is real, everything else is appearance. This sounds like the worst kind of metaphysical doctrine which is bound to be of no relevance to science. May I however ask the reader to be patient, suspend disbelief for the time being and let me examine how much of the role which mathematics and conservation laws play in physics can be derived from Meyerson's principle of identity? Note that I have spoken of this principle as of a tendency inherent in the human mind. If this tendency went unchecked, it would assert the existence of a unique Parmenidean Sphere which is undifferentiated in space and changeless in time; it would thus degrade all phenomena to the rank of mere sensory illusions. Becoming would thus be mere illusion and only Being would be real. The whole of science can be conceived as a sequence of attempts to rescue as much of Parmenides's sphere as is compatible with the phenomena. It is well-known that Greek Atomism postulates the void together with a plurality of particles each of which can be assimilated to a Parmenidean sphere. Both the atoms and the void are immutable in themselves. The flux of appearances is explained in terms of different configurations of the same particles within the same empty space. Let A and K stand for two consecutive states of affairs which, following Meyerson, will henceforth be referred to as the antecedent and the consequent respectively. Thus, according to Atomism, the differences between A and K are to be attributed to different distributions of the same atoms within an unchanging void. Classical Mechanics accentuates the parity between A and K by showing that the process leading from A to K is reversible. A reversal of all the velocities in K would lead back to A .

Every reduction in the number of different atoms, every theory proposing say that atoms differ only in mass and charge, are hailed by science as great steps forward. This is because each such step involves a reduction of diversity, *i.e.* an increase in homogeneity. In the same spirit Einstein proposed to define simplicity in terms of the paucity of the primitive elements occurring in a given theory. In Meyersonian language, the number of primitives is a measure of diversity. Already at this stage, we recognise the close affinity which exists between Meyerson's principle of identity on the one hand and, on the other, Einstein's unity principle which provided the heuristic of Unified Field Theories.

The progress of science is seen by Meyerson as a struggle between the mind which tries to impose identity and a differentiated reality which resists such an imposition. The mind cannot completely succeed in denying all diversity; but some aspects of reality allow of being subsumed under the principle of identity; what particular aspects can be determined only a posteriori, *e.g.* by trial-and-error methods. The general identity principle however is a priori.

I.1   Legal Explanations

The identity principle applies at two distinct levels: at the second-order level of laws and at the first-order level of things. This gives rise to what Meyerson calls legal explanations and causal explanations respectively. Legality means that the *same* law applies irrespective of place and time, *i.e.* the form of the law is both a-spatial and a-temporal. There is thus permanence or constancy of form: many diverse phenomena illustrate, or manifest, the same underlying law which constitutes their Platonic Form.[14]   For example, the falling stone, the moon revolving round the earth and the passenger thrown forward in a decelerating train are all instantiations of one and the same gravitational law. This permanence of the rule of law enables us to predict, hence has high survival value. According to Meyerson, every animal species needs for its survival this ability to anticipate facts. Thus he gives an evolutionary reason for the apriori character of the principle of identity as applied to laws. This principle is a priori in the sense of being inborn in a species which needs it for survival.

II.2   Causal Explanations and Conservation Laws

The human mind goes much further than assserting the permanence of natural laws; it searches for causal explanations, which Meyerson defines as
those under which certain substances, or things, are conserved; *i.e.* the
total quantum of these substances remains unchanged through time. Conservation laws thus flow from the principle of identity qua negation of
the diversity of things in time. I have said that Meyerson calls this
version of the principle the causal one: the effect is equal to, *i.e.* it
contains as much substance as, its cause.

Meyerson doubts that the causal principle, as distinct from the merely
legal one, has any survival value. He thinks that phenomenological laws
are all that is needed for a species to gain control over nature. The
causal tendency is a luxury which cannot be explained in biological
terms. It can however be argued that the postulation of substances which
perdure in time enables us to keep track of things, hence predict the
sensations which these things supposedly cause in us. To assume that
the tree exists even when I do not look at it enables me to explain, and
thus predict, what I shall perceive when I turn round and look at it
again. Mach claimed that the concept of material object can be replaced
by that of constant functional relationships between elements of sensation; but even he admitted that concepts like that of an electron having constant mass and charge effect great economy of thought and thus
have great pragmatic value. Moreover, the notion of constant functional
relationship, even if it could replace that of substance, is a cumber-
some notion introduced merely for reasons of philosophical purism; it
is far removed both from commonsense and from scientific realism which
remain our best guides for action. Thus both the legal and the causal
versions of the identity principle have great survival value.

Meyerson describes the way in which we construct conservation laws as
follows. The mind fixes on certain processes in nature which it hyposta-
sises; *i.e.* it turns these *processes* into *substances* or *things* whose
total quantity it then assumes to be constant. For example, commonsense
anchors sense-data in material objects which are supposed to perdure in
time. This commonsense hypothesis forms the basis of the classical law
of the conservation of mass. Another example is that of the principle
of inertia.[15]   For Aristotle motion was change and velocity a ratio
describing this process of change. Multiplying this ratio by mass,
Descartes turned it into a *thing*, which he called quantity of motion,
and whose sum total in the world remains constant. This constancy was
derived from God's immutability which is another name for the principle

of identity. Similarly, energy is hypostasised motion and position
within a field of force (kinetic and potential energies). This auto-
matically leads to a new conservation law, the conservation of energy
law. Note that in classical physics matter is the primary substance,
while momentum and energy derive their existence·from matter (and charge)
in motion. Special Relativity puts matter and energy on a par; so we can
suppose that there exists a unique substance which our senses apprehend
sometimes as radiant energy say and sometimes as hard impenetrable
matter. Special Relativity thus enhances the status of energy by making
it a constituent part of physical reality. This establishes a link between the Special Theory on the one hand and, on the other, General Relativity where energy considerations played an important part in the
discovery of the correct field equations. In connection with Relativity
however we should sound a note of warning: substances are generally
described not by single quantities but by tensors; the fact that tensor
equations are frame-independent reflects the absolute character of substances; conservation is expressed by the vanishing of the tensor's
divergence.

Summing up, we can say that the method of applying Meyerson's identity
principle to evolution in time is to single out some process common to
the antecedent A and to the consequent K ; then to hypostasise this
process, calling it say $\phi(A)$ and $\phi(K)$; finally to assert that
$\phi(A)$ = $\phi(K)$, *i.e.* to assert a conservation law. $\phi(A)$ may for example stand for the total mass in A , or the total momentum or energy.
We have seen that the human mind has a tendency to deny all forms of
change; if it could have its own way, it would affirm the identity of
A and K and thus abolish time (Note that, for Meyerson, time is
measured by change). Since this proves impossible, the mind resorts to
the next best thing, namely the assertion that A and K are equivalent under certain aspects. The relation $\phi(A) = \phi(K)$ is obviously an
equivalence relation.

This method of hypostasising processes is also used in mathematics. For
a long time functions were taken to be processes in which the value of
an independent variable x gave rise to the corresponding value of
$y = f(x)$. This process is clearly referred to by the word 'transformation' which is often used synonymously with 'function'. By defining
a function as a class of n-tuples, set theory reified this process into
a thing. Hermann Weyl wrote: 'Thus we are able to subordinate genetic
construction to the static existence of relations.[16] It is obvious that
the a-temporal character of mathematics makes it ideally suited to the
expression of theories based on the identity principle.

II.3   The Geometrisation of Nature

So far we have considered the identity principle as applied to change
in time, *i.e.* to temporal diversity. Let us now examine its application
to diversity in space, or more generally to diversity in some domain of
simultaneous events. Consider for example the simultaneous perception of
two qualities. Two so-called secondary qualities like colour and sound
strike us as being irreducibly different in a way that two frequencies
are not. There is continuity between two waves giving rise to two distinct sensations like those of colour and sound, whereas the sensations
themselves seem to be separated by an unbridgeable gap. This is why the
mind, which is hell-bent on eliminating diversity, tries to reduce all
secondary qualities to the primary ones. This however proves insuffi-
cient. Only spatio-temporal properties are deemed completely intelligible. Mass and charge are felt to be occult, or rather heterogeneous
relatively to space and time. Hence the Cartesian attempt to reduce
matter to extension and explain all physical phenomena in terms of figure and movement, *i.e.* in kinematical terms. As is well-known, this
Cartesian attempt failed. It was supplanted by the Newtonian system
which postulates space, time, mass and force as the primitive concepts.
With the emergence of electromagentism, charge and ether were added to
this list. The rise and development of the Relativity Programme can be
explained in Meyersonian terms as follows. General Relativity and the
sequence of Unified Field theories are the realisation of the old
Cartesian ideal. According to this view, the Cartesian enterprise failed
not because it was wrong-headed but because the underlying geometry was
too simple, too poor, to account for the diversity of the observed
phenomena. By complicating their geometry, Einstein, Weyl, Eddington
and Schrödinger proposed to construct global explanations of all physical phenomena. Meyerson's philosophical framework provides a rationale
as to why mathematics played such an important role in the creation of
the new systems, and why the Relativists tried time and again to force
physics into one geometry; they were trying to dissolve the qualitative
heterogeneity or diversity of the phenomena into the homogeneity of one
unified geometry.[17] This homogeneity does not of course mean that the
metric or the affine connections of the geometry do not change from one
point to the next; but this is the sort of continuous numerical change
which offends least against the mind's dislike of diversity. Anyway,
Hermann Weyl minimises the significance of this dependence of the metric
on each point of space-time in the following terms:

> *The nature of the metric is one and is absolutely given; only the
> mutual orientation in the various points is capable of continuous*

changes and dependent upon matter. Euclidean space may be compared to
a crystal, built up of uniform unchangeable atoms in the regular and
rigid unchangeable arrangement of a lattice; Riemannian space to a
liquid, consisting of the same indiscernible unchangeable atoms, whose
arrangement and orientation, however, are mobile and yielding to forces
acting upon them. [18]

Mach's definition of mass as negative inverse ratio of accelerations can
be seen as a first step towards the geometrisation of physics. [19] Einstein's explanation of gravity and inertia in Riemannian terms was of
course the major success of the Relativity Programme. The field equations
of GTR , namely $R_{mn} - \frac{1}{2} g_{mn} R = -kT_{mn}$ can be interpreted in two
very different ways. Reading them from right to left we can say that
energy determines, or at least modifies, geometry. However, we can also
read the equations from left to right and say that, once the geometry is
given, *i.e.* once the g_{mn}'s are given, then the energy content of space
is *defined*. This is one way of eliminating the right hand side of the
equations, *i.e.* the qualitative side, with which Einstein himself was
disatisfied. He wrote:

> *By this formulation one reduces the whole mechanics of gravitation to*
> *the solution of a single system of covariant partial differential equa-*
> *tions. The theory avoids all internal discrepancies which we have charged*
> *against the basis of classical mechanics. But, it is similar to a build-*
> *ing one wing of which is made of fine marble (left part of the equation),*
> *but the other wing of which is built of low grade wood (right side of the*
> *equation). The phenomenological representation of matter is, in fact, only*
> *a crude substitute for a representation which would correspond to all*
> *known properties of matter. (Out of my later years, p. 81.)*

Einstein was clearly giving the geometrical wing precedence over the
material one. In the same spirit Eddington had written:

> *According to the new point of view Einstein's law of gravitation does not*
> *impose any limitation on the basal structure of the world. $G_{\mu\nu}$ may*
> *vanish or it may not. If it vanishes we say that space is empty; if it*
> *does not we say that momentum or energy is present; and our practical*
> *test whether space is occupied or not - whether momentum and energy exist*
> *there - is the test whether $G_{\mu\nu}$ exists or not.* [20]

That this is not only Eddinton's idiosyncratic view is shown by the
follwing quotation from Schrödinger's 'Space-Time Structure':

> *I would rather you did not regard these equations $\left[that\ is$*
> *$- (R_{ik} - \frac{1}{2} g_{ik} \cdot R) = T_{ik} \right]$ as field equations, but as a definition of*
> *T_{ik} , the matter tensor. Just in the same way as Laplace's equation*
> *$div\ E = p$ (or $\nabla^2 V = - 4\pi p$) says nothing but: wherever the divergence*
> *of E is not zero we say there is a charge and call div E the densi-*
> *ty of charge. Charge does not cause the electric vector to have a non-*
> *vanishing divergence, it is this non-vanishing divergence. In the same*
> *way matter does not cause the geometrical quantity which forms the first*
> *member of the above equation to be different from zero, it is this non-*
> *vanishing tensor, it is described by it. (p. 99.)*

The Relativists tried to subject electromagnetism to the treatment which they had meted out to gravitation, *i.e.* they tried to imbed the electromagnetic field into geometry. Maxwell himself had looked upon the charge density p as the divergence of the electric field E ; and we have just seen that this was also Schrödinger's view. By trying to derive the electromagnetic field from a geometry which would also explain gravitation, the Relativists were, in Meyersonian terms, attempting to dissolve the specificity of electrical phenomena into the homogeneity of the same underlying kinematics. Matter and charge, which had appeared to us as possessing irreducibly different qualities, would now be fused into one global geometry. Putting it in theological language: God created *one* geometry, whose various aspects are apprehended by our *senses* in different ways. This is nothing but the culmination of the Cartesian programme. The Meyeronian scheme can be summed up as follows:

Principle of Identity:

(A)   *Identity of Form:* Legal Explanations (constancy of functional relationships, phenomenological laws)

(B)   *Identity of Substance:* Causal Explanations

(B1) *Identity of Substance in Time:* Classical Conservation Laws (Matter, Momentum, Energy)

(B2) *Identity of Substance in Space:* Unity of Matter (Cartesian Reduction of Matter to Space)

(B3) *Identity in Space-Time:* Geometrisation of Nature (General Relativity, Unified Field Theories, Block Universe)

Whatever one might otherwise think of it, Meyerson's philosophy explains why scientists tried to force their physical principles into a geometrical mould and consequently injected into these principles some of the surplus structure of the mould (See I.1 above). Before leaving Meyerson one final question should be asked: is it in principle possible for us to explain the whole of reality in geometrical terms? Meyerson admits that it is not. First the geometry itself, if it is to *define* matter, charge and energy, must contain arbitrary elements. If one adopts Eddington's or Schrödinger's approach, then one can give no reason why the metric g_{mn} or the affinity Γ^i_{mn} should assume any specific values at specific points; it is no longer possible to say it is because of the presence of matter or of charge, for this very presence is to be *defined* in terms of the g_m's or of the Γ^i_{mn}'s . One can for example, give no reason why space should be curved and not flat. Thus the values of the metric or of the affinity are contingently given, they are simply posited.

Secondly: although we have banished all qualities from physics, we have not thereby eliminated them but pushed them into a no man's land between physics on the one hand and physiology or psychology (or both) on the other. We have simply shifted the burden onto the so-called body-mind problem. In Aristotle's philosophy we could make sense of why certain bodies are *perceived* as hot say; according to Aristotle, this is because they *are* hot. In Relativity we have to *accept* that certain geometrical properties of the physical world surface as sensations at the level of consciousness. According to Meyerson, the transition from kinematics to sensation constitutes an irrational jump which will always remain incomprehensible to us. Moreover, physics does not completely succeed in banning all qualities from its domain: in testing physical hypotheses, we have to interpret certain propositions in operationalist terms. Hence there exist bridging principles between certain physical statements on the one hand and sense-data on the other. Finally, in its attempt to establish reversibility, and thus a kind of parity between part and future, physics has utterly failed. The entropy law underlines the fundamental asymmetry between past and future. Thus, according to Meyerson, nature sets definite limits to the applicability of the identity principle, *i.e.* to the principle of the negation of diversity, which remains an apriori tendency of the human mind. If one accepts that the identity principle has essential survival value, then nature has to comply with it only to the minimum extent that would enable the species to survive.

III   Continuity between STR and GTR.

I would now like to tackle an important problem whose solution turns out to be connected with the heuristic role of mathematics within physics; the problem namely of whether there exists any genuine continuity between the Special and the General Theories of Relativity. In other words: is there a unified Relativity Programme or does a *single word* denote a heterogeneous body of hypotheses simply because they were put forward by the same man?

III.1   The heuristic role of the Covariance Principle

One facile answer to the above question is that the Relativity Principle provides a continuous thread running through all the stages of the programme; the principle, namely, that all frames of reference belonging to some class K are physically equivalent. The General Theory simply extends this principle from the set of inertial frames to that of all possible - accelerated as well as unaccelerated - systems. The programme would thus be identified by a hard core, by one assumption shared by all scientists working in it. Although the Relativity Principle as it stands is both vague and metaphysical, it is well-known that it played an important regulative role in the Special Relativistic modifications of the laws of mechanics. However, when interpreted to mean only that all the laws of physics should be covariant with respect to some group of transformations, the principle turns out to be nearly empty. In 1917 Kretschmann pointed out that all physical laws can be covariantly re-written and Einstein accepted this criticism as valid. It looks there-fore as if the Relativity Programme had lost its most distinctive fea-ture since any hypothesis whatever can now be regarded as a relativistic theory. This first attempt to answer our question therefore seems to end in disaster. In his reply to Kretschmann Einstein asserted that the co-variance principle should be treated as a heuristic device rather than as an ordinary statement. Of all the covariantly formulated laws, we pick out the simplest one as the most suitable. The important point to note is that a theory which at first sight looks simple may, on covari-ant reformulation, turn out to be very complicated. Einstein wrote:

> *Although it is true that one can put every empirical law in a generally covariant form, yet principle a) [the principle of relativity] possesses a great heuristic power which has al-ready brilliantly proved its mettle in the case of gravitation and which is based on the following. Of two theoretical systems, both of which are in agreement with experience that one is to be preferred which, from the point of view of the absolute differential calculus, is the simpler (einfachere) and the more transparent (durchsichtigere) one. Let one express Newtonian gravitational mechanics in the form of generally covariant equations (four-dimensionally) and one will surely be convinced that principle a) excludes this theory from the practical if not from the theoretical point of view.*[21]

It is obvious from the above passage that Einstein meant 'simplicity' and 'clarity' in a pragmatic sense. Although words can be redefined in an arbitrary way, 'simplicity' is to my mind a misleading term, which ought to be replaced by something like 'compactness' or 'organic unity'. Even when covariantly formulated, Newton's theory cannot be said to be much more complicated than Einstein's;[22] its main disadvantage is that it contains two heterogeneous parts: the degenerate metric and the gravitational field. The latter is described by an affine connection not determined by the metric alone, so that we have to go beyond the metrical aspect of geometry in order to capture the field. In Einstein's theory the metric tensor and the gravitational potential are one and the same thing; the affinity $-\Gamma^{\alpha}_{\mu\nu}$ and the Christoffel quantities $\{^{\alpha}_{\mu\nu}\}$ are identical (in John Statchel's words, there is 'minimal coupling' between the field and the geometry). This it what is meant by saying that Einstein tried to geometrise physics, or equivalently, to make geometry empirical.[23]

III.2   Weyl, Eddington and Schrödinger

Let us now briefly consider various ways in which Einstein's programme was further developed. The following examples are designed to show that the attempt to imbed ever greater parts of physics into 'natural geometry' and thus construct unified theories was characteristic, not only of Einstein, but also of the other workers in the programme. This is precisely why one can legitimately speak of a *programme*. Let us note in passing that one speaks of 'unified', and not of 'simple', field theories. I have tried to explain above why 'unification' and 'simplicity' have, to my mind, very different meanings.

Einstein's and Maxwell's equations written in generally covariant forms are:

1. $\quad R_{\mu\nu} - \frac{1}{2} g_{\mu\nu} R = - k \; T_{\mu\nu}$ (k = constant).

2. $\left\{ \begin{array}{l} F_{[\mu\nu/\sigma]} \underset{\text{def}}{=} \frac{1}{3} (F_{\mu\nu/\sigma} + F_{\nu\sigma/\mu} + F_{\sigma\mu/\nu}) = 0 \qquad \text{and} \\[2mm] F^{\mu\nu}_{\;\;\;\;"\nu} = S^{\mu} \end{array} \right.$

where: $R_{\mu\nu}$ is the Ricci tensor, $g_{\mu\nu}$ the fundamental tensor, $T_{\mu\nu}$ the energy-momentum tensor, $F_{\mu\nu}$ the (anti-symmetric) electromagnetic tensor and S^{μ} the source vector.

In the Einstein-Maxwell theory the gravitational and electromagnetic fields play essentially asymmetric roles. The gravitational potentials are identical with the space-time metric. If electromagnetic energy is present, one term occurring in T_{μ}^{ν} is a multiple of $(\frac{1}{4}\delta_{\mu}^{\nu} F^{\alpha\sigma}F_{\alpha\sigma} - F^{\nu\sigma}F_{\mu\sigma})$. Einstein's equations for the case where only electromagnetic energy is present are obtained by applying a variational principle to the integral

3. $\quad I = \int [R + \frac{1}{2} h \; F^{\alpha\beta}F_{\alpha\beta}] \sqrt{-g} \cdot d\chi$ where: $dx = dx^0 \cdot dx^1 \cdot dx^2 \cdot dx^3$,

h = constant and $R = g^{\alpha\sigma}R_{\alpha\sigma}$.

In view of 1., the electromagnetic field contributes to the geometry but cannot, without further assumptions, be extracted from it; the electromagnetic field remains extraneous to the structure of space-time. Hermann Weyl tried to remedy this defect by constructing a geometry more general that Riemann's. Let us note that the first equation in 2. is equivalent to the existence of a vector potential K_{μ} such that:

$F_{\mu\nu} = K_{\mu/\nu} - K_{\nu/\mu}$.

Thus four functions, namely K_0, K_1, K_2 and K_3, completely determine the electromagnetic field. Weyl's solution consist's in extending Riemannian geometry so as to include four additional entities. In Einstein's theory, the gravitational field manifests itself in that a vector A^μ changes its direction when taken round an infinitesimal circuit in accordance with the formula:

$$\Delta A^\mu = \frac{1}{2} B^\mu{}_{\alpha\beta\sigma} \cdot A^\alpha \cdot H^{\beta\sigma} \qquad \text{where:} \quad B^\mu{}_{\alpha\beta\sigma} = \text{curvature tensor and}$$

$$H^{\beta\sigma} = \frac{1}{2} \oint (\xi^\sigma \cdot d\xi^\beta - \xi^\beta \, d\xi^\sigma) \ .$$

The length of A^μ however is not altered by the transport. Weyl constructs a new geometry which allows for a change of length given by:

4. $\frac{\delta 1}{1} = K^*_\mu \cdot dx^\mu$, $i.e.$ $\delta 1 = 1 \cdot K^*_\mu \cdot dx^\mu$ $(1^2 = g_{\mu\nu} \cdot A^\mu \cdot A^\nu)$,

where K^*_μ is some covariant vector which is later identified with the electromagnetic potential. Thus the change in length of A^μ manifests the presence of the electromagnetic field. By 4., $\delta 1$ is clearly proportional to 1 and is also independent of the direction of A^μ . These restrictions give us exactly the number of extra entities we want, namely the four functions K^*_i , i = 0,1,2,3 . The affinity $\Gamma^\mu_{\lambda\sigma}$ is now determined by the formula:

5. $\delta A^\mu = \Gamma^\mu_{\lambda\sigma} \cdot A^\lambda \cdot dx^\sigma$

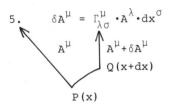

Consider the change of length that occurs when we transport a vector A^μ parallel to itselft from a point P(x) to a neighbouring point Q(x+dx). Then:

6. $\delta 1^2 = \left[g_{\alpha\beta}(x+dx) \cdot (A^\alpha + \delta A^\alpha) \cdot (A^\beta + \delta A^\beta) - g_{\alpha\beta} \cdot A^\alpha \cdot A^\beta \right]$

$\dot{=} \left[(g_{\alpha\beta} + g_{\alpha\beta/\sigma} \cdot dx^\sigma) (A^\alpha + \Gamma^\alpha_{\mu\nu} \cdot A^\mu \cdot dx^\nu) (A^\beta + \Gamma^\beta_{\sigma\gamma} \cdot A^\sigma \, dx^\gamma) - g_{\alpha\beta} \cdot A^\alpha \cdot A^\beta \right]$

$\dot{=} \left[g_{\alpha\beta/\gamma} \cdot A^\alpha \cdot A^\beta \cdot dx^\gamma + g_{\sigma\beta} \cdot \Gamma^\sigma_{\alpha\gamma} \cdot A^\alpha \cdot A^\beta \cdot dx^\gamma + g_{\sigma\alpha} \cdot \Gamma^\sigma_{\beta\gamma} \cdot A^\alpha \cdot A^\beta \cdot dx^\gamma \right]$

(by interchange of dummies and by the symmetry of $g_{\alpha\beta}$)

It follows that, if $\Gamma^\sigma_{\mu\nu} = - \{^{\ \sigma}_{\mu\nu}\}$, the last member of 6. vanishes, $i.e.$ the vector length is unaltered. This is why Weyl distinguished between the affinity $\Gamma^{\sigma}_{\mu\nu}$ and the Christoffel brackets $-\{^{\sigma}_{\mu\nu}\}$.

By 4. δl^2 can also be expressed as follows:

7. $\quad \delta l^2 = 2l\delta l = 2ll \cdot K^*_{\gamma} \cdot dx^{\gamma} = 2l^2 \cdot K^*_{\gamma} \cdot dx^{\gamma} = 2g_{\alpha\beta} A^{\alpha} \cdot A^{\beta} \cdot K^*_{\gamma} dx^{\gamma}$

$\qquad = 2g_{\alpha\beta} \cdot K^*_{\gamma} \cdot A^{\alpha} \cdot A^{\beta} \cdot dx^{\gamma}$

Comparing the two expressions for δl^2 and noting that the equality must hold for arbitrary values of A^{μ} and dx^{γ} , we deduce:

8. $\quad (g_{\alpha\beta/\gamma} - 2 g_{\alpha\beta} \cdot K^*_{\gamma}) + g_{\sigma\beta} \cdot \Gamma^{\sigma}_{\alpha\gamma} + g_{\sigma\alpha} \cdot \Gamma^{\sigma}_{\beta\gamma} = 0$

By circular permutation of α, β, γ, we obtain two further equations:

9. $\quad (g_{\beta\gamma/\alpha} - 2 g_{\beta\gamma} \cdot K^*_{\alpha}) + g_{\alpha\gamma} \cdot \Gamma^{\sigma}_{\beta\alpha} + g_{\sigma\beta} \cdot \Gamma^{\sigma}_{\gamma\alpha} = 0$

10. $\quad (g_{\gamma\alpha/\beta} - 2 g_{\gamma\alpha} \cdot K^*_{\beta}) + g_{\sigma\alpha} \cdot \Gamma^{\sigma}_{\gamma\beta} + g_{\sigma\gamma} \cdot \Gamma^{\sigma}_{\alpha\beta} = 0$

We take $\Gamma^{\alpha}_{\beta\gamma}$ to be symmetric in the indices β and γ . Subtracting the last 2 equations from the first one, we have:

$$g_{\sigma\gamma} \cdot \Gamma^{\sigma}_{\alpha\beta} = -[\alpha\beta,\gamma] + (g_{\gamma\alpha} \cdot K^*_{\beta} + g_{\gamma\beta} \cdot K^*_{\alpha} - g_{\alpha\beta} \cdot K^*_{\gamma})$$

where: $[\alpha\beta,\gamma] = \frac{1}{2} (g_{\gamma\alpha/\beta} + g_{\gamma\beta/\alpha} - g_{\alpha\beta/\gamma})$ is the Christoffel symbol of the first kind. Multiplying by $g^{\mu\gamma}$.

11. $\quad \Gamma^{\mu}_{\alpha\beta} = - \{^{\mu}_{\alpha\beta}\} + g^{\mu\gamma}(g_{\gamma\alpha} \cdot K^*_{\beta} + g_{\gamma\beta} \cdot K^*_{\alpha} - g_{\alpha\beta} \cdot K^*_{\gamma})$.

The affinity is thus uniquely determined and differs from the usual Christoffel quantity $-\{^{\mu}_{\alpha\beta}\}$ by the last term. This additional term is responsible for the change in length of a vector transported parallel to itself from one point of the manifold to another.

We now identify K^*_{μ} with the potential K_{μ} and claim that the electromagnetic field manifests its presence through the change of length experienced by a vector taken around a circuit. The electromagnetic field is thus imbedded in an extended geometric structure. Weyl's solution however is a little contrived. The two fields are not properly unified but forcibly brought together. Through conveniently assuming that $\delta l/l$ is independent of the direction of the vector, Weyl extends Riemannian geometry   just far enough to obtain the four extra quantities he needs. The unsatisfactory character of this unification of electromagnetism and gravitation is further brought out by the field equations which are obtained by varying the integral

12. $J = \int (R^2 + K\, F^{\alpha\beta} F_{\alpha\beta}) \cdot \sqrt{-g} \cdot dx$, where:

   $R = g^{\alpha\sigma} R_{\alpha\sigma}$, $F_{\alpha\beta} = K_{\alpha/\beta} - K_{\alpha/\beta}$ and $R_{\alpha\sigma}$ is the Ricci tensor

built out of the affine connections $\Gamma^{\sigma}_{\alpha\beta}$. Except for the fact that R is replaced by R^2 , the two integrals 3. and 12. are almost identical. Before commenting on the choice of the integrand in 12., it is necessary to explain the notion of gauge-invariance which plays a central role in Weyl's theory.

A gauge-transformation consists in a change of the unit of length at each point of space-time. This leads to the equations:

13. $d\hat{s}^2 = f\, ds^2$,   where   f   is some function of the coordinates.

14.   Hence $\hat{1} = \sqrt{f} \cdot 1$.

It follows from 13. that

15. $\hat{g}_{\mu\nu} \cdot dx^{\mu} \cdot dx^{\nu} = d\hat{s}^2 = f \cdot ds^2 = f \cdot g_{\mu\nu} \cdot dx^{\mu} \cdot dx^{\nu}$.   Thus:

16. $\hat{g}_{\mu\nu} = f \cdot g_{\mu\nu}$.

But we also have by 4.: $\frac{\delta 1}{1} = K_{\mu} \cdot dx^{\mu}$ $i.e.$ $\delta(\log 1) = K_{\mu}\, dx^{\mu}$.
Similarly: $\delta(\log \hat{1}) = \hat{K}_{\mu} \cdot dx^{\mu}$.   By 14: $\hat{1} = \sqrt{f}, 1$.   Thus:

17. $\hat{K}_{\mu} \cdot dx^{\mu} = \delta(\log \hat{1}) = \delta(\log(\sqrt{f}1)) = \frac{1}{2}\, \delta(\log f) + \delta(\log 1)$

   $= \frac{1}{2} \frac{\partial \log f}{\partial x^{\mu}} \cdot dx^{\mu} + K_{\mu} \cdot dx^{\mu}$.   Thus

18. $\hat{K}_{\mu} = \frac{1}{2} \frac{\partial \log f}{\partial x^{\mu}} + K_{\mu}$.

By 16. and 18. we have that a gauge-transformation consists of:

19. $\hat{g}_{\mu\nu} = f \cdot g_{\mu\nu}$ and $\hat{K}_{\mu} = K_{\mu} + \frac{1}{2} \frac{\partial \log f}{\partial x^{\mu}}$

where   f   is some function of the coordinates. Substituting in 11. we easily verify that:

20. $\hat{\Gamma}^{\mu}_{\alpha\beta} = \Gamma^{\mu}_{\alpha\beta}$.

We say of a function that it is gauge-invariant if it is unaffected by any transformation of the form 19. Thus, $\Gamma^{\mu}_{\alpha\beta}$ is gauge-invariant; this obviously also applies to every entity built out of the $\Gamma^{\mu}_{\alpha\beta}$'s (and of the coordinates). One important problem in Weyl's theory is to determine quantities having invariance properties with respect to both gauge and coordinate transformations; *e.g.* gauge-invariant tensors. With regard to equations 3. and 12., R^2 replaces R because $(R^2 + K\,F^{\alpha\beta}F_{\alpha\beta})\sqrt{-g}$, but not $(R + \frac{1}{2}h\,F^{\alpha\beta}F_{\alpha\beta})\sqrt{-g}$, is gauge-invariant. By the way, it should also be noted that $(R^2 + K\,F^{\alpha\beta}F_{\alpha\beta})\sqrt{-g}$ is gauge-invariant only in a 4-dimensional space. Thus Weyl's theory singles out four-dimensionality; or, more accurately, Weyl's theory explains why our world is four-dimensional, *provided* some rationale be given for the choice of $(R^2 + KF^{\alpha\beta}F_{\alpha\beta})\sqrt{-g}$ as the appropriate tensor density. Eddington strongly doubted the existence of such a rationale. He pointed out that, apart from the analogy which 12. bears to 3., there is no reason for stringing together two such heterogeneous functions as R^2 and $K \cdot F^{\alpha\beta}F_{\alpha\beta}$. The choice of $(R^2 + KF^{\alpha\beta}F_{\alpha\beta})\sqrt{-g}$ thus defeats Weyl's purpose, which is precisely to explain why gravity, which in Einstein's case is represented by $R_{\alpha\beta}$, should naturally coalesce with electromagnetism, which is represented by $F_{\alpha\beta}$. Eddington was to my mind right when he wrote: 'But the connection, though reduced to simpler terms, is not in any way explained by Weyl's action-principle. It is obvious that his action as it stands has no deep significance; it is a mere stringing together of two in-invarients of different forms. To subtract $F_{\mu\nu}F^{\mu\nu}$ from $^{*}G^2$ [*i.e.* from R^2] is a fantastic procedure which has no more theoretical justification than subtracting E^{ν}_{μ} from T^{ν}_{μ} . $[E^{\nu}_{\mu} = (\frac{1}{4}\,\delta^{\nu}_{\mu}F^{\alpha\beta}F_{\alpha\beta} - F^{\nu\sigma}F_{\mu\sigma})$ and $T^{\nu}_{\mu} = -\frac{1}{8\pi}(R^{\nu}_{\mu} - \frac{1}{2}\,\delta^{\nu}_{\mu} \cdot R + \lambda\delta^{\nu}_{\mu})]$.

At the moment we can only regard the assumed form of action A as a step towards some more natural combination of electromagnetic and gravitational variables'.[24]

Following Weyl, Eddington starts from the postulate that 'parallel displacement has some significance in regard to the ultimate structure of the world-it does not much matter what significance.'[25] He then goes on about setting up an affine geometry which *naturally* bifurcates into a gravitational part and an electromagnetic one. What 'naturalness' means in this context will become clear as we go along.

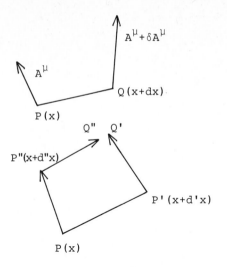

Let us again transport a contravariant vector A^μ from $P(x)$ to $Q(x+dx)$, constantly keeping A^μ parallel to itself. At Q the vector we thus obtain will have components $A^\mu+\delta A^\mu$, where δA^μ is bilinear and homogenous in A^α and dx^β. Hence: $\delta A^\mu = \Gamma^\mu_{\alpha\beta}\cdot A^\alpha\cdot dx^\beta$, the $\Gamma^\mu_{\alpha\beta}$'s being functions defined at each point of the manifold. Consider three neighbouring points: $P(x)$, $P'(x+d'x)$ and $P''(x+d''x)$. Let $\overrightarrow{P'Q'}$ and $\overrightarrow{P''Q''}$ be respectively the vectors obtained by the parallel transport of $\overrightarrow{PP''}$ and $\overrightarrow{PP'}$. The components of $\overrightarrow{P''Q''}$ are:

$d'x^\mu+\Gamma^\mu_{\alpha\beta}\cdot d'x^\alpha\cdot d''x^\beta$, so the coordinates of Q'' relatively to P are:

$d''x^\mu+d'x^\mu+\Gamma^\mu_{\alpha\beta}\cdot d'x^\alpha\cdot d''x^\beta$. Similarly, the coordinates of Q' with respect to P are: $d'x^\mu+d''x^\mu+\Gamma^\mu_{\alpha\beta}\cdot d''x^\alpha\cdot d'x^\beta$. If the parallelogramme law is to hold, Q' and Q'' must coincide, $i.e.$

$$\Gamma^\mu_{\alpha\beta}\cdot d'x^\alpha\cdot d''x^\beta = \Gamma^\mu_{\alpha\beta}\cdot d''x^\alpha\cdot d'x^\beta = \Gamma^\mu_{\beta\alpha}\cdot d'x^\alpha\cdot d''x^\beta .$$

Since $d''x^\beta$ and $d'x^\alpha$ are arbitrary infinitessimal displacements, $\Gamma^\mu_{\alpha\beta} = \Gamma^\mu_{\mu\alpha}$; so the affinity must be symmetric in its lower indices. Eddington assumes the parallelogramme law, pointing out however that such an assumption is by no means necessary on physical grounds. This underlines, once again, the importance of mathematical considerations in the construction of physical theories.

In view of the significace of parallel transport with regard to the structure of the world, Eddington considers the total variation which a vector A^μ undergoes as it is taken round a small closed curve at some point P. He finds that:

21. $\quad \Delta A^\mu = \frac{1}{2}\cdot B^\mu_{\;\lambda\sigma\nu}\cdot A^\lambda\cdot H^{\sigma\nu}$, where $H^{\sigma\nu} = \frac{1}{2}\oint(\xi^\nu\cdot d\xi^\sigma - \xi^\sigma\cdot d\xi^\nu)$ and

$\quad B^\mu_{\;\lambda\sigma\nu} = \quad$ curvature tensor $= \Gamma^\mu_{\lambda\sigma/\nu} - \Gamma^\mu_{\lambda\nu/\sigma} - \Gamma^\tau_{\lambda\sigma}\cdot\Gamma^\mu_{\tau\nu} + \Gamma^\tau_{\lambda\nu}\cdot\Gamma^\mu_{\tau\sigma}$

In view of 21., the Riemann-Christoffel tensor can be expected to play a fundamental role in General Relativistic theories. It is also well-known that the properties of a Riemannian space are closely connected with those of $B^\mu_{\;\lambda\sigma\nu}$; for example the vanishing of this tensor in a region is the necessary and sufficiant condition for the space to be flat. This is why we look for second-rank tensors built out of the curvature tensor.

The simplest method for finding such quantities constists in contracting μ with one of the indices λ, σ or ν . Since $B^{\mu}{}_{\lambda\sigma\nu}$ is anti-symmetric in σ and ν , $B^{\mu}{}_{\lambda\sigma\nu} = - B^{\mu}{}_{\lambda\nu\sigma}$ so $B^{\mu}{}_{\lambda\sigma\mu} = - B^{\mu}{}_{\lambda\mu\sigma}$. Hence we need only consider $B^{\mu}{}_{\mu\sigma\nu}$ and $B^{\mu}{}_{\lambda\sigma\mu}$. Let $B^{\mu}{}_{\lambda\sigma\mu} = R_{\lambda\sigma}$ and $B^{\mu}{}_{\mu\sigma\nu} = H_{\sigma\nu}$.

22. $R_{\lambda\sigma} = B^{\mu}{}_{\lambda\sigma\mu} = \Gamma^{\mu}{}_{\lambda\sigma/\mu} - \Gamma^{\mu}{}_{\lambda\mu/\sigma} - \Gamma^{\tau}{}_{\lambda\sigma} \cdot \Gamma^{\mu}{}_{\tau\mu} + \Gamma^{\tau}{}_{\lambda\mu} \cdot \Gamma^{\mu}{}_{\tau\sigma}$ and

$H_{\sigma\nu} = B^{\mu}{}_{\mu\sigma\nu} = \Gamma^{\mu}{}_{\mu\sigma/\nu} - \Gamma^{\mu}{}_{\mu\nu/\sigma} - \Gamma^{\tau}{}_{\mu\sigma} \cdot \Gamma^{\mu}{}_{\tau\nu} + \Gamma^{\tau}{}_{\mu\nu} \cdot \Gamma^{\mu}{}_{\tau\sigma} = \Gamma^{\mu}{}_{\mu\sigma/\nu} - \Gamma^{\mu}{}_{\mu\nu/\sigma}$

$= \Gamma_{\sigma/\nu} - \Gamma_{\nu/\sigma}$ where: $\Gamma_{\alpha} \overset{=}{\text{Def.}} \Gamma^{\mu}{}_{\mu\alpha} = \Gamma^{\mu}{}_{\alpha\mu}$.

Now note that, in view of the symmetry of $\Gamma^{\alpha}{}_{\beta\gamma}$ in β and γ , equation 22. entails:

23. $R_{\lambda\sigma} - R_{\sigma\lambda} = - \Gamma_{\lambda/\sigma} + \Gamma_{\sigma/\lambda}$.

24. Therefore $H_{\lambda\sigma} = R_{\sigma\lambda} - R_{\lambda\sigma}$. The second mode of contraction merely yields the antisymmetric part of $-2R_{\lambda\sigma}$. We are therefore *mathematically* led to write

25. $R_{\lambda\sigma} = g^{*}_{\lambda\sigma} + F^{*}_{\lambda\sigma}$, where $g^{*}_{\lambda\sigma} = \frac{1}{2} (R_{\lambda\sigma} + R_{\sigma\lambda})$, $F^{*}_{\lambda\sigma} = \frac{1}{2} (R_{\lambda\sigma} - R_{\sigma\lambda})$.

Remembering that the gravitational and electromagnetic fields are described respectively by a symmetric and an antisymmetric tensor $(g_{\lambda\sigma}$ & $R_{\lambda\sigma})$, we are driven to identify $g^{*}_{\lambda\sigma}$ with $g_{\lambda\sigma}$ and $F^{*}_{\lambda\sigma}$ with $F_{\lambda\sigma}$.

Commenting on Weyl's action-principle, Eddington[26] remarked that a natural invariant density, which however does not single out four-dimensionality, is $\sqrt{-\det[R_{\mu\nu}]}$. Schrödinger took up this Eddingtonian suggestion, but he dropped the symmetry condition on $\Gamma^{\alpha}{}_{\mu\nu}$, thus giving himself more latitude in that he now had 64 independent functions $\Gamma^{\alpha}{}_{\mu\nu}$ at his disposal.[27]

He wrote down his field equations in the form:

26. $\delta \left(\int \sqrt{-\det[R_{\mu\nu}]} \cdot d\chi \right) = 0$, which is equivalent to:

27. $R_{\mu\nu/\sigma} + R_{\lambda\nu} \cdot {}^{*}\Gamma^{\lambda}{}_{\mu\sigma} + R_{\mu\lambda} \cdot {}^{*}\Gamma^{\lambda}{}_{\sigma\nu} = 0$, where

${}^{*}\Gamma^{\alpha}{}_{\mu\nu} \overset{=}{\text{Def.}} \Gamma^{\alpha}{}_{\mu\nu} + \frac{1}{(N-1)} \delta^{\alpha}{}_{\mu} (\Gamma^{\tau}{}_{\nu\tau} - \Gamma^{\tau}{}_{\tau\nu})$, N = dimension of space.

Thus, like Eddington, Schrödinger tries to account both for gravitational and for electromagnetic phenomena in terms of one purely affine geometry.

The above derivations were given in some detail in order to show that each step in these developments, which covered a period of almost forty years, was largely determined by a priori mathematical considerations.

III.3 Conclusion

If we consider the sequence: Newton, Einstein, Weyl, Eddington, Schrödinger and Einstein, we see that each stage yields a more unified, but not necessarily simpler, theory than the previous ones. In Newton's case the background metric is independent of the field; in Einstein's 1916 theory the metric and the gravitational potentials are identified; the electromagnetic field occupies an intermediate position in that it affects the metric but is not imbedded in it. Weyl extends the geometry in order to accomodate both fields and Eddington sketches the outline of a theory in which an affine space unifies the two fields in a natural way; Schrödinger starts from where Eddington left off. This sequence is only meant to indicate a movement towards greater integration within the Relativistic Programme. The impression should not be given that each theory in the series was empirically better than the previous ones. In fact, both Weyl's hypothesis and Eddington's approach proved empirically unsuccessful. Moreover, Eddington's and Schrödinger's theories have a disturbing aspect. Einstein wanted both to geometrise physics and to turn geometry into an empirical science. In Weyl's theory the geometry and the two physical fields are co-extensive in that both are described by fourteen quantities at each point. In Eddington's and Schrödinger's cases we have forty and sixty-four independent functions $\Gamma^{\alpha}_{\mu\nu}$ at our disposal and only fourteen 'physical' variables to account for. It is true that the theories are still empirical in the sense of being experimentally testable, but the distance between hypothesis and observational statements has grown. This in itself is no defect, but it makes the independent testability of the theory imperative.

The historical and philosophical conclusions which can be drawn are:

1.   The General Covariance Principle, together with a compactness requirement, was part of the heuristic which guided the Relativity programme after 1906.

2.   Since the Lorentz-covariance principle played a similar heuristic role before 1908,[28] an important element of continuity is established between the Special and General Theories. Continuity obtains at the heuristic-mathematical rather than at the propositional level. It is a

dynamic type of continuity which has to do with a common approach to theory-construction rather than with a fixed set of shared assumptions.

3. 'Compactness' or 'degree of unification' is to be substituted for 'simplicity'. Weyl's theory for example is not simpler, but more uni-fied, than the conjunction of Einstein's and Maxwell's field equations.

Appendix: The Status of the Relativity Principle

The problem concerning the Relativity - or Covariance - principle is as follows. Let T be any statement. The Relativity Principle $R\boxed{T}$ is the proposition that, if T is any law of nature, then T is covariant. The main objection (first raised by Kretschmann) to the Relativity Principle is that any physical theory can be written in generally co-variant form. In other words: given any physical law T , there always exists a generally covariant physical law T^* such that $T \Longleftrightarrow T^*$. In other words: $R\boxed{T^*}$, since T^* is generally covariant.

If we consider $R\boxed{T}$ as an object-language statement in which T oc-curs, then from $T \Longleftrightarrow T^*$ and $R\boxed{T^*}$, there follows $R\boxed{T}$. Since T is an arbitrary physical law, $R\boxed{T}$ is vacuous. Even if we accept (Kretschmann's) thesis that, for any T , there exists a generally co-variant T^* such that $T \Longleftrightarrow T^*$, the above argument goes through only if $R\boxed{T}$ is treated as a statement in the object-language. But we know that the Relativity Principle is not vacuous, since we can exhibit state-ments (*e.g.* $\Sigma_{mo}\vec{v} = 0$) which are not Lorentz-covariant, let alone gen-erally covariant. Let me therefore propose the following - very tenta-tive - solution to the problem posed by the status of the Relativity Prinicple.

The very formulation of the Relativity Priniciple indicates that it is a proposition about the *statement* 'T' , in other words, the Relativity principle belongs to the metalanguage and should be written as $R('T')$. Instead of being an object-language statement $R\boxed{T}$ in which T *occurs*, the Relativity Principle is a metalinguistic statement *about* the state-ment 'T' . Thus, although $T \Longleftrightarrow T^*$ and $R('T^*')$, we cannot conclude that $R('T')$; since, in general, $'T' \neq 'T^*'$, we cannot conclude that $R('T') \Longleftrightarrow R('T^*')$.

I.e. if $R\boxed{T}$ were a statement in the object-language, then: from $T \Longleftrightarrow T^*$ follows $R\boxed{T} \Longleftrightarrow R\boxed{T^*}$; hence, since $R\boxed{T^*}$, we must also have $R\boxed{T}$. $R\boxed{T}$ however is not part of the object-language and should be written $R('T')$.

Thus, although 'T' can be written in the generally covariant form 'T*' , 'T' *itself* is not generally covariant.

Remember that R('T') stands for: if 'T' is a law of nature, then 'T' is generally covariant. Let R stand for: R('T') for all 'T' . Let R' stand for: for all 'T' , if 'T' is a law of nature, then 'T' is Lorentz-covariant. Obviously: R$\Rightarrow$R' . R' can be considered as part of the hard core of the Relativity Programme. General Relativity strengthened R' into R .

Notes

1. Cf. "Geometrie und Erfahrung" in: Einstein, *Mein Weltbild*, p. 119.

2. Cf. Oskar Becker, *Grundlagen der Mathematik*, pp. 144 - 167.

3. For the role which mathematical surplus structure plays in physics, cf. M. L. G. Redhead, "Symmetry in Intertheory Relations", in: *Synthese* 32 (1975).

4. Cf. Bernhard Riemann, *Collected Works*, (Dover) pp. 272 - 273.

5. As above, p. 286.

6. Cf. Mach, *Mechanics*, Introduction. Also: *Erkenntnis und Irrtum*, pp. 164 - 182.

7. H. Weyl, *Raum Zeit Materie*, § 40.

8. Eddington, *The mathematical Theory of Relativity*, § 97.

9. Schrödinger, *Space-Time Structure*, Chapter XII.

10. Duhem, *Aim and Structure of Physical Theory*, Part 2, Chapter 1.

11. For the important role which philosophical realism plays in the logic of discovery, cf. Popper, "Three Views Concerning Human Knowledge" in: 'Conjectures and Refutations' (Section 3).

12. For the fruitfulness of these equivalent reformulations, cf. Feynman, *The Character of Physical Law*, p. 168.

13. E. Meyerson, *Identité et Réalité*, Chapter I. Also: *De L'Explication das les Sciences*, Chapter V.

14. Cf. Popper, "The Aim of Science", in: *Objective Knowledge*, p. 191.

15. *Identité et Réalité*, Chapter III.

16. Weyl, *Philosophy of Mathematics and Natural Science*, p. 5.

17. Meyerson, *La Déduction Rélativiste*, Chapter XX.

18. Weyl, *Philosophy of Mathematics and Natural Science*, pp. 87-88.

19. Mach, *The History and the Root of the Principle of the Conservation of Energy*, Notes.

20.   Eddington, *The Mathematical Theory of Relativity*, p. 120.

21.   "Prinzipielles zur allgemeinen Relativitätstheorie", *Annalen der Physik*, Band 55, 1918. Also see Appendix 1.

22.   For this point I am indebted to Professor J. Stachel.

23.   Cf. above, see I.1.

24.   *Mathematical Theory of Relativity*, p. 212.

25.   *Mathematical Theory of Relativity*, p. 213.

26.   *Mathematical Theory of Relativity*, p. 206.

27.   Schrödinger, *Space-Time Structure*, p. 112.

28.   Cf. my "Why did Einstein's programme supersede Lorentz's?, *Brit. J. Phil. Sci.* <u>24</u> (1973) pp 95 - 123 and 223 - 62.

OVERTURN AND CONTINUITY OF THE HYPOTHESES IN THE FRAMING OF THE THEORY
OF RELATIVITY

An Attempt at Reconstructing the Transition from Classical to Relativistic Physics

Asaria Polikarov, Bulgarian Academy of Sciences, Sofia

> Clearly to recognize this axiom
> and its arbitrary character really
> implies already the solution of
> the problem.
>
> A. E i n s t e i n

1. Regarding the problem

We can tackle the problems of structure and dynamics of scientific theories in several ways, namely
- from the point of view of some ingredients or others of the theory
  (concepts, problems, methods, etc.) or
- complexly, bearing in mind all these ingredients;
- by juxtaposition with other scientific theories (of sequences of
  theories) or with certain philosophical conceptions, - what can also
  be accomplished complexly or along the lines of separate ingredients.

Here we shall attempt to juxtapose classical physics (mechanics and electrodynamics; theory of gravity) with the theory of relativity (special;
general) along the lines of their basic hypotheses. This, in our opinion,
is representative enough, *i.e.*, to a high degree - a substitute for a complex comparison.

1.1 Already the simple juxtaposition of the totalities of hypotheses (H)
characterizing, on the one hand, classical physics (C) and, on the other,
the theory of relativity (R), poses problems associated with philosophical
conceptions about the character of such transitions from C to R. One won-
ders: to what extent the *reconstruction* (along the lines of certain doc-
trines) prevails in the latter, the *refutation* of old H, the *incommensurability* or the *continuity* between classical and contemporary theories?

1.2 If we accept the view of the incommensurability of such successive
theories, then we must deal with two independent and incomparable sets of
H and hence the problem becomes to a great extent meaningless.

This corresponds neither to the logic of the construction of scientific
theories, nor to their history. This is expressed at least in the fact

that between the new and the old theory there not only consists a continuity, but that without the latter the new theory would not be conceivable. Even the radical refutation of certain statements presupposes their existence (and their assertion).[1]

1.3 On the other hand the attempts at explaining all changes in the growth of science within the framework of a continuity (Duhem, Brunschvicg) are untenable. They contradict not only the natural manner in which certain (revolutionary) changes are accepted, but quite generally, also the qualitative difference between improving within a given (classical or non-classical) theory and the transition from one theory to another.

Thus the problem arises to establish - possibly on the basis of acceptable criteria - the elements of continuity and of originality and their relationship of subordination (in every concrete case) and therefrom to draw general conclusions for the development of science.

2. What classes of hypotheses may be distinguished in physics?

2.1 We shall understand the hypotheses as a collective concept along the lines of the hypothetical-deductive view (as assumptions or presuppositions). Sometimes Einstein considers the postulate of the constancy of the velocity of light (law of light), as well as the principle of equivalence as hypotheses [Einstein, 1953; Einstein, 1912]. Also for Eddington, H. Thirring *et al*. the principle of equivalence is hypothetical [Eddington, 1960, § 54]. Sometimes the principle of relativity is also conceived as an hypothesis [Tonnelat, 1971, p. 153].

2.2 One can accept one or another classification of the hypotheses which was proposed by various authors. With a view to the problem under consideration, the classification of H in two classes with two subclasses seems suitable, namely

I. class: basic assumptions,
H_1 - presuppositions,
H_2 - principles (axioms, postulates) of the theory.

II. class: apparatus
H_3 - choice of the mathematical structure (calculus),
H_4 - empirical interpretation.

Correspondingly, the theory may be represented as a conjunction of these components, *i.e.*, as $<H_i>$ (i=1-4) [Polikarov, 1978]. In addition to these the supplementary class can be introduced

H_5 - consequences of the theory (which underlie an empirical proof).

3. Which hypotheses characterize classical physics?

Instead to juxtapose the set of basic hypotheses of classical physics (mechanics and electrodynamics) (C) and that of the theory of relativity (TR), we shall dwell mainly on the part which undergoes more or less essential changes. The changes in question are related (first of all) to the assumption of classical mechanics:

     I.   Kinematical

             C_{12} - of the absolute character of the length of bodies.
             C_{13} - of the absolute character of time.

     II.   Dynamical

             C_{14} - of constancy of mass.[2]

It must be emphasized that in classical mechanics these (and others) presuppositions are considered as true (certain) statements; their hypothetical character has been realized only later, especially on the basis of the theory of relativity.

     III.   Assumptions of electrodynamics:

             C_{15} - the microscopic electrodynamics (electron theory) is a theory of a constructive type (like the kinetic theory of gases).
             C_{16} - of the existence of the ether.
             C_{17} - of the immobility of the ether, *i.e.*, it represents a privileged frame of reference; the motion relative to it is an absolute one.

Additionally one assumes:

C_{18} - the contraction hypothesis (Fitzgerald, Lorentz, 1892). From C_{17} follows $\left[C_{21}\right]$ the invalidity of the principle of relativity in this domain, and from C_{18} - its validity only for effects of first order relative to $\frac{v}{c}$ $\left[C_{22}\right]$.

If we try to rank-order the C of classical physics, then C_{12}, C_{13} on the one hand, and C_{15} and C_{17}, on the other, come to the fore.

4. Which are the basic hypotheses of Special Relativity (SR)?

New hypotheses within the SR are

R_{11} (versus C_{15}) - the adequate theory of electromagnetic field is of principle type - as thermodynamics (for which Einstein states that it is the only physical theory of universal content which will never be overthrown within the framework of the applicability of its basic concepts $\left[\text{Einstein, 1957}\right]$.

R_{12} - the simultaneity of events is determined by light signals.

Here one admits that even in the moving reference system the light rays traveling from the clock to the observers eye behave more or less as we have always expected them to behave (Einstein) [Heisenberg, 1971].[3]

It is also presupposed the so-called standard synchronization, according to which the speed of light propagating in two opposite directions is the same, *i.e.*, that the coefficient (of simultaneity) $\varepsilon = 1/2$ [Reichenbach 1928].

There are considerations about the non-conventional character of such a definition (P. Bowman *e.a.*), whose convincing force has been questioned [Grünbaum, 1973, p. 670 ff; Melament, 1977].

R_{13} - clocks which are in slow transport synchrony are also in light signal synchrony [Ellis, 1978].

R_{14} - the so-called principle of the physical identity of the units of measure [Born, 1962].

R_{15} - Independence of the bahaviour of measuring rods upon their previous history [Einstein, 1953, p. 36, 63].

R_{21} (Antithesis of C_{21}) - of the universal validity of the special principle of relativity (first postulate of the SR).

R_{22} - of the constancy of the velocity of light (second postulate of the SR or law of light).

This may be formulated as an assertion of the existence of a fundamental velocity equal to that of light [Eddington, 1960, § 6].

It is interesting that R_{21}, which is new for electrodynamics, does not lead to important changes in its domain, whereas R_{22} (which is natural for the same) has radical consequences for mechanics. It is well-known that exactly from R_{22} follows the repudiation of the hypotheses C_{12} and C_{13} (as well as of the so-called ballistic hypothesis). In this sense Einstein maintains that there is a continuity between SR and electrodynamics, and its revolutionary impact concerns mechanics [Einstein, 1967].

This is displayed in the hypotheses

R_{31} - Lorentz-transformation (G_c), to which Einstein comes from a new point of view[4] and correspondingly

R_{41} (distinctly from Lorentz) - its non-metrogenic interpretation

R_{51} (Antithesis of C_{13}) - relativity of time,

R_{52} (Antithesis of C_{12}) - relativity of length.

R_{53} (Antithesis of C_{14}) - dependence of mass on velocity.

The velocity of light, as it follows from R_{31} and R_{53}, is at the same time a limiting velocity for the movement of material objects (tardions and luxons). Thus the principle of action by impact will be treated as universally valid.

R_{54} - Equivalence between mass (m) and energy (E): $E = c^2 m$ (α).

Einstein's interpretation of this relation is that mass and energy are essentially alike, *i.e.*, that they are different expressions for the same thing [Einstein, 1953, p. 47]. This interpretation cannot claim more than a limited validity [Polikarov, 1966, Chp. 3; Polikarov, 1973, § 8].

R_{55} (Antithesis of C_{16}) - Abandonment of the ether.

It deserves to be pointed out the following assumption

R_{16} - epistemological postulate, stating that the concepts and relations have justification only in so far as they are connected with sense impressions between which they form a mental connection [Einstein 1967].

In the framework of the SR other hypotheses have also been raised, among which first of all we have to mention that of Minkowski:

R_{42} - of the unitary examination of space and time.

The progress in method which electrodynamics owes to SR Einstein sees in this, that the number of independent hypotheses is diminished [Einstein, 1953, p. 41].

Furthermore we shall recall also the hypothesis

R_{56} - of the possible existence of the so-called tachions.[5]

All what was said hitherto may be summarized in the following table

Characteristic hypotheses

A. of classical physics

C_{11} - Euclidian space

C_{12} - absolute space

C_{13} - absolute time

C_{14} - m = Const

C_{15} - constructive theory of electromagnetism

C_{16} - existence of the ether

C_{17} - ether at rest

C_{18} - contraction hypothesis

C_{21} - invalidity of the principle of relativity

C_{22} - partial validity

B. of relativistic physics

R_{11} - theory on principle

R_{12} - light signal hypothesis

R_{13} - agreement between the two ways of defining simultaneity

R_{14} - identity of the units of measure

R_{15} - independence from the previous history

R_{16} - epistemological postulate

R_{17} - simplicity of the laws

R_{21} - principle of relativity (in the SR)

R_{22} - the law of light (c = Const)

R_{23} - preservation of the form of the laws

R_{31} - (G_c) - Lorentz transformation

R_{42} - Space-time continuum

R_{51} - relativity of time

R_{52} - relativity of space

R_{53} - m = m(v)

R_{54} - $E = c^2 m$ (α)

R_{56} - tachions

Lastly we should like to note the possibility of a variant construction of the SR, namely on the basis of the hypotheses R_{21} and R_{54} $\left[\textit{i.e.} \ (\alpha)\right]$; than R_{22} follows as a consequence.

5. Which are the essential changes in the basic hypotheses of physics brought about by the SR?

5.1 If we juxtapose C and R (by classes), we can state the following:

a) An important difference in the strategy or in the research programm. For electrodynamics it is characteristic what Poincaré said: "Let us take the Lorentz' theory, let us turn in all directions, let us modify it gradually and perhaps things will arrange themselves" $\left[\text{Tonnelat, 1971, p. 125}\right]$. Einstein carries out the transition from C_{15} to R_{11} while he is holding R_{21}. He compares the latter with Carnot's Principle $\left[\text{Einstein, 1921}\right]$.

Goldberg comments with good reasons that for Poicaré the theory of Einstein which does not proceed from a generalization of simple facts and which is not changeable, is not a good theory $\left[\text{S. Goldberg, 1970}\right]$.

The solution of SR, as understood by several authors (Lorentz, Weyl, Lanczos), belongs in a high degree to epistomology.

b) Radical changes concern C_1: C_{16} cancels out; C_{12} and C_{13} are replaced by R_{51} and R_{52}.

c) For the first time R_{22} is formulated. Instead of the refutation of the principle of relativity (C_{21}) or its partial acknowledgement (C_{22}), Einstein postulates R_{21}.

d) The Lorentz transformation (G_c) is substituted for the Galilei transformation (G_∞). Relevant is the new (Einstein) interpretation of the former. This shows that the gist of a scientific theory cannot be reduced to its mathematical apparatus.

e) It is worthwhile noting the four-dimensional approach which is based on R_{51} and R_{52} and at the same time leads to a new interpretation of the theory.

f) To the new consequences belongs also R_{54} $\left[\textit{i.e.} \ (\alpha)\right]$.[6]

It is remarkable that whereas in mechanics the Galilei transformation (G_∞) is deduced from C_{12} and C_{13}, in the SR one starts from R_{21} and R_{22} and obtains G_c (R_{31}) (and hence also R_{51} and R_{52}).

Hence the composition of the C hypotheses undergoes an important change. Along with this a couple of hypotheses (which are common with classical physics) are accepted as *e.g.*

C_{11} - Euclidian character of the space,

C_{18} - simplicity of the laws,

R_{23} - preservation of the concepts and the form of the laws, resp. the validity of the law of momentum and energy in the SR and the transition to classical values [Einstein, 1953, p. 45,46].[7]

The assumptions R_{14} and R_{15} are valid for classical physics too. Within the framework of the latter also R_{55} (non-existence of ether) instead of C_{16} is conceivable.

Let us note that the totality of hypotheses is not a stringent fixed set in the sense that the replacement of some statements by others is possible. So *e.g.* in his initial paper (1905) Einstein established the relation $\Delta E = c^2 \Delta m$ (β). Later he deduced the relation (α) wherefrom (β) follows [Einstein, 1907; Einstein, 1953, p. 46-47].

If during the framing of the theory of relativity the adduced statements may be viewed as hypotheses, later on some of them (*e.g.*, R_{22}, R_{42}, R_{31}) became certain statements, and others (the greater part of the group R_5) - facts.

5.2 Which hypotheses have a (mainly) heuristic and which a (mainly) logical function?

With a view to the problem of dynamics and structure of theories, it is interesting to distinguish, as far as possible, the hypotheses which played a considerable role in the framing of SR from those which are relevant to its logical foundations.

Important in heuristic respect is R_{11} and the theory of invariants [Einstein, 1953, p. 21]. A special role is assigned to R_{22} [Einstein, 1953, p. 28].

Of course a part of the heuristic assumptions is also important for the logical building of the theory. A foremostly logical role is played (within the framework of the ST) by R_{42} (Minkowski's interpretation).

One can assert that relative to dynamics or heuristics, the refutation of certain hypotheses and their replacement by new ones is decisive, *i.e.*, their change (overturn), whereas relative to the structure or foundation an essential role is played by the continuity. Einstein said once that Relativity as a deductive system is not distinct from the old physics [Einstein, 1928].

To be sure, the continuity is not alien to the dynamics of theories, especially relative to the requirement of preservation of the form of laws and their invariance. And vice versa, the change in the foundation is related to concrete non-classical statements, as *e.g.* to the definition of simultaneity, the law of light (R_{22}), etc.

6.   What is the hypothetical basis of General Relativity (GR)?

6.1 Without making a parallel examination of GR we can draw at once certain conclusions. The following turn out to be fundamental for GR:

   a) The idea of the generalization of the principle of relativity,
      or the framing of a relativistic theory of gravitation, respec-
      tively. In this connection Einstein becomes conscious of the
      meaning of the proportionality between heavy and inertial mass
      and he formulates the principle of equivalence. The requirement
      of a general covariance of the laws of nature has played an im-
      portant heuristic role.

   b) The principle of Mach has had a non-negligible relevance,

   c) Lastly, the idea to apply Riemannian geometry to these phenomena
      was of decisive importance.

In a logically reconstructed form one can accept Riemannian geometry and Mach's principle (which are connected by the principle of equivalence) as basic assumptions. This is represented by Wheeler by the following draw-ing (a threshing machine whose handle is the equivalence principle) [Wheeler, 1973].

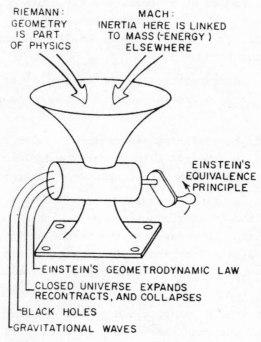

RIEMANN:
GEOMETRY
IS PART
OF PHYSICS

MACH:
INERTIA HERE IS LINKED
TO MASS (-ENERGY)
ELSEWHERE

EINSTEIN'S
EQUIVALENCE
PRINCIPLE

EINSTEIN'S GEOMETRODYNAMIC LAW
CLOSED UNIVERSE EXPANDS
RECONTRACTS, AND COLLAPSES
BLACK HOLES
GRAVITATIONAL WAVES

In the logical construction of GR, Riemannian geometry comes to the fore, whereas Mach's principle cancels out. According to some authors (V. Fock, J. Synge) the principle of equivalence is also superfluous. But in general the equivalence principle is conceived as unshakable both philosophically and experimentally.[8]

New postulates can be formulated for the logical foundation of GR. Thus according to Eddington the fundamental hypothesis of the TR runs as follows: "Everything connected with location which enters into observational knowledge - everything we can know about the configuration of events - is contained in a relation of extension between pairs of events". [Eddington, 1960, p. 10].[9]

Concerning the alternative theories of gravitation, it must be emphasized that the introduction of auxiliary fields of gravitation or/and prior geometries not only makes these theories become more complicated than the Einstein's theory, but also as containing elements which fail to have any experimental basis [Cp. Misner *e.a.*, 1973, § 17.6].

6.2   The novel elements in GR (relative to C and R of the SR) consist first of all in the following:

a) Relative to C_1 (as well as to C_2 and C_4):

   - in the physicalistic conception of geometry and in the giving up C_{11} (Euclidian character of space-time continuum) in favour of a Riemannian metrics of the same;

   - in the non-existence of extraneous (scalar) fields and of a prior geometry, and more generally in the local character of physics,

b) Relative to C_2 - in the generalization of R_{21} or in the requirement of covariance of the laws of nature relative to Gauss' transformation.
   Of great significance is also the principle of equivalence.

c) Relative to C_3, the four-dimensional approach is characteristic, as well as the use of tensor calculus and generally the idea of the prevalence of second order field equations throughout physics.
   The continuity of GR (relative to C and R of SR) is expressed by the fact that the following general assumptions are preserved:

      - the continuity of the field,
      - the law of conservation of momentum and energy,
      - the local validity of SR, etc.

7. What conclusions can be drawn about the dynamics of physical theories?

a) First of all we can state that - irrespective of the attempts
    at fitting Relativity in various philosophical (epistemological)
    conceptions - indeed it does not fit the scheme of any of these
    conceptions [Polikarov, 1966, Chp. 2; Polikarov, 1960; Tonnelat,
    1971, Chp. VIII, XII].

b) Relativity refutes the Duhem's thesis. This theory can not be
    reduced to an improvement whithin the fundamental classical hypo-
    theses. The methodology of Lorentz, to improve the theory in ac-
    cordance with the empirical data which conforms to the Duhem's
    view has failed. The necessity to advance new hypotheses every
    time new facts are known has been justifiably criticized as a
    weakness of the theory [Poincaré, 1900].
    Einstein's methodology which does not stop before the changes
    destroying the classical canons turns out to be doubtlessly su-
    perior. Especially characteristic in this respect is the GR
    which is created without a compulsion from the side of empirical
    knowledge.
    It is to the point to make a distinction between a change (im-
    provement) of certain hypotheses and their refutation and re-
    placement by essentially new ones. On this problem which is the
    subject of a special examination [Polikarov, 1973, § 6] we can
    limit ourserves to the general assertion that as a demarcation
    between these cases the domain of validity of the corresponding
    theoretical constructions can serve.

c) At the same time there is a continuity between the TR and clas-
    sical physics (mechanics and electrodynamics) - what makes
    Popper's conception (at least in its initial version - of so-
    called Naive Falsificationism) unreliable.
    It is worthwhile mentioning that the new theory is connected not
    only to the advancing of new hypotheses, but also to the aban-
    doning of other hypotheses. Indeed, the refutation of some hypo-
    theses goes hand in hand with the fixing of other (eventually
    modified) ones. In our opinion, the statement R_{23} (or the pre-
    servation of the form of laws) is characteristic.

d) Einstein's works played a considerable role in the framing and
    foundation of the so-called hypothetical-deductive view of scien-
    tific theories.
    Already in an early paper Einstein wrote of two stages in the

work of theoretical physics, namely the discovering of general
postulates or principles and the deducing of conclusions which
follow from them [Einstein, 1973, p. 221].

A complete system of theoretical physics, Einstein states, is
made up of concepts, fundamental laws which are supposed to be
valid for those concepts, and conclusions to be reachd by logical deduction. It is these conclusions which must correspond to
our separate experiences [Einstein, 1973, p. 272].

e) Attention has to be drawn to Einstein's formulation of the problem and method (programme, strategy or heuristics) and their ex-
plicite formulation along the lines (as Einstein himself said)
of what he is *de facto* doing.

A decisive moment in the framing of the SR is the selection of
the kind of theory, namely of a principal type.

In the creation of GR, Einstein was guided by the idea that this
theory can be constructed in a predominantly deductive way. If
it has been clear since the time of classical mechanics that all
knowledge of reality starts from experience and ends in it,
Einstein adds on the basis of his experience in the framing of
GR that in a certain sense pure thought can grasp reality
[Einstein 1973, p. 274].

As a matter of fact, in the constructing of GR Einstein proceeds
not simply from the fact of the proportionality of heavy and inertial mass, but he finds out this fact anew (if we can express
this so) from theoretical considerations and attaches a new meaning to it.

In general, the strategy of framing new theories has not a lawlike character [Feynman, 1965]. In TR, an arbitrary combination of elements (from
the old point of view) takes place, as well as an unusual interpretation.
This is especially displayed relative to the two solutions of the dilemma:
reduction of electrodynamics to mechanics or vice versa. Einstein not only
follows the nontraditional antimechanistic way, but after having trans-
fered the principle of relativity in electrodynamics (*i.e.*, enlarged its
validity) he applies the metrics he has established here in the domain of
mechanics and discoverd its more profound physical basis [Einstein, 1921].

When Einstein arrived at the scientific scene he already found several
elaborated theories: mechanics (M), the theory of gravitation (G), electrodynamics (E), thermodynamics (T), elements of the quantum theory (Q)
and certain branches of mathematics (Math) with their assumptions. It is
to be emphasized that in these domains variant theories or branches have

been framed which (partially) accept other hypotheses. Thus along with Newtonian M we also have Hertz's M, as well as a criticism of mechanism (AMs). Against phenomenological T stood statistical T. Likewise in the domain of E Lorentz built up a microscopic E. At the same time the quantum hypothesis shock the belief in the unlimited validity of E. Characteristic for the Math is the emergence of non-Euclidian (Riemannian) geometry which overcomes "the empirically oriented, natural-scientific mathematics" (A.A.Fraenkel) and at the same time creates the prerequisites for its physicalization.

Remarkable is the peculiar pathway followed by Einstein. For one thing (1) he continues the line of E and partially the work of Lorentz, while repudiating the conjectures C_{15} and C_{17} and putting forward R_{22} and R_{31}. Yet (2) he transfers this trend onto new rails, namely along the lines of phenomenological T (R_{11}). Then (3) he refutes C_{12} and C_{13} and retains certain statements of M (C_{11}, R_{21}, R_{23}). At the same time (4) he is leaning on AMs. Further (5) he abandons G and frames the GR, whereby the ideas (6) of the Riemannian geometry and partially (7) of Hertz's M find a creative development.

We have tried to visualize all that through a scheme, where the TR is outlined as a crossroad. Hereby the synthesis of the basic theoretical

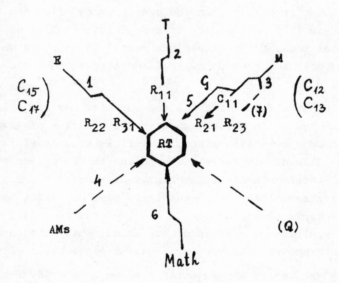

conceptions is brought about: 1 - 4 are essential for the framing of the SR, and 5 - 7 - for the GR.

This paper was prepared during my stay at the "Institut für Philosophie, Wissenschaftstheorie, Wissenschafts- und Technikgeschichte der Technischen Universität Berlin", sponsored by a special fellowship of the Alexander von Humboldt Foundation.

Notes

1. A propos, the conceptions of a negative type (like atheism, indeterminism, or refutation of the ether) have a meaning relative to the corresponding positive conceptions (theism, determinism, asserting the existence of the ether).

2. To this belong, according to Painlevé, also the postulate of the initial conditions, the axiom of causality, among others [Dugas, 1950, p. 478] .

3. Besides, it will be assumed the symmetry and the transitivity of the simultaneity.

4. Thre are known several (mathematical) ways which lead to a derivation of R_{31} [Cp. Dugas, 1950, p. 478].

5. Supplementary assumptions will be introduced in order to eliminate the so-called twins-paradox [Pokhrovnik, 1967; Tonnelat, 1971, Chp. VII, § 9]. The definition of the so-called 'Snapshots' (J. L. Synge) depends on certain assumptions too [Tonnelat, 1971, p. 169, 509].

6. This is the reason why P. G. Kard speaks about a violation of the principle of correspondence between classical and relativistic physics. We do not agree with this allegation. First of all a similar relation is known in electrodynamics; in the SR it undergoes a corresponding generalization. Besides there is a correspondence between the relativistic and non-relativistic formulae for momentum and kinetic energy [Cp. H. Goldstein, 1950].

   On the other hand it is argued that the relation (α) (which is suggested by SR and consistent with it) is not a deduction from this theory [Lindsey a. Margenau, 1957]. This allegation is not acceptable, because in the framework of the fourdimensional examination it is a natural consequence from SR.

7. To these postulates, according to Painlevé, belong also
   - the Kepler-Fresnel's postulate (which comprises the law of inertia and Fresnel's principle of the propagation of light),
   - the postule of the reversibility of the equations of motion, etc. [Dugas, 1950, p. 478,514].

   At the same time the hypothesis of the isotropy of space [Einstein, 1953, p. 24] is not necessary [Tonnelat, 1971, p. 135], as well as the assumption of the linear character of the transformation (Weyl) or of its continuity (A. Aleksandrov-V.Ovchinnikova; P. Suppes).

8. The principle of equivalence is substantiated by the gravitation red shift [Misner *e.a.*, 1973, § 7.4, 16.2] or, as Deser (1970) has shown, it follows from the requirement of consistensy [Misner *e.a.*, 1973, § 17.5].

9.   The additional postulates are:

- the principle of identification, resp. the principle of
  measurement
- the electromagnetic equations,
- the possibility to formulate the law of gravitation by
  differential equations of the second order [Eddington,
  1960, §§ 47, 74, 62 *e.a*] .

As a peculiar group of assertions of GR, Reichenbach states the
following: By the aid of light, solid bodies and natural clocks,
a space-time metric with certain properties can be constructed.
Further, as an assertion which is more general than the principle of equivalence he states that the coefficients $g_{\mu\nu}$ of the
metrics lead to a theory of gravitation [Reichenbach, 1965] .

It deserves mentioning the approach of R. Hojman, K. Kuchař and
C. Teitelboim, where the equation of Einstein-Hamilton-Jacobi is
basic and from which Einstein's field equation is derived
[Misner *e.a.*, 1973, § 17.5] .

References

M. Born, Einsteins Theory of Relativity, N.Y., (1962) 252

R. Dugas, Histoire de la mechnique, Neuchâtel, (1950)

A. S. Eddington, The Mathematical Theory of Relativity, Cambridge, (1960)

A. Einstein, Über die vom Relativitätsprinzip geforderte Trägheit der
    Energie, Ann. d. Phys., 23, (1907) 371

-, Zur Theorie des statischen Gravitationsfeldes, Ibid., 38, (1912) 443

-, A Brief Outline of the Development of the Theory of Relativity,
    Nature, 106, (1921) 782

-, A propos de "La déduction relativiste" de M. Emile Meyerson, Rev.
    philos. de la France et de l'étranger, 55, (1928) 161

-, The Meaning of Relativity, Princeton, NJ, (1953)

-, Autobiographical Notes, in: Albert Einstein - Philosopher-Scientist.
    Ed. P. A. Schillp, Evanston, IL, (1957) 3

-, Fundamental Ideas and Problems of the Theory of Relativity, in: Nobel
    Lectures. Physics 1901 - 1921, Amsterdam *e.a.*, (1967) 482

-, Ideas and Opinions, L. (1973)

B. Ellis, Is Signal Synchrony Independent of Transport Synchrony? Philos.
    Sci., 45, (1978) 309

R. Feynman, The Character of Physical Law, L. (1965)

S. Goldberg, Poincaré Silence and Einstein's Relativity, Brit. J. Hist.
    Sci., 5 (1970) 73

H. Goldstein, Classical Mechanics, Reading, Mas., Chp. 6, (1950)

A. Grünbaum, Philosophical Problems of Space and Time, Dordrecht-Boston, (1973)

W. Heisenberg, Physics and Beyond, N.Y. *e.a.*, (1971) 64

R. B. Lindsey and H. Margenau, Foundations of Physics, N.Y., (1957) 352

D. Melament, Causal Theories of Time and the Conventionality of Simultaneity, Noûs, 11, (1977) 293

Ch. W. Misner, K. S. Thorne and J. A. Wheeler, Gravitation, San Francisco, (1973)

H. Poincaré, Rapports du Congrès de physique de 1900, P., (1900) 22, 23, quoted after H. A. Lorentz, A. Einstein and H. Minkowski, Das Relativitätsprinzip, Lpz.-Bln., (1922) 7 - 8

S. J. Pokhrovnik, The Logic of Special Relativity, Cambridge, (1967)

A. Polikarov, Philosophische Diskussionen über die Relativitätstheorie, in: Forschen und Wirken. Festschrift zur 150-Jahr-Feier der Humboldt-Universität zu Berlin, Bd. III, B., (1960) 33

-, Relativity and Quanta, Moscow, (Russ.), (1966)

-, Science and Philosophy, Sofia, (1973)

-, On the Hypothetic-deductive Model of Scientific Theories, C. R. Acad. bulg. Sci., 31, (Russ.), (1978) 1097

H. Reichenbach, Philosophie der Raum-Zeit Lehre, Bln.-Lpz., (1928) 151

-, Axiomatik der relativistischen Raum-Zeit Lehre, Braunschweig, (1965), 110 - 111

M.-A. Tonnelat, Histoire du principe de relativité, P. (1971)

J. A. Wheeler, From Relativity to Mutability, in: The Physicist's Conception of Nature. Ed. J. Mehra, Dordrecht, (1973) 205

DIE ERKENNTNISTHEORETISCHEN QUELLEN EINSTEINS

Peter Janich, Universität Konstanz

Die Einladung, über die erkenntnistheoretischen Quellen A l b e r t
E i n s t e i n s zu referieren, habe ich angenommen, ohne mir die
besonderen Schwierigkeiten vor Augen zu halten, in die dieses Thema
führt. Freilich ist es nicht besonders schwierig, herauszufinden, was -
orientiert an einem Vorverständnis von Erkenntnistheorie - Einsteins
Lektüre war. Sofern also keine spektakulären Neuentdeckungen erwartet
werden, die Einsteins erkenntnistheoretischen Hintergrund in völlig neu-
em Licht erscheinen ließen, läßt sich ein Überblick wohl geben. Ohnehin
kann ich auf dem Gebiet, bisher unentdeckte Quellen zu finden, wenig leisten, da mir für diesen Vortrag nur bereits Gedrucktes als Grundlage zur
Verfügung stand. Mir sind also insbesondere noch nicht veröffentlichte
Teile des Einstein-Nachlasses, aber auch erkenntnistheoretische Quellen
in Form nicht veröffentlichter Gespräche mit Zeitgenossen unzugänglich
gewesen.[1] Die spezifische Schwierigkeit eines Referates über die erkenntnistheoretischen Quellen Einsteins rühren jedoch auch weniger von
der Quellensituation her, sondern ergeben sich aus der zu unterstellen-
den Erwartung, die hier aufgeworfene Frage würde eine *wirkungsgeschichtliche*
Antwort finden. Während die Präsentation einer Lektüreliste, eventuell mit
Inhaltsangaben, lediglich eine Präliminarie sein kann, fordern die groß-
artigen Leistungen Einsteins auf dem Gebiet der Physik die Frage nach
einem möglichen Einfluß erkenntnistheoretischer Meinungsbildung vor allem
beim jungen Einstein heraus.

Abgesehen von der Fülle kontingenter Schwierigkeiten, die im Falle gei-
stesgeschichtlicher Kausalbehaptungen mit der jeweils betrachteten Person und ihrer Situation zusammenhängen, erwachsen im Falle erkenntnistheoretischer Prämissen für physikalische Neuentwicklungen prinzipielle
- und deshalb zu diskutierende - Probleme, welche die Möglichkeit dra-
stisch beschränken, so etwas wie einen stringenten Nachweis philosophischen Einflusses auf die physikalischen Pionierleistungen Einsteins zu
führen. Hochgespannte Erwartungen an einen solchen Nachweis werden sich
also nicht erfüllen lassen.

1.   Die erkenntnistheoretische Lektüre Einsteins

Ich beginne bei einem Überblick über die von Einstein gelesenen Autoren
und, soweit bekannt, Titel. Einstein selbst äußert sich in seiner Autobiographie dazu und erwähnt auch einiges in seinen Briefen, wobei ich

vor allem an die Korrespondenzen mit M i c h e l e   B e s s o , M a x
B o r n   und A r n o l d   S o m m e r f e l d   denke. Außerdem sind
hier - quasi als halbe Korrespondenz - Einsteins Briefe an M a u r i c e
S o l o v i n e   zu nennen, deren Veröffentlichung in Buchform eine in-
formative Einführung Solovines, genauer ein Bericht über die gemeinsame
Berner Zeit, vorangestellt ist. (Die Briefe Solovines an Einstein sind
mir unbekannt.) Außerdem finden sich verstreut in der Literatur Hinweise,
und mit einem solchen möchte ich beginnen. In einem Brief an *Carl Seelig*[2]
schreibt Einstein:

> *Ich liebte mehr* (nämlich als poetische Bücher, Vf.) *Bücher
> weltanschaulichen Inhalts und im besonderen philosophische.
> S c h o p e n h a u e r ,   H u m e ,   M a c h ,   zum Teil K a n t ,
> P l a t o   und A r i s t o t e l e s .*

Welche Werke im Falle von Plato, Aristoteles und Schopenhauer gemeint
sind, konnte ich nicht feststellen. Ich kann nur vermuten, daß es sich
hierbei um keine sehr ausführliche und profunde Lektüre gehandelt hat.
Denn hinsichtlich Aristoteles, der von den drei Genannten für spätere
Einflüsse auf die Physik wohl der Einschlägigste wäre, schreibt Einstein
1948, also als Neunundsechzigjähriger, an Solovine, daß er als Abendlek-
türe seiner Schwester "einiges von den philosophischen Schriften des
Aristoteles" vorgelesen habe, und er fügt hinzu: "Es war eigentlich recht
enttäuschend; wenn es nicht so dunkel und konfus wäre, hätte sich diese
Art Philosophie nicht so lange halten können."[3] Diese Bemerkung scheint
mir dafür zu sprechen, daß Einstein bis dahin wohl keine genauen, ihm
irgendwie präsenten Kenntnisse von Aristoteles hatte, sondern allenfalls
einige Einschätzungen, die aus früherer und wohl eher peripherer Bekanntschaft mit der aristotelischen Physik durch Sekundärliteratur entstanden
waren. Ich habe auch keinen einzigen Hinweis finden können, inwiefern
Plato, Aristoteles oder Schopenhauer irgendeinen nachhaltigen Eindruck
auf Einstein gemacht haben könnten, sofern man nicht in der oberfläch-
lichsten Weise schon überall dort von Platonismus sprechen möchte, wo
die Mathematik für den theoretischen Physiker Bedeutung erlangt, ohne
unmittelbar mit Operationalisierung oder Erfahrung in Verbindung gebracht
werden zu können.

Bezüglich Einsteins Kant-Lektüre habe ich bei J. W i c k e r t   den
Hinweis gefunden, daß ein Mitschüler Einsteins aus der Aarauer Zeit,
H a n s   B y l a n d ,   berichtet, Einstein hätte sich damals an Kants
Kritik der reinen Vernunft "berauscht".[4] Abgesehen davon, daß Einstein
damals sechzehn oder siebzehn Jahre alt war, bin ich bezüglich einer
nachhaltigen Wirkung dieser frühen Kant-Lektüre sehr skeptisch. Denn in
seiner Replik zu den Aufsätzen in S c h i l p p ' s   Einstein-Band,

also im vorgerückten Alter, schreibt Einstein als eine Teilantwort auf
den Beitrag von H e n r y M a r g e n a u: "Ich bin nicht in Kant'-
scher Tradition aufgewachsen, sondern habe das Wertvolle, was neben heute
offenbaren Irrtümern in seiner Lehre steckt, erst spät begriffen."[5] Abgesehen davon, daß diese Bemerkung Einsteins einen expliziten Hinweis auf
eine spätere Lektüre enthält, bezieht sich die Auffassung von der "heute
offenbaren Irrtümern" wohl in erster Linie auf Kants Apriorismus, den
Einstein während seines ganzen Lebens - soweit sich dies aus den mir zugänglichen Quellen rekonstruieren läßt - psychologistisch mißverstanden
hat. Dabei hätte eine eingehende Kant-Lektüre, vielleicht unterstützt
durch eine Lektüre von Schriften des Neukantianismus, sehr wohl zeigen
können, daß die Notwendigkeit apriorischer Urteile von Kant keineswegs
als "denknotwendig"[6] im psychologischen Sinne gemeint war.

Ein weiterer Beleg dafür, daß eine erste ernstere Bekanntschaft Einsteins
mit Kant *nach* der Konzeption der Relativitätstheorien stattgefunden hat,
ist ein Brief an Max Born, der im Sommer 1918 geschrieben sein muß. Einstein berichtet dort aus einem Urlaub: "Ich lese hier unter anderem Kants
Prolegomena und fange an, die ungeheure suggestive Wirkung zu begreifen,
die von diesem Kerl ausgegangen ist und immer noch ausgeht."[7] Im Sinne
einer intensiven Lektüre Kants darf man also wohl davonausgehen, daß Einstein zumindest bei den Theorien, die heute bei vielen als eine Revolu-
tionierung unserer Vorstellungen von Raum und Zeit gelten, keinem Einfluß
durch die gerade auf diesem Gebiet einschlägige Philosophie Kants unter-
legen war. Die von Einstein oft beiläufig betonte Ablehnung der synthetischen Urteile apriori läßt sich zwanglos als Übernahme von anderen zeitgenössischen Autoren der Sekundärliteratur, vor allem der Wissenschaftstheoretiker des frühen Wiener und Berliner Kreises betrachten. Diese Einschätzung bestätigt meines Erachtens Einstein auch selbst: er soll auf
die ihm in der Société de Philosophie gestellte Frage, ob seine Theorien
mit der Lehre Kants im Widerspruch stünden, geantwortet haben: "Das ist
schwer zu sagen. Jeder Philosoph hat eben seinen eigenen Kant."[8] Wäre
Einstein wirklich an Kant orientiert gewesen, und sei es, daß ihn die Ablehnung Kants motiviert hätte, so wäre eine solche Antwort schwer verständlich.

Aus der ersten gegebenen Aufzählung bleiben also als Autoren Hume und
Mach. Weitere, ausführlichere Hinweise verdanken wir dem schon erwähnten
Vorwort, das Maurice Solovine zu den an ihn gerichteten Einstein-Briefen
geschrieben hat. Diese Angaben über die Lektüre der sogenannten *"Akademie
Olympia"* betreffen die gemeinsame Berner Zeit, genauer die Zeit von 1902
bis 1905, die für Einstein im Diskussionszirkel mit S o l o v i n e
und C o n r a d H a b i c h t wohl die ausgiebigste Beschäftigung

mit erkenntnistheoretischen Fragen mit sich brachte. Und dies ist die
Zeit vor den bahnbrechenden ersten Arbeiten, von der also eine Prägung
Einsteins durch erkenntnistheoretische Quellen am ehesten anzunehmen
wäre. Dem Bericht Solovines nach begann man mit K a r l P e a r s o n s
Die Grammatik der Wissenschaft, um dann bereits von Ernst Mach *Die Analyse der
Empfindungen* und die *Mechanik* zu lesen, von der Solovine behauptet, Ein-
stein hätte sie bereits früher einmal durchgearbeitet. Aus der Logik von
M i l l wurde das dritte Buch (Über die Induktion) gelesen, von
H u m e der Traktat über die menschliche Natur, sodann die Ethik von
S p i n o z a, die *Vorträge und Reden* von H. v. H e l m h o l t z,
einige Kapitel aus A m p è r e s *Essai sur la philosophie des sciences*,
B. R i e m a n n s *Grundlagen der Geometrie*, einige Abschnitte aus der
Kritik der reinen Erfahrung von R. A v e n a r i u s, außerdem J. C.
C l i f f o r d s Essai *Über die Natur der Dinge an sich*, schließlich R.
D e d e k i n d s Zahlentheorie und H. P o i n c a r é s *Wissen-
schaft und Hypothese*, dessen nachhaltiger Eindruck auf die drei Olympier
von Solovine besonders hervorgehoben wird.

Um diese Liste, in der Mach, Hume und Poincaré von Solovine als die wichtigsten bezeichnet werden, auch aus der Sicht Einsteins zu gewichten, zitiere ich aus zwei Briefen Einsteins an Besso. Am 6. März 1952 berichtet
er über

> *Bern, wo wir uns hauptsächlich mit D. Hume beschäftigen (in einer recht
> guten deutschen Ausgabe). Diese Lektüre war auf meine Entwicklung von
> ziemlichen Einfluß - neben Poincaré und Mach.*[9]

Und vier Jahre früher, am 6. Januar 1948, schrieb Einstein:

> *Was nun Machs Einfluß auf meine Entwicklung anlangt, so ist er sicher groß
> gewesen. Ich erinnere mich ganz gut daran, daß Du mich auf seine Mechanik
> und Wärmelehre hingewiesen hast während meiner ersten Studienjahre, und
> daß beide Bücher großen Eindruck auf mich gemacht haben. Wie weit sie auf
> meine eigene Arbeit gewirkt haben, ist mir offen gestanden nicht klar. So
> viel ich mir bewußt werden kann, war der unmittelbare Einfluß von D. Hume
> auf mich größer. Ich habe diesen zusammen mit Konrad Habicht und Solovine
> in Bern gelesen. Aber wie gesagt bin ich nicht imstande, das im unbewußten
> Denken Verankerte zu analysieren.*[10]

Im Hinblick auf die These dieses Referats ist hier nicht nur die
Einschätzung Einsteins von Bedeutung, daß Mach und Hume die für
ihn wichtigen Autoren waren, sondern auch, daß Einstein in einer
ernsthaften und nicht nur beiläufigen Bemerkung den Einfluß der
für ihn wichtigsten Autoren im Bereich des Unbewußten sieht,
nicht aber konkrete Einzelbehauptungen oder Theorien nennt, an
die er sich zustimmend oder ablehnend angeschlossen hätte.

Einen eher speziellen Hinweis über die Art der Wirkung, die Einstein
selbst von dieser Lektüre verspürt haben mag, gibt er in seiner Autobiographie: "Das kritische Denken, dessen es zur Auffindung dieses zentralen
Punktes bedurfte (Einstein spricht hier von dem Paradox, daß ein einem

Lichtstrahl nacheilender Beobachter dasselbe feststellen müsse wie ein auf der Erde ruhender), wurde bei mir entscheidend gefördert insbesondere durch die Lektüre von David Humes und Ernst Machs philosophische Schriften."[11] Philosophischer Einfluß wird also in einer allgemeinen Schulung kritischen Denkens, nicht im Sinne spezieller Thesen als konkreter theoretischer Vorleistungen gesehen. Im Sinne der wirkungsgeschichtlichen Frage würde ich allerdings entgegen der Selbsteinschätzung Einsteins den Einfluß Machs für stärker halten als den Humes, was mit der Kernthese dieses Referats zusammenhängt, die ich vorgreifend formulieren möchte: Es waren kaum erkenntnistheoretische als vielmehr physikalische Vorgaben, an denen sich Einstein bei seinen revolutionären Arbeiten zur Physik orientiert hat. (Diese These wird übrigens ähnlich von H. R e i c h e n - b a c h vertreten, der schreibt: "Alle Forschungsgegenstände Einsteins finden sich in erster Linie im Reich der Physik.")[12] Daß Machs Einfluß auf Einstein wohl stärker war als der Humes, liegt also nicht daran, daß Mach - etwa in seiner *Analyse der Empfindungen* - eine für Einstein attraktivere Erkenntnistheorie entwickelt hätte als Hume in seinem Treatise. In seinen späten Jahren hat Einstein die Mach'sche Erkenntnistheorie sogar explizit als *unhaltbar* bezeichnet.[13] Vielmehr war Mach eben auch Physiker und hat insbesondere in seiner *Mechanik* eine ausführliche Analyse und Kritik der Prinzipien N e w t o n s , insbesondere der Definitionen der Grundbegriffe der Mechanik gegeben. Daß letztlich diese Passagen des Gesamtwerks von Mach die für Einstein wichtigsten wurden, sehe ich unter anderem durch den ungewöhnlichen Umfang bestätigt, den Einstein in seinem Nachruf auf Ernst Mach in der Physikalischen Zeitschrift vom 1. April 1916 diesen Passagen widmet. Als weiteres Indiz dafür läßt sich anführen, daß Einstein (am 9. August 1909) an Mach geschrieben hat (in meiner Rückübersetzung aus dem Englischen; der Originalwortlaut ist mir nicht zugänglich.): "Ich kenne natürlich Ihre wichtigsten Schriften sehr gut, von denen ich am meisten Ihr Buch über die Mechanik bewundere."[14]

Die nach Solovine aufgezählten Autoren lassen sich grob in drei Gruppen einteilen. Die erste will ich die der großen Philosophen nennen, also Plato, Aristoteles, Hume, Kant und Schopenhauer, von denen uns hier nur noch Hume und Kant interessieren. Sodann die Gruppe der modernen Wissenschaftsphilosophen, nämlich Avenarius, Pearson, Clifford, Mach und Poincaré. Die dritte Gruppe von Autoren umfaßt die - freilich wissenschaftstheoretisch bedeutsamen - Fachwissenschaftler, nämlich Dedekind, Riemann, Helmholtz, K i r c h h o f f und die bei Solovine nicht erwähnten H e r t z und M a x w e l l .

Zur zweiten Gruppe, den Wissenschaftsphilosophen, mögen einige wenige Anmerkungen genügen. Der knapp vierunddreißigjährig verstorbene Clifford,

dessen Leistungen in Mathematik bekannt sind, ist in Einsteins Lektüre
mit einem einzigen Aufsatz vertreten. In diesem findet sich eine extra-
vagante Kant-Interpretation, die, salopp ausgedrückt, allenfalls zu einem
respektlosen Umgang mit der Philosophie Kants anregen kann. Avenarius, in
der Terminologie M. Schlicks zusammen mit Mach ein Immanenzpositivist,
kann als Vorläufer Machs angesehen werden und bedarf deshalb hier keiner
weiteren detaillierten Diskussion. Das Werk K. Pearsons wird erkenntnistheoretisch sozusagen von Hume und Mach abgedeckt, und in seiner lesens-
werten Analyse der Newtonschen Prinzipien, bei der die Newtonschen Bewegungsgesetze aus dem Prinzip der Relativität aller Bewegung rekonstruiert
werden, wird er an Einfallsreichtum und begrifflicher Prägnanz von Mach
übertroffen. Aus der Reihe der modernen Wissenschaftsphilosophen bleiben
demnach noch Mach und Poincaré als potentielle Kandidaten für einen Einfluß auf Einstein übrig. Den bislang noch nicht eingeordneten Ampère mit
seinem *Essai sur la philosophie des sciences* übergehe ich als enzyklopädisti-
sches Fossil, und auf Mill, bei dem ich gezögert habe, ihn ohne weiteres
unter die *großen Philosophen* einzureihen, komme ich noch beiläufig zu sprechen.

Spätestens anläßlich dieser hier gegebenen Einteilung von Autoren in drei
Gruppen ist eine terminologische Präzisierung nachzutragen. Während ich
Erkenntnistheorie im traditionellen und damit auch etwas unscharfen Sinne
für alle Bemühungen verwende, die sich mit Gegenständen und mit Formen
menschlicher Wissensbildung befassen, verstehe ich unter Wissenschaftstheorie eine wissenschaftliche, mit den Mitteln der Sprachphilosophie,
der Logik und einer terminologisch exakten Methodenlehre arbeitende Be-
mühung, die die Wissenschaft, in meinem Vortrag die Physik, zum Gegenstand hat. Mit dieser Unterscheidung läßt sich dann z.B. differenzieren
zwischen dem Erkenntnistheoretiker Mach, nämlich dem Autor der *Analyse der
Empfindungen*, dem Wissenschaftstheoretiker Mach, nämlich dem Kritiker Newtons, und schließlich dem Physiker Mach, etwa dem Entdecker der Schock-
wellen. Im folgenden betrachte ich die Frage nach einem erkenntnistheoretischen Einfluß auf die Physik Einsteins.

2. Das Problem der erkenntnistheoretischen Prägung

Eingangs wurde die Vermutung geäußert, ein Referat über die erkenntnistheoretischen Quellen Einsteins würde die Erwartungen wecken, eine wirkungsgeschichtliche Antwort zu finden. Weiter sei hier unterstellt, daß
es dabei um die Leistungen Einsteins geht, derentwegen er heute in erster
Linie gefeiert wird, nämlich um seine Beiträge zur Physik. Paradigmatisch
möchte ich mich hier auf die Relativitätstheorien, vor allem die spezielle
Relativitätstheorie, beschränken, weil dort der Bruch mit älteren Vor-

stellungen den prägnantesten Anlaß für eine wirkungsgeschichtliche Frage
bietet, und weil für ihren Gegenstand mehr als etwa bei den Arbeiten Einsteins zur Quantentheorie zu erwarten ist, daß einschlägige Fragen bereits von älteren Erkenntnistheoretikern diskutiert worden sind. Damit
hat sich die anstehende Frage darauf eingeschränkt, ob ein Einfluß von
Hume, Kant, von Mill oder Mach auf die Bildung der Relativitätstheorie
durch Einstein nachweisbar ist.

Bei Aufrechterhaltung auch nur bescheidener Standards an Argumentation
und Begründung ist diese Frage meines Erachtens unbeantwortbar, und zwar
prinzipiell. Diese Behauptung soll kurz begründet werden:

Die Ausgangssituation ist dadurch gekennzeichnet, daß zwei Sorten von
Texten vorliegen, nämlich erkenntnistheoretische auf der einen und physikalische auf der anderen Seite. Wie sind diese aufeinander zu beziehen?
Selbstverständlich sind die erkenntnistheoretischen Überlegungen, die
sich in Einsteins Lektüre finden, keine metasprachlichen Texte zu den
physikalischen Schriften Einsteins, wie dies etwa bei manchen Texten der
modernen analytischen Wissenschaftstheorie der Fall ist, sofern sie sich
konkrete Stücke der Physikgeschichte zum Gegenstand nehmen. Auch sonst
ist die Beziehung von erkenntnistheoretischen und physikalischen Texten
aufeinander - selbst im Falle von Äquivokationen, also wenn in beiden
dieselben Wörter wie z.B. Länge, Zeit, Körper, Raum usw. vorkommen -
nicht gegeben. Denn daß die alten Autoren keine Einsteinsche Terminologie
schreiben, ist trivial wahr, und daß sich Einstein - wohlgemerkt in seiner physikalischen Terminologie - nicht an Hume, Kant oder Mill anschließt,
ist empirisch wahr. Bedauerlicherweise gibt es auch darüber hinaus keinen
expliziten definitorischen, also logisch streng reformulierbaren Zusammenhang zwischen beiden Vokabularen. Dazu sind diese nämlich sowohl im
Falle der Erkenntnistheorie als auch der Physik zu unbestimmt. Damit will
ich sagen, was Fachleuten ohnehin klar ist, daß die erkenntnistheoretische Lektüre Einsteins und seine Physik nur über sehr weitreichende Interpretationen aufeinander beziehbar werden. Die Aufgabe, eine Wirkung
der Erkenntnistheorie auf Einsteins Physik nachzuweisen, wirft damit die
Frage nach einer geeigneten Interpretation der geannten Philosophen auf,
in der ein Zusammenhang zu Einstein herstellbar wird.

Formal scheint es hier die beiden Möglichkeiten zu gehen, die genannten
Erkenntnistheoretiker - wie man sich wohl üblicherweise ausdrückt -
im Sinne Einsteins zu lesen, d.h. so weit wie möglich in seine Terminologie
zu übersetzen, oder umgekehrt Einstein *erkenntnistheoretisch* zu lesen, d.h.
etwa auf erkenntnistheoretische Voraussetzungen zu befragen. Wer Fiktionen liebt, dem eröffnet sich hier ein weites und dankbares Betätigungsfeld.

Mir kommt es darauf an, daß beide Wege - und möglicherweise eine glückli-
che Mischung beider - die Wirkungsbehauptung über Erkenntnistheorie auf
Einstein zu einer Tautologie zu machen. Mit anderen Worten, das Vorgehen,
eine der beiden Textsorten mithilfe der anderen zu interpretieren, reduziert eine historische Frage auf analytisch wahre und damit historisch
uninteressante Sätze.

Weniger formal schließlich und konkreter bezogen auf die heutige Diskussion über die wissenschaftlichen Leistungen Einsteins wäre die weitere
Möglichkeit, Interpretationsgesichtspunkte von außen, etwa aus der modernen Wissenschaftstheorie oder dem gegenwärtigen Selbstverständnis der
Physiker, in Ansatz zu bringen. Auch hier gibt es kaum einen Zweifel, daß
sich auf diese Weise zahlreiche Ähnlichkeiten und Parallelen zwischen den
genannten Philosophen und Grundgedanken der Relativitätstheorie ergäben.
Es ist freilich zu befürchten, daß ein wendiger Kantianer z.B. mit Ver-
weis auf Einsteins kreative Phantasie für die Ausbildung neuer Begriffe
von Raum und Zeit für die eigene Partei reklamieren könnte, ein wendiger
Anhänger Humes seinen Lieblingsphilosophen im Empirismus Einsteins wie-
derfände und ein Historiker schließlich, der keinen Lieblingsphilosophen
hat, den alten Gegensatz von Empirismus und Rationalismus bei Einstein
z.B. in seinen Argumenten für die spezielle Relativitätstheorie aufs
Fruchtbarste aufgehoben fände. Alles dies erscheinen mir jedoch allenfalls geistreiche Kunststücke, aber an keiner Stelle streng überzeugende,
seriös argumentierende Belege.

Die in der Tat nächstliegende Interpretationshilfe, erkenntnistheoretische und physikalische Texte aufeinander zu beziehen, sei als letzte Va-
riante betrachtet: bekanntlich hat Einstein nicht nur physikalische Texte
geschrieben, sondern sich gern und oft zu philosophischen und wissenschaftstheoretischen Fragen geäußert. Es liegt deshalb der Ausweg nahe,
Einstein selbst als Einsteininterpreten zu bemühen. Bei diesem Verfahren
findet man sich außerdem in der glücklichen Lage, auf eine bekannte und
anerkannte Arbeit als Beispiel zurückgreifen zu können, nämlich auf den
Aufsatz *Mach, Einstein, and the Search for Reality* von G e r a l d   H o l-
t o n.[15] Dort geht es zwar nicht um Hume, Kant oder Mill, sondern um
Ernst Mach, und zwar, soweit Holtons Thesen betroffen sind, vor allem um
den Erkenntnistheoretiker Mach.

Der zitierte Aufsatz enthält als Kernthese die Behauptung, Einstein habe
während seines wissenschaftlichen Lebens eine philosophische Wandlung
durchgemacht, an deren Anfang - also etwa während der Berner Zeit und
damit auch während der Entstehungszeit der großen Arbeiten von 1905 - eine
sensualistisch-empiristische Position mit allergrößter Nähe zu Mach stand,

und die am Ende (das wohl schon in den dreißiger Jahren erreicht ist) auf
eine Position geführt hat, die Holten "rationalen Realismus" nennt. Holton unterscheidet dabei vier Phasen, nämlich den frühen Einfluß Machs auf
Einstein (vor allem bis etwa 1905); zweitens die Zeit des Briefwechsels
zwischen Mach und Einstein, von dem nur die Briefe Einsteins an Mach erhalten sind (die Zeit etwa von 1909 bis 1913, in die auch das persönliche
Zusammentreffen der beiden Gelehrten fällt); drittens die beginnende Di-
vergenz der Ansichten, die auf Seiten Machs mit der Abfassung des Vor-
worts zum Buch *Die Principien der physikalischen Optik* im Juli 1913 vollzogen
ist. (Das Buch erscheint posthum erst 1921 und wird auch erst dann Einstein bekannt.) Mach weist dort zurück, ein Vorläufer der Relativisten zu
sein, und lehnt ausdrücklich die Relativitätstheorie ab, der er wachsen-
den Dogmatismus vorwirft. Auf Seiten Einsteins vollzieht sich, so die
These Holtons, die Abkehr von Mach - zunächst von Einstein selbst unbe-
merkt, der noch von sich behauptet, ein Anhänger Machs zu sein, aber z.B.
über die einer Theorie zugrundeliegenden allgemeinen Erfahrungstatsachen
spricht und damit längst Behauptungen aufstellt, die mit Machs Position
unvereinbar sind. Diese Abkehr Einsteins von Mach kulminiert, nachdem
Einstein die Ablehnung der Relativitätstheorie durch Mach zur Kenntnis
genommen hatte, in der Behauptung, Mach sei zwar ein guter Mechaniker,
aber ein erbärmlicher Philosoph gewesen. In der vierten Phase schließlich
registriert Holton bei Einstein eine wachsende Ablehnung des Positivis-
mus, eine Kritik an der antimetaphysischen Haltung der Positivisten (die
ja deren Hauptmotiv seit den Tagen Machs, und so man will, seit den Tagen
Humes war), und eine Hinwendung zu einer "realistischen", die menschen-
unabhängige Existenz der Außenwelt hypostasierenden Position, die Holton
ohne Angabe von Gründen "rational" nennt.

Der Zusammenhang, in dem hier über Holton gesprochen wird, war mit der
Frage gegeben, ob vielleicht Einstein selbst der beste Einsteininterpret
im Hinblick auf seine erkenntnistheoretische Lektüre sei. Dabei darf für
Holton behauptet werden, daß dieser mit aller wissenschaftshistorischer
Sorgfalt vorgeht. Es seien deshalb die historischen Belege, die er an-
bietet, und die dann daran geknüpften Argumente näher betrachtet.

Holton zitiert einige Bemerkungen Einsteins über sein Verhältnis zu Mach,
die entweder aus der Autobiographie stammen, also im Alter von sechsund-
sechzig Jahren geschrieben wurden - sie lassen sich übrigens durch sinngemäß gleiche aus früheren Briefen vermehren -, oder Bemerkungen, die aus
zustimmenden Formulierungen Einsteins an Mach selbst stammen, so etwa,
daß Einstein seinen Brief vom 17. August 1909 schließt mit den Worten
"Ihr Sie verehrender Schüler". Alle diese Bemerkungen sind jeweils ganz
allgemeiner Natur, d.h. sprechen von Einfluß, Eindruck, Bewunderung und

Interesse, enthalten aber keine Hinweise auf bestimmte Thesen Machs.

Dann analysiert Holton die Arbeit über die Elektrodynamik bewegter Körper und bietet als Resultat an, daß die Machsche Komponente - "a strong
component, even if not the whole story" - sich in zwei Punkten zeige:
Erstens in Einsteins Insistieren, daß Grundlagenprobleme der Physik nur
gelöst werden können, nachdem eine erkenntnistheoretische Analyse durchgeführt ist (was selbstverständlich kein physikalischer Satz ist), und
zweitens, daß sich die Begriffe des Ereignisses bei Einstein und des Elements bei Mach nahezu vollständig überlappen ("overlaps almost entirely").

Dieses zweite Argument ist sicher falsch, weil die "Grundelemente" Machs
Empfindungen sind, die in einem Menschen stattfinden, während Einsteins
gleichzeitige Ereignisse am Ort als Schnitt zweier Weltlinien Außenbeob-
achtbares darstellen, also allenfalls Gegenstand einer Empfindung sein
können - aber auf diesen Fehler soll es hier nicht ankommen. Wichtig ist
vielmehr, daß einige allgemein gehaltene Bemerkungen Einsteins über sein
Verhältnis zu Mach für Holton den Grund dafür abgeben, daß eine physikalische Arbeit Einsteins unter der Perspektive eines Mach-Kenners gelesen
wird. Die Zuverlässigkeit der Interpretation hängt also davon ab, ob
Einstein sich selbst richtig sieht und versteht - und hieran zu zweifeln
begründet, wie oben beiläufig erwähnt, Holton selbst: Einstein hielt sich
nämlich auch dann noch für einen Anhänger Machs, als er unter Erfahrungs-
tatsachen (bei Mach noch einfache Empfindungen) abstrakte Sachverhalte,
ja korrekt gesehen sogar methodologische Normen faßte wie die Äquivalenz
von Inertialsystemen bezüglich optischer Erscheinungen. Es gibt - auch
bei Holton - weitere Belege dafür, daß die Selbstdefinition Einsteins bezüglich philosophischer Positionen sehr wohl fehlbar war. Die Schlußfolgerung daraus läßt sich vielleicht folgendermaßen verdeutlichen: Hätten
Einstein und Holton an jeder Stelle, an der über Einsteins erkenntnistheoretische Position gesprochen wird, statt "Mach" "Poincaré" geschrieben, so würde sich im gleichen Umfang überzeugend Einstein an seinen physikalischen Schriften als Konventionalist statt als Sensualist oder Empi-
rist erweisen.

Das diskutierte Beispiel mag zeigen, daß auch Einstein als Einsteininterpret nicht das Dilemma behebt, daß eine erkenntnistheoretische Prägung
Einsteins an seiner Physik nicht streng nachweisbar ist - außer man lese
sie hinein.

Die Untersuchung Holtons lassen aber auch einen anderen Schluß zu, der
freilich nicht mehr im Sinne Holtons liegt. Es ist sicher Holton zuzustimmen, daß Einstein in späteren Jahren über die Physik - etwa ihre Er-
fahrungsgrundlage - Auffassungen vertritt, die mit denen Machs unvereinbar

sind. Trennt man aber streng zwischen Behauptungen Einsteins zur Physik und Behauptungen über sein eigenes Verhältnis zu erkenntnis- und wissenschaftstheoretischen Lehrmeinungen oder Standpunkten, so erlauben die Untersuchungen Holtons, einen Wandel nur bei der zweiten Sorte von Behauptungen anzunehmen, und zwar in dem Sinne, daß Einstein zusammen mit dem Ausbau seiner physikalischen Theorien zunehmend bemerkt, daß er selbst in kein erkenntnistheoretisches Schema paßt. Auf dem Gebiet der Selbsteinschätzung also und nicht auf dem Gebiet der philosophischen Grundüberzeugung verändert sich nachweisbar etwas: Während sich der junge Einstein noch für einen Machianer oder für einen Empiristen hält, stellt der reife Physiker, der ja in zahllosen Debatten über seine Theorien zwangsweise ein Bild von der Physik entwickeln mußte, fest, daß dieses sein Bild auf keine der ihm bekannten Positionen paßt. Bei dieser Behauptung nun treten keine Interpretationsprobleme der oben genannten Art mehr auf, und Einstein kann unbedenklich beim Wort genommen werden:

> *Die gegenseitige Beziehung von Erkenntnistheorie und Science ist von merkwürdiger Art. Sie sind aufeinander angewiesen. Erkenntnistheorie ohne Kontakt mit Science wird zum leeren Schema. Science ohne Erkenntnistheorie ist - soweit überhaupt denkbar - primitiv und verworren. Hat aber einmal der nach einem klaren System suchende Erkenntnistheoretiker sich zu einem solchen System durchgerungen, so neigt er dazu, das Gedankengut der Science im Sinne seines Systems zu interpretieren und das abzulehnen, was in sein System nicht hineinpaßt. Der Scientist aber kann es sich nicht leisten, das Streben nach erkenntnistheoretischer Systematik so weit zu treiben. Er akzeptiert dankbar die erkenntnistheoretische Begriffsanalyse; aber die äußeren Bedingungen, die ihm durch die Erlebnistatsachen gesetzt sind, erlauben es ihm nicht, sich bei der Konstruktion seiner Begriffswelt allzusehr durch Festhalten an einem erkenntnistheretischen System beschränken zu lassen. Er muß dann dem systematischen Erkenntnistheoretiker als eine Art skrupelloser Opportunist erscheinen. Er erscheint als Realist insofern, als er eine von den Akten der Wahrnehmung unabhängige Welt darzustellen sucht; als Idealist insofern, als er die Begriffe und Theorien als freie Erfindungen des menschlichen Geistes ansieht (nicht logisch ableitbar aus dem empirisch Gegebenen); als Positivist insofern, als er seine Begriffe und Theorien   n u r   insoweit für begründet ansieht, als sie eine logische Darstellung von Beziehungen zwischen sinnlichen Erlebnissen liefern. Er kann sogar als Platoniker oder Phythagoräer erscheinen, insofern er den Gesichtspunkt der logischen Einfachheit als unentbehrliches und wirksames Werkzeug seines Forschens betrachtet.* [16]

Deutlicher hätte Einstein wohl kaum sagen können - und dies in seinen letzten Jahren - daß er selbst keine konsistente erkenntnistheoretische Position hat, und außerdem, daß sie ihm auch nicht fehlt. Danach fällt es auch leichter, im Hinblick auf die oben zitierten Bemerkungen Einsteins, er sei durch Hume, Mach und Poincaré beeinflußt, Einstein selbst zuzustimmen, wenn er sagt, daß dieser Einfluß nicht analysierbar sei. Damit soll freilich keineswegs ein solcher Einfluß gänzlich geleugnet werden. Aber worin genau ein solcher Einfluß besteht, läßt sich allenfalls in kühnen Interpretationen angeben, die mehr über den Interpreten als über Einstein aussagen.

Übrig bleibt lediglich auf demselben Allgemeinheitsniveau, auf dem Einstein Interesse an und Beeinflussung durch Erkenntnistheorie behauptet, einige Gemeinsamkeiten bei den von Einstein gelesenen Erkenntnistheoretikern auf der einen Seite und Einschätzungen Einsteins physikalischer Arbeiten auf der anderen Seite zu benennen. Allen Autoren gemeinsam ist ein aufklärerischer, gegen etablierte Autoritäten, vor allem in Form scholastisierender Systeme und Schulmeinungen gerichteter Impetus. In England wurde, um chronologisch bei Hume zu beginnen, die Aufklärung vom Empirismus getragen. Die Kritik, die Hume auf der Grundlage der Unterscheidung von *impressions* und *ideas* am scholastischen Kausalitäts- und Substanzbegriff geübt hat, ist insofern im Sinne der Aufklärung besonders attraktiv, als sie die Autonomie des einzelnen, wahrnehmenden und unvoreingenommenen über diese Wahrnehmungen reflektierenden Menschen betont - nicht anders, als die Kritik Machs an der Newtonschen Interpretation des Eimerversuchs und nicht anders als Einsteins Kritik am Gleichzeitigkeitsbegriff der klassischen Physik.

Sofern mein Anspruch ausdrücklich darauf beschränkt ist, Plausibilitäten, nicht aber strenge Argumentationen vorzutragen, läßt sich diese Betrachtung noch verlängern: wenn es historisch wahr ist - hierzu habe ich kein abschließendes Urteil -, daß im Falle der speziellen Relativitätstheorie für Einstein das Gedankenexperiment vom Beobachter, der einem Lichtstrahl hinterhereilt, eine wichtige Rolle gespielt hat, und wenn es historisch wahr ist, daß die Ununterscheidbarkeit von Schwere und Beschleunigung durch einen Beobachter in einem geschlossenen Raum ein Motiv für Einstein war, die allgemeine Relativitätstheorie zu entwickeln, dann sehe ich auffällige Parallelen zwischen der aufklärerischen Kritik Humes und Einsteins an Begriffen, die ihnen aus der Tradition entgegentraten. Kommt in dieser Situation Poincaré mit einem Nachweis zu Hilfe, daß viele vermeintlich empirische Sätze recht besehen Definitionen für einen oder einige in diesen Sätzen stehenden Termini sind, so ist dadurch der Weg für die Einsteinsche Überwindung der Physik Newtons schon geöffnet. In dieser Auffassung sehe ich mich von Reichenbach bestätigt, der schrieb:

> *Als logische Basis der Relativitätstheorie dient die Entdeckung, daß viele Aussagen, deren Wahrheit oder Falschheit als erweisbar angesehen wurde, bloße Definitionen sind.*[17]

Es wäre reizvoll, diese Parallelen noch weiterzuführen und glaubhaft zu machen, daß auch Kant ungezwungen zu den aufklärerischen Vordenkern Einsteins gezählt werden kann. Die zahlreichen ablehnenden Bemerkungen Einsteins gegen den Apriorismus zeigen eher Einsteins psychologistische

Interpretation Kants, wonach vor allem die Geometrie Euklids als "denknotwendig" bezeichnet wird, als daß hieraus eine Ablehnung der Kantischen
Philosophie im Kern spräche. Das Verhältnis Einsteins zur Philosophie
Kants scheint von einer nicht sehr kritischen Übernahme neopositivisti-
scher Kant-Kritik gekennzeichnet, wonach alle wissenschaftlichen Sätze
entweder analytisch oder empirisch seien. Einstein als Kantianer in Anspruch zu nehmen ließe sich an seine Bestimmungen physikalischer Grundbegriffe wie z.B. den der Gleichzeitigkeit anschließen. Dennoch würden
derlei Versuche einer erkenntnistheoretischen Klassifikation Einsteinscher
Beiträge zur Physik den Rahmen des Spekulativen kaum verlassen.

Oben war unterschieden worden zwischen Einsteins Bemerkungen über die
Physik und denjenigen über sich selbst im Hinblick auf erkenntnistheoretische Lehrmeinungen. Abschließend soll deshalb noch etwas über Einstein
als Wissenschaftstheoretiker gesagt werden. Immerhin hat Einstein 1933
in Oxford die Herbert Spencer Lecture zum Thema *Über die Methode der theoretischen Physik* gegeben; immerhin gibt es eine Reihe von Aufsätzen, die
Titel tragen wie *Physik und Wirklichkeit*, *Die Grundlagen der theoretischen Physik*,
Prinzipien der Forschung, *Über wissenschaftliche Wahrheit* und andere mehr. Wer in
Einstein den Philosophen oder zumindest den Wissenschaftstheoretiker sehen
möchte, wird geneigt sein zu fragen, warum die erkenntnistheoretischen
Quellen Einsteins nicht ins Verhältnis zu diesen Traktaten gesetzt werden,
wenn sich schon ein Einfluß auf die physikalischen Theorien selbst nicht
nachweisen läßt. Meine Antwort ist einfach, wenn sie auch manchen Ver-
ehrer Einsteins enttäuschen wird. Einsteins "gelegentliche Äußerungen erkenntnistheoretischen Inhalts", wie er sie selbst bezeichnet,[18] zählen
meines Erachtens nicht zu seinen großen Leistungen. Obgleich es selbstverständlich auch auf diesem Gebiet bereits Literatur gibt und Rekonstruk-
tionsversuche von Einsteins wissenschaftstheoretischen Meinungen noch ein
weites Betätigungsfeld bieten, ermutigen die bereits vorliegenden Analysen kaum, im Hinblick auf den jeweils zeitlich parallel zu Einsteins
Äußerungen erreichten Diskussionsstand in der Wissenschaftstheorie vom
Wissenschaftstheoretiker Einstein ein neues Physikbild zu erwarten.[19]
Es spricht manches dafür, daß Einstein dies selbst nicht viel anders gesehen hat. Sein Essai "Zur Methodik der theoretischen Physik" beginnt mit
den Worten:

> *Wenn Ihr von den theoretischen Physikern etwas lernen wollt über die von
> ihnen benutzten Methoden, so schlage ich Euch vor, am Grundsatz festzu-
> halten: Höret nicht auf ihre Worte, sondern haltet Euch an ihre Taten!*[20]

Wie berechtigt diese Empfehlung in seinem eigenen Falle war, dokumentieren
seine diesbezüglichen Äußerungen in vielfacher Weise. Dabei will ich noch
nicht einmal monieren, daß sich Einstein - gemessen jeweils an der zu

seiner Zeit verfügbaren Literatur - recht dilettantisch äußert und auch
Meinungen unberücksichtigt läßt, bei denen man annehmen könnte, er habe
sie über persönliche Kontakte, die historisch belegt sind, kennengelernt. Dilettantismus kann, wenn seine Resultate wahr, fruchtbar und
tragfähig sind, ein hervorragendes persönliches Qualitätsmerkmal sein,
aber die Resultate erkenntnis- und wissenschaftstheoretischen Dilettie-
rens bei Einstein sind dies alles nicht. Es fehlt ihnen durchweg an ter-
minologischer Erläuterung zu den entscheidenden Grundbegriffen wie Sin-
neserfahrung, Erfahrungstatsache, wissenschaftlich, wahr, bewährt, Theorie, Prinzip, Axiom, Natur, Wirklichkeit, Intuition, Begriff, Sprache
usw. Einstein bedient sich hier einfach einer Bildungssprache, deren
Fallstricke an entscheidender Stelle ihm offensichtlich nicht bewußt
sind. Schon dieses terminologische Defizit macht die wissenschaftstheoretischen Bemerkungen Einsteins zu vagen, unkontrollierbaren Äußerungen, an
denen z.B. häufig noch nicht einmal feststellbar ist, ob sie beschreibend
oder empfehlend gemeint sind. Vom heutigen Standpunkt der Diskussion in
der Wissenschaftstheorie hält kein wissenschaftstheoretischer Beitrag
Einsteins der Überprüfung stand. Dieser Sachverhalt ist, wenn man die
Entstehung und die Funktion dieser Essais vor Augen hat, weder besonders
überraschend, noch verdient er besondere Betonung. Diese Essais sind nämlich - wie übrigens auch diejenigen vieler anderer berümter Physiker -
sowohl methodisch als auch historisch immer n a c h den physikalischen
Leistungen entstanden, von denen sie explizit oder implizit handeln. Mir
ist kein erkenntnis- oder wissenschaftstheoretischer Aufsatz eines berühmten Physikers bekannt, der programmatisch wäre u n d dem die Durchführung des Programms gefolgt wäre. Vielmehr ist die Nachträglichkeit
solcher Äußerungen ein Zeichen dafür, daß post factum noch einige Auf-
räumarbeiten geleistet werden, die sich daraus ergeben, daß neue (und in
ihrer Anerkennung durch Physiker von erkenntnistheoretischen Sätzen völlig unabhängige) physikalische Ansätze oder Resultate mit den persönlichen, meist alltagssprachlich formulierten Meinungen ihres Urhebers zur
Physik kollidieren. Aus einem dann eher psychologisch zu erklärenden Kon-
sistenzbedürfnis heraus wird das Physikbild nachträglich zurechtgerückt.
Terminologische oder logische Prägnanz wären hierbei hinderlich. Zu Gun-
sten Einsteins sei hinzugefügt, daß ihn nicht das geringste Verschulden
trifft, wenn eine autoritätsfixierte Öffentlichkeit ihren Idolen Kompe-
tenzen zuschreibt, die diese nicht beansprucht haben, mehr noch, Einsteins bekannt antiautoritäre Einstellung würde dies verbieten.

Als Resümee läßt sich deshalb vielleicht ziehen: es ist zumindest nicht
nachweisbar, daß es erkenntnistheoretische Quellen waren, die eine neue
Physik durch Albert Einstein haben entstehen lassen. Möglicherweise hat

Einstein an diesen Quellen Ermutigung und Bestätigung für seine ohnehin vorhandene Haltung gefunden, möglicherweise auch gesucht. Einstein als Erkenntnistheoretiker und Philosoph aber ist meines Erachtens Teil der Einstein-Legende. Möglicherweise hätte er selbst dazu gesagt, was er 1948 tröstend an Besso geschrieben hat:

Ein Schmetterling ist kein Maulwurf; das soll aber kein Schmetterling bedauern.[21]

Anmerkungen

1. So wies in der an diesen Vortrag anschließenden Diskussion z.B. I. Stachel (Princeton) daraufhin, daß die Hume-Lektüre Einsteins nicht auf den Treatease beschränkt war, was er mit einem Zitat aus dem noch nicht veröffentlichten Einstein-Nachlass belegte.

2. cf. Carl Seelig, Albert Einstein. Eine dokumentarische Biographie. Zürich/Stuttgart/Wien 1954, s. 134.

3. A. Einstein, Lettres a Maurice Solovine, Paris 1956, S. 90.

4. Johannes Wickert, Albert Einstein in Selbstzeugnissen und Bilddokumenten. Hamburg 1972, S. 22.

5. Paul Arthur Schilpp (Hrsg.), Albert Einstein als Philosoph und Naturforscher, engl. 1951, dt. Stuttgart 1955, S. 505.

6. Für die zahlreichen Stellen, an denen Einstein "apriorisch" gleichsetzt mit "denknotwendig", möge stellvertretend stehen:

   Es ist deshalb nach meiner Überzeugung eine der verderblichsten Taten der Philosophen, daß sie gewisse begriffliche Grundlagen der Naturwissenschaft aus dem der Kontrolle zugänglichen Gebiete des Empirisch-Zweckmäßigen in die unangreifbare Höhe des Denknotwendigen (Apriorischen) versetzt haben.

   Albert Einstein, Grundzüge der Relativitätstheorie, Braunschweig 1965, S. 2.

7. Albert Einstein, Hedwig und Max Born, Briefwechsel 1916-1955, München 1969, S. 19.

8. J. Wickert, loc. cit. S. 23.

9. Max Flückiger, Albert Einstein in Bern. Bern 1974, S. 197.

10. Albert Einstein, Michele Besso, Correspondence 1903-1955, dt. u. franz., Paris 1972, S. 391.

11. P. A. Schilpp, loc. cit., S. 20.

12. Hans Reichenbach, Die philosophische Bedeutung der Relativitätstheorie, in: P. A. Schilpp, loc. cit., S. 188-207, S. 189.

13. *Ich sehe Machs wahre Größe in der unbestechlichen Sekpsis und Unabhängigkeit; in meinen jungen Jahren hat mich aber auch Machs erkenntnistheoretische Einstellung sehr beeindruckt, die mir heute als im wesentlichen unhaltbar erscheint.*

Albert Einstein, Autobiographisches, in: P. A. Schilpp, loc. cit. S. 8.

14. Gerald Holton, Mach, Einstein, and the Search for Reality, in: Daedalus 1968, S. 636-673, S. 643.

15. cf. Anmerkung 14.

16. cf. P. A. Schilpp, loc. cit. 507-508.

17. H. Reichenbach, loc. cit. S. 192.

18. Albert Einstein, Bemerkungen zu den in diesem Bande vereinigten Arbeiten, in: P. A. Schilpp, loc. cit. S. 507.

19. Stellvertretend hierfür mögen die beiden im Schilpp-Band enthaltenen Beiträge gelten. Cf. Victor F. Lenzen, Einsteins Erkenntnistheorie; und F. S. C. Northrop, Einsteins Begriff der Wissenschaft.

20. Carl Seelig (Hrsg.), Albert Einstein, Mein Weltbild, Frankfurt 1955, S. 113.

21. Brief an Michele Besso, loc. cit. (cf. Anmerkung 10), S. 392.

THE GENESIS OF GENERAL RELATIVITY

John Stachel, Dept. of Physics, Boston University, USA

Responding to a letter inquiring about what influences had led him to
his discoveries, Einstein replied that he could not indicate anything
which had served as an external motive. He characterized his proper
life's work as the search to answer three questions:

1) how the representation of a light ray depends on the state of
   motion of the coordinate system to which it is referred;
2) the basis for the equality of inertial and gravitational mass;
3) whether the gravitational and electromagnetic fields can be
   comprehended in a unified scheme.

The attempt to answer the first question led Einstein to the special
theory of relativity. The search had its beginning in his teens (when
he first wondered about what would be observed if one succeeded in running alongside a light ray), but started in earnest during his final
years at the ETH (where he conceived the idea of measuring the motion
of the earth through the ether by an optical experiment). It continued
for seven years until 1905, when he was able to rapidly complete the
theory after realizing that simultaneity must be a relative concept.

The second question was the starting point of the general theory of
relativity which occupied Einstein, with varying degrees of intensity,
between 1907 and 1915, when the final version of the theory was completed.

The third question was already briefly raised in the 1916 review paper
on general relativity. Einstein's search for a satisfactory unified field
theory started in earnest a few years later, and was to continue until
the very last days of his life.

In another letter, Einstein elaborated on the nature of his motivation
in attempting to answer such questions. He described it as the striving
for a logically simple interpretation of empirically known connections.
He described the state of psychic tension generated by what he felt to
be fundamental incompleteness in the theoretical system of physics, and
the striving to overcome this incompleteness, spurred on by his intense conviction of the existence of a logically simple solution. He compared his motivation to that of someone trying to solve a puzzle or
chess problem, spurred on by the knowledge that the solution must exist.
These comments shed some light on the significance of Einstein's cosmic

religiosity: it was a deep emotional conviction of the lawful nature
of the universe which supplied the psychic energy needed to keep working
on such "puzzles" over a decade or more. Even in the case of the unified
field theory, where Einstein felt far from certain of his own ultimate
success, he felt that some more fundamental solution to the "puzzle"
must exist than that provided by quantum mechanics.

Both at the time of its creation, and looking back in later years, Einstein saw his work on general relativity as something quite unique in
his life. He felt that if he had not worked out the special theory of
relativity, somone else - perhaps Paul Langevin - would have done so.
His approach to the general theory was entirely his own (apart from
mathematical help from Marcel Grossmann), carried through with enormous
efforts in the face of skepticism if not active hostility from physicists
he respected, such as Planck and Abraham. He characterized his efforts
on special relativity as mere child's play compared to what was needed
to complete the general theory. His feeling of satisfaction at its completion marked the high point of his career. His whole later outlook on
science was colored by the experience of the search; and the terms of
his quest for a unified field theory were set by the nature of the generally covariant theory of gravitation that was its outcome.

In this paper I shall try to briefly outline the steps in this search
to solve that second "puzzle" - the central episode in Einstein's scientific activity. I hope to publish a larger study, with full references,
at a later date.

Einstein himself, in material he prepared for Erwin Freundlich's book
- the first popular book on general relativity, published in 1916 - divided the story of his search into three parts (I attach no mystical
significance to this recurrence of triads in our story):

1) In 1907, he had the basic idea for a generalized theory of relativity, based upon the attempt to find a fundamental explanation
   for equality of gravitational and inertial mass.
2) In 1912 came the recognition of the non-Euclidean nature of the
   space-time metric, and its physical determination by gravitation.
3) In 1915 Einstein arrived at the correct field equations for gravitation, and the explanation of the hitherto anomalous precession
   motion of the perihelion of Mercury.

The Equivalence Principle

In 1907, Einstein was asked by Johannes Stark to contribute a review
paper on the principle of relativity to Stark's *Jahrbuch*. In this paper

Einstein first published his reflections on the relationship between
the relativity principle and gravitation. From later comments, however,
we know that this was not his first attempt at a relativistic theory of
gravitation. He had apparently tried to set up a special-relativistic
gravitational theory by some such obvious generalization of Newton's
theory as transforming Poisson's equation for the Newtonian gravitational
potential into d'Alembert's equation. But then he had been struck by the
fact that such a theory violated the equivalence between gravitational
and inertial mass. This feature of the gravitational interaction had
been known for centuries, but appeared to have no deeper explanation
within Newtonian theory. Einstein became convinced that this "coincidence" must hold the key to any fundamental explanation of gravitation:
gravitation and inertia must be two aspects of the same phenomenon. He
held to this conviction throughout the eight years it took him to arrive
at his final formulation of the field equations of general relativity
- years during which other colleagues tried to set up theories of gravitation which abandoned the strict equivalence between gravitational and
inertial mass. Luckily, it was not until his 1914 discussion of Nordström's theory that Einstein realized that a scalar theory of gravitation was compatible with the equivalence principle. By that time, he
was firmly convinced of the need for a (metrical) tensor theory of gravitation. Since it is sometimes implied that Einstein was unaware of the
anomalous precession of the perihelion of Mercury at the time, let me
note that in a letter of 1907 he states that he is working on a new
theory of gravitation and hopes to be able to explain the anomalous precession with its help. He added that unfortunately it did not seem to
want to come out right - as indeed it did not until 1915! This comment
suggests, however, that he might have arrived at a gravitational analogue of the Sommerfeld special-relativistic orbit precession as early
as 1907.

At any rate, he abandoned such special-relativistic attempts to generalize the Newtonian theory in 1907. From then on he pinned his hopes for
a relativistic theory of gravitation on the attempt to generalize the
relativity principle from inertial frames to some wider class of accelerated frames of reference. The 1907 review paper gives us the first
fruits of this attempt. Einstein had apparently been troubled even earlier by the question of why inertial frames should be singled out by
the (special) relativity principle. One major source of his doubts came
from his reading of Mach's *Mechanics*, with its critique of Newton's
concepts of inertia and absolute space - a topic to which we shall return near the end of this paper. Of course, if one accepted the exis-

tence of one privileged frame of reference (ether frame) there was no problem of relativity. But once one gave that up, there seemed no logically compelling reason why one should stop at the relativity of inertial frames. The conventional answer to such doubts was that the laws of nature took a simpler form in the inertial frames of reference, because in these frames one did not have to introduce the fictitious so-called "inertial forces" (which would be better called "non-inertial", for our purposes) in writing down Newton's laws of motion. But Einstein saw that this argument really did not hold, in at least the simplest case, if one took into account the equivalence between gravitational and inertial mass. Consider a uniformly accelerated (with respect to any inertial frame) frame of reference, without any gravitational field, on the one hand; and an inertial frame of reference with a uniform gravitational field of the same magnitude and opposite direction, on the other. Because of the equivalence of gravitational and inertial mass, all bodies fall with the same acceleration in a uniform gravitational field (Galileo effect), so there is no way by a mechanical experiment to tell the two situations apart. Einstein made the bold assumption that this equivalence between the two frames - one inertial with uniform gravitational field, the other accelerated without gravitational field - held for all physical phenomena, in particular for electromagnetic effects (the parallel with his extension of the mechanical relativity principle in special relativity is obvious). Using this as a heuristic guide, he was able to draw several quantitative conclusions, which he regarded as at least a first approximation to results which any complete theory of gravitation should yield. Among these was the prediction that a light ray must be deflected by a gravitational field, but at that time he could suggest no experimental test.

For several years after that, the gravitational problem receded into the background of his efforts, although never dropped entirely. For example, in a letter to Sommerfeld in 1909, he remarks on the importance of generalizing his work on uniformly accelerated translation to uniformly rotating rigid bodies, but he does not seem to have worked on this problem until 1912. One can suggest several reasons for this:

1) Although absolutely convinced of the correctness of his basic approach, he was unsure of just how to proceed to develop his ideas further. He was much troubled that, whereas giving the space and time coordinates direct physical significance in inertial reference frames had played a crucial role in the development of special relativity, the coordinates seemed to lose their direct physical interpretation in accelerated frames

of reference.

2) During these years he became deeply involved in the attempt to find a fundamental explanation of the quantum effects he had been so instrumental in bringing to light. Indeed, his first attempt at a "unified theory" came at about this time: he was looking for a unified quantum theory of electrons and electro-magnetic fields. However, after several cycles of elation and defeat as one attempt after another proved a failure, he decided a fundamental theory was beyond his efforts. He returned to his earlier approach of taking the existence of the quantum (provisionally) for granted, and elucidating the consequences of its existence.

3) An even more prosaic explanation may be that during this period he was inducted into the academic world, first as a part-time Privatdozent in Bern, and then as Associate Professor at Zürich. His bibliography shows a definite fall-off in his (always extraordinary) rate of publication in 1908-1909, quite consistent with the acclimatization period of a new junior faculty member.

At any rate, it was only when he moved to Prague in 1911 as a full Professor that he returned to the problem of gravitation with full vigor. In a preface to the Czech edition of his popular book on *Relativity*, he remarks that it was within the quiet halls of the Prague Institute of Theoretical Physics that he found the necessary equanimity to slowly give his earlier ideas a more definite form. The first product of Einstein's Prague ruminations was his 1911 paper "On the Influence of Gravitation on the Propagation of Light". This he regarded as basically an improved version of his earlier considerations on the equivalence principle. He stressed two conclusions, one experimental, the other theoretical, which proved important to the later history of general relativity.

The experimental conclusion was new: it would actually be possible to test the idea that light was deflected by a gravitational field. One should look for a shift in the apparent position of a star near the edge of the sun during an eclipse, compared with its position on an ordinary night. A Prague colleague put Einstein in contact with the Berlin astronomer Erwin Freundlich, who became extremely enthusiastic about the possibility of testing this and other astronomical consequences of Einstein's ideas about gravitation (e.g., gravitational red shift), and devoted much of the next decade to such efforts. Bad weather and world war conspired to twice keep Freundlich from getting the needed

eclipse photographs. Since Einstein's 1911 paper predicted only half
the deflection of the final 1915 theory, it is interesting to speculate
on what might have been the reception of Einstein's final theory if it
had appeared *after* an experimental test had shown the deflection was
twice what Einstein had earlier predicted.

The theoretical conclusion that was to prove important for Einstein's
further work was his treatment of the velocity of light - actually an
improved derivation of a result obtained in the 1907 paper. Measured
with the global time coordinate of a uniformly accelerated reference
frame, c is no longer a constant, but depends on the acceleration.
By Einstein's equivalence principle, this meant that the velocity of
light in a gravitational field must be variable; and Einstein gave a
formula showing how it varied with the gravitational potential. Using
this, it was easy to derive the deflection of a light ray in a gravitational field by analogy with that of a light ray in a medium with
variable index of refraction.

It is clear from some comments at the beginning of his next paper that
Einstein had been thinking about the application of his equivalence principle to the uniformly rotating frame of reference; but this was not the
problem to which he turned his attention in the body of the paper. The
reason perhaps lies in the fact that he could hardly accept his own
conclusions, as indicated by the tentative nature of his comments about
the rotating disk, and did not know what to do with the paradoxical result to which he was led - a result to which we shall return shortly.
In line with his step-by-step approach to the gravitation problem (as
he characterized it in a letter to Ehrenfest), he devoted two 1912 papers to the problem of the static gravitational field, a problem which
he felt able to handle on the basis of his concept of the variable speed
of light. If the speed of light depends on the gravitational potential
in a static gravitational field, then we may replace the gravitational
potential by the (variable) speed of light. Gravitational field equations for the speed of light are set up in the first 1912 paper, and
modified in the second to accord with the conservation laws when gravitational momentum and energy are taken into account. Naturally, this
leads to a non-linear theory of the gravitational field. But the establishment of equations of motion for a particle in the gravitational
field proved to be the most important result of these papers. In an
Addendum to the second of the papers, added in proof, Einstein notes
that the equations of motion can be derived from a variational principle
which is formally the same as Planck's special-relativistic variational
principle for particle motion, except that the speed of light is now a

function of position (not time, of course, since only the static case is treated).

The Metric Tensor

Before turning to the next and crucial step in the story, it is important to recall that during the years between 1907 and 1911 Minkowski had presented his four-dimensional formulation of special relativity theory, and it had been brought to the attention of the theoretical physics community largely by the expository efforts of Sommerfeld and Laue. At first, Einstein's reaction had been rather negative. From his student years on, he later remembered, he had felt that the physicist's mathematical requirements were quite modest; and that any toying with advanced mathematics was pure luxury - or worse, needless pedantry. Jests are reported in which Einstein spoke of the Göttingen mathematicians as making relativity so hard that the physicists wouldn't be able to understand it. This skeptical attitude was only reinforced by his 1912 work on the static gravitational field. The concept of a variable speed of light when gravity was taken into account seemed to undercut the mathematical basis for the symmetry between space and time which underlay Minkowski's approach to special relativity; there are a number of comments to this effect in Einstein's 1912 correspondence. However, his notebooks from this period seem to indicate that he was studying Minkowski's approach, possibly via Sommerfeld's influential exposition.

Now we have the elements that went into the next step in Einstein's search, the crucial step: his adoption of a non-flat four-dimensional metric tensor as the correct mathematical representation of the gravitational field. An attempt to understand how Einstein arrived at this truly revolutionary viewpoint is complicated by the fact that there is not a hint of it in his 1912 papers on gravitation; while the very next paper in 1913 starts off from this viewpoint. Therefore, one can only speculate on exactly what happened in the half-year between. Here is my (incomplete) outline of the story:

After completing his treatment of the static case, Einstein planned to treat stationary gravitational fields, as he informed Ehrenfest. This may have returned his attention to the rigidly rotating disk, the simplest stationary problem. What had troubled Einstein about his earlier analysis was that it seemed to show that measuring rods at rest on such a rotating disk would show a ratio of its circumference to its diameter different from π. Einstein seems to have confirmed this earlier, tentative conclusion. He had good reason to believe the result because it could be based on reasoning from the viewpoint of a non-rotating inertial frame (i.e., purely special relativistic viewpoint), plus the assumption that small measuring rods at rest on the disk are unaffected by their acceleration. A non-Euclidean geometry must hold in the disk frame of reference. But then, by the equivalence principle, this rotating frame of reference should be equivalent to an inertial frame of reference plus a certain gravitational field. Einstein concluded that, in the presence of a gravitational field, measuring rods will map out a non-Euclidean geometry.

Einstein already knew something about non-Euclidean geometry. About the only mathematical lectures that stood out favorable in his remembrances of his years at the ETH were those by the geometer Karl Friedrich Geiser, who had lectured on infinitesimal geometry - another name for differential geometry. From the notebooks of Einstein's classmate and later collaborator Marcel Grossmann it is known that these lectures included at least some elements of the Gaussian theory of surfaces, including the use of arbitrary curvilinear coordinates on a two-dimensional surface.

Einstein seems to have put this idea of curvilinear coordinates for two-dimensional non-Euclidean geometries together with his earlier result on the variational principle for a particle in a static gravitational field, to conclude that his variational integrand could be regarded as the line element of a four-dimensional non-flat space-time, invariant under arbitrary transformations of the space-time coordinates. What was required by the break-down of the invariance of special relativity under the Lorentz group when gravitation was taken into consideration was not to drop Minkowski's four-dimensional point of view. Rather, one had to generalize the group of coordinate transformations and apply them to a non-flat space-time. From this point of view it became obvious that the correct mathematical representation of the gravitational field was the array of ten coefficients of the coordinate differentials in the expression for the line element - the metric tensor.

Two problems remained: to reformulate the non-gravitational parts of physics in the light of this new non-Euclidean metrical approach; and to find the correct field equations for the gravitational field itself. But the crucial step had been taken: the recognition of the dual role of the metric tensor as the representation of both the space-time structure and of the gravitational field. Once this mathematical structure had been identified, it was only a matter of time until the two problems just mentioned were solved. Einstein was not certain just how wide the new group of coordinate transformations should be under which the resulting theory would be invariant. But clearly, any answer short of the most general coordinate transformations would only raise the question again: why stop here? So he seems to have felt strongly that the final theory must be invariant under general coordinate transformations.

The Field Equations

Shortly after his return to Zurich in October 1912, as Professor at the ETH, Einstein had more or less clearly formulated this program. He must have felt that his mathematical background was not adequate to work out these problems alone. He turned for help once more to his former ETH schoolmate Marcel Grossmann. The latter had placed his careful note-books at Einstein's disposal when final examination time came around, and had urged his father to intervene with the head of the Swiss Patent Office when Einstein was desperate for a job after graduation. Now Grossmann was a Professor of Mathematics at the ETH. Perhaps remembering Grossmann's earlier work on non-Euclidean geometry, Einstein asked Grossmann whether mathematical methods existed for formulating equations which were invariant under arbitrary coordinate transformations. Grossmann found that the work of Riemann, Christoffel, Ricci and Levi-Civita, from the mid-nineteenth century to the early years of the twentieth, had already provided the tools needed. Tensor analysis proved to be the mathematical formalism adapted to setting up generally covariant equations. Einstein and Grossmann soon succeeded in writing generally covariant equations incorporating the effect of the gravitation field on other physical systems; actually, much of this work had already been done by others and only had to be adapted to the new point of view. It also quickly became clear that the ten-component stress-energy tensor was the correct relativistic generalization of the Newtonian mass as the source of the metric-gravitational field.

But the search for the correct field equations for the gravitational field, relating this stress energy tensor to the metric tensor, proved a much longer story. As mentioned above, Einstein had hoped that the entire theory would be generally covariant. Grossmann pointed out that there was a certain tensor, formed from the metric tensor and its first and second derivatives and now known as the Ricci tensor, which was the practically unique candidate for generally covariant field equations. But calculation seemed to show that a theory based on these field equations did not yield Newton's theory of gravitation as a first approximation for weak gravitational fields. Such a "correspondence limit" was correctly seen as a criterion for any new theory of gravitation, but not until 1914 did Einstein realize that the calculation had been incorrectly

interpreted. But, Einstein being Einstein, he had not remained satisfied with rejecting general covariance on the basis of a mere calculation. He had developed a simple meta-argument to show that the equations
for the gravitational field *could* not be generally covariant. By the
time he realized the Newtonian limit argument was in error, the meta-
argument held him in its sway. What was that argument? Einstein stated
that the causality principle demanded that a given distribution of matter
and energy - a given stress-energy tensor mathematically speaking -
should result (modulo appropriate boundary conditions) in a unique gravitational field. So far so good; but he felt that a unique gravitational
field should correspond to a unique metric tensor. Now whether this is
correct or not depends on how we interpret "unique". If we mean physically
unique, that is fine (there are subtle mathematical problems even here,
which we need not discuss). But Einstein interpreted it to mean unique
as a mathematical function of the coordinates. Clearly no generally covariant field equations can have that property, precisely because their
form as a function of the coordinates must change under coordinate transformation. Just because a theory was generally covariant, mathematically
different-looking solutions corresponding to the same sources could be
physically identical. But it was apparently only late in 1915 that Einstein realized this flaw in his meta-argument.

Under the sway of first the erroneous Newtonian limit argument and then
the erroneous meta-argument, Einstein developed a non-covariant set of
equations for the gravitational field. Yet he was never really comfor-
table with giving up general covariance. His correspondence from early
1913, when he did the original work with Grossmann, until late 1915 is
filled with justifications of the non-covariant equations and statements
of how satisfied he was with them - probably as much to convince himself as his correspondents - and equally passionate statements that
somehow the final theory must be generally covariant.

During this period he was also engaged in discussions - and sometimes
polemical interchanges - with the authors of other theories of gravitation, notably Abraham, Mie and Nordström. Einstein, working with
Fokker, was happy to find that one of Nordström's scalar theories of
gravitation could be put into generally covariant form; they remark at
the end of their 1914 paper on this subject that one should perhaps
reexamine the question of general covariance for the Einstein-Grossmann
theory (it is here that Einstein acknowledges the error in his earlier
work). But somehow this did not stop him from publishing the meta-argument against general covariance in another 1914 paper. Since Nordström's
theory did not predict any deflection of light by a gravitational field,

Einstein looked upon future eclipse results as providing the critical test between the two theories. Abraham, after his efforts to create a special-relativistic theory of gravitation with a variable speed of light had been criticized by Einstein, abandoned the relativity concept altogether (at least verbally) and delivered himself of stinging attacks on Einstein's efforts. He did not fail to note that Einstein had first abandoned Lorentz invariance and then general covariance. Einstein rather welcomed these polemics - at least someone was publicly paying attention to his work on gravitation. Although his public comments were much more restrained in tone than Abraham's, in his private correspondence he gave vent to such gems as a reference to one of Abraham's theories as a stately horse, but lacking three legs. He referred privately to Mie's special relativistic gravitational theory as fantastic and having a vanishingly small inner probability.

When Einstein joined the Prussian Academy of Sciences in 1914, he included a brief discussion of the basis of his hopes for a generalized relativity theory in his Inaugural Address. Planck, in his reply for the Academy, took the rather unusual step of criticizing these hopes, probably because Einstein's remarks aligned the latter with Mach's views (Planck and Mach were then engaged in a bitter controversy over the philosophical foundations of their views on physics) on the relativity of inertia.

Difficulties with his non-covariant theory continued to pile up. He found that a mathematical argument, developed to show that his Lagrangian was unique, actually permitted almost any Lagrangian. He found that the rotating disk metric was not a solution to his field equations. He found that the value calculated for the anomalous precession of the perihelion of Mercury was not in agreement with the observed value. So in mid-1915 he started a fundamental reexamination of the whole problem, in the course of which he must have found the flaw in his meta-argument, and returned happily to the generally covariant approach.

In speculating on the reasons for the long delay between the initial formulation of the covariant program in late 1912-early 1913 and its final vindication in late 1915, one must bear in mind, of course, the revolutionary nature of what Einstein was trying to do. What seems obvious, from a contemporary viewpoint, about mathematical properties of covariant equations was clearly not so obvious to one working his way towards such equations from primarily physical viewpoints and criteria. In addition, one may speculate on the influence of various non-scientific events on Einstein's work during this period. It was the period of his move to Berlin; of his final separation from his first wife, who returned to Zurich with the children to whom he was deeply attached shortly after the move to Berlin; and of the outbreak of the first World War which so deeply affected Einstein's whole outlook. In any case, he did succeed in overcoming most of his confusions about general covariance by late 1915, and the development of the theory proceeded quite rapidly from that point on.

Quite rapidly, but not without another mis-step which, Einstein later regretted, he had immortalized in print. He first readopted the 1913 generally covariant field equations, in which the Ricci tensor was set equal to the gravitational coupling constant times the stress-energy tensor. This results in field equations that are satisfactory outside of matter, where the stress-energy tensor vanishes. Indeed, Einstein was extremely pleased when he was able to finally derive a value for the Mercury perihelion precession in good agreement with the observed one. But where matter did not vanish, his field equations were not mathematically consistent, he found, unless the trace of the stress-energy tensor vanished. His first response was to make a virtue of necessity and to argue that, although the trace of the phenomenological stress-energy tensor of ordinary matter might be non-vanishing, the trace of the electromagnetic stress-energy tensor did vanish. Perhaps a fundamental theory of ordinary matter would show that it was basically electromagnetic in nature (such speculations were not uncommon at the time, notably in the work of Mie), so that at a fundamental level the trace of the stress-energy tensor did vanish.

Shortly after publishing this justification of his equations, Einstein realized that they could be slightly modified so that they remained generally covariant without requiring any condition on the trace of the stress-energy tensor. Thus, Einstein finally arrived at the field equations which today we know as the Einstein equations for the gravitational field. In 1916 he published a review article on the general theory of relativity in which he pointed out that the principle of general covariance did not force any definite assumptions about the nature of matter (i.e., non-gravitational sources of the gravitational field). He left it as an open question whether a combined theory of gravitation and electromagnetism could shed further light on the nature of matter. Here we may see an oblique comment on his earlier error, as well as on Hilbert's continuation of the Mie program; and the germ of Einstein's interest in a unified field theory.

Cosmology and Epistemology

But Einstein did not regard his work as complete for another, and to
him more pressing, reason. We have mentioned that one of the strong mo-
tivations for Einstein's search for a generalization of the relativity
principle was his reading of the critique of Newton in Mach's *Mechanics*
There have been many varying interpretations of Mach's text; but let it
suffice here to say that Einstein interpreted Mach as demanding a theory
in which the inertial properties of a body would not come from its re-
sistance to motion with respect to absolute space, but rather from its
interaction with all the rest of the matter in the universe. On such an
interpretation, the difference in shape between a rotating and non-rotating liquid (Newton's bucket experiment) must be due to one liquid's
rotating relative to the distant matter in the universe, while the other
does not (the first argument for an extension of the relativity principle
in Einstein's 1916 review paper is a version of this Machian epistemological argument). During the long years of his search for the general
theory, Einstein had often consoled himself with the strength of this
argument - several times in direct correspondence with Mach in which he
defended the latter against Planck, at that time the best-known opponent
among physicists of Mach's epistemological credo. Without wishing to
imply that Einstein was a full-fledged Machian - which he was not, or
that he was an opponent of Planck, to whom he was otherwise very close
- he certainly drew much of his inner conviction in the search for a
generalized relativity principle from this aspect of Mach's doctrine.

Yet when his theory was finally completed, it did not embody Einstein's
version of Mach's principle. He had been able to show to his satisfaction
that the inertia of a body was influenced by the presence of other bodies
in its neighborhood; but it was by no means the case that the total inertial of one body was determined by the mutual interactions between all
the bodies in the universe. Indeed, from Einstein's point of view it
was a scandal that a solution to his field equations should exist which
corresponded to the presence of a single body in an otherwise "empty"
universe (i.e., no other bodies present). He made this clear in a number
of letters in 1916-17 to Schwarzschild, Klein, de Sitter and others.
The way out for him was to construct a cosmological model such that the
matter entirely determined the metric. This was the basic motivation for
his 1917 static closed cosmological model, which he looked upon as the
completion of his general-relativistic program. He was even willing to
modify his field equations once more, by introducing the cosmological
constant term, in order to achieve this end. De Sitter soon found an
"empty" solution to the new field equations, but it was not static.

Einstein was disturbed by the existence of this solution, and found various reasons for rejecting it.

In later years, of course, he was forced to give up his static model when it became clear that a non-static expanding model gave the most natural explanation for the observations of galaxy recession. By then he was also no longer wedded to Mach's principle. He realized that it was ultimately tied to a matter (in the conventional, particulate sense of the word) ontology, and that Mach's arguments were irrelevant in the context of a field ontology, basic to his approach to the unified field problem. But it took some time until the observational and ontological arguments finally convinced Einstein to give up his static model. Regardless of the original motivation for (and the fate of) this particular model, it must be recognized that Einstein's basic approach to cosmology - setting up a metrical model and attempting to correlate its features with observations - has dominated theoretical work in the field for over half a century.

The development of general relativity not only revolutionized theoretical physics, it also changed Einstein's views on epistemology. As he stated in a letter to Lanczos, his work on gravitation turned him from a skeptical empiricist into a believing rationalist. Neither of these extremes should be taken literally as a completely adequate characterization of Einstein's early or late philosophical views. But they do correctly indicate the trend of development of his views, and the crucial role that his work on general relativity played in this evolution. He more and more came to see the search for mathematical simplicity as the source of progress in theoretical physics - even if it was in constant need of control by experience.

On the other hand, it would not be correct to say that Einstein regarded what he had done as a reduction of physics - even the gravitational part of physics - to geometry. He explicitly repudiated such an interpretation of his work in his comments on Myerson's book on relativity in 1927, as well as on numerous other occasions. He sought a generally covariant field description for physical reality, but did not regard such a description as any more or less of a geometrization of physics than the fact that distance occurs in Newton's law of gravitation, or that Maxwell's equations can be written in terms of vectors and tensors.

Einstein reacted sharply against any attempt to extract a philosophy from his theories. He stated that a mathematical-scientific theory such as relativity might well be the *object* of philosophical investigation, but could never form the *basis* of a philosophy.

He was aware, even during the earliest days of his triumph, of the limitations of what he had done. He realized almost at once that general relativity did not solve the problem of the structure of matter, as we have seen. He discussed the possibility, as early as 1917, that no continuum theory might be able to do justice to the quantum nature of matter. Yet he always felt that the possibility that it *could* also was not excluded, even by the triumph of quantum mechanics; and that this possibility should be explored to the end, particularly in view of the lack of any adequate non-continuum mathematical model. He often acknowledged the mortality of all physical theories, including his own.

ZUM PRODUKTIVEN DENKEN BEI EINSTEIN. EIN BEITRAG ZUR ERKENNTNISPSYCHO-LOGIE

Johannes Wickert, Universität Tübingen

Wer Einstein verstehen will, kann seine physikalischen Schriften studieren, sich mit seinen Traktaten und Aphorismen zur Erkenntnis- und Wissenschaftstheorie, zur Ethik und Politik befassen, biographische Ereignisse in Rechnung stellen, Gestalt und Werk historisch einordnen und Kommentare aus ganz verschiedenen Wissenschaften lesen: ein vielseitiges Material wartet und hilft zum Verständnis; aber e i n Zugang ist vergleichsweise selten beachtet worden: ein erkenntnispsychologischer. Erkenntnispsychologie befaßt sich mit Erlebnissen und Verhaltensweisen, die etwas mit der Hervorbringung von Wissenschaft zu tun haben.

I. Gewinnung der Fragestellung

Überall da, wo Menschen beteiligt sind - und wo sind sie es nicht? - verschafft sich der Psychologe Zutritt und redet mit. So befindet er sich in einem weiten Arbeitsfeld. Doch was Chance hat, in methodischer wie in inhaltlicher Hinsicht offen zu sein, kann leicht umschlagen in Verzettelung. In einigen Gegenden nämlich, in denen er sich aufhält, ist die Luft ein wenig dünn, in anderen übt er methodologische Askese in positivistischer Einseitigkeit (Sander). - Wo sollen wir denn unsere Argumente verankern? Ist es denn nicht beliebig, was wir hervorheben und wie wir es interpretieren?

1. Wertheimers Beitrag. Es gibt ein einzigartiges Dokument, auf das wir uns beziehen können. Im Jahre 1916 führte der Gestaltspsychologe M. W e r t h e i m e r Gespräche mit Einstein. Die beiden gingen schrittweise die Entstehung (vor allem) der Speziellen Relativitätstheorie durch. Wertheimers Begriff des 'Produktiven Denkens' wurde dadurch wesentlich geformt. "Das waren wunderbare Tage", teilt Wertheimer mit, "als ich... das Glück hatte, Stunden und Stunden mit Einstein zusammenzusitzen, allein in seinem Arbeitszimmer, und von ihm die Geschichte der dramatischen Entwicklungen zu hören, die in der Relativitätstheorie gipfelten. Während dieser langen Gespräche richtete ich an Einstein sehr ins Einzelne gehende Fragen über die konkreten Ereignisse bei seinen Überlegungen. Er beschrieb sie mir, nicht in Allgemeinheiten, sondern in einer Erörterung des Entstehens jeder der Fragen." [37,S.194] Diese psychologische Untersuchung - sie erschien 1943 in Wertheimers letztem Buch "Productive Thinking" - ist für uns die wichtigste Orientierung.

2. <u>Charakteristik des Themas</u>.   Auch wenn wir Wertheimers Forschungsarbeit in den Mittelpunkt stellen - und damit die Gestalttheorie - müssen
wir uns freihalten und andere Sichtweisen zulassen. Eine eklektische Be-
trachtungsweise, die in allen Punkten klar begründet sein muß, ist bei
unserem Problem das kleinere Übel (gegenüber einer Deutung, die bereits
in den Begriffen und Denkgewohnheiten einer Schule fixiert ist). "Ver-
nehmende Offenheit" (H. Kunz) gebietet besonders die Kreativitätsforschung. Denn ihr ist es nicht gelungen, die Zuständigkeit   e i n e s
bestimmten Fähigkeitsbereiches, Umstandes oder Prozeßverlaufes für kreative (gedankliche) Höchstleistungen nachzuweisen. Viele Variablen kommen
in Betracht (neben kognitiven Fähigkeiten emotionale Dimensionen, körper-
liche Konstitution, allgemeine und spezielle äußere Bedingungen, Lern-
felder u.s.f.). Wird versucht, einen oder wenige Faktoren für sich zu
sehen, um dadurch Höchstleistungen zu erklären, so bleibt immer ein Rest;
nimmt man alle zusammen, so entsteht ein riesiger dunkler Komplex -
'Genialität' -, der aus sich heraus zu existieren und sich einer von außen kommenden Analyse zu widersetzen scheint. Etwa Unfaßbares schwingt
mit: darf man es, kann man es überhaupt analysieren? Wie immer Genialität verstanden wird (als etwas qualitativ ganz anderes oder als bloße
Steigerung dessen, was in jedem Menschen ist) - über Möglichkeiten der
Erforschung entscheiden die Methoden. Auf sie müssen wir zunächst achten.

Wir finden Zugang zur Spezifität unseres Problemes, indem wir die Zeit-
verhältnisse vergleichen, die der Untersuchungssituation eigen sind (Z_1)
mit denen des zu untersuchenden Gegenstandes (Z_2). Sind etwa beide, wie
im Experiment,   s y n c h r o n ,   so kann das Geschehen durch eine
Funktion beschrieben werden: $x = f(y)$. Beginnend zum Zeitpunkt t
wird die die Variable y regierende unabhängige Variable x manipuliert
- sie erhält unterschiedliche Werte - und die Wirkung bei y wird beobachtet. Im experimentellen Tun wird bewußt eine zeitliche Ordnung geschaffen, bei der Z_1 und Z_2 identisch sind. Den aufeinanderfolgenden
Zeitpunkten des gewählten Intervalls lassen sich Ursachen- bzw. Wirkungs-
rollen zuordnen.

Diese Umittelbarkeit ist oft nicht zu erreichen. Die meisten historischen
Themen machen eine derartige Vergegenwärtigung unmöglich. Das Geschehen
tritt dann auf in der Gestalt zeitlich zurückliegender Wirkungen. Die
Glieder einer Kausalbeziehung sind gleichsam vertauscht. Die Ursachefak-
toren können nun nicht durch systematische Variation und damit auch nicht
durch Ausschluß unbeteiligter Faktoren entdeckt werden. Der zeitliche
Abstand von Z_1 und Z_2 schafft Raum für Möglichkeiten und Deutungen.

Von Einsteins Hand liegen etwa 400 wissenschaftliche Arbeiten vor, ein überdurchschnittliches Werk hinsichtlich seines Umfanges, neuartig generös hinsichtlich seiner Bewertung durch die Fachwelt. - Dies ist also der Tatbestand, den wir als Wirkungen begreifen - wo liegen die Ursachen? - Der Historiker wird versuchen, die dokumentierten Umstände zu sammeln und sachlogisch sowie in ihrer zeitlichen Folge ordnen. Für den Psychologen ist dieses Programm unvollständig. Zwar geben objektive Ereignisse Hinweise, Anekdoten, den großen Leistungen einen menschlichen Puls, aber alles ist vieldeutig. Man kann sogar an Einsteins Beispiel demonstrieren, daß gerade durch die Betrachtung objektiver Begebenheiten Höchstleistungen unerklärlich werden.

Es muß also noch etwas geben, denkt man sich, das es zu erschließen gilt, will man überhaupt zu Antworten kommen. - Konnte nicht Einstein viele Jahre hindurch gedankliche Widersprüche aushalten? Dies ist objektiv belegt. - Aber eine solche Eigenart entspricht nicht einer oft beobachteten psychischen Tendenz, Widersprüche ohne Lösung aufzuheben. Man könnte also schließen, Einstein habe über eine ganz besondere 'Ambiguitätstoleranz' verfügen können. Nach diesem Beispiel könnte man historisch gesichertes Wissen mit psychologischer Erfahrung vergleichen und verknüpfen. Man würde dann Einsteins Persönlichkeit durch Merkmale charakterisieren können. Angenommen, dies gelänge, eine Frage muß zuvor erörtert werden: wie verhalten sich solche Merkmale zum kreativen Produkt? - Hierzu stellen wir uns vor: A (ein Physiker) denkt über 'Masse' nach (z.B. über 'Masse' in der Allgemeinen Relativitätstheorie) - B (ein Philosoph) denkt über das Denken nach (z.B. über Induktion). C (ein Psychologe) versucht beides: er denkt über 'Masse' nach und will gleichzeitig - experimentell - sein Denken denkend beobachten.

Wird C seine Absicht ausführen können? Hierzu lassen sich viele Antworten sammeln, eine davon lautet: Nimmer kann dies Kunststück gelingen! Die Gedanken sind der Denker (James). Was im Denken geschieht, kann immer nur in der R ü c k s c h a u begriffen werden - aber eben nicht wie ein historisches Thema! Denn bei diesem ist ja ein unmittelbares Dabeisein (eine Identität von z_1 und z_2) prinzipiell möglich. Daraus folgt ein fundamentaler Unterschied zu einem Gegenstandsbereich, der sich funktional darstellen, zeitlich ordnen und durch Kausalbeziehungen charakterisieren läßt.

> Hier steckt ein großes Geheimnis: An einer fertigen Vase erkenne ich untrennbar verbunden Form u n d Materie, aber auch in allen Stadien ihrer Entstehung. Gibt es je reine Form (reines Denken)? - Teile lassen sich austauschen: für Materie ganz unterschiedliche Materialien einsetzen! Durch diese Transformation wird eine Einsicht möglich (die sich z.B. in der Geschichte des Trägheitsbegriffes als fruchtbar erwies): Man denkt sich Materialien als

Träger, deren Beschaffenheiten in Rechnung zu stellen
sind. Widerstrebt nicht, so kann man dann fragen, die
Materie der Form? Und kann nicht, umgekehrt, eine positive Wirkung von der Materie zur Form übergehen, indem sie sogar formspendend wirkt?

Um dies kurz anzuwenden, wollen wir eine Episode ernster
nehmen als sie vielleicht war: Einstein sollte einen Vortrag halten und bot stattdessen der Zuhörerschaft ein
Geigenkonzert (und wir erinnern uns an den von ihm ge-
lebten Aphorismus: Wissenschaften und Künste seien Zweige
desselben Baumes [vlg. 7,S.14]). Man sieht einen Materie-
anteil (logisches Schließen, begrifflich-physikalische
Formung, mathematische Ausdrücke und Zusammenhänge ...
Töne...). An solchen Transfereffekten spürt man, daß es
sich um zwei Bereiche handelt, kann sie aber nicht prä-
zisieren und auswerten (denn immer ist die Vase da).
Aber man kann das Geigenspiel und die Vorträge untersuchen und den Akteur bitten, seine Taten zu erläutern.

Kein anderer moderner Naturforscher hat mehr über Wahrnehmung, Denken,
Sprache, Logik, Erkenntnis- und Wissenschaftstheorie ausgesagt als Einstein. Seine   n a c h t r ä g l i c h e n   Erklärungen sind für die
Erkenntnispsychologie von großem Wert. (Mit berechtigten Erwartungen
wendet sich also C an A .) Aber Einstein befaßte sich mit erkenntnis-
psychologischen Themen auch in der Vorbereitung für seine Forschung: und
dies macht seine Überlegungen noch wertvoller. Im allgemeinen nämlich
leben Erkenntnistheorie und Wissenschaftspraxis getrennt. Bei Einstein
sind beide untrennbar miteinander verbunden: "Erkenntnistheorie ohne
Kontakt mit Science wird zum leeren Schema. Science ohne Erkenntnistheorie ist - soweit überhaupt denkbar - primitiv und verworren." [6,S.507]
Und wiederum hat dieses Aufeinanderangewiesensein einen doppelten Aspekt:
Wichtig ist nämlich nicht nur das, was Einstein hierüber gesagt, gedacht
und geschrieben hat, sondern auch das, was er getan hat. Denn das Werk
selber verhält sich zu vielen psychologischen Problemen gleichsam wie
Form und Materie: es trägt Psychologie in sich, aber nicht so unbequem
und verschlüsselt wie sonst. In der Relativitätstheorie ist nämlich die
Psychologie in der Gestalt physikalischer Aussagen und Zusammenhänge so
aufgearbeitet, daß man sagen kann: die Gedanken sind Denken. Grundlage
dieser These ist Einsteins "erkenntnistheoretischer Hauptsatz": Der
Physiker, so erklärt er, könne gar nicht weiterarbeiten, ohne ein viel
schwierigeres Gebiet kritisch zu betrachten, d.h. die Natur unseres alltäglichen Denkens zu analysieren. Unentbehrlich ist für Einstein diese
Analyse - warum? Er antwortet: "Die ganze Naturwissenschaft ist nichts
weiter als eine   V e r f e i n e r u n g   unseres alltäglichen Denkens"
[7,S.63] (Oder: "Das wissenschaftliche Denken ist eine Fortbildung des
vorwissenschaftlichen." [8,S.138])

Produktives Denken wirkt demnach im Vorgang der 'Verfeinerung'. Erkenntnispsychologisch ist dieses mit M a c h s Überlegungen eng verwandte Prinzip aufschlußreich. Wir folgern, daß Einsteins große Tat nicht nur darin bestand, die absolut gesetzten Raum-Zeit-Begriffen aufgehoben, neu durchdacht und definiert zu haben, sondern er hat auch etwas ganz anderes absolut Gesetztes aufgehoben: die Denkstruktur. Vor Einstein - und nach ihm wohl wieder - sieht man nämlich die Naturgesetze als unabhängig von der Struktur des menschlichen Denkens. (In den Vorträgen dieses Kongresses zum Themenkreis "Kosmologie" wurden sehr viele Varationen von Raumvorstellungen vorgestellt. So differenziert dieses Denken auch war, man setzte den "Observer" als etwas gänzlich Unbeteiligtes an.) Ich weiß wohl, daß damit das Anthropomorphismus-Probleme wieder hervorgerufen wird und finde es auch ärgerlich, das Objektive der Naturwissenschaft mit subjektivem Durcheinander zu verbinden - ärgerlich zwar, aber durch viele Einstein-Zitate zu belegen, durch eine Analyse der Entstehung von Einsteins Theorien zu demonstrieren und in seinem erkenntnistheoretischen Gedankenkreis auf eine höhere Ebene zu bringen.

> Da draußen - von unserem Denken und Fühlen unabhängig - gibt es die objektive Natur [vgl.6,S.2] . Sie ist als eine reale Welt 'vernünftig'. Denn wäre sie nicht 'vernünftig', wie könnte sonst Wissenschaft gelingen?
>
> Was ist Wissenschaft? Naturwissenschaft "... ist der Versuch einer nachträglichen Rekonstruktion alles Seienden im Prozeß der begrifflichen Erfassung" [7,S.29] . Was kennzeichnet diese Rekonstruktion? Sie orientiert sich an dem Prinzip, die Sätze mit Sinnesdaten zu verknüpfen; sie folgt einer formalen Forderung logischer Einfachheit; sie ist nie Zustand, Sein, immer Werden: In der Neuordnung bekannter Rekonstruktionselemente oder in der Neugewinnung bislang verborgener. Der Forscher ist Schöpfer, nie Verwalter.
>
> Eigenartig ist, daß es für sein Tun keine externen Gütekriterien gibt. Der ganze Prozeß durchläuft einen Kreis. Alle Möglichkeiten der Rekonstruktion liegen nur in ihm. Gäbe es eine logische Verbindung (derart, daß Verknüpfungsprinzipien, die im Kreis gelten, auch Brücken nach Bereichen außerhalb schlagen lassen) zwischen Rekonstruktion und Konstruktion (ebenso zwischen theoretischem Satz und Sinnendatum), so wäre die Denkstruktur und alles damit psychisch verbundene in einem anderen Sinn als hier zu verstehen. Im populären Verständnis wäre dann ein Satz verifizierbar. Er aber erborgt - nach Einstein - seinen Wahrheitsgehalt einzig von dem System, dem er angehört, und dieses wiederum ist 'wahr' nur insofern sich alle Sätze v o l l s t ä n d i g der Erlebnisgesamtheit zuordnen lassen [vgl.6,S.5] . Die vollständige Zuordnung ist das eigentlich rationale Kriterium (nicht, wie oft zitiert, die Sicherheit der Zuordnung, denn diese ist alogisch, intuitiv [vgl.11,S.129]). Eine vollständige Zuordnung aber ist nie zu erreichen: immer wird zum eben noch Vollständigen Neues hinzukommen.
> Einsteins geistiges Werk kommt mir vor wie ein Netz mit unterschiedlich großen Maschen: es umfaßt alle wesentlichen Lebensbereiche (Religion, Ethik, Politik), geht - enger werdend - in weite Bereiche philosophischen und naturphilosophischen Denkens

und zentriert sich, dicht geknüpft, in der Physik, aber alles ist miteinander verbunden. Man möchte gerne die Ergebnisse Einsteinscher Naturwissenschaft für sich haben - ohne jene ein wenig verdächtigen Zutaten - aber gerade dies wäre eine Forderung, die seinem Verständnis von Wissenschaft im Kern widersprechen würde.

Von Einstein legitimiert, werden wir und in der Folge mit der Struktur des Denkens befassen, als ob dies ein Teil der Physik wäre: durch Einbeziehung erkenntnistheoretischer Vor- und Nacharbeit des Forschers selbst (hierzu besonders Wertheimers Beitrag), aber auch durch Auswertung des physikalischen Werkes, in welchem alltägliches Denken in einer anderen Gestalt - verfeinert - wiederzufinden ist. Der Psychologe unternimmt dies in enger Kooperation mit dem Physiker, dem Wissenschaftstheoretiker und Historiker, allerdings aus einem spezifischen Blickwinkel: er sieht primär nicht auf inhaltliche Definitionen und Zusammenhänge, er versucht nicht die Rekonstruktionselemente durch formalgedankliche Vorlagen der Wissenschaftstheorie zu begutachten, sondern achtet auf das lebendige Zusammenspiel der Teile - welche Teile sind da gemeint und wie lauten die Spielregeln? Ein winziges Stück dieser anspruchsvollen Frage werden wir hier beantworten.

II. Ein Programm für produktives Denken: Strukturarbeit

1. <u>Anagramm</u>.   Die literaturhistorische Forschung erkannte lange nicht, daß German Schleifheim von Sulsford ein Pseudonym ist, ein Anagramm, das bei anderer Anordnung der Buchstaben den Namen Christoff von Grimmelshausen ergibt. Diese ars combinatoria des Barockdichters ist e i n   Grundmuster für einen Denkvorgang. Die Ausgangslage ist der eigentliche Name, ein Ganzes, das aus 28 Buchstaben - nennen wir sie Elemente - besteht. Man findet in ihm eine Anzahl von Beziehungen zwischen den Teilen (z.B. nachbarliche Folge der Buchstaben, Plazierung der Worte) und zwischen den Teilen und dem Wortganzen. Diese Beziehungen ergeben eine Struktur. Um ein Anagramm zu bilden, kann man die Buchstaben nach Zufall, nach den Regeln der Kombinatorik oder auch nach Maßgabe einer bestimmten Strategie vertauschen. Was heißt 'bestimmte' Strategie? Angenommen, es soll ein Anagramm gebildet werden, das wie eine geographische Bezeichnung erscheinen soll. Über geographische Namen besitzen wir allgemeine Erfahrungen (Hornberg klingt eher so als Kulismus). Man orientiert sich also an Merkmalen, die unentbehrlich zu einem Zielberich zu gehören scheinen, die bei allen Einzelerscheinungen dieses Bereiches vorkommen und auch für eine Neubildung zutreffen sollen. Die Möglichkeiten der Umstellung der 'Elemente' lassen sich auf diese Weise reduzieren, umso mehr, je geringer die Anzahl individueller Merkmale gegenüber den allgemeinen ist. Wesentlich ist hier nicht, wie ein Zielbereich auf den

Denkprozeß einwirkt (z.B. auch vor-rational), sondern daß er überhaupt eine selektive, richtungsgebende Funktion erhält.

Wertheimer würde dem zustimmen: Produktives Denken heißt Umstrukturieren, es wird eine Neugestaltung der Beziehungen einer Anzahl von Elementen angestrebt. Gut, aber nach welcher Strategie? Im Falle Grimmelshausen handelte der Denker nach   Z i e l n o t w e n d i g k e i t e n : das Material, die Elemente, müssen sich in diesem instrumentellen Denken einer fremden Absicht unterordnen. Der Denkvorgang ist beendet, wenn das gewünschte Ziel erreicht ist. Wertheimer setzte hier anders an: Der Denker muß die Struktur des Gegebenen abtasten, die der Elemente. Er wird dann eine Strategie finden, die sich aus   S t r u k t u r n o t w e n - d i g k e i t e n   ergibt.

Der springende Punkt liegt bei der Untersuchung von 'fremden' Zielnotwendigkeiten und 'eigenen' Strukturnotwendigkeiten. Wie kann man überhaupt darüber entscheiden? - Eigentum dem, der etwas gebraucht (so eine Auffassung aus der Geschichte der Rechtssprechung); wir müssen darauf achten, was produktives Denken benötigt: auf den Prozeß. Er hat folgenden Verlauf: Der Denkende sucht Übersicht (hier: er hat alle Buchstaben in der Gestalt irgendeiner Anordnung im Blick). Von diesem Ganzen geht er aus. Ob es 'richtig' oder 'falsch' sei, diese Frage stellt sich hier gar nicht. Er merkt sich das 'Ausgangs-Ganze' gut. Nun ordnet er die Teile neu (hier: er stellt Buchstaben um) und prüft sofort, wie diese Neuordnung dem Ganzen steht. Produktives Denken vollzieht sich im raschen Vergleich zwischen Überblick und je neuen Bedeutungsnuancen der Teile. Immer wieder werden diese nach der Korrektur durch den Blick aufs Ganze gesetzt, bis das Strukturbild klarer geworden ist.

2. Probleme mit dem Anfang. Eine Wissenschaft, die lange und bei ganz Unterschiedlichen in Bearbeitung war, stellt sich dar als ein Sammelsurium methodischer Möglichkeiten, begrifflicher Fassungen, theoretischer Modelle, erfahrungswissenschaftlicher Daten, Wertungen, Denkgewohnheiten, Mischung von Grundlagen und Anwendung u.s.f. Wer in diesem Umkreis produktiv denken will, muß erst Grundvoraussetzungen schaffen.

Wie findet man sie? Wir haben mit Absicht ein Anagramm gewählt, um ein Schema für produktives Denken vorzustellen, auch wenn in der wissenschaftlichen Forschung eine festgesetzte Anzahl von Elementen gerade nicht vorgegeben ist. Ein Ausgangsganzes zu schaffen und zu fragen, ob die Elemente vollständig sind, kommt beim Forscher immer hinzu. Aber dies ist (gegenüber den Prozessen des Umstrukturierens) kein eigenständiger kognitiver Vorgang. Für Vollständigkeit gibt es kein absolutes Maß. Um etwas zu vervollständigen, muß eben erst eine Ausgangslage da sein.

Denn erst in Bezug auf diese kann man von einer Gerichtetheit sprechen,
der das Denken folgt.

> Meist scheitert produktives Denken ganz am Anfang, weil die
> Strukturgewinnung nicht gelingt. Ein Student erklärte mir in
> einer Diplomprüfung: "Ich würde schon gerne nachdenken, wie
> Sie wollen, aber ich weiß nicht worüber!" Er war über Fakten
> und Theorien informiert, konnte aber nur reproduzieren.

> Zwei Hauptschwierigkeiten fallen auf: Erstens: Eine endlose
> Reihe von Bedeutungen. Wird an irgendein Gegenstand gedacht,
> so bildet sich beim Informieren eine Summe unterschiedlicher
> Akzentuierungen. Kleine Konflikte können sich dabei anhäufen:
> wird gerade   e i n   Aspekt beachtet und hervorgehoben, so
> meldet leicht ein zweiter, dritter ... seine Berechtigung an.
> Der Denker müßte jetzt gewichten, doch nach welcher Maßgabe?

> Fast immer wird diese ermüdende Arbeit nur dadurch unterbro-
> chen, daß (abgesehen vom Zeitdruck) ein eigener Standort oder
> eine bestimmte Absicht   e i n e   Sichtweise beleuchten,
> alles andere aber ausblenden. Man hat eine Aufgabe erfüllt,
> aber was geschieht mit dem unbeachtbaren Rest? - Zweitens:
> Interferenzen. Zwei Wellen löschen einander, die um eine
> halbe Wellenlänge gegeneinander verschoben sind. - So kann
> die Strukturgewinnung dadurch verhindert werden, daß ein
> Denkimpuls den anderen aufhebt. Ein derartiger Effekt -
> als Grundmuster - wirkt wie eine Blockade und führt zu kog-
> nitiver Hilflosigkeit. (Es wird in einem späteren Abschnitt
> zu zeigen sein, wie Einstein diese Probleme bewältigte.)

3. Gerichtetheit als Strukturnotwendigkeit.   Denken kann als 'Funktion'
verstanden werden (etwa im Sinne Rohrachers): als Werkzeug, dessen Beschaffenheit zwar zu beachten ist, das aber nur funktioniert, wenn es
durch eine Kraft in Gang gebracht und ausgerichtet wird. Sinngemäß stellt
sich auch Wertheimer die 'strukturelle Spannung' als Vektor vor. Er argumentiert: Zwar kann man in Einsteins Theorien Resultate und traditionell
logische Operationen erkennen (wie z.B. Abstraktionen, abgeleitete Folgerungen, Syllogismen), aber was erhalten wir durch eine solch stückhafte
Analyse? - "Wir bekommen einen Haufen..." [37,S.210] . Nur integriert -
ausgerichtet - sind die Sätze der Logik sinnvoll. Für sich genommen zer-
schneiden sie lebendige Denkvorgänge. Sie kommen von außen her ins Bild
aufgrund von Rückbesinnung. Das   s i n n v o l l e   'Wenn-dann' verlangt, unentbehrlich, innere Zusammenhänge mit einem Ganzen, also strukturelle Bezogenheit. Wertheimer erinnert an die Erörterung des Linné-
schen Systems in der Pariser Akademie: Exakt logisch geschlossen, schließe
nicht aus, völlig verschiedene Gegenstände zusammenzubringen oder andere
zu trennen, die zusammengehören. Was (traditionell) logisch gesehen, in-
tentisch erscheint, kann strukturell Verschiedenes bedeuten. Und wenn
auch zwei Verschiedenes tun, so können ihre Taten doch strukturell dieselben sein. "Um in einer veränderten Situation dasselbe zu tun, muß man
es anders machen." [37,S.244] Und Veränderungen kennzeichnen doch die
Natur! Wenn also die traditionelle Logik sich auf statische Eigenschaften

beschränkt, so ist sie sachblind. Denn in der Wissenschaft habe jeder einzelne Satz immer etwas Gerichtetes in seinem Zusammenhang. In Wertheimers Untersuchung wird Einstein zitiert, der die siebenjährige Bearbeitungszeit der Speziellen Relativitätstheorie wie folgt charakterisiert: Es habe nie einen "blinden Schritt" gegeben. Immer das Problem vor Augen, habe das Denken eine Spur verfolgt. Wörtlich protokolliert Wertheimer: "Während all dieser Jahre hatte ich ein R i c h t u n g s - g e f ü h l , das Gefühl, gerade auf etwas Bestimmtes zuzugehen. Es ist natürlich sehr schwer, dieses Gefühl in Worten auszudrücken, aber es war ganz entschieden der Fall und klar unterscheidbar von der Art der späteren Überlegungen über die rationale Form der Lösung. Natürlich ist hinter solch einer Gerichtetheit immer etwas Logisches; aber ich habe es in einer Art von Überblick, gewissermaßen sichtbar vor Augen." [37,S.212,213].

Erkenntnispsychologisch ist diese These außerordentlich wichtig, ja sie scheint das eigentliche Kernproblem zu formulieren. Nun aber gibt es die Frage zu bearbeiten, wie eine Ausrichtung zustandekommt. Wertheimer antwortet: Produktives Denken befinde sich im Einflußbereich des sogenannten Prägnanzprinzips. Dieses fordert, "daß eine Organisation des Feldes so klar und einfach zu werden strebt, wie es die gegebenen Bedingungen gestatten" [37,S.225] .

a)   Ein Postulat?   Die wohl einfachste Lösungsstrategie ist die, ein Problem in ein Postulat zu verwandeln. Dies angenommen, könnte Einsteins "formale Forderung" an eine physikalische Theorie zur inhaltlichen Präzisierung des Prägnanzprinzips dienen: es wird dabei gesagt, was in Bezug auf eine Theorie 'Klarheit' und 'Einfachheit' bedeutet, logische Sparsamkeit an Begriffen, die die Basis bilden, aber dennoch noch mit der Erfahrung verknüpfbar.

b)   Eine "Heimlich arbeitende unterirdische Macht" (Metzger).   Wertheimers Anspruch geht aber über den eines Postulates hinaus. Gerichtetheit ist eine Konsequenz bestimmter Strukturverhältnisse. Erhebt sich nämlich in einer Struktur ein Problem (z.B. eine Lücke wird entdeckt), so entspringen die folgenden Denkoperationen der jeweiligen strukturellen Spannung. Ob nun die Denkakte in Sonderung, Gruppierung, Zentrierung oder struktureller Transformierung bestehen, entscheidet das Feld selber. Gut verständlich wird eine solche Strukturnotwendigkeit im Falle der sogenannten 3. Prägnanzstufe. Man stellt sich ein Quadrat vor, das an einer Stelle ein wenig verbeult ist. (Metzger sagt dazu: es ist nicht 'heil'). Gemessen an einem sauber gezeichneten Quadrat liegt es auf der Hand, was zu tun ist. Aber dies gibt Wertheimers Anliegen ungenau wieder. Strukturnotwendigkeit heißt, daß die 3 Seiten des Quadrates eine A u f -

f o r d e r u n g   ergeben, die Beule in der 4. Seite zu entfernen.
Ähnlich könnte es sich auch in einer Theorie verhalten: eine Störung,
ein strukturfeindliches Teilchen wird entdeckt und in der Folge struk-
turgemäß bestimmt. - Solange man deskriptiv solche Operationen beschreibt,
sind die Forderungen des Prägnanzprinzips einleuchtend. Dunkel aber ist
der theoretische Unterbau.

> Wertheimer - und in der Nachfolge Metzger - gehen von
> immanent wirkenden Kräften aus. Das Feld hilft sich
> gleichsam selber. Das erinnert an Aristoteles, in dessen
> Physik die Körper von sich aus ihren natürlichen Ort zu
> erreichen streben. Doch sei jene ordnende Kraft nicht mit
> einem entelechialen Prinzip gleichzusetzen, weil darin ein
> noch nicht verwirklichter Zielbereich zu stark betont werde
> (Metzger). Der Plan des Denkens ergebe sich nach Maßgabe
> des Prägnanzprinzips in kleinsten Schritten aus dem jewei-
> ligen Strukturzusammenhang. - Auch lehnt Metzger Platons
> Erkenntnisideal (nach welchem Erkennen Wiedererkennen sei
> und demzufolge das Denken durch eine frühere Einheit ge-
> lenkt werden könnte) ab und verweist auf Kant, dessen
> "Kompromiß" - Formen der Anschauung stellen einen subjek-
> tiven Anteil, die Anschauung selber sind als etwas Objek-
> tives, das Feld Betreffendes, zu verbuchen - am ehesten
> als Vorlage dienen könnte. Doch ist m.E. dieser Zusammen-
> hang ungenau und im Einzelnen nicht nachgewiesen.
>
> Da die im Feld wirkenden Kräfte immer nur in der Form von
> Wirkungen auftreten (Metzger), dürften sich kaum Kriterien
> zur Lokalisierung der Ursachen finden lassen - im Feld, im
> Erkenntnisvermögen oder in beidem? (Letzteres hat Köhler
> nachzuweisen versucht; er ging von Wertheimers Theorem
> "was innen, das ist außen" (ein Goethe-Aphorismus) aus und
> postulierte eine Gleichgestaltigkeit psychischer und nicht-
> psychischer Welt (Isomorphie). - Anhand der publizierten
> Äußerungen Einsteins läßt sich diese Frage   s y s t e -
> m a t i s c h   ebensowenig erörtern wie seine Beziehung
> zu großen Denkern. Kant, um ihn als Beispiel zu erwähnen,
> fragte, wie Wissenschaft möglich sei; Einstein   m a c h -
> t e   Wissenschaft. Das führt zu einem prinzipiellen Unter-
> schied: Der Forscher Einstein interessierte sich für Er-
> kenntnistheorie lebenslang, aber nie anders als im Zusammen-
> hand mit seiner wissenschaftlichen Arbeit. Er sei "skrupel-
> loser Opportunist" und überall macht er erkenntnistheore-
> tische Anleihen. Trotzdem sind   T e n d e n z e n   er-
> kennbar; so sagt er: "Das Wirkliche ist uns nicht gegeben,
> sondern aufgegeben nach Art eines Rätsels". [6,S.505] Da-
> bei ist der Forscher mit sich allein (das ist ein carta-
> sianischer Zug). Über das Gelingen der nach freien Stücken
> vorgenommenen Rekonstruktion kann man sich 'wundern', aber
> es gibt kein Darüberhinaus. Zwar kommt von den Sinnerfahrun-
> gen das Rohmaterial, wodurch Denken inhaltlich wird, doch
> die 'V o r s c h r i f t e n' für das Denken sind nur im
> menschlichen Geist begründbar, durch sich selbst begründbar,
> d.h. durch ihre Ganzheit: wenn sich die Erlebnisgesamtheit
> dem Ganzen aller Einzelerlebnisse vollständig und sicher zu-
> ordnen läßt [vgl.8,S.44 ff. und 6,S.5] .

Ist hierdurch nicht eine Entscheidung getroffen, die Prägnanztendenz im
Erkenntnisvermögen zu suchen? Wenn man noch hinzunimmt, daß - wie oben

zitiert - Einstein das 'Richtungsgefühl', das zeitlich vorher war, von
der 'späteren rationalen Form der Lösungen' trennt - haben wir damit
nicht eine Rahmenhypothese? Es wäre von einer fühlenden Weltvergegen-
wärtigung auszugehen, aus der sich rationale Formen ableiten lassen. Man
ist an einen Satz von W. F. Otto erinnert: Mozart konnte sagen, daß ihm
ein sinfonischer Satz mit dem ganzen Reichtum seiner in zeitlichem Verlauf sich entfaltenden Formen und Bewegungen wie eine ruhende Gestalt
auf einmal gegenwärtig sein." [15,S.147] - Ehe wir dies ausführen, steht
noch der Nachtrag eines prinzipiellen Aspektes der Strukturarbeit an,
dessen Spuren wieder eindeutiger zu Wertheimer führen.

4.   'Äquivalenz' von Kognition und Emotion.
Indem eine strukturelle Spannung entsteht, ist auch ein Bedürfnis da.
Nur bei der Sache, mit anderen Worten: innerhalb eines Kreises zu sein,
in welchem das Problem sich erhob, begünstigt eine besondere Art von
kognitiver Arbeit. Ihr Kennzeichen ist u.a. dies, daß sie direkt vom
Problem her motiviert wird und zwar so lange, wie die strukturelle Spannung besteht. In reiner Form handelt es sich hier um intrinsische Motivation - so verstehen wir Wertheimers Anspruch.

Dies ist insofern eine ungewöhnliche Konstellation, als meist externe
Verstärker in den Sachbezug treten. Die so formulierte Äquivalenz ist
dann aufgehoben, wenn etwa, ein Zweck die Forschungsarbeit leitet und
motiviert oder auch, wenn ein sozialer Verstärker den Sachkontakt unter-
bricht (ein Lehrer lobt den Schüler, der gerade ganz bei der Sache ist.
In diesem Fall kann die soziale Anerkennung stören, ablenken und wie ein
Surogat sättigen).

Der Sachbezug hält gefangen. Es entsteht ein Verlangen, eine Erkenntnis-
lust, w i r k l i c h aufzuklären - man will das Ganze erkennen, man
sucht den Überblick. Darauf zu verzichten wird unerträglich. Man fragt
(immer und immer wieder): Ist das wirklich so? Ist mir das klar, habe
ich das wirklich verstanden? Ist das nicht noch einfacher zu fassen? Im
Falle Einstein ist es nicht schwer, die 'Äquivalenz' von Kognition und
Emotion zu belegen: Ein Widerspruch tat ihm weh wie sonst Schmerzen.

a)   Er verlangt die "Suche nach Wahrheit um ihrer selbst willen" -
Wissenschaft verkümmert sonst! Diese Überzeugung hat aber nichts mit
'wertfreier Wissenschaft' zu tun. Gemeint ist ein erkenntnispsycholo-
gisches Moment: daß nämlich Wahrheit überhaupt nur gesucht werden darf
und gefunden werden kann, wenn alle Möglichkeiten menschlicher Existenz
in den Bereich, der wahr werden soll, involviert sind.

b)   Zahlreiche biographische Momente wie auch Äußerungen von Zeitgenossen belegen, daß für Einstein alles Wesentliche auf wissenschaftliches

Denken reduziert wird. Er lebt, d.h. er denkt. Viele Formen des Daseins
sind in einer. So erscheint alles außerhalb des Denkens, als ob es un-
entwickelt oder aufgelöst sei. I n f e l d , der langjährige Freund,
berichtet: "Ich kenne niemanden, der so allein und so losgelöst war wie
Einstein. Seine außerordentliche Freundlichkeit, seine absolute Bescheidenheit, seine Geradheit in der Behandlung von Menschen und sozialen
Ideen war trotz allem gegenteiligen Anschein unpersönlich und losgelöst."
[19,S.467] Der 'Einspänner' Einstein selbst sagt: "Der wahre Wert eines
Menschen ist in erster Linie dadurch bestimmt, in welchem Grad und in
welchem Sinn er zur Befreiung vom Ich gelangt ist." [8,S.11] Befreiung
wovon und wofür? Vom 'Blind-Triebhaften', den 'Fesseln des Nur-Persönli-
chen' (und damit von einem Weltteil, der für viele der wichtigste ist).
Frei wofür: für das Denken. Eng mit Spinoza verwandt, erkennt Einstein
in der "Nichtigkeit des Daseins" eine Gefangenschaft, aus der nur die
Tätigkeit des Denkens führt. - Ohne diese den Kern seiner Persönlichkeit
bezeichnende Auffassung bleiben seine denkerischen Höchstleistungen un-
verständlich.

c)   Wie stark emotionale Anteile die Strukturarbeit im Zusammenhang mit
der Entstehung der Speziellen Relativitätstheorie mitbestimmen, zeigt
folgender Bericht Wertheimers: Früher hätten Fragen beunruhigt wie: "Wie
wäre es, wenn man hinter einem Lichtstrahl herliefe? Wie, wenn man auf
ihm ritte? Wenn man einen Lichtstrahl auf seiner Reise verfolgte, würde
seine Geschwindigkeit dann abnehmen? Wenn man schnell genug liefe, würde
er sich dann überhaupt nicht mehr bewegen?" [37,S.195] - Am Anfang war
also die Beunruhigung, ein "gewisser Zustand der Verwirrung". Sonderbar,
alles schwer beschreibbar - aber das Interesse wurde gefangengenommen.
Er las nicht nach oder befragte andere, er dachte nach und überlegte sich
einfache Experimente, um die Erdbewegung zu messen. Doch die Beunruhigung
blieb erhalten, bis erste Überzeugungen heranwuchsen. Eine absolute Bewegung, das fühlte er, gibt es nicht. Ein Restunbehagen blieb zurück,
vielleicht Anlaß zur nächsten Überzeugung: Es ist doch ziemlich unwahr-
scheinlich, daß sich 'c' vom Bewegungszustand des Beobachters beeinflussen
läßt! - Erste Arbeit: er versuchte jahrelang Maxwells Gleichungen zu verändern. Und damit wurde er auch sicherer in der Meinung, daß die Lage
beim Licht nicht wesentlich anders sei als bei mechanischen Vorgängen.
Das Michelson-Experiment schien seine Gedanken zu bekräftigen - doch die
Idee von Fitzgerald befriedigte nicht. "Die Schwierigkeit war zwar 'weg-
geschafft'. Aber... die Lage war nicht weniger beunruhigend als zuvor;
er hatte das Gefühl, daß die Hilfshypothese eine Annahme ad hoc sei, die
am Kern der Sprache vorbei gehe." [37,S.200]
Er habe sich gefragt: Ist das wirklich klar? Habe ich das wirklich ver-

standen? "Während dieser Zeit war er oft niedergeschlagen, manchmal verzweifelt, aber von den stärksten Vektoren angetrieben." [37,S.200] Er
empfindet ein leidenschaftliches Verlangen zu verstehen. "Er fühlte irgendwo eine Lücke, ohne daß er imstande war, sie zu klären, ja, sie auch
nur zu benennen. Er fühlte, daß die Schwierigkeit tiefer ging als bis zu
dem Widerspruch zwischen Michelsons erwartetem und erzieltem Befund. Er
spürte, daß ein bestimmter Bereich in der Struktur der Gesamtsituation
ihm in Wirklichkeit nicht so klar war, wie er sollte, obwohl er bisher
von jedermann, ihn selbst eingeschlossen, ohne jede Frage hingenommen
worden war." [37,S.200] In der letzten Phase der Strukturgewinnung, kurz
bevor der konstruktive Schritt einer Strukturverbesserung gelingt, sagte
er sich: "Wenn zwei Ereignisse am selben Ort stattfinden, verstehe ich
klar, was Gleichzeitigkeit bedeutet... Aber... bin ich mir wirklich klar
über das, was Gleichzeitigkeit bedeutet, wenn sie sich auf Ereignisse an
zwei verschiedenen Orten bezieht? ... Ist es mir im ersten Fall so klar
wie im zweiten? ... Nein!" [37,S.201]

III. "Talent ist Charaktersache: Ausdauer und Geruchssinn"

1. Das Rätsel 'Intuition'. Während seines Studiums fühlte sich Einstein belastet von dem Vielen, das den Geist erfülle und vom Wesentlichen
ablenke. Er sei damals ganz überwältigt gewesen von der Masse des er-
fahrungsmäßig Gegebenen und ungenügend Verbundenen. Wie konnte es gelingen, eine Struktur zu finden? Er habe bald gelernt, antwortet er,
"dasjenige herauszuspüren, was in die Tiefe führen konnte, von allen anderen aber abzusehen." [6,S.6] Auf mathematischem Gebiet sei die Intuition nicht stark genug gewesen, wie im Bereich der Physik, um das fundamental Wichtige, Grundlegende, sicher vom Rest der mehr oder weniger ent-
behrlichen Gelehrsamkeit zu unterscheiden. Und zu Studenten sagt er: Es
sei ganz sonderbar mit den wissenschaftlichen Bestrebungen: oft sei
nichts von größerer Wichtigkeit, als zu sehen, wo es nicht angezeigt ist,
Zeit und Mühe anzuwenden. Man müsse einen Instinkt darüber erlangen, was
unter Aufbietung der äußersten Anstrengung gerade noch erreichbar sei.

Wir befinden uns jetzt an einer außerordentlich schwierigen Stelle. Die
Rede ist vom Gefühl, vom Instinkt und Intuition. In vielen Mitteilungen,
die Einstein über sich und seine wissenschaftlichen Anstrengungen macht,
finden wir diese Begriffe. Wie soll man sie verstehen? Einstein hat wohl
bei der Verwendung des Begriffes 'Instinkt' nicht an Lorentz, beim Begriff 'Gefühl' nicht an Krueger und beim Wort 'Intuition' nicht an Bergson
gedacht. Eine Klärung wird nicht erreicht durch Orientierung am handels-
üblichen Begriffsverständnis. Aber was hilft es: auf der Landkarte, die
Einsteins kreatives Denken und Werk abbildet, gibt es weiße Flecke: Das

fundamental Wichtige, lehrt er, den Anfang einer Wissenschaft muß man er-
spüren. Was die Richtung betrifft, die das Denken bestimmt, so gibt es
Richtungsgefühl. - Was unterscheidet wissenschaftliche Wahrheit von
"leerer Phantasterei"?: Der Grad der Sicherheit der intuitiven Verknüpfung von Begriffen und Sinnenerlebnissen; an der Logik ist logisch die
Einhaltung ihrer Regeln, alles andere ist alogischer Natur (z.B. sie auf-
zufinden). Noch 1948 teilt Einstein in einem Brief an Solovine mit: er
lese der Schwester aus einem Buch vor, in welchem Ptolomäus gegen Ari-
starch Argumente vorbringe. "Ich habe dabei an manche Argumente der heutigen Physiker denken müssen: gelehrt und raffiniert, aber instinktlos.
Das Abwägen von Argumenten in theoretischen Dingen bleibt eben Sache der
Intuition." [11,S.88] - Wie geht man mit diesen unbequemen Ausdrücken
um? Ist es denn nicht für die Ratio schwindelerregend, überall mit dem
Gefühlsbereich zusammenzustoßen? Selbst die heutige Psychologie spart
diesen Themenkreis weitgehend aus. Und doch, sollte man nicht neugierig,
vorurteilsfrei, kurz über die Grenzen schauen?

2. Zwei Erkenntniskräfte. Er sei ein Spätentwickler gewesen, erzählt
Einstein, und so habe er sich noch wundern und staunen können, als sich
die Altersgenossen bereits an alles gewöhnt hatten, auch an die Phänomene
Raum und Zeit. Früh aber war Mißtrauen gegen alles Selbstverständliche erwacht; es wuchs, lebenslang, anstatt sich zu beruhigen. Die Haltung
des alten Forschers aus Princeton, der eine "Einheitliche Feldtheorie"
sucht und von sich sagt, er sei bald der einzige, der noch daran glauben
könnte, ist dem jungen Physiker der "Speziellen Relativitätstheorie" sehr
ähnlich. (Auch) in geistiger Hinsicht ließ sich Einstein nicht soziali-
sieren.

Aufschluß gibt eine berühmte Episode. Man schenkte dem Jugendlichen einen
Kompaß. "Das diese Nadel in so bestimmter Weise sich benahm, paßte so gar
nicht in die Art des Geschehens hinein, die in der unbewußten Begriffswelt Platz finden konnte (an 'Berührung' geknüpftes Wirken)." [6,S.3]
Aus diesem Bericht läßt sich ein Schema gewinnen: ein Sinneneindruck -
eine Wahrnehmung - erschien mit dem vorhandenen 'unbewußten Begriffsfeld'
(oder, wie Einstein auch sagt: mit einem "gut gefügten Gefühlssystem")
oder mit einem 'hinreichend fixierten Begriffssystem' unvereinbar. Die
Gefühlsstruktur - so nennen wir diesen Bereich - charakterisiert, daß alle Geschehnisse phänomenal-kausal geordnet erscheinen
(an Berührung geknüpftes Wirken), nur jener neue Eindruck nicht. Und
weiter: jene Gefühlsstruktur muß gut gefügt sein, so erklärt Einstein
ausdrücklich, weil sonst ein Zusammentreffen mit einem derart denkwürdi-
gen Gegenstand gar nicht "hart und intensiv" erlebt werden kann. Ein
Konflikt ist unvermeidbar - wie kann man ihn lösen?

Ehe wir im Sinne Einsteins antworten - ein so nicht erwartetes Erlebnis
läßt sich prinzipiell auf ganz unterschiedliche Weise assimilieren. Es
gibt viele Techniken, um Dissonanzen aufzugeben, um ein inhomogenes Wahr-
nehmungsfeld zu vereinheitlichen (z.B. durch Formen des Verdrängens).
Aber Einstein erhielt sich den Konflikt und versuchte eine ganz besondere
Art der Bewältigung d u r c h   d a s   D e n k e n . - Was ist die
Entwicklung des Denkens anders als eine beständige Flucht aus diesem Diskrepanzerlebnis!(so er selbst). Was sucht der Mensch im Tempel der Wissenschaft? "Der Mensch sucht in ihm irgendwie adäquater Weise ein ver-
einfachstes und übersichtliches Bild von der Welt zu gestalten und so die
Welt des Erlebens zu überwinden, indem er sie bis zu einem gewissen Grad
durch dieses Bild zu ersetzen strebt. Dies tut der Maler, der Dichter,
der spekulative Philosoph und der Naturforscher, jeder in seiner Weise."
[8,S.142] Hier geht es um den Ursprung von Forscher- und Künstlertum:
warum suchte Einstein Ersatz, warum strebte er danach, die Welt des alltäglichen Erlebens zu überwinden und wie konnte der neue selbstgewählte
Entwurf gelingen? Was Freud an Leonardos Werdegang zu erkennen glaubte
(er forschte, anstatt zu lieben) könnte vielleicht ähnlich auch bei Einstein nachgewiesen werden. Auch wären Eriksons Analysen fortzusetzen -
wir aber wollen das psychoanalytische Instrument nicht einsetzen. Mit nur
geringer Kenntnis der Einzelheiten wird die Sicht auf das Ganze - nicht
unbedacht - etwas unscharf bleiben.

3. Gefühlsstruktur. a) Formale Merkmale. Wird wie beim Einsatz
aktualgenetischer Methoden ein Bild zunächst nur kurzzeitig, dann immer
länger bis zur ausführlichen Betrachtung dargeboten, so steht am Anfang
eine gefühlsartige Komplexqualität, aus deren Diffusität sich allmählich
eine differenzierte Endgestalt herausbildet. Dauer und Art eines solchen
Vorganges - von der Vorgestalt oder dem Gestaltkeim bis zur differenzier-
ten Wahrnehmung - hängt von vielen Einzelbedingungen ab (z.B. von der Entropie der dargebotenen Figur). Sie interessieren hier nicht. Wir stellen
uns vor, es gäbe ein Wesen I, das nur Vorgestalten hätte, gegenüber einem
anderen (Wesen II), das sich im Besitz von Endgestalten wähnte. Es würde
in einer stark auf eine einfache Struktur reduzierten Welt leben. - Was
heißt da reduziert und einfach? Erstens: Einer alten Unterscheidung nach
ist Empfindung mehr auf die Spezifität der Reize gerichtet (II), Gefühl
eher auf den Aktus des Empfindens. Jene ist hochdifferenziert und viel-
fältig, dieses einheitlich (wie bei I). Viele unterschiedliche Reize
können ein und dasselbe Gefühl hervorrufen. - Zweitens: Wie steht es,
wenn wir vom Gefühl aus nach der Reizvielfalt fragen? Wir antworten zunächst mit einer grundsätzlichen Überzeugung, daß nämlich die Quelle
aller menschlichen Fähigkeiten daraus entspringt, sich in alles, Belebtes

458

und Unbelebtes,   e i n z u f ü h l e n   (Volkelt). Einfühlen? Wer vor
einer Plastik steht, um sie zu betrachten, läßt gleichsam zu, daß seine
Gefühle von ihr   v e r k ö r p e r t   werden. Indem der Betrachter die
Figur so beseelt, erfährt er, was diese (die Reizspezifität) daraus
macht. Gefühle erhalten einen Inhalt: sie treten jetzt - z.B. in der
Gestalt einer ästhetischen Empfindung - auf (Herder). Unsere Überlegung
schildert einen aktualgenetischen Verlauf. Strukturen werden verwandelt.
Man kann sich einen Kreisbogen vorstellen, der vom Betrachter zur Figur
verläuft, und einen zweiten, der zu ihm - 'beinhaltet' - zurückkehrt.
Der Gegenstand spricht mit. So wandert das Gefühl in seiner Weltoffen-
heit von Reiz zu Reiz. Dort, beim Reiz, wird nicht mehr gefühlt, da wird
gewußt. Das Wesen II findet da seine Wissensstruktur. Drittens: Wie verhält sich die Gefühlsstruktur zur Endgestalt (Wissensstruktur)? - Ist
das eine im anderen enthalten und wenn ja, wie soll man sich das denken?
Gibt es doch, so fragen wir weiter mit Skepsis, eine Art von Isomorphie?
- Erinnern wir uns an Einsteins Kompaß! Bei diesem Beispiel ist nämlich
nicht ohne weiteres klar, wie ein Diskrepanzerlebnis - eine inhomogene
Wahrnehmungsstruktur - zustandekommt. Eingedenk einiger Wahrnehmungs-
theorien (z.B. von Hebb) spielt beim 'etwas für Wahr-Nehmen' die Erwartung eine entscheidende Rolle. Im Falle des Kompasses dürfte die Erwartung nicht auf Eigentümlichkeiten des Gegenstandes gerichtet sein (auf
dessen Farbe, Größe, Beschaffenheit der Teile), sondern auf etwas, das
gar nicht zu sehen ist: auf die Nichterfüllung des Kausalprinzips. (Damit
befinden wir uns nicht auf der Denkebene Kants! - Wir setzen später an.)
Auch wenn wir zu Analysezwecken Vor- und Endgestalt getrennt haben. Beide
sind kontinuierlich verbunden. Die eine ist eine Verfeinerung der anderen. Verfeinert wird 'alltägliches Denken', das ist zum Beispiel der Umgang mit solchen Dingen, die sich berühren und gegenseitig bewegen...

> Ein ausgezeichnetes Beispiel wird von Einstein selbst
> detailliert ausgeführt: die Genese des Raumbegriffes.
> Beim freien Assoziieren bilden sich Reihen deren Glie-
> der einander wachrufen. Kommt ein Glied in verschiede-
> nen Reihen, immer wieder vor, so kann dies der Aufmerk-
> samkeit nicht entgehen. Es wird von Einstein als "ord-
> nendes Element" bezeichnet. Je weiter wir nun in das
> Feld wissenschaftlichen Denkens eindringen, desto do-
> minierender wird jenes Ordnungszeichen (das übrigens
> nicht notwendig mit einem reproduzierbaren Symbol ver-
> knüpft sein muß.)

> Wesentlich ist das Prinzip der Kontiguität (im Falle
> des Raumbegriffes): das gleichzeitige Auftreten von
> Gesichts- und Tasteindrücken. Durch diese Assoziation
> entstehen 'Körper'. Im weiteren Verlauf wird ein Unter-
> scheidungskriterium gewonnen, das der räumlichen Di-
> stanz aufgrund unterschiedlicher Lagerung. Körper kön-
> nen sich berühren oder - bei unterschiedlichem Abstand
> voneinander - entfernt sein. Berühren sie sich nicht, so

kann man andere, sogar ganz unterschiedliche Körper,
dazwischen legen, ohne ihre räumliche Distant dadurch
zu verändern. Der Begriff des Zwischenraumes entsteht.
Wenn man nun geneigt ist, die Körper als etwas 'Reales'
einzustufen, so gibt es keinen Grund, dieses Prädikat
dem Zwischenraum abzusprechen. - Die 'Verfeinerung' des
Raumbegriffes läßt sich bis zur Allgemeinen Relativitätstheorie darstellen. Wir aber, lieber Leser, müssen
aus Raumgründen abbrechen. Ich empfehle, das Studium
folgender Quellen: 7, S.46-53, S.58-140; 8, S.156-166,
S.180-193, S.207-212.

b) <u>Richtungsweisende Strukturmerkmale</u>. Die Relativitätstheorie sei,
so Einstein, ein "schönes Beispiel", um die Anwendung eines "unentbehr-
lichen und wirksamen Mittels der Forschung" aufzuzeigen: das Leitprin-
zip der L o g i s c h e n   E i n f a c h h e i t . Zwar werden die
Ausgangshypothesen immer erlebnisferner, aber dafür wird das "vornehmste
Ziel der Wissenschaft" erreicht: ein Mindestmaß and Hypothesen bei einem
Maximum an Erlebnisdaten, die logisch deduktiv zu umspannen seien
[vgl.8,S.111] . Ähnlich Kopernikus: gegenüber anderen astronomischen
Erklärungen konnte er kein einziges neues Datum finden, wodurch sein
Weltsystem sich ergeben hätte. Er fand es durch Umstrukturierung, Rollen-
tausch und Neuzentrierung - wie im Falle der Relativitätstheorie - nach
der Leitidee des Logisch-einfachen.

Ein zweites Vorbild der Gefühlsstruktur für die Ausgestaltung der begrifflichen Rekonstruktion ist das bereits erwähnte   K a u s a l p r i n -
z i p . Gegen den Widerstand fast aller Physiker hielt Einstein "radikal" daran fest. Er könne sich nämlich, so erklärt er, ohne den Glauben,
die Welt sei logisch- kausal zu rekonstruieren, gar keine Naturwissenschaft vorstellen. Die gegenwärtige "Wahrscheinlichkeitsmystik" sei ihm
"unerträglich". Zwar sei der mathematische Formalismus (Heisenbergs) gewiß
richtig, aber ein Fragment, das erst in einer zukünftigen Physik einen
Platz gewinnen müsse. Freilich, wenn die Zahl der mitwirkenden Faktoren
zu groß werde (wie beim Wetter), lasse Einem die wissenschaftliche Methode leicht im Stich; aber man könne doch daraus nicht schließen, daß
das Kausalprinzip nicht gelte! [Vgl.7,S.33; 6,S.414]

Ein drittes Moment ist die   A n s c h a u l i c h k e i t .   Einsteins
Prinzip, verwickelte Probleme durch Analysen anschaulicher Bilder (Lichtstrahl, Fahrstuhl), an die "einfache Fragen" gestellt werden, zu lösen,
dürfte für die Erkenntnispsychologie zu einem besonderen Forschungsge-
genstand werden. Vielfältige, lang andauernde gedankliche Verläufe,
zahllose emotionale Vorgänge, Lektüre einschlägiger Literatur: all diese
Anstrengungen brauchen einen Ort (eine Art Sparkasse) an welchem sie
leicht, auch nach Unterbrechungen, gesammelt werden können. - Zweitens
aber ist die Technik der "einfachen Fragen" ein Dokument dafür, daß

Einsteins begriffliche Rekonstruktion wirklich vom Alltäglichen ausgegangen ist.

c)   Zur Stabilität der Gefühlsstruktur

1.   Der innere Kontrollort.   Die Analyse von Entwicklungsverläufen ergibt oft einen übergroßen Anteil an Fremdbestimmung. In allen Regionen
des Lebensraumes haben 'Ursachen' ihren Sitz, deren 'Wirkungen' sich im
Individuum abspielen. So entsteht 'Soziale Identität', aber keine Selbst-
identität, wie sie der Produktive braucht und wie sie - ausgeprägt - von
Einstein erworben wurde. Studiert man nämlich seine Entwicklung, so läßt
sich kurz gesagt eine Umkehrung des Kausalverhältnisses beobachten: der
Ort, von dem aus das Geschehen kontrolliert wird, ist nach innen ver-
lagert. Er attribuierte - eindeutig im kognitiven Bereich - internal.
Die Gründung und Stabilisierung einer 'inneren Institution' läßt sich
bis ins Einzelne bei Einstein verfolgen.

2.   Sättigung - Wachstum.   Ein Muskel atrophiert, wenn er nicht gebraucht wird. Ebenso trocknen psychische Fähigkeiten aus, wenn sie nicht
in Funktion sind. Sie erhalten und verbessern sich in permanentem Trai-
ning. Dieses Entwicklungsprinzip ist zu ergänzen: Viele Verläufe endigen
(auf der Subjektseite) mit S ä t t i g u n g oder (auf der Objekt-
seite) mit G e s t a l t z e r f a l l (diesen Aspekt wollen wir nicht
beachten). Immer aber ist die Frequenz (!) ausschlaggebend.

> Motivationale Sättigung versuchen W a c h s t u m -
> m o d e l l e aufzuschieben oder gar aufzuheben. Man
> denk sich, daß ein motivierender Reiz (Hunger) nicht
> durch den Vorgang (Essen) gelöscht wird, sondern daß
> der Vorgang selbst auf den Reiz verstärkend wirkt. Aus
> einem Kreis entsteht eine Spirale. Das Motiv greift
> immer wieder nach. Dies sind grobe Umschreibungen, das
> eigentliche Problem ist noch verdeckt. Es wird sicht-
> bar, wenn man die Arten der Verstärker genau unter-
> sucht, die Wachstum bewirken sollen. Ein Aspekt lautet:
> Nachdem etwas getan ist, ergibt sich aus dem Werk ein
> neuer Anreiz, der vom ursprünglichen unabhängig ist
> (autonome Motivation).

Einstein hatte Freude am Denken und Forschen (auch wenn er einmal sagte,
wer die geistige Arbeit kenne, der reiße sich nicht um sie). Ja er zeigt
einer wissenschaftsmüden Zeit des Durchschnitts, daß Forschen viel mit
Freude und Begeisterung zu tun haben kann. Für diese positive Einstellung
gibt es zunächst eine simple Erklärung. Die Denkarbeit ist - wir haben
dies beschrieben - aus Diskrepanzerlebnissen hervorgegangen. Denken wird
wie eine Konfliktreduktion erlebt - und dies übt einen positiven Effekt
auf das konfliktreduzierende Moment - das Denken - aus.
- Wie aber soll das Denkresultat einen neuen motivierenden Effekt erhalten (wenn doch das Diskrepanzerlebnis, das Anlaß gab, aufgehoben ist)?

Hier zeigt es sich: Es kommt darauf an, daß das Denkresultat auf eine
gut gefügte, breit angelegte, g a n z h e i t l i c h e Gefühlsstruk-
tur bezogen werden kann. An irgendeiner Stelle wird sich dann ein neuer
Kontrast ergeben, eine neue Diskrepanz, die weitermotiviert - und dies
immer und immer wieder. Das Gefühl ist Nährboden alles Psychischen
(Krueger), auch des Denkens, der bei Einstein nie aufgezehrt wurde. Die
Gefühlsstruktur löste sich nie zugunsten einer Informationsstruktur auf.
In allen Stationen seines Schaffens spricht er vom Geheimnisvollen, vom
Staunen, vom Sich-Wundern (mirari): "Das Schönste, was wir erleben können,
ist das Geheimnisvolle, das an der Wiege wahrer Kunst und Wissenschaft
steht. Wer es nicht mehr kennt und sich nicht mehr wundern kann, nicht
mehr staunen kann, der ist sozusagen tot und sein Auge erloschen."
[8,S.10]

3. Spiel.   Wie viele Große spielte Einstein - er hat dies oft bezeugt.
Im Menschen schlummert ein großes Potential, das nur auf der Ebene des
Spiels mit dem Merkmal der Irrealität aktual wird. Man kennt dies beim
Kind und muß es auch dem Erwachsenen wünschen, der produktiv denken will.
Wo sind Assoziationen 'flüssiger' als im Spiel? ("Das ist ein Küchen-
stuhl", sagt ein Erwachsener - "Das ist ein Thron, ein Schaukelstuhl, ein
Schaukelpferd... ein richtiges Pferd", sagt ein Kind über den gleichen
Gegenstand). Solche Transferierungen gelingen im Irrealen - im Spiel,
weil keine realen Folgen die Phantasie belasten. Doch kommt es darauf an,
die Irrealitätsebene für die Realitätsebene fruchtbar zu machen. Dies
gelang, so behaupte ich, Einstein. Er schuf sich eine Denktechnik, die
es ihm ermöglichte, die Nieten wegzuwerfen (sie im Irrealen zu belassen),
den Gewinn aber real auszuwerten. - Wodurch konnte eine Selektion nach
Nieten und Treffern zustande kommen? Wir erinnern an die beiden Erkenntniskräfte 'Gefühlstruktur' und 'Denken'. Es spielte - in Eile formuliert
- die eine 'Kraft' der anderen etwas vor; Spielen muß Vorspielen sein!
Dann lassen sich nämlich im reichhaltigen Versuch-Irrtum-Material des
Spiels Aspekte entdecken, die für sich gar nicht aufgekommen wären.

4. Sinn.   Es gibt auch im Psychischen kein perpetuum mobile. Das
System Mensch verschleißt nur um zu sein. Jeder Historiker, der die Werke
der Großen studiert (und jeder Mensch, der selber eine wichtige Aufgabe
hat), weiß, daß anspruchsvollere Aktivitäten ohne Verbindung mit sinn-
spendenden Idealen erstarren. Doch sind die besten Überzeugungen wertlos,
wenn man im faktischen Tun das ganz Große vergißt. Nicht so bei Einstein!
Seine Arbeitsweise kennzeichnet Durchlässigkeit: Was er zum Beispiel über
'Kosmische Religiosität' fühlt, ist wirklich relevant für die denkeri-
sche Theoriearbeit. Dies ist folgerichtig, weil Ideale der Gefühlsstruktur entstammen, die vom Wissen-Schaffen nicht ausgeschlossen ist.

5. Die Welt drum herum. Wir haben individuelle Aspekte beachtet und können nicht sagen: "I djin" - der Sinn ist erschöpft -, zumal wir kulturelle und soziale Bezüge unbeachtet ließen. Aber das ist ein neues Thema. Es gab - fachlich - den Dialog mit den Besten und - emotional - viele Formen des Ruhms. So sehr hat die zivilisierte Welt ihn angenommen, das man ihm das Regierungsamt des neuen Palästina anbot und einmal sogar meinte, er sei das Weltgewissen. Die Gegengabe Einsteins ist nicht nur ein großes physikalisches Werk - wir wissen nicht, wie lange es dem steten Gang der Wissenschaft wird folgen können - sondern ein Beispiel, w i e W i s s e n s c h a f t g e m a c h t w i r d : man braucht einen ganzen Menschen, der alle Möglichkeiten entwickelt, durchlebt, verbindet und abrufen kann im Augenblick eines kurzen Gedankens.

Literatur

[1] R. Bergius, Psychologie des Lernens, 1971.

[2] D. E. Berlyne, Konflikt, Erregung, Neugier, 1974.

[3] H. Drüe, E. Husserls System der phänomenologischen Psychologie, 1963.

[4] A. Einstein, Grundgedanken der allgemeinen Relativitätstheorie und Anwendung dieser Theorie in der Astronomie. Preußische Akademie der Wissenschaften, Sitzungsberichte, 1. Teil, S. 315, 1915.

[5] A. Einstein, Vier Vorlesungen über Relativitätstheorie, gehalten im Mai 1921 an der Universität Princeton, 1922.

[6] A. Einstein, "Nekrolog" und "Bemerkungen zu den in diesem Bande vereinigten Arbeiten, in: Albert Einstein als Philosoph und Naturforscher (Hrsg. von Schilpp) 1951.

[7] A. Einstein, Aus meinen späten Jahren, 1952.

[8] A. Einstein, Mein Weltbild, 1953.

[9] A. Einstein, Grundzüge der Relativitätstheorie, 1956.

[10] A. Einstein, Briefwechsel mit M. Solovine, Paris 1956.

[11] A. Einstein, Briefe an Maurice Solovine, 1961.

[12] A. Einstein und A. Sommerfeld, Briefwechsel, 1968.

[13] Einstein-Born, Briefwechsel, 1969.

[14] Einstein-Besso, Briefwechsel, 1972.

[15] F. Förster, Von Glück und Leid des Erfinders, in: Mitteilungen der Deutschen Patentanwälte, 68. Jahrgang, Heft 8/9, 1977, S. 141 - 177.

[16] S. Freud, Eine Kindheitserinnerung des Leonardo da Vinci, 1976.

463

[17] C. F. Graumann, (Hrsg.), Denken, 1971.

[18] A. Hermann, Physik als Philosophie und Weltgeschichte. Zu Leben
      und Werk von Einstein, Hahn, Laue und Meitner, in: Gedächtnis-
      ausstellung zum 100. Geburtstag von Einstein, Hahn, Laue und
      Meitner, 1979.

[19] L. Infeld, Albert Einstein - seine Persönlichkeit, sein Werk und
      unsere Zeit, in: Universitas, 23. Jahrg., Heft 5, Mai 1968.

[20] M. Jammer, Der Begriff der Masse in der Physik, 1974.

[21] P. Janich, Die Sprache der Physik und die Wirklichkeit der Natur-
      wissenschaften, in: Dialectica, Vol. 31, Nr. 3 - 4, 1977,
      S. 301 - 312.

[22] G. Kaminski, Zur Analyse von Urteilssequenzen, Psych. u. Prax. 3,
      1959, S. 184 - 192.

[23] S. Kreuzer, Beiträge psychologischer Spieltheorien zur Aktivi-
      täts- vs. Disengagement-Diskussion in der psychologischen Geron-
      tologie, Diplomarbeit an der Univ. Tübingen 1978.

[24] U. Laucken, Naive Verhaltenstheorie, 1974.

[25] K. Lewin, Die Entwicklung der experimentellen Willenspsychologie
      und die Psychotherapie, 1970.

[26] E. Mach, Die Mechanik, Historisch-kritisch dargestellt, 1963.

[27] E. Mach, Erkenntnis und Irrtum, 1968.

[28] W. Metzger, Psychologie, 1968.

[29] G. Révész, Talent und Begabung, Grundzüge einer Begabungspsycho-
      logie, 1952.

[30] K. Rossmann, Wert und Grenze der Wissenschaft, Zur Symbolik von
      Dürers Kupferstich 'Melencolia I', in: Offener Horizont, Fest-
      schrift für K. Jaspers zum 70. Geburtstag, 1953 (S. 126 - 48).

[31] F. Sander und Hans Volkelt, Ganzheitspsychologie, Grundlagen -
      Ergebnisse - Anwendung, 1967.

[32] W. Schulz, Philosophie in der veränderten Welt, 1972.

[33] C. Seelig, A. Einstein, 1954.

[34] G. Ulmann, Kreativität, 1968.

[35] G. Ulmann (Hrsg.), Kreativitätsforschung, 1973.

[36] M. Wertheimer, Drei Abhandlungen zur Gestalttheorie, 1925.

[37] M. Wertheimer, Produktives Denken, 1964.

[38] J. Wickert, A. Einstein, 1971.

[39] J. Wickert, Psychologie der Zeit (unveröffentl. Manuskript) 1978.

[40] J. Wickert, Isaac Newton (erscheint 1979).

EINSTEINS DIALOG MIT DEN KOLLEGEN

Karl von Meyenn, Universität Stuttgart

Einleitung

Der Begriff vom Menschen als gesellschaftliches Wesen wurde zur Zeit der
Aufklärung geprägt und ist seither Ansatzpunkt jeder erklärenden Entwicklungsgeschichte des Individuums.

Will man den einzigartigen Werdegang eines Einsteins verstehen, so muß
auch hier seinen Beziehungen zur geistigen Umwelt besondere Aufmerksamkeit geschenkt werden.

Vielfach wurde geäußert, daß Einstein ungeachtet seiner Freundlichkeit
und Menschenliebe einen geringen Bezug zu seinen Mitmenschen hatte, ja,
daß er in voller Zufriedenheit auch ohne sie hätte existieren können.[1]
Er selbst bezeichnete sich des öfteren als   " t y p i s c h e r   E i n -
s p ä n n e r "[2]   und schwärmte einmal von der Tätigkeit eines einsamen
Leuchtturmwärters.

Zahlreiche Augenzeugen berichten demgegenüber von lebhaften Diskussionen,
die Einstein mit den Kollegen und im Kreis seiner Freunde geführt hat.

In diesem Lichte erscheinen solche Bemerkungen Einsteins auf den ersten
Blick paradox. Wir wollen versuchen, zu verstehen, wie dieser Hang zur
Isolierung und eine ausgesprochene Diskussionsfreudigkeit dennoch mit-
einander verträglich sein können.[3]

Eine Diskussion ist nicht notwendigerweise ein Dialog, denn dazu gehören
ebenbürtige Partner mit der Fähigkeit, jeweils differenziert auf die
Argumente des anderen einzugehen. Die Gruppe der in Frage stehenden Personen wird durch diese Forderung beträchtlich eingeschränkt.

Die geistige Auseinandersetzung zwischen Menschen kennt aber neben dem
Gespräch noch die Lektüre. Einstein selbst hat seit seiner Jugend viel
gelesen und ihre Wichtigkeit anerkannt. "Was einer selbständig denkt ohne
Anlehnung an das Denken und Erleben anderer", heißt es in einem seiner
Zeitschriftenaufsätze, "ist auch im besten Falle ziemlich ärmlich und
monoton".[4]

Im folgenden wollen wir nun einige der wichtigsten Dialogformen und
Partner aus Einsteins Leben herausgreifen und mit ihrer Hilfe die Besonderheiten seiner Entwicklung beleuchten.

Die prägenden Jahre

Wir wissen, daß Einstein in seiner Jugend viele Enttäuschungen erlebte.
Als Kind jüdischer Eltern mußte er in der Schule viele Demütigungen erdulden und einen wirklichen Klassenfreund hatte er nicht. Als Ersatz ver-
tieft er sich in die Bücher; diese ersetzen ihm den Freund, und hier
fand er seine ersten Anregungen. Wie wichtig Einstein solche Jugender-
fahrungen wertete, hat er später sehr deutlich in dem Nachruf auf seinen
Freund   P a u l   E h r e n f e s t   gesagt:

> *Es scheint mir, daß die Neigung zu übermäßiger Selbstkritik mit
> Erlebnissen im Knabenalter zusammenhängen. Demütigung bzw. gei-
> stige Unterdrückung durch verständnislose und egozentrische Leh-
> rer tut schweren, untilgbaren Schaden im kindlichen Gemüte, der
> gar oft das spätere Leben verhängnisvoll beeinflußt.*

Das ihm die Lektüre keineswegs nur zur Unterhaltung diente, zeigen die
Lehren, die er schon im Alter von 12 Jahren aus ihrem Inhalt zu ziehen
vermochte: Aus einem braven Kind wird ein aufsässiger Knabe, der die
Religion und den Staat als Erfindungen des Menschen zum Betrug seines-
gleichen erblickte.

Diesen negativen Eindruck kann der junge Einstein jedoch kompensieren
durch eine wachsende Liebe zu der Wissenschaft. "Zunächst glaube ich mit
Schopenhauer", sagte er später, "daß eines der stärksten Motive, die zu
Kunst und Wissenschaft hinführen, eine Flucht ist aus dem Alltagsleben
mit seiner schmerzlichen Rauheit und trostlosen Öde, aus den Fesseln der
ewig wechselnden eigenen Wünsche."[5]

Die gleiche Konsequenz beweist Einstein im praktischen Leben. Mit 15
Jahren verläßt er die ihm verhaßte Schule[6] und sucht sich nach einigem
Umherirren die ihm geeignetere Umgebung in der friedlichen Schweiz.

Ein zweiter Abschnitt seines Lebens beginnt. In der Schweiz schließt
Einstein seine ersten Freundschaften und entfaltet sich zu einem aufge-
schlossenen und geselligen Menschen. Seinen Freunden   B e s s o ,
S o l o v i n e   und   H a b i c h t   ist er sein Leben lang treu geblieben. Seine Professoren dagegen empfanden seine allzu große Offenheit und seinen nachlässigen Vorlesungsbesuch als einen Mangel an Respekt und Interesse.

Als einziger seiner Gruppe erhält er 1900 nach erfolgreichem Abschluß am
Züricher "Poly" keine Assistentenstelle, und er landet schließlich nach
zweijähriger Irrfahrt beim Berner Patentamt.

Ohne Zweifel hat Albert Einstein in diesen ersten Schweizer Jahren viele
Freunde gefunden und intensiv mit ihnen diskutiert. Sie gründeten sogar
zu diesem Zweck ihre vielgeliebte "Akademia Olympia". Zu einem echten

Dialog in unserem Sinne ist es aber hier nicht gekommen.[7] So hat es
auch Einstein im Rückblick empfunden, wenn er über sein Verhältnis zu
Besso einmal sagte: "Einen besseren Resonanzboden hätte ich in ganz
Europa nicht finden können."[8] Aufschlußreich ist auch Bessos bedeutungs-
volle Äußerung über diese Zeit in seinem Brief vom 17. Januar 1928:
"Meinerseits war ich in den Jahren 1904 und 1905 Dein Publikum; habe ich
bei der Fassung Deiner Mitteilungen zum Quantenproblem Dich um einen
Teil Deines Ruhmes gebracht, Dir dafür in Planck einen Freund verschafft."[9]

Die ersten Arbeiten und ihre Aufnahme in der Fachwelt

Noch im gleichen Jahre 1900 nach Abschluß seinerStudien reichte Einstein
seine erste Arbeit bei den Annalen der Physik ein. Es handelt sich um
den Versuch, aus der Oberflächenspannung einer Flüssigkeit die Größe der
molekularen Wechselwirkungen zu bestimmen.[10] Als Literatur hatte er
lediglich das Ostwaldsche Lehrbuch der Allgemeinen Chemie benutzt.

Einstein sandte einen Sonderdruck seiner Arbeit an W i l h e l m
O s t w a l d   und bat ihn um eine Assistentenstelle in seinem Leipziger
Institut.

Weshalb Einstein sich gerade an Ostwald wendet, ist uns heute schwer
verständlich. Wilhelm Ostwald gehörte als prominentester Vertreter der
Energetik einer der Machschen Naturphilosophie nahestehenden Richtung
an und war damit ein entschiedener Gegner der B o l t z m a n n schen
Atomistik. Auf der Lübecker Naturforscherversammlung war es 1895 zwischen
beiden Partein zu heftigen Auseinandersetzungen gekommen, ähnlich wie
sie sich ein viertel Jahrhundert später in Bad Nauheim wiederholen sollten.

Einsteins Untersuchung ging aber gerade von einer atomistischen Vorstellung der Materie aus und mußte deshalb den Unwillen Ostwalds hervorrufen.

Entweder war Einstein damals über Ostwalds Position schlecht unterrichtet,[11] oder er überschätzte zu sehr die Überzeugungskraft seiner Argumente. Ostwald hat jedenfalls auf Einsteins Brief und ein gleichzeiti-
ges Gesuch des Vaters[12] nicht reagiert.

Ein gutes Jahr später ist eine zweite Arbeit fertig, die sich inhaltlich
direkt an die erste anschließt. Auch in ihr findet man keine Literatur-
hinweise. Später hat Einstein sie als "wertlose Erstlingsarbeiten" betrachtet.[13]

Sobald Einstein eine feste Anstellung im Patentamt gefunden hat, beginnt
er eine Reihe von systematischen Untersuchungen über die Grundlagen der
statistischen Thermodynamik, die ihre Krönung durch die Arbeiten des

Jahres 1905 erhalten.

Alle diese Arbeiten sind praktisch im Alleingang entstanden. Weder ein Briefwechsel noch ein Gespräch mit den zuständigen Fachleuten hatte stattgefunden. Um so ausgiebiger hatte sich Einstein dafür mit Boltzmanns Gastheorie, der P l a n c k schen Wärmestrahlung und der L o r e n t z schen Elektronentheorie auseinandergesetzt.[14]

Die großen Arbeiten aus dem Jahre 1905 erscheinen so als das Ergebnis selbständigen Nachdenkens über den in den Lehrbüchern der Zeit dargebotenen Stoff. Mit der schon im Leben gezeigten Konsequenz vermochte Einstein jetzt auch auf geistigem Gebiet mit den gewohnten Denkstrukturen zu brechen.

Der große Durchbruch

Man hat das Jahr 1905 oft in Anlehnung an Newton Einsteins Wunderjahr genannt.[15] Das sollte aber nicht darüber hinwegtäuschen, daß alle diese Arbeiten bereits in den vorangehenden angelegt sind. Die Arbeiten von 1905 bedeuten aber insofern eine Wende in Einsteins Leben, weil sie ihm den Zugang zur Fachwelt eröffneten. Die erste von ihnen enthält Einsteins Lichtquantenhypothese. Durch sie berührte Einstein die Interessensphäre zweier Physiker, die später seine härtesten Opponenten werden sollten. Dennoch gehören P h i l i p p L e n a r d und J o h a n n e s S t a r k [16] zu den ersten Gelehrten, mit denen Einstein eine wissenschaftliche Korrespondenz anbahnte.

Lenard hatte die experimentellen Vorarbeiten zu Einsteins quantentheoretischer Interpretation des photoelektrischen Effektes geliefert.[17] Wahrscheinlich sandte ihm Einstein seine Veröffentlichung, in der er die "bahnbrechende Arbeit" von Lenard gebührend gewürdigt hatte. Einen Brief vom 16. November 1905, in dem sich Einstein für eine ihm zugeschickte Arbeit bedankt und eine Frage stellt, läßt Lenard zunächst unbeantwortet.

Angesichts dieser Mißerfolge ist es um so wichtiger, daß gerade der angesehenste Physiker des Deutschen Reiches zu denjenigen gehörte, die sich zuerst für Einsteins Relativitätstheorie interessierten.[18] Im WS 1905/ 1906 hat Planck persönlich im Berliner Physikalischen Kolloquium darüber referiert und die Theorie später zur Deutung der K a u f m a n n schen Ablenkungsversuche von Elektronenstrahlen herangezogen.[19]

Diese Meinung vertritt auch Einstein in einem Aufsatz aus dem Jahre 1913:[20]

> *Der Entschiedenheit und Wärme, mit der P l a n c k für diese Theorie eingetreten ist, ist wohl zum großen Teil die Beachtung zuzuschreiben, die diese Theorie bei den Fachgenossen so schnell gefunden hat. Planck hat als erster die Gleichungen der Bewegung*

des materiellen Punktes nach der Relativitätstheorie aufgestellt...
Auch entwickelte er in einer Untersuchung... den wichtigen Zusam-
menhang, welcher nach der Relativitätstheorie die Energie und die
träge Masse verknüpft.

Interessant ist, daß Einstein an dieser Stelle Planck an der Entdeckung
der Masse-Energie-Äquivalenzrelation teilhaben läßt, obwohl er früher
gegenüber Stark seine Priorität in dieser Angelegenheit betont hatte.[21]

Es wird berichtet, daß Planck damals Einstein zu seiner neuen Theorie
gratulierte.[22] "Meine Arbeiten finden viel Würdigung", schreibt Einstein
im Mai 1906 seinem Freund Maurice Solovine, "und geben Anlaß zu weiteren
Untersuchungen. Prof. Planck schrieb mir neulich darüber."[23]

Daß Planck sich an diesen Untersuchungen selbst beteiligte, hat er damals aber noch nicht gewußt.

Für Johannes Stark, mit dem Einstein spätestens seit Anfang 1907 regel-
mäßig Briefe austauschte, hat Einstein seinen ersten Übersichtsartikel
über die Relativitätstheorie verfaßt, der nun im Jahrbuch der Radioaktivität und Elektronik erschien. Bei dieser Gelegenheit informierte
ihn Stark über Plancks Bemühungen. "Es ist gut", erwiderte Einstein,
"daß Sie mich auf die Plancksche Arbeit über die Kaufmannschen Versuche
aufmerksam gemacht haben; ich wußte nichts von einer Untersuchung des
Herrn Planck über diesen Gegenstand."[24]

Immer drückender wurde die Last der Arbeit im Patentamt. Zu seinen eigenen Forschungen war seit dem Erscheinen der großen Abhandlungen von
1905 eine immer mehr anwachsende Korrespondenz hinzugekommen.

Obwohl man inzwischen auch im Patentamt auf die außergewöhnlichen Talente
des neuen Mitarbeiters aufmerksam geworden war und ihm neben einem besseren Gehalt Zeit für seine eigenen Untersuchungen ließ, war Einstein
"entschlossen, unter die Privatdozenten zu gehen."[25] Vorläufig mußte
er sich mit anderen Mitteln behelfen.

Nachdem Planck die Aufmerksamkeit auf Einsteins Arbeiten gelenkt hatte,
begannen sich auch die anderen Forscher für den Autor der Relativitätstheorie zu interessieren. Man darf aber nicht vergessen, daß Einstein
für die meisten Ordinarien als sonderbarer Außenseiter und kleiner An-
gestellter sozusagen nicht gesellschaftsfähig war. So waren es dann
besonders die jüngeren unvoreingenommenen Nachwuchskräfte, welche die
Initiative zur Herstellung eines persönlichen Kontaktes ergriffen.

M a x   L a u e , Assistent von Planck, besuchte Einstein 1906 während
eines Ferienaufenthalts in der Schweiz.[26] Obwohl ihn das Gespräch mit
Einstein stark beeindruckte und er sich dem Studium der Relativitätstheorie widmete, kam es damals zu keiner näheren Beziehung zwischen

den beiden.

Die Kunde von Plancks Interesse an Einsteins Arbeiten machte auch den Breslauer Kreis auf ihn aufmerksam. Zu diesem Kreis gehörten damals F r i t z   R e i c h e ,   M a x   B o r n   und   R u d o l f L a d e n b u r g .   Im Sommer 1908 besuchte Ladenburg daraufhin den Gelehrten im Schweizer Patentamt und berichtete darüber seinen Freunden.[27]

Frühe Kontakte mit den Kollegen

Die erste fruchtbare Zusammenarbeit mit Einstein kam Anfang 1908 mit dem jungen Österreicher   J a k o b   J o h a n n   L a u b   zustande. Nach absolviertem Studium der Mathematik und der mathematischen Physik unter   H i l b e r t   und   M i n k o w s k i   in Göttingen war L a u b   Anfang 1907 zu   W i l h e l m   W i e n   nach Würzburg gegangen. Dort verteidigte er als These seiner Dissertation (November 1906) die spezielle Relativitätstheorie.

Im Juni des folgenden Jahres veröffentlichte Laub seine ersten beiden Arbeiten auf diesem Gebiet. Nach einem Vortrag im physikalischen Kolloquium regte Wien, den das Thema interessierte, einen Besuch bei Einstein an.

Im Februar 1908 schreibt Laub an Einstein[28]   und bitte ihn um einen kürzeren Aufenthalt in Bern. In den folgenden drei Monaten war Laub zu Besuch bei Einstein. Aus dieser Gemeinschaftsarbeit und der Diskussion auf dem Nachhauseweg vom Patentamt und in Einsteins Wohnung entstanden in Kürze zwei Abhandlungen, zu denen Laub im wesentlichen die mathematischen Ideen beisteuerte. Aber auch in persönlicher Beziehung haben sich die beiden gut verstanden.[29]

Durch Laub, der viel umherreiste, erhielt man in Deutschland die ersten Nachrichten über Einsteins Persönlichkeit und seine Arbeitsweise. Umgekehrt war Einstein durch diese Zusammenarbeit mit den neuesten mathematischen Methoden der Göttinger Schule in Berührung gekommen, und er erhielt durch den anschließenden Briefwechsel mit Laub einen Einblick in die Interna der deutschen Hochschulen.[30]

Ebenso wie später   H e i s e n b e r g   bei seiner Entdeckung der Quantenmechanik Erfolg hatte, ohne den ihr zugrunde liegenden Matrizenkalkül zu kennen, so hatte sich auch Einstein den Zugang zur speziellen Relativitätstheorie allein mit Hilfe physikalischer Betrachtungen eröffnet. Hilbert soll einmal geäußert haben, daß Einstein das Originellste und Tiefste über Zeit und Raum zu sagen vermochte, weil er nichts über die Philosophie noch Mathematik von Zeit und Raum gelernt hat.[31]

Die Göttinger Mathematiker, mit Rudolf Minkowski an ihrer Spitze, inter-
essierten sich vorwiegend für die mathematischen Seiten der neuen Theorie.
Schimpfte Heisenberg später über den Göttinger Mißbrauch mit seiner
Quantenmechanik, so klagte Laub, der bei Einstein dessen physikalische
Denkweise kennengelernt hatte, jetzt in einem Brief vom 18. Mai 1908
Einstein seinen Kummer:[32]

> *Es ist ganz merkwürdig, was dem M a t h i a s   C a n t o r   an
> der   M i n k o w s k i schen Arbeit gefällt. Er schätzt nur die
> Behandlung der Zeit und Koordinaten als gleichartige Größen
> (x_1, x_2, x_3, x_4), daß man das als Drehung behandeln kann... Ich glaube,
> er hat sich durch die nichteuklidische Geometrie imponieren lassen.
> Cantor und der Mathematikus   v.   W e b e r   werden die Arbeit im
> physikalischen Kolloquium vortragen... Wäre nicht Ihre Arbeit vor-
> handen, so wären wir mit der Minkowskischen Transformations-Glei-
> chung für die Zeit höchstens auf demselben Standpunkt (was die
> physikalische Deutung betrifft) wie mit der Lorentzschen "Local
> Zeit".*

Seine nächste Arbeit - eine Anwendung der Relativitätstheorie auf die
Dispersionserscheinungen - ließ Jakob Johann Laub 1909 durch Lenard der
Heidelberger Akademie vorlegen. 1908 wurde er sein Assistent.

Bald traten die ersten Meinungsverschiedenheiten mit Lenard auf. Einstein, davon unterrichtet, riet, Lenards "Schrullen" zu ertragen, denn
er sei "ein großer Meister, ein origineller Kopf".[33] Durch Laubs Vermittlung erhält nun auch Lenard Informationen über Einstein aus erster
Hand, und er entschließt sich, die seit langem ruhende Beziehung durch
einen Brief wieder aufzunehmen.[34]

Einsteins Ansehen in der Fachwelt war im Jahre 1908 erheblich gestiegen.
Wichtige Beiträge zur speziellen Relativitätstheorie von namhaften Gelehrten waren bei den Zeitschriften eingegangen.

Zum erstenmal äußerte Einstein Anfang 1908 den Wunsch, "auf dem dies-
jährigen Naturforschertag [in Köln] - wenn irgend möglich - zugegen [zu]
sein."[35] Aber auch ohne Einsteins persönliche Teilnahme wurde diese Veranstaltung zu einem großen Triumph für die Relativitätstheorie.

Im Mittelpunkt der Kölner Tagung stand der glanzvolle Vortrag von
Hermann Minkowski über Raum und Zeit, der vielen Teilnehmern (darunter
Max Born,   A r n o l d   S o m m e r f e l d   und Max Planck) in ewiger
Erinnerung blieb, und das allgemeine Interesse an dieser Theorie auch
über die speziellen Fachkreise hinaus erweckte. Die Messungen an
B e c q u e r e l   strahlen von   A l f r e d   B u c h e r e r   gaben
nun eindeutig der "Einsteinschen Fassung" des Relativitätsprinzips den
Vorzug gegenüber der "Lorentzschen Fassung".[36]

Dem fachlichen Erfolg folgten jetzt die ersten äußeren Anerkennungen.

Den Höhepunkt aber bildete Einsteins berühmter Vortrag auf der nächsten Naturforscherversammlung in Salzburg, der ihm neben der Anerkennung den ersten persönlichen Kontakt zu vielen seiner späteren Fachkollegen brachte, darunter mit Max Planck, Arnold Sommerfeld, Max Born, L i s e M e i t n e r und O t t o H a h n .[37] Die persönliche Bekanntschaft mit Johannes Stark, der damals zu den wenigen Befürwortern der Lichtquanten gehörte, hatte allerdings eine allmähliche Entfremdung dieser beiden so verschiedenartigen Menschen zur Folge.

Aber auch in einer anderen Hinsicht war diese Veranstaltung bedeutungsvoll, indem sie einen Wendepunkt in der Physik herbeiführte. Einstein hatte in seinem Vortrag das durch seine Lichtquantenhypothese aufgeworfene Problem, das Welle-Teilchen-Problem, in den Vordergrund gestellt. In sehr geschickter Weise wurde dadurch das Interesse, das man seiner Relativitätstheorie entgegenbrachte, für die Quantentheorie ausgenutzt.

Er selbst aber hatte zu diesem Zeitpunkt längst die s p e z i e l l e R e l a t i v i t ä t s t h e o r i e hinter sich gelassen und wendete sich dem Kernstück seines Systems, der A l l g e m e i n e n R e l a - t i v i t ä t s t h e o r i e zu, die von nun an zur zentralen Aufgabe seines Lebens werden sollte.[38]

Die Allgemeine Relativitätstheorie ist noch mehr wie die Lichtquantentheorie und die spezielle Relativitätstheorie das Werk eines einzigen Mannes.

Dennoch sollte man nicht vergessen, daß das Gravitationsproblem auf eine alte Tradition zurückblicken kann und bereits im vergangenen Jahrhundert viele populärwissenschaftliche Verarbeitungen aufzuweisen hat.[39] Um 1900 hatte auch H e n d r i k A n t o o n L o r e n t z eine Gravitationstheorie veröffentlicht.[40]

Bedeutsam dürfte in diesem Zusammenhang die Bekanntschaft mit W a l t h e r R i t z gewesen sein. Zwischen Ritz und Einstein hatte eine Meinungsverschiedenheit über die Ursache der Irreversibilität bestanden, die Ritz mit der Unmöglichkeit von einlaufenden Wellen in Zusammenhang brachte.[41] Als es zu einer Einigung kam, veröffentlichten sie im April 1909 eine gemeinsame Note in der Physikalischen Zeitschrift.

Ritz war ein gebürtiger Schweizer und nur wenig älter als Einstein. Er hatte ebenso wie dieser an der ETH in Zürich studiert und sich später in Göttingen habilitiert. Ritz gehörte zu den großen Hoffnungen der damaligen Zeit, er war mit E h r e n f e s t befreundet und auch Lorentz und Sommerfeld schätzten ihn sehr.

Ritz stand einem Kreis von Physikern nahe, die an einer Weiterbildung der Gravitationstheorie arbeiteten, darunter insbesondere die Tübinger R i c h a r d   G a n s   und   F r i e d r i c h   P a s c h e n . Ritz selbst hatte 1909 eine Emissionstheorie der Gravitation veröffentlicht und die   P e r i h e l v e r s c h i e b u n g   des   M e r k u r berechnet.

Die oben angeführte Note ist von beiden Autoren unterzeichnet, und wir müssen annehmen, daß Ritz zur Beilegung der Differenzen Einstein in Zürich aufgesucht hat. Bei dieser Gelegenheit dürfte, wie L. P y e n s o n zuerst bemerkt hat, eine ausführliche Diskussion über ihre Standpunkte zur Gravitationsfrage stattgefunden haben.[42]   Leider ist Ritz, der schon seit vielen Jahren an einem schweren Lungenleiden litt, drei Monate nach diesem Treffen mit Einstein im Alter von 31 Jahren gestorben.

Die nächste Zwischenstation des akademischen Aufstiegs von Einstein war die Lehrkanzel für theoretische Physik an der Karls-Universität in Prag. Die Berufung dort hatte   A n t o n   L a m p a   in die Wege geleitet.[43] Die Kunde davon hatte sich schon rumgesprochen.   P a u l   E p s t e i n berichtete aus München seinem ehemaligen Studienkollegen Paul Ehrenfest:[44]

> *Eigentlich hatte ich die Absicht nach Zürich zu Einstein zu gehen; nun hat er aber einen Ruf nach Prag, dem er folgen will, und es fehlt nichts als die Bestätigung vom Kaiser... Prag stelle ich mir als einen ungemütlichen Ort vor... Übrigens soll Einstein die Elementarquanten überwunden und wieder glänzende Sachen haben...*

In der Prager Zeit fällt auch der 1. Solvay-Kongreß, zu dem Einstein geladen war: "Der dortige Kongreß sah überhaupt einer Wehklage auf den Trümmern Jerusalems ähnlich... Gefordert wurde ich wenig, indem ich nichts hörte, war mir nicht bekannt gewesen wäre."[45]

Die große Freundschaft

Im Februar 1912 besuchte ihn dort Paul Ehrenfest, mit dem er schon länger korrespondierte.[46]   Ehrenfest kam aus Rußland und hatte zuvor München, Berlin und Wien aufgesucht. (Mit Einsteins thermodynamischen Arbeiten war er vertraut. In seinen Notizbüchern taucht der Name Einsteins seit 1911 auf.) Dort hatte er viele Kollegen und Freunde getroffen (in München war er bei Sommerfeld, in Wien begegnete er dem jungen Schrödinger). Er brachte viele Neuigkeiten mit in die abgelegene Stadt. In seinem Tagebuch vermerkte er die Eindrücke seiner ersten Begegnung mit Einstein, die den Beginn einer langen Freundschaft markierte und die wir deshalb hier etwas ausführlicher wiedergeben:[47]

> *23. Februar: Endlich Einfahrt in Prag - grau. Heraus! Gehe zum Ausgang. Dort Einstein (mit Zigarre im Mund) und Frau. Gleich in's Kaffeehaus ... Reden über Wien, Zürich, Prag ... Schon im Café Ergodenhypothese. - Auf dem Weg zum Institut schon Streit*

über alles. Regen auf der Gasse - Schmutz - immer zu Diskussion.
Institut: Aula - Stiege hinauf in [die] theoretische Physik.

Dann folgt eine kleine Skizze von Einsteins Arbeitszimmer mit der Bemerkung:

Lackierter Boden, sehr große Zimmer. Mit Einstein noch weiter
gestritten.

Anschließend hatte Einstein eine Verabredung ins Quartett und ließ Ehrenfest mit Anton Lampa zurück. Erst am späten Abend traf man sich wieder
in Einsteins Wohnung. Ehrenfest fertigte auch eine Skizze der Wohnung an.

Gegen 12 h kommen Einsteins. - Noch Tee. Von 12 - 2:30 h.- Mit
Einstein gestritten. Sehr spät schlafen.

Am nächsten Morgen, 24. Februar:

8 h auf ... Frühstück im Vorzimmer mit Buberl... [Später] ins
Institut. Dort schon Einstein. Wir kommen sofort wieder ins
Streiten (rücklings auf Sesseln am Mitteltisch und stehend)...
Später erzählt mir Einstein seine Gravitationsarbeit... Zentri-
fugieren der Strahlung. Später schlafen.

Der 25. Februar ist ein Sonntag:

Morgens weckt mich Albert' [Einsteins Sohn] auf... Gemeinsames
Frühstück ... Mit Kindern. Erzähle Einstein ... Er sucht Feh-
ler... (falsch). Ich gebe "Flohrechnung". (Albert' freut sich).

Vorher spielten wir Klavier. Violin-Sonate von B r a h m s
(Junge singt!) - Ja wir werden Freunde sein. - [Ich] war furcht-
bar froh. -

[Später auf einem Spaziergang.]

Einstein mit Kinderwagen und Kleinen ... Dann weiter in enormem
Schritt ... N.B. Einstein spricht oft ein derbes Wörtlein vor
den Ohren des Jungen ... In Tram Nr. 6 bis nahe Palatsky-Brücke.
(Er hat im Mantel [ein] Loch!) ... Einstein und Frau gehen in[s]
Konzert. Ich allein ... Einsteins kommen (schenken Bilder). Reden
über   J o f f e -Dispersion.

Am folgenden Morgen geht Ehrenfest allein ins Institut und verbringt den
Morgen in derBibliothek, während Einstein bei Lampa ist. 26. Februar:

Nach dem Essen Spaziergang (Sonne!!!). Frau mit Einstein und
kleinem Jungen, ich hinten an: Betrübt - letzten Floh zerquetscht
... gehe sehr geknickt hinter ihnen her. Dann mit Einstein ins
Institut. Auf [dem] Weg erzählt er mir [über die] Dispersions-
frage (Rubens). [Wir] sprechen dann im Institut einigermaßen faul
über spezifische Wärme von H_2.[48]

[Ich] mache [eine] "Galgenbemerkung". Einstein ganz elektrisiert.
Später zum Abendessen nachhause. Zuhause setzt er mir die Gravi-
tationsarbeit auseinander.

Am Dienstag, dem 27. April hält Ehrenfest von 7:30 h bis 9 h seinen
Strahlenvortrag.   K o w a l e w s k y ,   R o t h m u n d , P i c k ,

E i n s t e i n ,   L a m p a ,   C z u r d a   sind unter den Zuhörern.
"Schafte Abgrenzung gegen Einstein", notierte Ehrenfest. "Dann ging es
ins Restaurant und später wieder ins Kaffeehaus." Mittwochnachmittag
ist Seminar. Im "Stiegenzimmer" (Tafel, lange Bänke, bemerkte Ehrenfest)
trägt Einstein über die   N e r n s t sche Arbeit vor (wahrscheinlich
über das Nernstsche Wärmetheorem, über welches Nernst auf der letzten
Naturforscherversammlung in Karlsruhe vorgetragen hatte).[49]

Am letzten Besuchstag werden nochmals gemeinsam Brahmslieder (Mädchen-
lied, Geh' schlafen, mein Kind, Liebestreu, Nachtwandler) gespielt. Dann
am 29. Februar:

> *Mit Einstein in [die] Konditorei (Schokolade), dann längs der*
> *Moldau ... Einstein schimpft über G o e t h e . Stadtbesich-*
> *tigung. Auf [den] Bahnhof - also los - Eingang. Einstein hilft*
> *tragen. Am Zug. - Bis zum letzten Wagen. Einstein ... sehr lieb*
> *und innig - Abschied. Was bringt die Zukunft?*

Hier spricht Ehrenfest das eigentliche Motiv seiner Rundreise an: Er
suchte eine neue Stellung. Nachdem er hörte, daß Einstein inzwischen
einen Ruf and die ETH in Zürich erhalten hatte, erschien ihm die Prager
Nachfolge eine Möglichkeit. Da er aber jüdischer Abstammung und obendrein
konfessionslos war, erfüllte er nicht die Voraussetzungen für eine Be-
werbung; er war auch nicht zu einem Kompromiß bereit, den ihm Einstein
und Lampa unterbreiteten.[50]

An Stelle dessen machte sich Ehrenfest jetzt große Hoffnungen auf eine
bescheidenere Stellung an der ETH in Einsteins unmittelbarer Nähe. Die
Aussichten dafür waren aber schlecht.

Zu dieser Zeit beabsichtigte Lorentz seine Lehrtätigkeit in Leiden aufzugeben und wählte - zur Überraschung aller - Ehrenfest an zweiter Stelle nach Einstein zu seinem Nachfolger. Da Einstein damals nicht zu haben
war - er hatte bereits in Zürich zugesagt - wurde diese Stelle für Ehrenfest frei.[51] Den Ausschlag dazu hatte wohl ein sehr lobendes Gutachten
von Sommerfeld gegeben,[52] der selbst mit dem Gedanken spielte, Ehrenfest als Privatdozenten nach München zu holen.

Ehrenfest gehörte von nun an zu den treuesten Freunden Einsteins, ein
regelmäßiger Briefwechsel[53] und häufige Besuche zeugen davon.

Vor Antritt seiner Berliner Stellung verbrachte Einstein im März 1913
eine Woche mit den Ehrenfests in Leiden. Von dort aus besuchten sie gemeinsam Lorentz, den Einstein hoch verehrte und jetzt auch als Menschen
kennenlernte.

Die Besuche bei den holländischen Freunden waren für Einstein ein will-
kommener Ausgleich für die ihm persönlich weniger nahestehenden Kollegen

in Berlin und die vielen Anfeindungen, die er hier erdulden mußte.
Während des 1. Weltkrieges schrieb er seinem Freund Besso:[51]

> *Unterdessen habe ich wunderschöne Tage in Holland verlebt. Dort
> ist die allgemeine Relativität schon ganz lebendig geworden ...
> Mit Ehrenfest und besonders mit Lorentz verbrachte ich unver-
> geßliche Stunden ... Ich spüre überhaupt, daß ich diesen Men-
> schen unvergleichlich näher stehe.*

Die politischen Verhältnisse in Deutschland hatten sich nach dem Kriege
zugespitzt. "Not und Hunger sind fürchterlich in der Stadt. Wohin wir
politisch steuern, weiß kein Mensch", heißt es in einem seiner Briefe
Anfang 1920.[55]

Einstein setzte damals noch große Hoffnungen auf die junge Republik.
Trotz der ersten offenen Angriffe gegen die Juden[56] versprach er 1919
Max Planck, "Berlin nicht den Rücken zu kehren, bevor nicht Verhältnisse
eintreten, die ihm einen solchen Schritt als natürlich und richtig erscheinen lassen."[57] Diesem Versprechen fühlte sich Einstein verpflichtet, auch als die holländischen Freunde ihm die großzügigsten Angebote
machten.[58]

Beiläufig hatte Einstein in einem Brief einmal erwähnt, daß er finanzielle Sorgen habe. Das war für Ehrenfest Anlaß genug, sofort eine Kopie
von Einsteins Brief mit folgender Bemerkung an Lorentz zu senden: "Wenn
man ihm wenigstens seine Geldschwierigkeiten nehmen könnte!! Der N o -
b e l   P r e i s ?!"[59] Als sich im Sommer 1920 die politischen Angriffe
gegen Einstein richteten, schlug auch Lorentz die Verleihung des Nobel-
preises zur Überwindung der Schwierigkeiten vor.[60]

Ehrenfest gehörte wohl zu den ganz wenigen Personen, denen Einstein
seelisch und geistig wirklich nahestand. Keinem seiner anderen Kollegen
hat Einstein je so herzliche Worte gesagt wie in seinem Nachruf für
Ehrenfest.[61]

Noch deutlicher wird diese Verbundenheit in zwei Briefen vom 6. Juni und
24. August 1920 an Ehrenfest ausgedrückt:[62]

> *Das öftere Zusammensein tut uns außerordentlich gut und es ist,
> als hätte die Natur uns füreinander erschaffen.*
>
> *Wir werden von jetzt an in engem persönlichen Kontakt bleiben.
> Ich weiß, wie gut uns das tut und daß sich ein jeder weniger
> fremd in dieser Welt fühlen wird.*

Ein weiteres Beispiel für ihre Geistesverwandtschaft ist ein längerer
fachlicher Briefwechsel aus dem Jahr 1913 über das Ehrenfestsche Adia-
batenprinzip. Zum erstenmal begegnen wir hier einer Methode der Wahr-
heitsfindung, die wir als den   E i n s t e i n s c h e n   D i a l o g'
bezeichnen möchten, und die darin besteht, den Partner durch einfache

Gegenbeispiele zu einer neuen Antwort zu nötigen. Am schönsten hat uns
Einstein diese Dialogform in seinen berühmten Diskussionen mit N i e l s
B o h r   über die Interpretation der Quantentheorie vorgeführt.

Einstein in Berlin

Doch bevor wir auf diese letzte große Phase der physikalischen Auseinandersetzung in Einsteins Leben eingehen, möchten wir ein wenig über
seine Beziehungen zu seinen Berliner Kollegen erfahren.

"Nicht ohne gewisses Unbehagen sehe ich das Berliner Abenteuer näher-
rücken ...", heißt es in einem Brief an Besso. [63] Dennoch stellt ohne
Zweifel die Berliner Periode von fast zwanzig Jahren den Höhepunkt seiner wissenschaftlichen Laufbahn dar, auch wenn hier seine höchsten Triumphe durch große menschliche Enttäuschungen geradezu kompensiert wurden.
"Ich werde also in Dahlem wohnen und in Habers Institut ein Zimmer haben... Es wird ohne Zweifel in Berlin sehr interessant werden, schon
zum Zusehen"; [64] und "ich habe einen sehr erfreulichen wissenschaftlichen Verkehr, kurz es geht mir gut ...", lautet sein Stimmungsbericht
Anfang 1918. [65]

Mit seiner a l l g e m e i n e n   R e l a t i v i t ä t s t h e o -
r i e   von 1916 hatte sich Einstein gleichsam in das Abseits der physikalischen Entwicklung begeben. [66] Die großen Fortschritte wurden jetzt
auf dem Gebiete der von ihm selbst in Bewegung versetzten Quantentheorie
gemacht. Die Methode der stufenweisen Anpassung von Theorien an die
experimentellen Tatsachen unter vorläufigen Verzicht auf logische Ge-
schlossenheit hatte sich besonders in Gestalt der B o h r - S o m m e r -
f e l d s c h e n   Q u a n t e n t h e o r i e   bewährt. Einstein, der
sich mit derartig provisorischen Gebilden nicht anfreunden konnte, hat
nur noch sporadisch - wenngleich äußerst erfolgreich - in diese Entwicklung eingegriffen. Sein hoher Anspruch wurde ihm hier letztlich zum
Verhängnis.

Wenn wir Einsteins Verdienste in gerechter Weise beurteilen wollen,
müssen wir neben seinen rein wissenschaftlichen Leistungen aber noch
eine zweite Großtat würdigen, die ebenfalls einen revolutionären Um-
bruch in dem Bewußtsein unserer Zeit einleitete.

Einstein hat nämlich durch den Einsatz seines wissenschaftlichen Ansehens
in die politischen und weltanschaulichen Auseinandersetzungen der zwanziger Jahre - besonders in Deutschland - ein Tabu altherkömmlicher Traditionen gebrochen. [67] Er hat damit ein der neuen Zeit gemäßeres Umdenken
in der Ethik der Wissenschaftler beschleunigt, das wir heute mit dem
Schlagwort von der Verantwortung der Wissenschaft kennzeichnen.

Hören wir dazu ein Urteil Einsteins über seine Kollegen aus dem Jahr
1917:[68]

> ... ich kann es nicht fassen, daß in ihrem persönlichen Verhalten
> grundanständige Menschen in Bezug auf die allgemeinen Angelegen-
> heiten einen so ganz anderen Standpunkt einnehmen ... Nur ganz
> selten selbständige Charaktere können sich dem Drucke der herr-
> schenden Meinung entziehen. In der Akademie scheint kein solcher
> zu sein.

Unter diesem doppelten Aspekt muß man also Einsteins Aktivitäten in den
Berliner Jahren betrachten.

Das distanzierte Verhältnis zu den Kollegen äußerte sich auch darin,
daß keine einzige seiner Arbeiten aus dieser Periode durch Kooperation
mit den deutschen Wissenschaftlern entstand.

Seine Abneigung gegen das ' P r e u ß i s c h e ' hat er einmal in
einem Brief an   R o m a i n   R o l l a n d   mit den folgenden Worten
charakterisiert:[69]

> Dies Land ist durch den Waffenerfolg von 1870, durch Erfolge auf
> dem Gebiete des Handels und der Industrie zu einer Machtreligion
> gekommen, die in T r e i t s c h k e   einen adäquaten, gar nicht
> übertriebenen Ausdruck gefunden hat. Diese Religion beherrscht fast
> alle Gebildeten; sie hat die Ideale der Goethe-Schiller-Zeit fast
> vollkommen verdrängt.

In einem späteren Brief vom 5. Januar 1929 an Besso heißt es dann:[70]

> Politisch sieht es nicht erfreulich aus - die Preußen haben sozu-
> sagen nur ihr Fleisch vertauscht mit anderem.

Einstein betrachtete die deutschen Gelehrten der Wilhelminischen Zeit
als unbescheiden und befürwortete deshalb sogar ihren zeitweiligen Ausschluß vom internationalen geselligen Verkehr.[71] Als dann die ersten
Ergebnisse der Untersuchungskommission für Kriegsverbrechen vorlagen,
nahm er wieder die Position der Gerechtigkeit ein und schrieb die weitsichtigen Worte:[72]

> Schon a priori, schrieb er an Lorentz, ist es unglaublich, daß
> die Bewohner eines ganzen großen Landes moralisch minderwertig seien!
> Nach meiner Überzeugung sind es eben die Verhältnisse, die ' P r e u -
> ß e n ' wachsen lassen; und meine Angst ist, solche möchten nun unter
> den ganz veränderten Verhältnissen anderwärts nachwachsen. Hoffentlich
> ist diese Furcht nicht berechtigt.

Als man dann Einstein unter Ausschluß seiner Kollegen zum 4. Solvay-
Kongreß 1924 einlädt, zeigt er sich mit diesen solidarisch.

> Denn es entspricht meiner Überzeugung, daß man in die wissenschaftlichen
> Bestrebungen keine Politik hineintragen und daß man überhaupt die ein-
> zelnen Menschen nicht für den Staat verantwortlich machen sollte, dem
> sie zufällig angehören.[73]

Einstein und Planck

An dieser Stelle müssen wir auch auf das besondere Verhältnis zwischen
Einstein und Max Planck eingehen. Wir hatten gesehen, wie sehr Planck
sich für die junge Relativitätstheorie eingesetzt hatte, und vor allem
Planck war es, der Einstein in Berlin haben wollte.[74] Das Verhältnis
der beiden war von der gegenseitigen Hochachtung ihrer wissenschaftlichen
Leistungen bestimmt. Ihren Höhepunkt erreichte dieses Verhältnis, als
Einstein 1918, damals als Vorsitzender der Deutschen Physikalischen Gesellschaft (DPG), eine ergreifende Festrede zu Plancks 60. Geburtstag
hielt. "Der Gefühlszustand, der [Planck] zu seinen Leistungen befähigt",
sagte er dort, "ist dem des Religiösen oder Verliebten ähnlich."[75]

Aus einem Brief vom 9. Januar 1925 an Lorentz hören wir aber auch, worin
die beiden nicht übereinstimmen:[76]

> Ich habe ... Herrn Planck gebeten, eine nationale Kommission für
> intellektuelle Zusammenarbeit zu begründen, bzw. zu bemuttern.
> Ich sah gleich, daß diese meine Aufforderung ihn unglücklich machte,
> und der Arme kämpfte nun ein Vierteljahr ... Gestern aber kam er ...
> und erklärte, daß er die ihm zugedachte Funktion nicht übernehmen
> könne ... Ich beurteile Plancks wirkliche Situation so: Er selbst
> täte es eigentlich gerne, aber die Gebundenheit an seine Gemeinschaft
> erlaubt es ihm nicht.

Die spätere Haltung Plancks schien Einsteins Ansicht zu bestätigen, ihre
Wege trennten sich 1933.

Auch seine Beziehung zu Sommerfeld, mit dem Einstein eine langjährige
fachliche Korrespondenz unterhielt, war nicht frei von einigen Vorbehalten. "Sommerfelds Buch [Atombau und Spektrallinien] ist hübsch",
schrieb er am 27. Januar 1920 an seinen Freund Born, "wenn ich auch
offen sagen muß, daß diese Persönlichkeit für mich aus Gott weiß was
für einem unbewußten Grunde etwas nicht ganz Reines in ihrem Klang hat".[77]

Besser haben - auch in ihren politischen Ansichten - Einstein und Born
übereingestimmt. Beide unterhielten seit ihrer Bekanntschaft in Berlin
(1915) einen stetigen Briefwechsel, und besonders Max Born war es, der
in den frühen Jahren durch übersichtliche Aufsätze und Vorträge über die
spezielle und allgemeine Relativitätstheorie wesentlich zur Anerkennung
der Einsteinschen Theorien beitrug.

Das Verhälnis zu Bohr

Weit interessanter ist natürlich die Frage, wie sich die beiden antago-
nistischen Hauptfiguren der modernen Physik Albert Einstein und Niels
Bohr verstanden haben. Beide hatten gleichzeitig den Nobelpreis empfangen
(Einstein für 1921, Bohr für 1922). Ihre erste Begegnung fand anläßlich
eines Berliner Besuches von Niels Bohr im April 1920 statt, den Einstein

seinerseits am 26. Juni des gleich Jahres erwiderte. In einem Brief an Bohr aus dieser Zeit heißt es:[78]

> *Ich begreife jetzt, warum Ehrenfest Sie so liebt. Ich studiere jetzt Ihre großen Arbeiten und habe dabei – wenn ich gerade irgendwo stecken-bleibe – das Vergnügen, Ihr freundliches Jungen-Gesicht vor mir zu sehen, lächelnd und erklärend. Ich hab' viel von Ihnen gelernt, haupt-sächlich auch, wie Sie den wissenschaftlichen Dingen gegenüberstehen.*

Die Einladung zu diesen Berliner Vorträgen hatte allerdings Arnold Sommerfeld angeregt, als er den Vorsitz der DPG übernahm.[79]

"Bohrs Arbeiten flößen mir große Hochachtung ein durch den sicheren Instinkt, der sie leitet", schrieb Einstein zwei Jahre später.[80] Ein-stein war es auch, der den Wahlvorschlag für Niels Bohr zum korrespon-dierenden Mitglied der Preußischen Akademie aufsetzte.[81]

Trotz dieser freundschaftlichen Beziehungen darf man nicht übersehen, daß Einstein und Bohr von ihrem methodischen und erkenntnistheoretischen Standpunkt her grundsätzlich verschiedenartigen Richtungen angehörten.[82] Einsteins Bemühungen konzentrierten sich im wesentlichen auf die Errich-tung eines naturwissenschaftlichen Weltbildes auf klassisch-determini-stischer Grundlage, wobei die Relativitätstheorie als ein erster Schritt in dieser Richtung anzusehen ist.[83] Einstein glaubte außerdem, daß es möglich sein sollte, durch reines Denken ein solches Weltbild aufzu-finden.[84]

L é o n R o s e n f e l d hat Einstein einen Hang zum Mystizismus vorgeworfen, weil er – seiner Meinung nach – keine Begründung dafür zu geben vermag, wie diese wunderbare Harmonie zwischen unseren Gedanken-Konstruktionen und den Erscheinungen der Außenwelt zustande kommen sollte.[85]

Aber gerade das Bestreiten einer solchen Beziehung ist für Einstein ein Zeichen von mystischer Einstellung: In einem Brief an Schrödinger sagt er:[86] "Es gibt auch noch den Mystiker, der ein Fragen nach etwas unab-hängig vom Beobachteten Existierenden, ... überhaupt als unwissenschaft-lich verbietet (Bohr)."

Aber nur die wenigsten seiner Zeitgenossen sind Einstein in dieser An-sicht gefolgt. "Ich teile vollständig Ihre Auffassung", schreibt Pauli 1954, "daß Einstein sich 'in seiner Metaphysik verrannt' hat."[87]

In dem Spannungsfeld dieser beiden naturphilosophischen Gegensätze hat sich dennoch, dank der großen Persönlichkeit ihrer Vertreter, eine kon-struktive Diskussion entwickelt, die schließlich zu einer Klärung der Standpunkte führte.[88] Da Einstein nicht geneigt war, von seiner Grund-einstellung abzuweichen,[89] versuchte er durch eine Fülle von geistreichen

Gedankenexperimenten, die Ende der zwanziger Jahre zu lebhaften Auseinandersetzungen führten, die Endgültigkeit des durch die quantentheore-
tische Ungenauigkeit geschaffenen statistischen Charaktere physikalischer
Aussagen zu bestreiten.

Trotz ihres freundschaftlichen Verhältnisses zueinander herrschte zwischen
Einstein und Bohr auch eine gewissen Distanziertheit, die vielleicht bei
Menschen ihrer Art durch die verschiedenartige Einstellung zu ihrer Wissenschaft bedingt ist.[90] Aber auch andere entscheidende Wesensunter-
schiede ließen sich dafür anführen. Wir hatten eingangs erwähnt, daß
Einstein sich öfters als "typischer Einspänner" bezeichnete, was sich
auch darin äußerte, daß er einen großen Teil seiner Erkenntnisse in
völliger Abgeschiedenheit gewann. Einstein hat keine Schule im eigentlichen Sinne hinterlassen.[91]

Bohrs späteren großen Erfolge sind z.T. nur durch den Kreis seiner vielen
Schüler und Mitarbeiter, die er um sich versammelte, möglich gewesen.
Pauli sprach von einer 'Hauptstadt der Atomphysik' in Kopenhagen.[92]

Dennoch hat Einstein gerade auf die jüngeren Physiker eine ungewöhnliche
Faszination ausgeübt und sie durch seine Arbeiten und seine Ansichten
stark geprägt.

Pauli

Interessant ist hier das besondere Verhältnis zu Wolfgang Pauli, demjenigen Physiker der jüngeren Generation, der Einstein am besten verstand und zugleich Bohr sehr nahe war.

Wolfgang Pauli ist in Wien aufgewachsen. Er ist in seiner Jugend unter
dem direkten Einfluß von E r n s t   M a c h - dem bekanntlich auch
Einstein nahestand[93] - groß geworden. Mach war ein enger Freund des
Vaters und hat die Lektüre für sein frühreifes Patenkind, Pauli jun.,
ausgewählt.[94]

In der Schule studierte Pauli die W e y l sche Modifikation der Einsteinschen Gravitationstheorie und hatte bei Antritt seines Studiums
bereits eine Arbeit zu diesem Thema bei der Physikalischen Zeitschrift
eingereicht. Im Alter von 20 Jahren begann Pauli seine Arbeit an dem
berüzmten Relativitätsartikel für die mathematischen Enzyklopädie, der
bei Einstein größte Bewunderung hervorrief.

Wenn Pauli sich nach Abschluß dieser Arbeit auch den atomphysikalischen
Problemen zuwandte, so machten ihn seine frühen und tiefen Kenntnisse
der Relativität zu einem geeigneten Vermittler zwischen dem Einsteinschen und Bohrschen Gedankensystem.[95] Es ist schwer zu sagen, wem dieser
beiden Gelehrten sich Pauli in diesen Jahren mehr verpflichtet fühlte.

Wie stark sich dieser Einfluß Einsteins und Bohrs auf ihre jungen Zeitgenossen und Mitarbeiter auswirkte, äußerte sich auch in einer ähnlichen
Ausdrucksweise, die diese zuweilen wohl ganz unbewußt übernahmen. Einstein z.B., der sich mit Vorliebe einer bildhaften Ausdrucksweise bediente, drückte 1924 seinen Unwillen über die Quantensprünge mit folgenden Worten aus:[96]

> *Wenn schon, dann möchte ich lieber Schuster oder gar Angestellter*
> *in einer Spielbank sein als Physiker.*

Vielleicht ist es kein Zufall, wenn auch Pauli ein Jahr später seinen
Unmut über die unbefriedigende Situation in der Quantentheorie ähnlich
formulierte:[97]

> *... ich wollte, ich wäre Filmkomiker oder so etwas und hätte nie*
> *etwas von Physik gehört.*

Umgekehrt bemerkte der 21jährige Heisenberg nach einem etwas zu lang
geratenen Satz in Anspielung auf Bohrs lange und komplizierte Satzbil-
dung:[98]

> *Man darf Bohr nicht in allen Dingen nachahmen.*

Die Beeinflußbarkeit der jungen Physiker durch ihre hochverehrten Lehrer
war natürlich nicht nur auf diese verbalen Äußerungen beschränkt. Was
Einstein dachte, interessierte damals Göttingen und München ebenso wie
Kopenhagen.

Einstein hatte am 28. Mai 1924 im Berliner Kolloquium über die B o h r -
K r a m e r s - S l a t e r   T h e o r i e   referiert, die damals
großes Aufsehen erregte und von vielen Physikern als die Lösung des
Quantenproblems angesehen wurde.[99]

"Diese Arbeit ist ein alter Bekannter von mir,[100] kommentierte Einstein
und führte einen ganzen Katalog von Einwänden auf. Schon eine Woche darauf informiert Heisenberg Pauli:[101]

> *Gestern sprach ich Einstein, der z.Z. hier [in Göttingen] ist, über*
> *die Bohrsche Theorie und Einstein hat hundert Einwände...*

und Rudolf Ladenburg unterrichtete   H e n d r i k   A n t h o n y
K r a m e r s   ebenfalls davon.[102]

Im September 1924 traf Pauli Einstein auf der Naturforscherversammlung
in Innsbruck. Auch er gibt Bohr einen ausführlichen Bericht über Einsteins Einwände gegen die neue Strahlentheorie:[103]

> *Sie sagten schon zu Ostern wiederholt zu mir, Sie möchten gerne*
> *wissen, was Einsteins Argumente sind,*

und schließt mit den Worten:

*Sie sehen, selbst wenn es für mich psychologisch möglich wäre, mir
meine wissenschaftlichen Meinungen auf Grund irgendeiner Art von
Autoritätsglauben zu bilden, so wäre dies doch logisch unmöglich,
da die Meinungen zweier Autoritäten einander hier so sehr widersprechen.*

Diese Beispiele illustrieren zur Genüge, welches Gewicht in den zwanziger Jahren Einstein bei der Meinungsbildung unter den führenden Physikern zukam.

Mit Einsteins Fortgang im Jahre 1933 wurde der geistige Dialog in der
europäischen Physik grundlegend gestört, und das goldene Zeitalter der
Physik fand damit sein Ende. Für die Genese der Wissenschaft bedarf es
einer 'kritischen Masse' an Geist, welcher unter sich und mit den anderen kulturtragenden Faktoren seiner Umwelt über das Gespräch und den
Meinungsaustausch die notwendigen Verbindungen herstellt.

Auch Einsteins schöpferische Kraft überdauerte diese Verpflanzung nicht;
sein Ruhm als Forscher und als Bahnbrecher eines neuen Bewußtseins bleibt
dadurch aber unangetastet.

==============================

Für die Erlaubnis, in die Notizbücher und in die Ehrenfest-Korrespondenz
im Boerhaave-Museum in Leiden einzusehen, bin ich sehr dankbar. Ebenso
hat mir John J. Stachel aus Princeton freundlicherweise ein Verzeichnis
der wissenschaftlichen Korrespondenz der Albert-Einstein-Sammlung zur
Verfügung gestellt.

Anmerkungen

1. Vgl. den Brief [73] von Einstein an Born in: Albert Einstein -
   Hedwig und Max Born - Briefwechsel; München 1969, und Borns
   Kommentar dort auf S. 135.

2. Diesen Ausdruck hat Einstein öfters gebraucht. So. z.B. in seiner
   Autobiographie, die er für den Schilpp-Band schrieb und in
   seinem Aufsatz "Mein Weltbild" von 1934.

3. Vgl. hierzu die Bemerkung in Einsteins 'Selbstporträt' aus dem
   Jahr 1936: "Heute lebe ich in jener Einsamkeit, die in der
   Jugend so schmerzlich, aber in den Jahren der Reife so köst-
   lich ist." In: Albert Einstein: Aus meinen späten Jahren.
   Stuttgart 1979. (Bei der früheren 1. deutschen Auflage von
   1952 handelt es sich um eine Rückübersetzung ins Deutsche!)

4. In Karl Seelig: Albert Einstein; Eine dokumentarische Biographie.
   Zürich 1954, dort S. 136.

5. A. Einstein: Prinzipien der Forschung. Rede zum 60. Geburtstag von
   Max Planck. Auch abgedruckt in Albert Einstein: Mein Weltbild.
   Amsterdam 1934.

6. Auch in diesem Punkt hatte er ein ähnliches Schicksal wie sein
    späterer Freund Paul Ehrenfest, der sogar aufgrund seiner
    schrecklichen Schulerlebnisse so weit ging, die Schulerziehung
    seiner Kinder selbst zu überwachen. Vgl. z.B. M. J. Kleins
    Ehrenfest-Biographie, Amsterdam 1970, Vol. 1.

7. Siehe hierzu auch G. J. Whitrow: Einstein, the Man and his
    achievement. New York 1967, S. XI.

8. In C. Seelig, op. cit. (Fußnote 4), S. 85.

9. Möglicherweise veranlaßte Besso Einstein damals zur Streichung
    einer Kritik an der Planckschen Strahlungstheorie.- Zietiert
    nach Einstein-Besso-Correspondence 1903 - 1955, herausgegeben
    von P. Speziali, Paris 1972.

10. Vgl. hierzu den Aufsatz von H. Ezawa: Einsteins Beitrag zur sta-
    tistischen Mechanik; in Albert Einstein, sein Einfluß auf Phy-
    sik, Philosophie und Politik. Herausgegeben von P. C. Aichel-
    burg u. R. U. Sexl, Braunschweig 1979. Ebenso enthält die weni-
    ger bekannte Darstellung von C. Lanczos: The Einstein Decade
    1905 - 1915, New York 1974, einen guten Überblick dieser Zeit.

11. Interessant ist in diesem Zusammenhang, daß Einstein in seinem
    Aufsatz 'Max Planck als Forscher' (Naturwiss. $\underline{1}$, 1077 - 1079,
    1913) "eine 1896 in Wiedemanns Annalen erschienene polemische
    Arbeit Plancks 'Gegen die neuere Energetik', weil sie zweifel-
    los einen bedeutenden Einfluß auf die Fachkollegen ausübte",
    gelobt hat.

12. Hermann Einstein an W. Ostwald, 13. April 1901. Zitiert in Banesh
    Hoffmanns und Helen Dukas Einstein-Biographie.

13. Brief von Einstein an Stark, 7. Dezember 1907. Abgedruckt bei
    A. Hermann: Albert Einstein und Johannes Stark. Briefwechsel
    und Verhältnis der beiden Nobelpreisträger. Sudhoffs Archiv
    $\underline{50}$, 267 - 285 (1966).

14. Möglicherweise hatte Einstein damals versucht, auch mit Boltzmann
    Kontakt aufzunehmen. Von einer Korrespondenz ist bisher noch
    nichts bekannt.

15. Vgl. z.B. H. G. Schöpf: Albert Einsteins annus Mirabilis 1905,
    Gesch. Naturwiss., Technik, Med., Leipzig $\underline{15}$, 1 - 17 (1978).

16. Vgl. A. Kleinert und Ch. Schönbeck: Lennard und Einstein. Ihr
    Briefwechsel und ihr Verhältnis vor der Nauheimer Diskussion
    von 1920. Gesnerus $\underline{35}$, 318 - 333 (1978). Der Briefwechsel mit
    Stark wurde von A. Hermann (vgl. Fußnote 13) publiziert.

17. Lenard erhielt dafür den Nobelpreis für das Jahr 1905.

18. Unzutreffend ist z.B. die Äußerung von C. P. Snow, der in seinem
    Artikel 'Variety of Man' (auszugsweise abgedruckt in 'Einstein,
    A Centenary Volume' 1979, S. 3) behauptet, man habe damals in
    Deutschland erst vier Jahre später das Genie Einstein erkannt.
    Dagegen sprechen die Briefe von Einstein an Solovine (z.B.
    3. Mai 1906) und sein früher Briefwechsel mit Planck. (Vgl.
    Fußnote 23.) Im April 1908 haben Minkowski und Lorentz auf
    einem internationalen Kongreß in Rom die Verdienste des Sechs-
    undzwanzigjährigen gewürdigt. Selbst die Tatsache, daß Einsteins
    ungewöhnliche Arbeiten damals in den Annalen aufgenommen wurden,

484

hat A. Hermann (u.a. in seinem neuen Buch 'Die Neue Physik', Fußnote 19) erstaunlich gefunden. Planck selbst gibt dafür eine Erklärung während der Festsitzung zu seinem 80. Geburtstag. Über seine frühere Mitwirkung bei der Annalen-Redaktion sagte er: "Mir schwebte immer das Schicksal der berühmten Erstlingsarbeit von Julius Robert Mayer vor..., die bekanntlich von dem damaligen Annalenredakteur I. C. Poggendorff zurückgewiesen wurde. Das hat der sog. zünftigen Physik ungemein geschadet ..." (Abgedruckt in Verh. d. DPG 19, 57 - 76 (1938)).

19. Vgl. hierzu A. Hermann: Die Neue Physik, München 1979, und die detailliertere Untersuchung von S. Goldberg: Max Planck's Philosophy of Nature and His Elaboration of the Special Theory of Relativity. Hist. Stud. Phys. Sci. 7, 125 - 160 (1976).

20. Op. cit. (Fußnote 11).

21. Brief Einsteins an Stark, 13. Februar 1908. (Enthalten in A. Hermanns Aufsatz, Fußnote 13).

22. Vgl. A. Moszkowski: Einstein, Einblicke in seine Gedankenwelt. Hamburg/Berlin 1921, S. 227.

23. Einstein an Solovine, 3. Mai 1906. Enthalten in 'Albert Einstein: Lettres à Maurice Solovine', Paris 1956.

24. Brief Einsteins an Stark, 11. November 1907, (vgl. A. Hermann, op. cit. Fußnote 13).

25. Brief Einsteins an Besso, Januar 1903. [Der Brief 01 der Einstein-Besso-Correspondence; op. cit. (Fußnote 9)].

26. Vgl. z.B. A. Hermann, op. cit. (Fußnote 19), S. 17.

27. A. Hermann: Frühgeschichte der Quantentheorie (1899 - 1913), Mosbach 1969, dort S. 79.

28. Brief Laub an Einstein, 2. Februar 1908, zitiert nach L. Pyenson, op. cit. (Fußnote 30).

29. Vgl. C. Seelig, op. cit. (Fußnote 4), S. 86.

30. Vgl. L. Pyenson: Einstein's Early Scientific Collaboration, Hist. Stud. Phys. Sci. 7, 83 - 123 (1976).

31. Wiedergegeben in Philipp Franks Einsteinbiographie (Neudruck 1979), S. 335.

32. Abgedruckt bei L. Pyenson, op. cit. (Fußnote 30), S. 99 - 100.

33. Brief Einstein an Laub, 1908, vgl. A. Kleinert und Ch. Schönbeck, op. cit. (Fußnote 16).

34. Brief Lenard an Einstein, 5. Juni 1909, vgl. A. Kleinert und Ch. Schönbeck, op. cit. (Fußnote 16).

35. Brief Einstein an Stark, 22. Februar 1908, vgl. A. Hermann, op. cit. (Fußnote 13).

36. A. H. Bucherer: Messungen an Becquerelstrahlen. Die experimentelle Bestätigung der Lorentz-Einsteinschen Theorie. Physik. Z. 9, 755 - 761 (1908). Vgl. hierzu auch A. Hermann und U. Benz:

Quanten- und Relativitätstheorie im Spiegel der Naturforscher-
versammlungen 1906 - 1920. in: Wege der Naturforschung 1822 -
1972 im Spiegel der Versammlungen Deutscher Naturforscher und
Ärzte. Herausgegeben von H. Querner und H. Schipperges. Springer 1972, dort S. 125 - 137.

37. Vgl. hierzu A. Hermann: 'Einstein auf der Salzburger Naturforscher-
versammlung 1909'. Phys. Bl. 25, 433 - 436 (1969); und 'Zur
Frühgeschichte der Quantentheorie'. Einleitung zu Bd. 7 der
Dokumente der Naturwissenschaften. Stuttgart 1965.

38. Vgl. Brief Einstein an K. Habicht, 24. Dezember 1907. [Zitiert bei
C. Seelig (op. cit., Fußnote 4) S. 91] und Einsteins Artikel im
Jahrbuch der Radioaktivität und Elektronik 4, 411 - 462 (1907),
Teil V, der am 4. Dezember 1907 abgeschlossen war.

39. Vgl. hierzu die Zusammenstellung bei C. Isenkrahe: Das Rätsel von
der Schwerkraft. Braunschweig 1879.

40. Vgl. hierzu H. A. Lorentz: La Gravitation. Scientia 16 (1914).
Abgedruckt in Collected Papers, Vol. VII, S. 116 - 146.

41. Vgl. hierzu P. Jordan: Albert Einstein; Sein Lebenswerk und die
Zukunft der Physik. Stuttgart 1969, dort S. 220 und 236.

42. Vgl. L. Pyenson, op. cit. (Fußnote 30). Eine genauere Untersuchung
der Korrespondenz von Ritz könnte vielleicht weitere Hinweise
über seine Beziehungen zu Einstein geben.

43. Vgl. Brief A. Lampa an Mach, 18. Dezember 1910. (Zitiert bei
K. D. Heller: Ernst Mach, Wegbereiter der modernen Physik.
Wien und New York 1964, dort S. 146). Ein Teil des Briefwechsels
zwischen Anton Lampa und Albert Einstein wurde von A. Kleinert
in Gesnerus 32, 285 - 292 (1975) veröffentlicht.

44. Brief Epstein an Ehrenfest, 1. Dezember 1910. Das Original dieses
Briefes befindet sich im Paul Ehrenfest Archiv des Boerhaave-
Museums in Leiden.

45. Brief Einstein an Besso, 26. Dezember 1911. (Publiziert in der
Einstein-Besso-Correspondence, op. cit. Fußnote 9.)

46. Zum erstenmal wird Ehrenfest in einem Brief Einsteins vom 21.
Oktober 1911 an Besso genannt.

47. Insgesamt 190 Notizbücher von Ehrenfest befinden sich im Ehren-
fest-Archiv des Boerhaave Museums in Leiden. (Die Datierung
der Aufzeichnung erfolgte nach dem russischen Kalender!) Diese
Notizbücher wurden z.T. von M. J. Klein für seine Ehrenfest-
Biographie (op. cit., Fußnote 6) ausgewertet. Auf dem Deckel
des Notizbuches (ENB: 4-10), dem wir unsere Auszüge entnehmen,
steht neben einigen Fragen, die Ehrenfest an Einstein stellen
wollte, auch die Bemerkung: "Einstein-Tintenfeder ideal." In
seinem Nachruf "In memoriam Paul Ehrenfest" (abgedruckt in:
'Aus meinen späten Jahren', Stuttgart 1979), hat Einstein über
seine erste Begegnung mit Ehrenfest berichtet, und es ist be-
merkenswert, wie genau seine Darstellung mit Ehrenfests Notizen
übereinstimmt.

48. Auf dem 1. Solvay-Kongreß hatte man im vergangenen Herbst gerade
Nernsts neue Vorstellungen über die gequantelten Zustände von
Gasmolekülen diskutiert.

49. W. Nernst: Über ein allgemeines Gesetz, das Verhalten fester Stoffe bei sehr tiefen Temperaturen betreffend. Verh. d. D.P.G. 13, 921 - 925 (1911).

50. Vgl. Brief Einstein an Ehrenfest, 25. April 1912, und Ehrenfest an Lorentz, 6. Mai 1912. Zitiert bei M. J. Klein, op. cit. (Fußnote 6) S. 179 und 180.

51. Brief Lorentz an Ehrenfest, 13. Mai 1912. Vgl. M. J. Klein, op. cit., S. 186.

52. Brief Sommerfeld an Lorentz, 24. April 1912. Vgl. M. J. Klein, op. cit., S. 184 f.

53. Über hundert Briefe aus der Einstein-Ehrenfest-Korrespondenz (1912-1933) werden in den Verzeichnissen des Ehrenfest-Archivs in Leiden (ein Katalog wurde von Bruce Wheaton angefertigt und 1977 vom Museum Boerhaave herausgegeben) und des Einstein-Archivs in Princeton aufgeführt.

54. Brief Einstein an Besso, 31. Oktober 1916. Einstein-Besso-Correspondence, op. cit. (Fußnote 9).

55. Brief Einstein an Ehrenfest, 7. April 1920. Zitiert nach: O. Nathan und H. Norden: Albert Einstein. Über den Frieden. Weltordnung oder Weltuntergang? Bern 1975.

56. Vgl. z. B. den Brief Einstein an Ehrenfest, 4. Dezember 1919: "Hier ist starker Antisemitismus und wütende Reaktion, wenigstens bei den 'Gebildeten'. Ich bemerkte es besonders in Rostock". In Nathan-Norden, op. cit. (Fußnote 55).

57. Brief Einstein an Ehrenfest, 12. September 1919. Vgl. M. J. Klein, op. cit., S. 311 f. und Nathan-Norden, op. cit. (Fußnote 55), wo der deutsche Originaltext wiedergegeben ist.

58. Brief Ehrenfest an Einstein, 2. September 1919. Vgl. M. J. Klein, op. cit., S. 310 f.

59. Ehrenfest an Lorentz, 12. September 1919. M. J. Klein, op. cit., S. 312.

60. Vgl. Nathan-Norden, op. cit. (Fußnote 55).

61. Vgl. Fußnote 47.

62. Aus M. J. Klein, op. cit., S. 323 und 314. (Rückübersetzung aus dem Englischen).

63. Brief Einstein an Besso, Ende 1913. Brief [9] der Einstein-Besso-Correspondence, op. cit. (Fußnote 9).

64. Brief Einstein an Besso, Anfang März 1914. Brief [10] der Einstein-Besso-Correspondence.

65. Einstein an Besso, 5. Januar 1918, ibid.

66. Schon gegen Ende 1913 heißt es in einem Brief an Besso: "Zur Gravitationsarbeit verhält sich die physikalische Menschheit ziemlich passiv. Das meiste Verständnis hat wohl Abraham dafür. Er schimpft zwar... kräftig..., aber mit Verstand... Laue ist den prinzipiellen Erwägungen nicht zugänglich, Planck auch nicht, eher Sommer-

feld. Der freie unbefangene Blick ist dem (erwachsenen) Deutschen überhaupt nicht eigen. "ibid., Brief [9].

67. Anläßlich einer Beratung zur Gründung der hebräischen Universität in Palästina schrieb er am 12. Dezember 1919 an Besso: "Ich gehe nicht deshalb hin, weil ich mich für besonders sachverständig hielte, sondern deshalb, weil mein seit der englischen Sonnenfinsternis-Expedition hoch im Kurs stehender Name der Sache nützen kann." ibid.

68. Brief Einstein an Lorentz, 3. April 1917. Nathan-Norden, op. cit.

69. Brief Einstein an Rolland, 22. August 1917. Nathan-Norden, op. cit., S. 40.

70. Einstein-Besso-Correspondence, op. cit.

71. Brief Einstein an Rolland, 1. August 1919. Nathan-Norden, op. cit.

72. Brief Einstein an Lorentz, 21. September 1919. ibid.

73. Brief Einstein an Lorentz, 16. August 1923. ibid.

74. Vgl. M. Planck: Wahlvorschlag für Albert Einstein zum Ordentlichen Mitglied der Preußischen Akademie der Wissenschaften, Berlin, 12. Juni 1913.

75. A. Einstein: Zu Max Plancks 60. Geburtstag 1918. Ansprache in der DPG in Berlin. (Vgl. Fußnote 5).

76. In Nathan-Norden, op. cit.

77. Einstein-Born-Briefwechsel, op. cit. (Fußnote 1).

78. Brief Einstein an Bohr, 2. Mai 1920. Abgedruckt in: Niels Bohr. Collected Works. Vol 3, edited by J. Rud Nielson. Amsterdam, New York, Oxford 1976. Vgl. auch das Schreiben von Ehrenfest vom 14. Mai 1920 an Bohr: "Seit einer Woche ist Einstein bei mir - Herrlich! - Wann werde ich Euch beide einmal zugleich hier haben? - Er hat mit mit außerordentlicher Wärme über Sie erzählt, wie er Sie in Berlin hörte und sah."

79. Vgl. Brief Sommerfeld an Einstein, Juni 1918. Enthalten in A. Hermann: Einstein-Sommerfeld-Briefwechsel. Basel/Stuttgart 1968.

80. Brief Einstein an Born, o.D. (1922). (Brief Nr. [42] des Einstein-Born-Briefwechsels. Vgl. Fußnote 1).

81. Datum der Wahl, 1. Juni 1922.

82. Einer Einladung zu den berühmten Göttinger Bohr-Festspielen (vgl. Brief Born an Einstein, 12. Februar 1921) war Einstein trotz seiner guten Beziehungen zu Bohr nicht gefolgt.

83. Vgl. z.B. Einsteins Brief an Besso vom 5. Januar 1929: "Aber das Beste, an was ich fast die ganzen Tage und die halben Nächte gegrübelt und gerechnet habe, ist nun fertig vor mir und auf zwei Seiten zusammengepreßt unter dem Namen 'einheitliche Feldtheorie'. Das sieht altertümlich aus ... Den in diesen Gleichungen kommt kein Plancksches h vor. Aber wenn man an die Leistungsgrenzen des statistischen Fimmels deutlich gelangt

sein wird, wird man wieder zur zeiträumlichen Auffassung reue-
voll zurückkehren..." Einstein-Besso-Correspondence.

84.　　"Ich zweifle nicht mehr an der Richtigkeit des ganzen Systems, mag
die Beobachtung der Sonnenfinsternis gelingen oder nicht. Die
Vernunft der Sache ist evident." Brief Einstein an Besso, Anfang März 1914. (ibid.)

85.　　L. Rosenfeld: Conflict between Einstein and Bohr. Z. Phys. 171,
242 - 245 (1963). (Abgedruckt in: Selected Papers of Léon Rosenfeld. Edited by R. S. Cohen and J. J. Stachel. D. Reidel Publ.
Co. 1979, dort S. 517 - 521).

86.　　Brief Einstein an Schrödinger, 9. August 1939. Enthalten in K.
Przibram: Briefe zur Wellenmechanik. Wien 1963.

87.　　Brief Pauli an Born, 15. April 1954, Pauli-Letter-Collection
(PLC), CERN.

88.　　Vgl. N. Bohr: Diskussion mit Einstein über erkenntnistheoretische
Probleme in der Atomphysik. Beitrag zu: Albert Einstein. Philosopher Scientist, London 1949.

89.　　In diesem Sinne äußerte er sich später einmal in seinem Brief vom
11. November 1940 an Besso: "Ich selber verfolge mein wissenschaftliches Ziel unbeirrt und eigensinnig festhaltend an dem
Weg, den mich mein Instinkt weist". Einstein-Besso-Correspondence.

90.　　Vgl. z.B. Bohrs Charakterisierung der Diskussion mit Einstein
(vgl. Fußnote 88) "Und doch blieb ein gewisser Unterschied in
der Einstellung..."

91.　　Otto Stern gehörte zu den ganz wenigen, die man als Einstein-
Schüler bezeichnen kann.

92.　　Vgl. W. Pauli: Niels Bohr zum 60. Geburtstag. Rev. Mod. Phys. 17,
97 - 101 (1945).

93.　　Vgl. z.B. Einsteins Nachruf auf Mach in der Physikalischen Zeitschrift 17, 101 - 104 (1916).

94.　　Wolfgang Pauli jun. erhielt von Mach ein Exemplar seiner Mechanik
(1912) mit der Widmung "Meinem lieben Patenkinde Wolfgang in
freundlichem Gedenken. München-Vaterstetten, 17. Oktober 1913."
Mit 13 Jahren hatte Pauli Eulers Analysis studiert.

95.　　Vgl. z.B. Born-Einstein-Briefwechsel, S. 217 ff.

96.　　Brief-Einstein an Hedwig und Max Born, 29. April 1924. (Born-
Einstein-Briefwechsel).

97.　　Brief Pauli an Kronig, 21. Mai 1925. In: Wolfgang Pauli: Wissenschaftlicher Briefwechsel mit Bohr, Einstein, Heisenberg, u.a.
Bd. 1. Herausgegeben von A. Hermann und K. v. Meyenn. Springer
1979.

98.　　Brief Heisenberg an Pauli, 6. März 1922. Ibid.

99.　　Vgl. Brief Pauli an Kramers, 27. Juli 1925: "Ich halte es überhaupt für ein ungeheures Glück, daß die Auffassung von Bohr,
Kramers und Slater durch die schönen Experimente von Geiger

und Bothe... so schnell widerlegt worden sind... Aber viele
ausgezeichnete Physiker (wie z.B. Ladenburg, Mie, Born) hätten
daran festgehalten und diese unglückselige Abhandlung... wäre
vielleicht für lange ein Hemmnis des Fortschritts der theoretischen Physik geworden." Ibid.

100. Brief Einstein an Ehrenfest, 31. Mai 1924. Zitiert nach K. Stolzenburg: Die Entwicklung des Bohrschen Komplementaritätsgedankens
in den Jahren 1924 bis 1929. Stuttgarter Dissertation 1976.

101. Brief Heisenberg an Pauli, 8. Juni 1924. Pauli-Briefwechsel.

102. Brief Ladenburg an H. A. Kramers, 8. Juni 1924. Zitiert nach
Stolzenburg, op. cit. (Fußnote 100).

103. Brief Pauli an Bohr. 2. Oktober 1924. Pauli-Briefwechsel.

BIOGRAPHIES OF EINSTEIN

David C. Cassidy*, Historisches Institut, Universität Stuttgart

P h i l i p p   F r a n k   once wrote: "To comprehend Einstein is to comprehend the world of the twentieth century."[1] Einstein and his accomplishments are an integral part of our century and of the world in which we live. Yet today -- a hundred years after his birth and nearly a quarter century since his death -- the nature, development, and contribution of the man and his work are still only partially understood. Part of the reason lies in their complexity, diversity, and influence. Seen from the outside, much of what Einstein said and did is highly technical, confusing, and often paradoxical. In addition, few scientists have stimulated the public imagination and interest the way Einstein has, and any attempt to comprehend his life and work must first penetrate the layers of wonderment and myth still surrounding them.

Most importantly, many of the crucial documents on Einstein and his work and views are not yet widely accessible to scholarly study, and the treatment of topics for which documents appear non-existent is yet to be defined. At present, only Einstein's correspondence with   M a x   B o r n ,   A r n o l d   S o m m e r f e l d ,   M i c h e l e   B e s s o,   and   M a u r i c e   S o l o v i n e   has been published in its entirety.[2] If all goes well, these volumes will be joined in the near future by the complete Einstein Collection in Princeton.[3]   Other scientific collections, such as the papers of   N i e l s   B o h r[4]   and the correspondence of   W o l f g a n g   P a u l i ,[5]   are now emerging from the press. Additional microfilmed documents and interviews on twentieth-century physics have been deposited at five locations in the last decade,[6]   and a project to locate and survey world holdings on the history of twentieth-century physics and its cultural relations is now nearing completion.[7]

This flurry of recent activity suggests that the appropriate time for the encompassing biographical study of Einstein is only just arriving, but it is not too early to consider what biography can contribute to a comprehension of Einstein and what it has contributed so far. It is well known that Einstein himself discouraged biographies, not only because of his modesty, the ironic accusations of publicity seeking by his enemies, and what he saw as attempts to cash in on his fame, as he told   D a v i d   R e i c h i n s t e i n ,[8]   but also because he regarded the "merely personal" as simply unimportant. In his "Autobio-

graphical Notes" he declared: "The essential in the being of a man of
my type lies precisely in *what* he thinks and *how* he thinks, not in what
he does or suffers."[9]   In his case, Einstein wrote in a just-published
remark, his internal thoughts originated and played themselves out in
a way independent of his personal doings and sufferings. "Therefore,"
he declared, "no biography!"[10]

To a certain extent Einstein was right. It is his "thought" and its product in science and in human affairs -- not his personal life and sufferings -- that formed the essence of what he was to us and to himself.
Moreover, much of what went on in his thought, especially in physics,
probably did occur independently of outside factors. But to understand
what and *how* this thought contributed to our world: the problems handled,
the solutions found or not found, the motives, insights, controversial
stands, and observed development, one can hardly isolate it from the
broader historical context, social-professional interactions, and life
circumstances within which it occurred. In so far as another human being
will ever be able to penetrate Einstein's inner thoughts on science,
culture, and society, his doings and sufferings are important in helping
us to comprehend the nature, development, impact, and even origins of
these thoughts formed amidst a real life lived within a specific historical context. This is where biography can make its greatest contribution.

What a particular biography actually does contribute is strongly dependent upon its technical components. It is an explicit function of the
type and capacity of the sources consulted, the skills and sensitivity
of the biographer, his conception of the biographical form, and the
audience for which he is writing. Such factors have strongly determined
the contributions of the voluminous literature on Einstein's life and
work produced so far. My remarks here are limited to the over fifty books
and book-length articles dealing with Einstein's life or life and works
produced since Einstein's literary friend, the pseudo-Boswellian
A l e x a n d e r   M o s z k o w s k i ,[11] published his perceptions
of Einstein in 1921. Biographies since then can be placed into four
major groups according to their outstanding technical factors: those
authored by Einstein's friends, family, friends of the family, and
colleagues; those permeated by ideology and social critique; those directed at a wide audience, including young people; and finally those
that, by their contribution, approach and influence, have become standard works, or claim to be.

Nearly all biographies authored by individuals close to Einstein were
written before any collected documents or scholarly research on his

work and times were available. Since most of these authors were also
unfamiliar with Einstein's physics, their perceptions of his professional
side were severely limited. Yet their works are valuable documents in
themselves. Einstein's family, friends, and family friends provide inside information on his life and activities and those of others close
to him that is not obtainable elsewhere. Einstein's son-in-law
R u d o l f K a y s e r ,[12] one of the few biographers to receive
Einstein's blessing, included an account of his youth presumably as
Einstein himself recalled it around 1930. However, because Kayser and
other such authors, when not speaking for Einstein directly, see him
from only one close-up perspective and often through the filter of their
own personalities, the information in their works must be carefully
evaluated before it can be extracted and applied to any future work.
A n t o n i n a V a l l e n t i n ,[13] a journalist and friend of
Einstein's second wife Elsa, provides unique glimpses of Einstein seen
through female eyes, as well as impressions of Elsa herself, but this
is her only perspective and her perceptions are not always distinguished
from Elsa's. P e t e r M i c h e l m o r e [14] interviewed Einstein's
son Hans Albert at length, obtaining his views of Einstein's first wife
Mileva and their marriage. This makes for interesting reading, but it is
the only perspective of Einstein provided, and, again, Michelmore's own
conjectures are often indistinguishable from Hans Albert's remarks. I
hope he saved his notes.

Einstein's colleagues and assistants are somewhat disappointing.
D a v i d R e i c h i n s t e i n ,[15] a professor of physical chemistry in Zürich who often asked Einstein for professional advice, uses
anecdotal vignettes of his faculty colleague and what he has deduced
from them as excuses to expound at length upon his own views.
B a n e s h H o f f m a n n ,[16] one of Einstein's assistants in the
thirties who had access to the Einstein Collection and the collaboration
of Einstein's secretary H e l e n D u k a s , provides a readable
account of the life and physics spiced with interesting archival documents and photographs. But he does not come close to achieving the
depth of insight possible with his qualifications and sources. Although
he worked with Einstein, he only hints at what he actually did and what
he learned about Einstein from it. This is evident in his handling of
Einstein's "genius". He calls it a "magic touch" containing "powerful
intuition" (pp. 7 - 8, 80) and "that intuitive sense of communion with
the universe ... that defies our attempts at definition" (p. 195). He
could at least explain these expressions, if not attempt a definition,
however unsuccessful.[17] Perhaps it is a law of human nature that one's

wonderment at the man increases with the inverse square of one's near-
ness to him.

By far the greatest number of Einstein biographies are directed at the
general buying public. Since the authors of most of these works are un-
familar with the physics or the history and derive all of their information from a few "standard works" already known to be inadequate, the
result is often an inaccurate description of Einstein's science and a
superficial chronology of his life with little appreciation of the difficult
problems either has posed. Many of these works may be the residue of
what   L e o p o l d   I n f e l d   called "an era of hucksterism in
popular science,"[18] yet the general public is certainly entitled to a
readable -- though accurate -- biography of the man who transformed much
of the world in which we live, even if he has not yet fully succumbed
to scholarly probing. Moreover, any biographer who wishes to be read
must make Einstein intelligible to non-specialists without becoming
superficial or inaccurate.   J e r e m y   B e r n s t e i n ,[19] who
discussed Einstein with   P h i l i p p   F r a n k ,   is perhaps the
most insightful and careful of the biographers writing for a wide audience. He offers reasonable accounts of the physics and includes contemporary developments of general interest, such as black holes and
nuclear energy, but even he does not always escape errors, over-simpli-
fications, and superficialty.

Four biographies merit special attention because of their positive or
negative contributions, methodology, and influence: those by
P h i l i p p   F r a n k ,[20]   C a r l   S e e l i g ,[21]   B o r i s
K u z n e t s o v ,[22] and R o n a l d   C l a r k .[23] Published in
1947, Frank's was the first of these and it is still the most inclusive
and reliable. Frank was a physicist and member of the Vienna Circle of
logical positivists. He succeeded Einstein in Prague and encountered
him later in Berlin and Princeton; all of which placed him in a position
to begin comprehending the man, his physics, and their cultural relations. This he does, but no more. Frank produces not a "biography," but
a still-life portrait of the man framed at the moment he is writing,
what he calls "a description of Einstein's personality and his relation
to his environment" (pp. 55 - 56). When Frank wrote this book around
1940, there were no documents or recollections other than published
works available to him, nor the historical analyses produced since then.
His main sources are his own recollections, readings, and insights
supplemented by those of his subject, which make the work of value in
themselves. The first words of the text are: "As far back as Einstein's
memory extends..." (p. 3), and on the same page we hear traces of

dialect in Einstein's later speech. By page 11 the precocious youngster is already "aspiring to learn the laws of the universe." He never grows or develops; he is always the Einstein of later years.

But Frank's portrait is a good portrait, especially considering when it was written. We see the subject as professor, colleague, family man, and public figure. Frank provides intelligible summaries of his physics and handles most of the necessary social and cultural themes, although many historical remarks have since been eclipsed, the period after 1941 is of course missing, and Frank's own philosophy sometimes gets in the way. He offers the proper diversity of topics which takes him through many of the important themes of this century, but they are distributed over separate, descriptive, often superficial accounts, rather than absorbed into an encompassing study adding up to the totality of the living, thinking man.

Seelig took a different approach in three editions of what he called a "documentary biography." By this he meant the "documents" speak for themselves with little interference from the author. Since Seelig strings together his raw material in chronological order, some passages running over five pages,[24] his portrayal excludes the resolution of contradictions in the documents and the handling of themes, such as pacifism or religion, that defy chronological order or do not appear in the documents. But because he hardly tampers with his sources, his work is a valuable source book for Einstein studies.

Seelig, a Swiss writer, journalist, and art critic, was unable to tackle Einstein's physics and wisely -- but unfortunately -- stayed clear of it. When he began his biography in the early fifties, there were still only scant biographical documents available, and many of those persons who knew Einstein during his most successful years in Switzerland were at least as old as he. Seelig traced many of these people and gathered their written recollections. He also corresponded with Einstein himself, as well as with many physicists, politicians and others who knew him. The material thus generated, combined in subsequent editions with some archival holdings and published recollections on the later years in Berlin and Princeton, formed the main bulk of Seelig's documents. Other sources since made available are of equal or greater importance, but the early collection and publication of this material was a valuable service. It even enabled Seelig to attain new biographical insights. The effort is marred only by the lack of notes or bibliography indicating the origins and locations of the complete documents. They now constitute the core of the Einstein-Sammlung at the Eidgenössische Technische Hochschule in

Zürich.[25]

Frank and Seelig were both confronted with a lack of biographical sources. Frank accordingly drew upon his personal acquaintance with Einstein,
Einstein's recollections, and his own insights; Seelig set about collect-
ing documents. In addition, Frank understood Einstein's physics and
provided readable accounts of it. Seelig understood little of the physics, but penetrated parts of Einstein's early life when many of his
great works were produced.

Although Kuznetsov and Clark had the advantages of fuller sources and
recent scholarly studies of their subject, they fall short of Frank and
Seelig in their treatment of both the life and works -- the former because of his methodology, the latter because of his skills. Kuznetsov
was apparently trained in physics and probably reads German and English,
although he speaks neither . Clark, a former war correspondent, is un-
skilled in physics or social analysis and apparently reads little German.

Kuznetsov never applies his skills at Seelig's documentary level because
he finds it unnecessary to do so. He is interested in the effect of genius on the grand flow of intellectual-scientific history, not in individual geniuses. A genius, he argues, is he who accelerates or alters
this flow, and "a scholar's biography records not the upshot of his
scientific achievement, but the gradient of scientific progress associated with it..." (p. 7). The contribution of Einstein's finished, genial
work seen as a retrospective unit is of primary importance, Einstein the
genius is not.

Kuznetsov draws support from Einstein's "Autobiographical Notes." That
Einstein said he tried to free himself from the "merely personal" encouraged Kuznetsov to suppress "'personal' biography" in favor of
"'extra-personal' history of science" (p. 50). Einstein's "Notes" also
served as his main source on Einstein's science. Kuznetsov is not alone
in failing to appreciate that Einstein wrote an autobiography, not a
biography.

Kuznetsov accordingly produced two editions of a "biography of a genius"
(p. 3) in which the genius himself is missing. Instead we are given
generalizations about the place of his work and views as a logical unit
within Kuznetsov's version of the broad course of intellectual and scientific history since the Renaissance. Einstein's actual life and work are
so secondary that a unique "Einstein" is unnecessary. Were he a school
or generation of scientists, it would alter few essentials of the book.

Kuznetsov's historiography of the individual in history is certainly not new, but still deserves some attention. Einstein's various works -- not the logical unit Kuznetsov sees -- must indeed be evaluated as contributions to the science of their time. Moreover, generalizations about scientific knowledge before and after the work of a great scientist are interesting and valuable, especially if the science afterward is our own, but they do not resolve the historical problems of how and why history occurred as it did, or the biographical problems of how and why this particular individual was able to contribute to it.

Because he was unable to handle the physics, society, or Einstein's difficult language, Clark relied upon various technical advisers in producing a wealth of information on Einstein's later public life gleaned from numerous archival sources. Clark's wealth of information enables him to make several biographical assertions, but his often superficial approach and derivative interpretations prohibit his achieving Frank's reliability, sensitivity, or diversity. The biographer is certainly rare who can handle well all aspects of Einstein's life and work, but every biographer should be able to handle at least one of those aspects. Nevertheless, Clark is one of the few to assert that Einstein the student possessed a "prickly arrogance" that made him quite different from the "gentle philosopher" of later years (p. 26). Clark looked through a large number of documents in search of Einstein's role in the atomic bomb project and concluded that he knew much more about it than is generally believed. Finally, Clark's Einstein is a tragic figure who is overcome by his real or apparent paradoxes, particularly the paradox of a man of high political ideals forced onto the center stage of an often barbaric century. All of these valuable and plausible assertions need to be explored in greater detail and depth before they can be fully accepted.

Clark manages to ask some of the right questions about Einstein's nonscientific life, but his skills and approach prevent the careful, penetrating analysis of the total man needed to answer them. He rightly sees that Einstein's formative years up to age 16 are crucial to understanding him, but it is precisely these years for which solid sources are unavailable. Clark accordingly devotes only 20 of 631 pages to them (pp. 3 - 23) and in mechanical fashion, but tries to argue that Einstein's life-long suspicion of authority and his "obsession with exploring and understanding the physical world" (p. 32) grew out of them. He rightly notes that the popular interest in relativity theory has not been adequately explained (p. 246), but only repeats the conjectures of others. He knows he should explain why Einstein remained in Berlin amidst the anti-Semitic campaign against him, but seems unaware that he provided two explanations separated by 40 pages (pp. 264 - 265, 305), neither of which mentions Einstein's own explanation to Sommerfield.[26] Clark calls Einstein an "imaginative genius" in one sentence, then quotes him in the next saying "I have no particular talent" (p. 6). That no attempt is made to resolve at least this paradox arising between Einstein and his biographer is indicative of the distance still separating Clark from his subject.

Each of these biographies contributes something to the comprehension of Einstein, particularly within their own time and setting, yet Einstein the living, thinking, working man is still missing. He is still missing because the essentials of sound biography are still missing: the unique biographer skilled in physics, language, history, and culture who possesses the sensitivity needed to comprehend this complex, multi-faceted man; secondly, the proper diversity and use of biographical and historical documents; and thirdly, the proper timing that allows one to draw upon scholarly studies of the period, its science, and the existing sources. Yet to be answered are such questions as: what happened and why during that *annus mirabilis* of 1905? How does one explain Einstein's tremendous popularity, as well as the tremendous hate evoked in some circles, and what effects did they or did they not exert upon him? When and how did his social-political views originate, how did they develop, and what was their actual influence? How were his science, politics, and religion all related or unrelated in his own mind at different periods of his life?

To answer such questions requires -- in addition to the above essentials -- a different biographical form than represented so far, a form that encompasses the total Einstein but places at its center what makes him of interest to begin with: his work and "thought" and the way they contributed to our century. Though such a form exists in most types of biography, it has been developed perhaps most fully among literary biographers who, seeking to understand a writer's works, have looked beyond them to the life and culture within which they were produced.[27] As a result they have developed a genre that is more inclined to view the heroic genius as a normal, intelligent, hard-working, neurotic human being possessed of certain well-developed talents -- hence approachable by mere mortals -- and that examines his creative life and work, as Dostoevsky's biographer writes, "in the context of a massive reconstruction of the social-cultural life of his period."[28]

There are of course limits to such a biographical form applied to Einstein, especially to Einstein the scientist. On the other hand, the history of science, notoriously plagued in the past by heroic biography, has recognized that science is a complex group endeavor that often transcends the work of an individual genius and that science too must often be understood within a changing social-cultural context. Owing to its notoriety and inability to move far from the great individual, biographical history of science has fallen into some disrepute, despite several excellent contributions.[29] But in so far as the right combination of biographer, sources, and approach "permits us to watch a great mind at work,"[30] without losing sight of that mind's actions and reactions with the life, community, culture, science, and society within which it operates, it will contribute immensely to our comprehending, as Einstein himself wrote, "how the efforts of a life hang together and why they have led to expectations of a definite form."[31]

Notes

*   The author thanks the Alexander von Humboldt-Stiftung for support of this work, and Gerald Holton, John Michel, and the session participants for their comments.

1.   Philipp Frank, *Einstein: Sein Leben und seine Zeit* (Munich, 1949; Braunschweig, 1979), p. 7. The "Einleitung" (pp. 7-14) was not included in the English edition: *Einstein: His Life and Times*, trans. from manuscript by G. Rosen (New York, 1947).

2.   A. Einstein, Hedweg and Max Born , *Briefwechsel 1916 - 1955* (Munich, 1969); A. Einstein and Arnold Sommerfeld, *Briefwechsel*, ed. by Armin Hermann (Basel and Stuttgart, 1968); A. Einstein and Michele Besso, *Correspondance 1903 - 1955*, edited with simultaneous French translation by Pierre Speziali (Paris 1972); A. Einstein, *Lettres à Maurice Solovine* (Paris, 1956).

3.   Edited by John Stachel, Institute for Advanced Study, Princeton, New Jersey. See David Dickson, "Einstein: disagreement delays publication of collected works," *Nature, 278* (1979), 294-295.

4.   Niels Bohr, *Collected Works*, general editor L. Rosenfeld (Amsterdam, beginning in 1972).

5.   Wolfgang Pauli, *Wissenschaftlicher Briefwechsel mit Bohr, Einstein, Heisenberg u.a.*, Volume 1: 1919 - 1929, ed. by Armin Hermann and Karl von Meyenn (New York, 1979).

6.   American Institute of Physics, New York; American Philosophical Society, Philadelphia; University of California, Berkeley; University of Minnesota, Minneapolis; Niels Bohr Institute, Copenhagen. A catalogue was published by Thomas S. Kuhn, *et.al.*, *Sources for History of Quantum Physics: An Inventory and Report* (Philadelphia, 1967); Supplement (Berkeley, 1973).

7.   Directed by Bruce Wheaton, Office for History of Science and Technology, 470 Stephens Hall, University of California, Berkeley, California.

8. Einstein to David Reichinstein, 2 May 1932, reproduced in David
    Reichinstein, *Albert Einstein, sein Lebensbild und seine Welt-
    anschauung* (Prague, 1935), pp. 279 - 281.

9. A. Einstein, "Autobiographical Notes" (German with simultaneous
    English translation), in *Albert Einstein: Philosopher Scientist*,
    ed. Paul Arthur Schilpp (New York, 1949), 1 - 95; on p. 33, his
    emphasis.

10. Einstein, "Vorwort" written in 1942, first published in the 1979
    edition of Philipp Frank, *Einstein*; remark on second unnumbered
    page.

11. Alexander Moszkowski, *Einstein: Einblicke in seine Gedankenwelt*
    (Hamburg and Berlin, 1921).

12. Anton Reiser (Rudolf Kayser), *Albert Einstein: A Biographical
    Portrait* (New York, 1930; London, 1931).

13. Antonina Vallentin, *The Drama of Albert Einstein*, trans. Moura
    Budberg (Garden City, 1954).

14. Peter Michelmore, *Einstein: A Profile of the Man* (London, 1963).

15. Reichinstein, *Albert Einstein*.

16. Banesh Hoffmann, with the collaboration of Helen Dukas, *Albert
    Einstein: Creator and Rebel* (New York, 1972).

17. If it reached him in time, Gerald Holton's study, "On Trying to
    Understand Scientific Genius", *American Scholar*, 41 (1971)
    95 - 110, may have helped.

18. Leopold Infeld, *Albert Einstein: His Work and Its Influence on Our
    World* (New York and London, 1950), p. 5.

19. Jeremy Bernstein, *Albert Einstein* (New York and London, 1973).

20. Frank, *Albert Einstein*. Page numbers in the text refer to the
    English edition.

21. Carl Seelig, *Albert Einstein und die Schweiz* (Zürich, 1952);
    second edition: *Albert Einstein, eine dokumentarische Biographie*
    (Zürich, 1954); third edition: *Albert Einstein, Leben und Werk
    eines Genies unserer Zeit* (Zürich, 1960).

22. Boris Kuznetsov, *Einstein*, trans. into English by V. Talmy (New
    York, 1970); second edition: *Einstein, Leben-Tod-Unsterblich-
    keit*, trans. Siegfried Wollgast and Helmut Fuchs (Basel and
    Stuttgart, 1977). Page numbers in the text refer to the English
    version.

23. Ronald W. Clark, *Einstein: The Life and Times* (New York, 1971).
    The British edition (London, 1973) was revised and altered for
    copyright reasons. Page numbers in the text refer to the New
    York edition.

24. E.g., Seelig, *Einstein*, second edition, pp. 41 - 46 and 119 - 125.
    The corresponding passages in the third edition are pp. 56 - 63
    and 170 - 179.

25. Catalogue of the collection: *Die Einstein-Sammlung der ETH-Biblio-thek in Zürich, ein Überblick für Benützer der Handschriften-Abteilung* (Zürich, 1970).

26. Einstein to Sommerfeld, 6 September 1920, in *Briefwechsel*, p. 69. See also, Einstein, "Self-Portrait," in Albert Einstein, *Out of My Later Years* (New York, 1950), p. 5.

27. See, for example, Leon Edel, *Literary Biography*, The Alexander Lectures, 1955 - 1956 (Toronto, 1957).

28. Joseph Frank, *Dostoevsky*, Volume 1: *The Seeds of Revolt, 1821-1849* (Princeton, 1976), p. xii.

29. Thomas L. Hankins offers a response, "In Defense of Biography: The Use of Biography in the History of Science," *History of Science*, 17 (1979), 1 - 16. I thank Professor Hankins for providing a pre-publication copy of his manuscript.

30. J. Z. Fullmer, "Davy's Biographers: Notes on Scientific Biography," *Science*, 155 (1967), 285 - 291; on p. 291.

31. Einstein, "Autobiographical Notes," p. 95.

NATIONALISTISCHE UND ANTISEMITISCHE RESSENTIMENTS VON WISSENSCHAFTLERN GEGEN EINSTEIN

Andreas Kleinert, Fachhochschule für Bibliothekswesen, Stuttgart

Als der Physiker und Nobelpreisträger Wilhelm Wien vor fast 60 Jahren hier in Berlin vor den Mitarbeitern der Firma Siemens und Halske über die Relativitätstheorie sprach, hielt er es für erforderlich, in seinem Vortrag folgendes zu bemerken:

> [Es] ist um die sogenannte allgemeine Relativitätstheorie ein Streit entbrannt, wie er in der Geschichte der Wissenschaften wohl noch nicht vorgekommen ist. Der Kampf hat den wissenschaftlichen Boden ganz verlassen und ist ins politische und dogmatische Gebiet übertragen worden, wobei naturgemäß die bei solchen Streitigkeiten sich einstellende Erbitterung nicht gefehlt hat.[1]

Es gehörte in den zwanziger Jahren beinahe schon zum guten Ton, sich von den hier angesprochenen Methoden der Auseinandersetzung mit der Relativitätstheorie zu distanzieren, wenn man sich kritisch zu Einstein äußern wollte, und so finden wir entsprechende Bemerkungen in den Schriften zahlreicher Einstein-Gegner, die um ihre wissenschaftliche Reputation besorgt waren. Mit Politik oder Antisemitismus, so versichern sie, habe ihre Kritik nichts zu tun.[2]

An dem von Wien erwähnten Kampf um die Relativitätstheorie, der "den wissenschaftlichen Boden ganz verlassen" hatte, wirkten in erster Linie Leute mit, die, wie sie zumeist selbst zugaben, in physikalischer Hinsicht Dilettanten und Laien waren. Widerlegungen der Relativitätstheorie von Autoren wie einem Regierungsrat Fricke[3] oder einem schwer Kriegsbeschädigten und alten Hitler-Getreuen Heinsohn[4] erschienen in so großer Zahl, daß es durchaus zutreffen dürfte, wenn Philipp Lenard 1933 schrieb, daß er schon seit Jahren etwa alle 14 Tage eine solche Schrift zur Begutachtung erhalte.[5]

Kritische und polemische Äußerungen zur Relativitätstheorie, auf die die Charakterisierung Wiens zutrifft, kamen aber auch von Physikern. Die Zahl derer, die sich so äußerten, war zwar relativ klein, aber was sie sagten, erhielt durch das Ansehen, das sie aufgrund von wissenschaftlichen Leistungen, hohen Auszeichnungen oder einfach wegen ihrer Stellung als Hochschullehrer besaßen, ein ganz anderes Gewicht als die Ausführungen der Regierungsräte, Kriegsteilnehmer und Hitler-Getreuen.

Über solche, von Physikern stammende Äußerungen zur Relativitätstheorie, die von nationalistischen und antisemitischen Vorurteilen geprägt sind, will ich im folgenden sprechen.

Wien irrte sich, wenn er seine Bemerkungen über wissenschaftsfremde Elemente in der Auseinandersetzung mit Einstein auf die Allgemeine Relativitätstheorie beschränkte. Bereits im Jahre 1915, also zu einem Zeitpunkt, als nur die Spezielle Relativitätstheorie vorlag, ist diese zur Zielscheibe einer von übersteigertem Nationalismus geprägten Polemik geworden.

Der größere Rahmen, in dem sich diese erste Polemik gegen Einstein abspielte, war der später als "Krieg der Geister" bezeichnete Nebenkriegschauplatz,[6] auf dem sich kurz nach Ausbruch des ersten Weltkriegs die Intellektuellen der kriegführenden Mächte betätigten. Dieser mit Tinte und Druckerschwärze geführte Krieg hatte damit begonnen, daß zahlreiche deutsche Intellektuelle, allen voran die Universitätsprofessoren, Aufrufe und Manifeste veröffentlichten, in denen sie sich ohne Einschränkung zur deutschen Kriegspolitik bekannten und insbesondere die Verletzung der Neutralität Belgiens zu rechtfertigen suchten. Besonderes Aufsehen erregte in diesem Zusammenhang der von 93 Gelehrten und Künstlern unterzeichnete "Aufruf an die Kulturwelt" vom 4. Oktober 1914.[7] Der Umstand, daß zu seinen Unterzeichnern viele Naturwissenschaftler gehörten (u.a. die Physiker Lenard, Nernst, Röntgen und Planck), führte in Frankreich zu einer heftigen Gegenreaktion: zahlreiche französische Gelehrte griffen nun ihrerseits zur Feder, und sie beschränkten sich dabei nicht auf die Auseinandersetzung mit dem politischen Standpunkt ihrer deutschen Fachgenossen, sondern gingen gewissermaßen selbst zum Angriff über. Das Manifest der deutschen Intellektuellen war ihnen eine willkommene Gelegenheit zur Abrechnung mit der deutschen Wissenschaft, die ihnen ohnehin schon längst, und nicht ganz ohne Grund, als ein Instrument des wilhelminischen Imperialismus suspekt geworden war.[8]

Zu den französischen Gelehrten, die sich in Büchern, Vorträgen und Zeitungsartikeln über die deutsche Wissenschaft verbreiteten, gehörte auch der bekannte Physiker Pierre Duhem (1861 - 1916). Er sah nach seinen eigenen Worten in dieser Tätigkeit seinen bescheidenen Beitrag zur nationalen Verteidigung; seiner Ansicht nach war nicht nur der Boden des Vaterlandes besetzt worden, sondern auch das französische Denken war unter die Knechtschaft des ausländischen, sprich deutschen Gedankengutes geraten: "La pensée étrangère a réduit en servitude la pensée francaise."[9] Duhem versuchte, die wissenschaftliche Erkenntnis unter nationalen Gesichtspunkten zu sehen: rein wissenschaftsimmanente Aussagen, so behauptete er, seien geprägt vom Nationalcharakter des Gelehrten, der sie ausgesprochen hat.

Auch andere französische Gelehrte waren der Meinung, es gebe eine typische deutsche Denkweise, und wir begegnen Formulierungen wie "structure de l'esprit germanique" und "mécanisme cérébral des Allemands".[10] Duhem aber hat sich als einziger auch um den Nachweis bemüht, daß sich diese angeblichen germanisch-deutschen Gehirnstrukturen auch in den wissenschaftlichen Arbeiten deutscher Gelehrter niederschlagen: spezielle Merkmale der deutschen Wissenschaft, so schreibt er, könne man ableiten aus gewissen Anlagen in der deutschen Intelligenz ("les faire dériver de quelques dispositions essentielles de l'intelligence allemande").[11]

Duhem bezieht sich vor allem auf die theoretische Physik, und seine Argumentation sieht etwa folgendermaßen aus.[12]

Es geht in der Physik darum, die Fülle der Einzelaussagen über die Natur aus wenigen, allgemeinen, mathematisch formulierten Aussagen (Axiomen, Prinzipien) logisch abzuleiten. Es kommt also einmal darauf an, die richtigen Axiome zu finden, die am Anfang der Argumentation stehen, und es geht darum, aus den Axiomen durch mathematische Deduktion zu den daraus folgenden Einzelaussagen zu gelangen, die dann dem Experiment bzw. der Beobachtung nicht widersprechen dürfen.

Die Fähigkeit zum Auffinden der richtigen Axiome und die Fähigkeit, daraus die richtigen Schlüsse zu ziehen, bezeichnet Duhem nun mit einem Begriffspaar, das er von Pascal entlehnt hat: zur Deduktion, also zum Ableiten der richtigen Schlußfolgerungen aus den Axiomen, bedarf es des "esprit de géométrie", während zum Finden von Axiomen oder Prinzipien der "esprit de finesse" erforderlich ist. Der "esprit de géométrie" folgt festen, ihm von außen auferlegten Regeln, während der "esprit de finesse" intuitiv, oft sprunghaft und eher gefühlsmäßig vorgeht, um zu den Erkenntnissen zu gelangen, die nicht durch logisches Schließen aus Bekanntem abgeleitet werden können. "Esprit de géométrie" und "esprit de finesse" sollen nun bei den verschiedenen Völkern unterschiedlich ausgeprägt sein, und Duhem wird nicht müde, zu versichern, daß die Deutschen sich durch ein Übermaß an "esprit de géométrie" auszeichneten, während der "esprit de finesse" bei ihnen stark verkümmert sei.

Diese Einseitigkeit des deutschen Geistes hänge dann wieder mit den bekannten deutschen "Tugenden" zusammen: wer sich von den Regeln der deduktiven Logik beherrschen läßt, der muß offenbar arbeitsam, unterwürfig, gewissenhaft und vor allem diszipliniert sein.[13]

Der Mangel an "esprit de finesse" zeigt sich nach Duhem bei den deutschen Physikern vor allem durch Mangel an gesundem Menschenverstand. Die deutschen Physiker, so sagt er, gehen bei dem Versuch, Hypothesen Prinzipien oder Axiome aufzustellen, von völlig willkürlichen Postulaten aus,[14] die dem gesunden Menschenverstand, dem "sens commun", Hohn sprechen.

Wie es nicht anders zu erwarten ist, wenn der gesunde Menschenverstand ins Spiel kommt, muß die Relativitätstheorie dazu herhalten, um Duhems Auffassung von der deutschen Wissenschaft zu belegen. Vor allem die Aussage, daß die Lichtgeschwindigkeit eine Grenzgeschwindigkeit sein soll, die kein bewegter Körper erreichen kann, ist für ihn eine unerträgliche Zumutung, die nur auf dem Boden des deutschen "esprit de géometrie" entstehen konnte.

In Anbetracht der Parallelen, die diese erste Kampagne gegen die Relativitätstheorie zu der Polemik aufweist, die später mit veränderten nationalen Etiketten in Deutschland gegen Einstein geführt wurde, will ich zitieren, was Duhem zur Speziellen Relativitätstheorie schrieb, die er, dem damaligen Sprachgebrauch entsprechend, als Relativitätsprinzip ("pincipe de la relativité") bezeichnete:

> Die Tatsache, daß das Relativitätsprinzip alle Empfindungen des gesunden Menschenverstandes durcheinander bringt, erweckt nicht das Mißtrauen der deutschen Physiker - ganz im Gegenteil. Es [das Prinzip] zu akzeptieren bedeutet gleichzeitig, alle Lehrsätze umzustoßen, in denen von Raum, Zeit und Bewegung die Rede war, alle Theorien der Mechanik und der Physik. Eine solche Verwüstung hat nichts an sich, das dem germanischen Denken mißfallen könnte. Auf dem Gebiet, auf dem die alten Lehrsätze beseitigt wurden, wird der geometrische Verstand der Deutschen voller Freude eine ganze Physik neu errichten, deren Grundlage das Relativitätsprinzip sein wird. Wenn diese neue Physik unter Mißachtung des gesunden Menschenverstandes allem widerspricht, was aufgrund von Beobachtungen und Erfahrungen in der Mechanik des Himmels und in der irdischen Mechanik aufgebaut worden war, so werden die Anhänger der rein deduktiven Methode nur um so stolzer sein auf die unbeugsame Strenge, mit der sie die zerstörerischen Konsequenzen ihres Postulats bis zum Ende verfolgt haben werden. (...) So ist die Relativitätsphysik aufgebaut; so schreitet die deutsche Wissenschaft voran, stolz auf ihre algebraische Strenge und voller Mißachtung für den gesunden Verstand, den alle Menschen mitbekommen haben.[15]

In Jean-Pierre Achalme fand Duhem bald einen Mitkämpfer, der seine Gedanken aufgriff und weiter entwickelte. Achalme brachte 1916 unter dem bezeichnenden Titel "La science des civilisés et la science allemande" ein über 200 S. starkes Buch heraus, in dem es über die Relativitätstheorie unter anderem heißt, sie sei eine wissenschaftliche Entwicklung, die man am besten mit dem Futurismus und dem Kubismus in der Kunst vergleiche, und die einzig darauf bedacht sei, bewährte Traditionen zu zerstören. Im gleichen Stil wird dann bei Achalme noch ein anderer Bereich der modernen Physik als typisch deutsch abqualifiziert, an dessen Entwicklung Einstein ebenfalls beteiligt war: die Theorie der Lichtquanten. Einstein wird hier freilich nicht genannt, sondern nur Planck, der sich als Mitunterzeichner des "Aufrufs an die Kulturwelt" bei einer solchen Verquickung von Nationalismus und Physik noch besser als Zielscheibe

eignete. Plancks Quantentheorie, so heißt es, sei ein mathematisch-
metaphysisches Delirium, bei dem Abstraktionen als Realitäten angesehen
würden, und es ist für Achalme der Gipfel der Absurdität, wenn die Atome
der Wärme, des Lichtes, ja der Energie schlechthin aufgrund der Relativitätstheorie sogar eine Masse besitzen sollen.[16]

Die vom übersteigerten Nationalismus der ersten Kriegsjahre geprägte
Polemik gegen die Relativitätstheorie und die frühe Quantentheorie, die
ich hier skizziert habe, ist eine kurze Episode geblieben, die schnell
vorüber ging und an die sich bald niemand mehr erinnerte. Vor allem in
Deutschland scheint sie kaum bekannt geworden zu sein, denn während den
deutschen Gelehrten noch lange nach Kriegsende ihre Unterschriften unter
den "Aufruf an die Kulturwelt" vorgeworfen wurden, der sogar als Begründung für einen langjährigen internationalen Boykott der deutschen
Wissenschaft herhalten mußte, fand sich in Deutschland niemand, der den
Franzosen ihre Torheiten von 1915 ins Gedächtnis gerufen hätte.

Stattdessen wurde bald darauf in Deutschland gegen die moderne Physik
polemisiert, wobei die Struktur der Argumentation genau dieselbe sein
sollte: die physikalischen Theorien, die man wegen ihrer Unanschaulich-
keit ablehnte, wurden als typisches Produkt artfremden, feindlichen
Denkens angesehen, das aus der eigenen Gedankenwelt entfernt werden
mußte. Aus der "Science allemande" von 1915/16 wurde die "jüdische
Physik" der zwanziger und dreißiger Jahre; der gesunde, an der Erfahrung geschulte Menschenverstand, den Duhem und Achalme für Frankreich
reklamiert hatten, wurde nun zum Markenzeichen der in Deutschland ge-
pflegten arischen oder eben deutschen Physik, und wieder waren Relativitätstheorie und Quantentheorie betroffen.

An der weiteren Entwicklung der Quantentheorie war Einstein nicht mehr
beteiligt; ich will deshalb im folgenden vor allem schildern, wie die
Relativitätstheorie in Deutschland von einigen Physikern bekämpft wurde, die dabei physikalische und ideologische, d.h. im Antisemitismus
wurzelnde Argumente miteinander verknüpften.

Im Zusammenhang mit dem vom Antisemitismus geprägten Kampf gegen die
Relativitätstheorie ist als erster Philipp Lenard zu nennen. Seine Einwände gegen die Relativitätstheorie hatten mit Antisemitismus oder mit
persönlichen Vorbehalten gegen Einstein zunächst nichts zu tun. Den
wissenschaftlichen Werdegang Einsteins hat Lenard von Anfang an verfolgt, und zwar durchaus mit wohlwollendem Interesse. Zwischen 1905
und 1909 tauschten die beiden Sonderdrucke aus und führten eine kurze
Korrespondenz; Lenard nennt Einstein in einem Brief vom Juni 1909 einen
tiefen, umfassenden Denker und bezeichnet seine Theorie des Photoeffekts

als "umfassende, wunderbare, von Ihnen gefundene Beziehungen".[17]

Lenards Auseinandersetzung mit der Relativitätstheorie begann im Jahre 1910 mit einem vor der Heidelberger Akademie der Wissenschaften gehaltenen Vortrag "Über Äther und Materie". Er diskutiert hier die Schwierigkeiten, die durch die von ihm akzeptierte Spezielle Relativitätstheorie für die traditionelle Vorstellung vom Äther als dem Medium der optischen und elektrischen Erscheinungen entstanden sind, und hofft, daß sich diese Schwierigkeiten durch gewisse Korrekturen an der bisherigen Äthervorstellung beheben lassen. Persönliche Angriffe gegen Einstein enthält dieser Vortrag nicht, und noch 1913 spielte Lenard mit dem Gedanken, Einstein auf eine Professur für theoretische Physik nach Heidelberg zu berufen.

Lenards nächste Veröffentlichung zur Relativitätstheorie erschien 1918 unter dem Titel "Über Relativitätsprinzip, Äther, Gravitation"; auch dieser Artikel ist frei von jeder Polemik. Lenard hebt die Bedeutung der Speziellen Relativitätstheorie ausdrücklich hervor und vergleicht sie in ihrer Tragweite mit dem Satz von der Erhaltung der Energie; lediglich die Allgemeine Relativitätstheorie kritisiert er in einigen Punkten.

Auf Lenards Einwände antwortete Einstein noch im selben Jahr 1918 in einem Aufsatz in den "Naturwissenschaften";[18] darauf antwortete wieder Lenard in einer Neuauflage seiner Schrift von 1918, erschienen im März 1920. Die Diskussion zwischen den beiden Physikern nahm allmählich an Schärfe zu, blieb aber durchaus im wissenschaftlichen Rahmen. Lenards Einwände gegen die allgemeine Relativitätstheorie kreisten letztlich immer mehr um einen Punkt, der schon Duhem an der Relativitätstheorie gestört hatte: der "gesunde Verstand", so Lenard, sträube sich gegen diese Theorie, worauf Einstein antwortete, der gesunde Menschenverstand sei als Schiedsrichter in physikalischen Fragen ungeeignet.

Es läßt sich recht genau angeben, ab wann bei Lenards Kampf gegen die Relativitätstheorie der Antisemitismus eine Rolle spielte. Anfang 1921 erschien die oben genannte Abhandlung von 1918 ("Über Relativitätsprinzip, Äther, Gravitation") in der dritten Auflage, und hier finden wir in einer als Ergänzung hinzugefügten Fußnote über die angebliche Kühnheit des theoretischen Physikers, der neue Hypothesen formuliert, den Satz "Deutsche Eigenschaft ist *diese* Kühnheit nicht". Im selben Jahr erschien von Lenard eine weitere Veröffentlichung zur Relativitätstheorie unter dem Titel "Über Äther und Uräther", in der er sich ausdrücklich von seinen früheren, zustimmenden Äußerungen zur Speziellen Relativitätstheorie distanziert, und die 1922 erschienene zweite

Auflage dieser Schrift enthält ein "Mahnwort an deutsche Naturforscher"
mit unverkennbar antisemitischen Angriffen gegen Einstein. Da ist die
Rede von der "vor nicht Rassekundigen versteckten Begriffsverwirrung,
welche um Herrn Einstein als deutschen Naturforscher schwebt",[19] und
auch der Satz "Es ist bekannte jüdische Eigenschaft, sachliche Fragen
sofort aufs Gebiet von persönlichem Streit zu verschieben"[20] ist eine
deutliche Anspielung auf seine Auseinandersetzung mit Einstein. Auch
die Relativitätstheorie wird hier zum erstenmal als typisches Produkt
des jüdischen Geistes bezeichnet, wenn es heißt: "Lebt gesunder deutscher Geist - der freilich Pflege und Schutz nötig hätte - wieder auf,
so wird von selbst der Fremdgeist weichen müssen, der als dunkle Macht
überall auftaucht und der auch in allem, was zur 'Relativitätstheorie'
gehört, so deutlich sich ausprägt."[21]

Man muß Lenard zugute halten, daß es nicht ganz aus der Luft gegriffen
war, wenn er das Gefühl hatte, daß Einstein ihn in einen persönlichen
Streit hineingezogen hatte. Einstein hatte ihn im August 1920 in einem
Artikel im "Berliner Tageblatt" in einer Weise angegriffen, die er durchaus als persönliche Beleidigung auffassen mußte, und da er bis zu diesem Zeitpunkt nur in wissenschaftlichen Fachzeitschriften über die Relativitätstheorie geschrieben hatte, war er durch diese öffentliche Attacke des 17 Jahre jüngeren Einstein tief gekränkt.[22]

Erst 1933 finden wir bei Lenard wieder eine antisemitische Äußerung
gegen Einstein: es ist jener oft und meist falsch zitierte Artikel im
"Völkischen Beobachter", in dem er Einstein als "Relativitätsjuden"
bezeichnet, dessen "mathematisch zusammengestoppelte 'Theorie' (..) nun
schon allmählich in Stücke zerfällt, wie es das Schicksal naturfremder
Erzeugnisse ist."[23]

Bis zu diesem Zeitpunkt hat Lenard sich darauf beschränkt, die von ihm
abgelehnte Relativitätstheorie und das Verhalten Einsteins pauschal als
jüdisch zu bezeichnen, ohne daß er näher darauf eingegangen wäre, wie
sich nun genau jüdisches von nicht-jüdischem Denken unterscheide und
welches die speziell jüdischen Merkmale der Relativitätstheorie seien.
Eine solche, detaillierte Auseinandersetzung mit der Relativitätstheorie
auf dem Boden der nationalsozialistischen Rassenlehre setzte nach 1933
ein, und erst jetzt äußerten sich außer Lenard auch andere Physiker in
dieser Weise.

Ich komme nun zu den Angriffen gegen Einstein und die Relativitätstheorie, die auf der Grundlage der nationalsozialistischen Rassenlehre nach
1933 erfolgt sind. Die Physiker, die sich dieser Bewegung anschlossen,
sahen sich als Vertreter einer "Deutschen Physik", die das Gegenstück

zur jüdischen Physik bilden sollte.[24] Im Vorwort zu seinem 1935 erschienenen Lehrbuch, das eben diesen Titel "Deutsche Physik" trägt, hat
Lenard zum erstenmal ausgeführt, wie sich die Zugehörigkeit zum Judentum bei einem Physiker zeigen soll. Wie alles, was Menschen hervorbrin-
gen, so argumentiert Lenard, sei auch die Wissenschaft rassisch, blut-
mäßig bedingt, und so habe sich "sehr breit eine eigentümliche Physik
der Juden entwickelt", die "ein auffallendes Gegenstück zur deutschen
Physik ist." Was Lenard dann zur jüdischen Physik schreibt, ist sehr
allgemein:

> *Dem Juden fehlt auffallend das Verständnis für Wahrheit, für mehr als*
> *nur scheinbare Übereinstimmung mit der von Menschen-Denken unabhängig*
> *ablaufenden Wirklichkeit, im Gegensatz zum ebenso unbändigen wie*
> *besorgnisvollen Wahrheitswillen der arischen Forscher. Der Jude hat*
> *kein merkliches Fassungsvermögen für andere Wirklichkeiten als etwa*
> *die des menschlichen Getriebes und der Schwächen seines Wirtsvolkes.*
> *Dem Juden scheint wunderlicherweise Wahrheit, Wirklichkeit überhaupt*
> *nichts Besonderes, von Unwahrem Verschiedenes zu sein, sondern gleich*
> *irgendeiner der vielen verschiedenen, jeweils vorhandenen Denkmöglich-*
> *keiten.*

Ferner soll es eine "dem jüdischen Geist eigene Eiligkeit, mit uner-
probten Gedanken hervorzutreten" geben, kurzum: "Die jüdische Physik
ist somit nur ein Trugbild und eine Entartungserscheinung der grundlegenden arischen Physik." Lenard sagt zwar, um diese jüdische Physik zu
charakterisieren, könne "am gerechtesten und besten an die Tätigkeit
ihres wohl hervorragendsten Vertreters, des wohl reinblütigen Juden
A. Einstein, erinnert werden", dessen "Relativitäts-Theorien" die ganze
Physik umgestalten und beherrschen wollten; er lehnt es aber dann ausdrücklich ab, in seinem Lehrbuch näher auf dieses "verfehlte Gedanken-
gebäude" einzugehen.[25]

Was an der Relativitätstheorie typisch jüdisch sein soll, haben andere
zu zeigen versucht. Es waren dies J o h a n n e s   S t a r k ,
W i l h e l m   M ü l l e r   und B r u n o   T h ü r i n g , ferner
einige jüngere Lenard-Schüler. Ich beschränke mich hier auf die Äußerung
von Stark, Müller und Thüring, denn alles, was andere in diesem Zusammenhang gesagt haben, ist in den Schriften dieser drei Physiker enthalten.

Zunächst zu Stark. Im Gegensatz zu Lenard, der etwa ab 1920 die gesamte
Einsteinsche Physik ablehnte,[26] beschränkte sich Stark auf die Allgemeine Relativitätstheorie; gegen die Spezielle Relativitätstheorie hat
er sich an keiner Stelle ausgesprochen. Im übrigen galt sein Interesse
viel mehr der Quantenmechanik, wie sie von Bohr und Sommerfeld, später
dann von Heisenberg und Schrödinger entwickelt worden ist, und wenn er
zur Relativitätstheorie Stellung nahm, so geschah dies stets im Zusam-

menhang mit sehr viel ausführlicheren Äußerungen zur Quantentheorie.

Starks erste Schrift dieser Art, "Die gegenwärtige Krisis in der Deutschen Physik", stammt aus dem Jahre 1922. In dieser Schrift taucht zum erstenmal eine Formulierung auf, die von den Gegnern der modernen theoretischen Physik noch oft aufgegriffen werden sollte: der Vorwurf des Dogmatismus. So wie das kirchliche Dogma Behauptungen aufstellt, die vom Menschen aufgrund seiner bisherigen Erfahrungen nicht verstanden werden können, so ist nach Stark auch die Quantentheorie "in ihren Voraussetzungen durchaus dogmatisch",[27] ihre Dogmen sind für ihn zu diesem Zeitpunkt (1922) die Bohrschen Postulate mit den Ergänzungen von Sommerfeld. Auch der Allgemeinen Relativitätstheorie wirft er einen solchen Dogmatismus vor: sie sei physikalisch inhaltslos und wolle aus geistreichen Fiktionen mit Hilfe mathematischer Operationen physikalische Erkenntnisse gewinnen; sie sei charakterisiert durch eine Übertreibung ins Abstrakte und Formale und beschränke sich auf das intellektuelle Spiel mit mathematischen Definitionen und Formeln.

Der Antisemitismus spielt in dieser Schrift keine Rolle, und politische Bemerkungen betreffen nicht die Relativitätstheorie selbst, sondern die Art und Weise, wie sie propagiert wurde. Diese Propaganda, so schreibt Stark, fand in der Zeit der politischen und sozialen Revolution einen fruchtbaren Boden, da sie auch von einem Umsturz redete, nämlich dem unserer bisherigen Anschauungen von Raum und Zeit. Einstein wirft er in diesem Zusammenhang seine internationale Gesinnung vor, die sich in seinen Vortragsreisen ins Ausland manifestiere.[28]

Erst nach 1933 hat Stark seine Vorstellungen vom Dogmatismus der modernen theoretischen Physik weiter präzisiert, am ausführlichsten in einem Vortrag "Jüdische und deutsche Physik" im Jahre 1941.[29] Den Gegensatz zwischen theoretischer und experimenteller Physik bezeichnet Stark wieder als dogmatische und pragmatische Einstellung:

> *Die dogmatische Einstellung sucht die wissenschaftlichen Erkenntnisse aus dem menschlichen Geist herauszuholen. (...) Die pragmatische Einstellung holt ihre Erkenntnisse aus der sorgfältigen Beobachtung und aus zweckmäßig angestellten Experimenten. (...) Die dogmatische Einstellung glaubt, neue Erkenntnisse durch mathematische Operationen am Schreibtisch gewinnen zu können. (...) Die pragmatische Einstellung sucht die Erkenntnis der Wirklichkeit in geduldiger, oft jahrelanger Laboratoriumsarbeit.*[30]

Neu ist, daß dogmatisch für ihn jetzt gleichbedeutend mit jüdisch und pragmatisch soviel wie deutsch ist:

> *Es erscheint notwendig, den jüdisch-dogmatischen Geist und die deutsch-pragmatische Einstellung in der Physik scharf zu kennzeichnen. (...) Die dogmatische Einstellung ist dem jüdischen Geist artgemäß. Denn die Juden sind überwiegend dogmatisch veranlagt; auf sie geht die theolo-*

gische Dogmatik zurück; die Schöpfer und Vertreter der soziologischen
Theorien in der neueren Zeit waren auch überwiegend Juden. Die dog-
matische Theorie, welche in der Physik der neueren Zeit am meisten
propagiert worden ist, stammt von einem Juden.[31]

Auch in dieser Schrift geht es Stark in erster Linie darum, die Quantentheorie als dogmatisch und damit jüdisch zu kennzeichnen; bei dem, was
er zur Relativitätstheorie schreibt, hat man wie schon 1922 den Eindruck, daß ihn mehr ihre Propagierung störte als die Theorie selbst.
"Die weltweite aufdringliche Propaganda für Einsteins Relativitäts-
theorien"[32] ist für ihn ein Beispiel für dogmatische Einstellung, aber
welches eigentlich die Dogmen der Relativitätstheorie sind, erfahren
wir (im Gegensatz zur Quantentheorie) nicht.

Die zahlreichen Schriften, in denen Wilhelm Müller gegen die Relativitätstheorie Stellung nimmt, bringen inhaltlich wenig Neues zu dem, was
bereits von Lenard und Stark gesagt wurde. Seine Veröffentlichungen,
die zum großen Teil in der "Zeitschrift für die gesamte Naturwissenschaft" erschienen sind,[33] sind vor allem in sprachlicher Hinsicht be-
merkenswert: im Vergleich zu der polemischen Diktion Müllers sind die
Ausführungen Lenards und Starks geradezu von vornehmer Zurückhaltung.
Magischer Atheismus, standpunktslose Pseudophysik, Schwindel, talmu-
distische Inflationsphysik, skrupelloseste Wirklichkeitsverfälschung,
großer jüdischer Weltbluff – das sind nur einige der Attribute, mit
denen Müller die Relativitätstheorie belegt. Wie Lenard kritisiert er
die "Abseitigkeit und Verstiegenheit des den gesunden Menschenverstand
verleugnenden theoretischen Formalismus",[34] und von Stark übernimmt er
den Vorwurf, die Physik werde durch die Relativitätstheorie gewissen
Dogmen unterworfen, die obendrein "seinerzeit während der Episode der
deutschen Schmach ausgebrütet und ausgebreitet wurden".[35] Ein solches
Dogma ist für Müller auch "die dogmatisch festgesetzte Grundtatsache der
konstanten Lichtgeschwindigkeit".[36] Natürlich weiß er, daß die Grundlage für dieses angebliche Dogma der Michelson-Versuch ist, doch das
hindert ihn nicht daran, sich auch hier auf den gesunden Menschenverstand zu berufen: "Wenn man sich ganz unbefangen diese Dinge überlegt,
so muß man sagen, daß es viel vernünftiger wäre, zu vermuten, daß der
Michelsonsche Versuch keine unbedingt sichere Beweiskraft besitzt, eine
Theorie von so allgemeiner Gültigkeit aufzubauen, die sofort in sich
zusammenstürzen würde, wenn ein ähnlicher Versuch mit wesentlich ver-
feinerten Methoden einmal die Veränderlichkeit der Lichtgeschwindigkeit
zeigen würde".[37] Müllers Bemühungen, die Relativitätstheorie als typisch jüdisch darzustellen, sind über weite Strecken rein deklamatorisch.
Immer wieder behauptet er, die Theorie habe "eine tiefe Beziehung zum
jüdischen Wesen und Schicksal", sie sei "eine spezifische jüdische

Angelegenheit, eine Formel, hinter der sich der jüdische Wille verbirgt", man finde hier "die ganze Entwurzeltheit und Wesenlosigkeit des Judentums in einer mathematisch verkleideten Form wieder". In der Relativierung von Raum und Zeit sieht er den "steten Drang [des Juden], alle artlichen Verschiedenheiten in der Welt zu Fall zu bringen und seinem wurzellosen Wesen anzugleichen, alle Dinge miteinander zu vertauschen und für einander einzusetzen und zu schicksallosen Elementen eines entseelten Materialismus herabzumindern."[38] Durch die Relativitätstheorie werde "das Weltbild, das die nordische Menschheit geschaffen hat, zersetzt und zerrüttet",[39] und "in der Fassung des relativistischen Weltbildes, in der Raum und Zeit ihre durch Kant gesicherte Stellung eingebüßt haben, liegt eine tiefe und letztgültige Zerstörung der moralischen Selbstherrlichkeit und aller objektiven Maßstäbe, die allein die Freiheit und Zurechnungsfähigkeit des Menschen begründen können".[40] Gekennzeichnet ist das relativistische Weltbild nach Müller ferner durch einen "abstrakten Materialismus, der durch Verkehrung aller durch die Natur vorgegebenen Rangordnung ausgezeichnet ist und immer in irgendeiner versteckten Form dem Marxismus in die Hände arbeitet".[41]

B r u n o   T h ü r i n g s   Schrift "Albert Einsteins Umsturzversuch der Physik" ist erschienen in den "Forschungen zur Judenfrage" des Reichsinstituts für Geschichte des neuen Deutschlands.[42] Laut Verlagsankündigung "ragt diese Untersuchung dadurch hervor, daß in ihr mit den Mitteln der strengen Wissenschaft die rassischen Hintergründe aufgedeckt werden, die das Verfahren Einsteins kennzeichnen". Im Vorwort schreibt Thüring, seine Abhandlung verfolge das Ziel, "die in der Relativitätstheorie lebendige Denkmethode in ihrer Beziehung zum talmudischen Denken klarzulegen"; der größte Teil der 65 Seiten starken Broschüre ist jedoch eine Wiederholung dessen, was auch bei Lenard, Stark und Müller steht: seit der Antike seien "beinahe bis zum Jahre 1900 die exakten Naturwissenschaften das Tätigkeitsfeld einzig und allein arischer Menschen gewesen",[43] und den Zusammenbruch der klassischen Physik hätte nur ein Jude herbeiführen können, "der mit dieser jahrtausende langen Arbeit von Natur aus keinen inneren Kontakt hatte".[44] Viel Raum nehmen die sattsam bekannten Ausführungen ein, die nicht der Relativitätstheorie, sondern der Person Einsteins gelten: er sei Zionist, Pazifist, Internationalist und Bundesgenosse der Marxisten. Im Zusammenhang mit der Relativitätstheorie wiederholt Thüring den von Stark erhobenen Vorwurf des Dogmatismus, und wie Müller sieht er ein solches Dogma im Postulat von der Konstanz der Lichtgeschwindigkeit, ein, so Thüring, willkürliches Urteil, dessen Begründung, der Versuch des "jüdischen Physikers Michelson", kein eindeutiges Ergebnis geliefert habe.[45]

Der Kern von Thürings Untersuchung ist der Versuch, talmudisches Denken und Relativitätstheorie miteinander zu verknüpfen, und hier geht er in der Tat über seine Vorgänger hinaus, die, wie z.B. Müller,[46] einen solchen Zusammenhang zwar behauptet hatten, den Beweis aber schuldig geblieben waren.

Thürings Argumentation ist die folgende. Die jüdische Religion, so sagt er, kennt einmal die Thora, die 5 Bücher Mose, die als absolutes Gotteswort gelten, und außerdem eine Sammlung von mündlich überlieferten Verhaltensvorschriften, den sog. Halachot, "von denen dogmatisch behauptet wird, daß sie in der Thora implizite enthalten seien".[47] Aufgabe des Rabbiners sei es nun, eine Methode zu entwickeln, mit der sich beweisen läßt, daß diese Überlieferungen (Halachot) tatsächlich in der Thora enthalten sind; diese Beweismethode sei der von den jüdischen Schriftgelehrten gemachte Midrasch. Nach dieser Charakterisierung der jüdischen Religion fährt Thüring fort:

> *Es ist nach dem Vorangegangenen nicht allzuschwer, die Identität des talmudischen Denkens mit den Grundlagen und Methoden der relativistischen Physik in seiner ganzen Ausdehnung zu erkennen. Was dort die Thora ist, ist hier in der relativistischen Wissenschaft "die Natur". (...) Daneben bestehen – wie dort die Halachot – hier die Prinzipe und Postulate, die als Verhaltensvorschriften oder naturwissenschaftliche Gebote weiter überliefert und festgehalten werden sollen. Um sie aber in ihrer Richtigkeit zu "beweisen", muß die Verbindung zur Thora, d.h. hier der "Natur" hergestellt und gezeigt werden, daß sie eigentlich schon in der Natur, in jenen Effekten enthalten sind. Es muß also ein Midrasch gemacht werden, d.h. es muß die Denkmethode gesucht werden, die es ermöglicht, formal die Postulate (Halachot) mit irgendwelchen Natureffekten (Thoraworten) zu verbinden.*[48]

"Die Relativitätstheorie", so faßt Thüring seine Erkenntnisse zusammen, "ist lediglich eine Denkmethode, welche ihrer inneren Herkunft und Struktur nach mit dem talmudischen Midrasch identisch ist."[49]

Wie Stark ist übrigens Thüring der Ansicht, daß auch die Quantentheorie dem jüdischen Denken entspricht, daß sie, wie er sagt, "sich dieser talmudischen Methode des formalen Verbindens von Halachot und Thora durch einen Midrasch" bediene. Die Tatsache, daß hier auch eine Reihe von Nichtjuden mit am Werk gewesen sind, zeige nur, "daß sich auch Nichtjuden talmudische Methoden zu eigen machen können."[50]

Die Vorgänge, die ich hier beschrieben habe, zeigen in aller Deutlichkeit, wie schwer es Wissenschaftlern fallen kann, vertraute Denkgewohnheiten aufzugeben, und wie weit sie unter Umständen gehen bei dem Versuch, ein wissenschaftliches Weltbild, das sie einmal akzeptiert haben, zu verteidigen und zu erhalten.

Einstein hat einmal gesagt, daß die Relativitätstheorie wohl ein ganz

anderes Echo gefunden hätte, wäre er nicht Jude von freiheitlicher und
internationaler Gesinnung, sondern Deutschnationaler mit oder ohne
Hakenkreuz.[51] Das trifft sicher zu für die Reaktionen, die sich in der
Presse und in der breiten Öffentlichkeit zeigten, aber für die ablehnenden Reaktionen auf die "neue Physik" aus dem Lager der Physiker war die
Person Einsteins wohl eher von untergeordneter Bedeutung. Mit welch
heiligem Ernst die Physiker jener Zeit bestrebt waren, an ihrem Weltbild festzuhalten, dafür gibt es zahlreiche Beispiele; ich erinnere nur
an die pathetischen Äußerungen Plancks, der im Zusammenhang mit der
atomistischen Interpretation des 2. Hauptsatzes von einem "Akt der Verzweiflung" und einem "Opfer an seinen bisherigen physikalischen Überzeugungen" sprach.[52] Dieses Gefühl, daß man seinen physikalischen
Überzeugungen so wie seinem Glauben oder seiner politischen Haltung
treu bleiben müsse, war auch bei den Gegnern Einsteins vorhanden, und
es war bei ihnen so ausgeprägt, daß sie sich schließlich zu Schritten
bereit fanden, durch die sie im Kreis der Wissenschaftler zu Außen-
seitern wurden. Während bei den meisten der "Akt der Verzweiflung" darin bestand, daß sie das Neue dann doch akzeptierten und ihre alten Vorstellungen aufgaben, sah bei einigen Physikern ein solcher "Akt der
Verzweiflung" am Ende so aus, daß sie im Kampf um die Bewahrung kon-
servativer Positionen in der Wissenschaft Verstärkung im Bereich der
Politik und der Ideologie suchten. Bei der Relativitätstheorie boten
sich natürlich Einsteins Pazifismus, sein Judentum und seine freiheit-
liche Gesinnung geradezu an, um sie in den Kampf mit einzubeziehen,
aber das Beispiel Duhem zeigt, daß notfalls auch der Kriegsgegner
Deutschland dazu dienen konnte, um eine physikalische Theorie zu diskreditieren. Die Tatsache schließlich, daß auch die Quantentheorie,
deren Urheber größtenteils keine Juden waren und politisch eher rechts
standen, mit denselben politisch-ideologischen Mitteln bekämpft wurde
wie die Relativitätstheorie, macht deutlich, daß die Rasse, die Natio-
nalität oder der politische Standpunkt eines Physikers höchstens willkommene Vorwände waren, um seine Theorie zu diskreditieren. Das Ausschlaggebende bei den Gegnern der modernen Physik war die tief in
ihrer Persönlichkeitsstruktur wurzelnde Aversion gegen die neuen Denk-
formen.

Anmerkungen

1. Wilhelm Wien, Die Relativitätstheorie vom Standpunkte der Physik
   und Erkenntnislehre, Leipzig 1921, S. 13.

2.   So z.B. Friedrich Reinhard Lipsius, Wahrheit und Irrtum in der
     Relativitätstheorie, Tübingen 1927, S. 3; Oskar Kraus, Offene
     Briefe an Albert Einstein und Max von Laue, Wien und Leipzig
     1925, S. X.

3.   H. Fricke, Der Fehler in Einsteins Relativitätstheorie, Wolfen-
     büttel 1920.

4.   Johannes Heinsohn, Einstein-Dämmerung. Kritische Betrachtungen zur
     Relativitätstheorie, Leipzig 1933

5.   Vorwort zu der Schrift von Heinsohn (Anm. 4).

6.   Der Krieg der Geister. Eine Auslese deutscher und ausländischer
     Stimmen zum Weltkriege. Hrsg. von Werner Kellermann, Dresden
     1915.

7.   IB. S. 64-69

8.   Vgl. dazu Harry W. Paul, The Sorcerer's Apprentice, The French
     Scientist's Image of German Science 1840-1919, Gainesville
     1972; Brigitte Schroeder-Gudehus, Les scientifiques et la
     paix, Montréal 1978, S. 63 - 97.

9.   Pierre Duhem, La science allemande, Paris 1915, S. 4

10. Albert Dastre, Du rôle restreint de l'Allemagne dans le progrès
     des sciences, in: Gabriel Petit et Maurice Leudet (Hrsg.): Les
     Allemands et la Science, Paris 1916, S. 75 - 97, hier S. 80.

11. Pierre Duhem, Science allemande et vertus allemandes, in: Petit/
     Leudet, a.a.O., S. 137 - 152, hier S. 138.

12. Um die nach Duhem typisch deutschen Merkmale mathematischer und
     physikalischer Theorien geht es in seinem Aufsatz "Quelques
     réflexions sur la science allemande", in: Revue des deux mon-
     des, 25 (1915) S. 657 - 685.

13. "Le Germain est laborieux, (...) le Germain est minutieux, (...)
     le Germain est discipliné, (...) le Germain est soumis."
     Duhem, Science allemande . . . (Anm. 11), S. 140.

14. "a partir de postulats librement formulés", Duhem, Quelques ré-
     flexions ... (Anm. 12), S. 669.

15. Que le principe de relativité déconcerte toutes les intuitions
     du sens commun, ce n'est pas, bien au contraire, pour exciter
     contre lui la méfiance des physiciens allemands. Le recevoir,
     c'est, par le fait même, renverser toutes les doctrines où il
     etait parlé de l'espace, du temps, du mouvement, toutes les
     théories de la Mécanique et de la Physique; une telle dévasta-
     tion n'a rien qui puisse déplaire à la pensée germanique; sur
     le terrain qu'elle aura déblayé des doctrines anciennes, l'es-
     prit géométrique des Allemands s'en donnera à coeur joie de
     reconstruire toute une Physique dont le principe de relativité
     sera le fondement. Si cette Physique nouvelle, dédaigneuse du
     sens commun, heurte tout ce que l'observation et l'expérience
     avaient permis de construire dans le domaine de la Mécanique
     céleste et terrestre, la méthode purement déductive n'en sera
     que plus fière de l'inflexible rigueur avec laquelle elle aura
     suivi jusqu'au bout les conséquences ruineuses de son postu-
     lat. (...) Ainsi s'est faite la Physique de la relativité;

ainsi la science allemande progresse, fière de sa rigidité
algébrique, regardant avec mépris le bon sens que tous les
hommes ont reçu en partage. Ib. S. 681 - 683.

16. Le principe de relativité est la base d'une évolution scientifique
qui ne peut être mieux comparée qu'au futurisme et au cubisme
dans l'art. Tout ce qui peut se réclamer de la tradition ou du
bon sens est inexorablement piétiné. Le seul souci de cette
école est de heurter et détruire. Plus les déductions semblent
absurdes, plus elles lui apparaissent kolossales et supérieures
aux vérités admises, et l'on se trouve bientôt en présence d'un
chaos où fusionnent en hurlant le concret et l'abstrait.

Nous emprunterons un bon exemple de ce délire mathématico-méta-
physique à la théorie des quanta de Max Planck, professeur de
physique à Berlin, un des 93 intellectuels d'outre-Rhin. Planck
a été amené à concevoir que les différentes formes de l'énergie,
la lumière, la chaleur, etc., sont réductibles en parties
aliquotes indivisibles, les quanta. Prenant ces abstractions
pour des réalités, il introduit aussitôt la notion essentielle-
ment concrète d'atome et accorde à la lumière, à la chaleur,
etc., une structure atomique. Il y aurait donc des atomes de
chaleur, de lumière, d'énergie mécanique (!), enfin d'énergie
en général; en vertu du principe de relativité, ces atomes
posséderaient même und masse douée d'inertie (!!). Pierre-
Jean Achalme, La science des civilisés et la science allemande,
Paris 1916, S. 162.

17. Zu den Beziehungen zwischen Lenard und Einstein vgl. Andreas Kleinert und Charlotte Schönbeck, Lenard und Einstein, Ihr Briefwechsel und ihr Verhältnis vor der Nauheimer Diskussion von
1920, in: Gesnerus, 35 (1978) S. 318 - 333.

18. Albert Einstein, Dialog über Einwände gegen die Relativitätstheorie, in: Die Naturwissenschaften, 6 (1918) S. 697 - 702

19. Philipp Lenard, Über Äther und Uräther, 2. Aufl. Leipzig 1922, S. 3

20. Ib. S. 9

21. Ib.

22. Zu Einsteins Zeitungsartikel und Lenards Reaktion vgl. Kleinert/
Schönbeck (Anm. 17), S. 328 ff.

23. Völkischer Beobachter, 13. 5. 1933, Falsch wiedergegeben wird
dieser Artikel bei Philipp Frank, Albert Einstein. Sein Leben
und seine Zeit, München 1949, S. 376; ferner bei Edgar Lüscher:
Experimentalphysik I, Mannheim 1967, S. 47.

24. Zur Geschichte der "Deutschen Physik" vgl. Alan D. Beyerchen,
Scientists under Hitler. Politics and the Physics Community in
the Third Reich, New Haven u. London 1977.

25. Alle Zitate dieses Abschnitts aus Philipp Lenard, Deutsche Physik,
2. Aufl. München u. Berlin 1938, S. IX-X.

26. Lenards erste ablehnende Äußerung gegenüber der Speziellen Relativitätstheorie findet sich in dem Artikel "Über Äther und Uräther", in: Jahrbuch der Radioaktivität und Elektronik, 17
(1920) S. 307 - 356, hier S. 309.

27. Johannes Stark, Die gegenwärtige Krisis in der deutschen Physik, Leipzig 1922, S. 19

28. Ib. S. 14 ff.

29. In: Jüdische und deutsche Physik, Vorträge ... hrsg. von Wilhelm Müller, Leipzig 1941, S. 21 - 56.

30. Ib. S. 22 - 23.

31. Ib. S. 22 - 24.

32. Ib. S. 23.

33. Judentum und Wissenschaft, Leipzig 1936; Jüdischer Geist in der Physik, in: Zeitschrift für die gesamte Naturwissenschaft, 5 (1939) S. 162 - 175; Die Lage der theoretischen Physik an den Universitäten, Ib. 6 (1940) S. 281 - 298; Grundsätzliches zur Eröffnung des Kolloquiums für theoretische Physik an der Universität München, in: Jüdische und deutsche Physik (Anm. 29) S. 5 - 20.

34. Grundsätzliches ..., S. 9.

35. Die Lage der theoretischen Physik ... S. 281.

36. Jüdischer Geist ... S. 162; ebenso in: Grundsätzliches ... S. 13.

37. Ib.

38. Alle Zitate in: Judentum und Wissenschaft, S. 52 - 53.

39. Jüdischer Geist ... S. 166.

40. Judentum und Wissenschaft, S. 52.

41. Jüdischer Geist ... S. 165.

42. Berlin 1941.

43. Thüring, S. 7.

44. Ib. S. 19.

45. Ib. S. 33 - 37.

46. "talmudische Inflationsphysik", in: Die Lage der theoretischen Physik, S. 292.

47. Thüring, S. 38.

48. Ib.

49. Ib. S. 42.

50. Ib. S. 39 - 40.

51. Meine Antwort, Über die anti-relativitätstheoretische GmbH, in: Berliner Tageblatt, 26. August 1920.

52. Armin Hermann, Frühgeschichte der Quantentheorie, Mosbach 1969, S. 31.

ISOLATION UND KOOPERATION DER NATIONALEN SCIENTIFIC COMMUNITIES

Brigitte Schroeder-Gudehus, Inst. d'histoire et de sociopolit. des
                    sciences, Univ. de Montréal, Canada

Gegen Ende der zwanziger Jahre - im September 1928 - schrieb A l b e r t
E i n s t e i n a n H u g o K r ü s s, den Direktor der Staatsbibliothek, daß er sich in der Völkerbundskommission für geistige Zusammenarbeit eigentlich nie recht am Platze gefühlt habe. Krüss, der Mitglied eines Unterausschusses der Kommission war, hatte Einstein schon
öfter auf den Jahressitzungen in Genf vertreten. Einstein hielt sich
einfach nicht geschaffen für die Aufgaben, wie die Kommission sie sich
gestellt hatte. Und er erklärt Krüss, warum er überhaupt, 1922, die
Ernennung zum Mitglied angenommen habe: er sei in die Bresche gesprungen,
weil doch "bei der damaligen Mentalität unserer Geistigen" keiner zu
finden gewesen wäre, der sich mit internationaler Gesinnung hätte kom-
promittieren wollen. Daß er, Einstein, wirklich eine Verbindung hätte
herstellen können, sei ihm nie eingefallen. Dazu sei seine Verwurzelung
in der deutschen geistigen Elite viel zu oberflächlich gewesen.[1]

Aus diesen paar Sätzen schon läßt sich entnehmen, daß die sogenannte
"internationale Gelehrten-Republik" in den zwanziger Jahren in einem
nicht sehr erfreulichen Zustand war. Ein wichtiger Aspekt dieser Situation, der in Einsteins Brief natürlich stillschweigend vorausgesetzt ist,
muß hier ergänzt werden: gewiß wäre es praktisch unmöglich gewesen, in
den maßgebenden Kreisen der deutschen Professorenschaft und unter denen,
die sich sonst als Kulturträger verstanden, jemanden zu finden in den
frühen zwanziger Jahren, der sich mit einem Eintreten für den Völker-
bundsgedanken hätte bemerkbar machen wollen. Andererseits wäre aber eine
derartige Geste auch in einem grossen Teil des Auslands sehr schlecht
angekommen.

1922 - das war eine Zeit, in der der sogenannte "Boycott" der deutschen
Wissenschaft empfindlich wirksam war: internationale wissenschaftliche
Gremien, in denen die Siegermächte vertreten waren (und das waren sie
eigentlich in den allermeisten Fällen), dachten gar nicht daran, deutsche
Gelehrte zu ihren Versammlungen einzuladen, geschweige denn, sie als Mitglieder aufzunehmen. Für die Völkerbundskommission bestätigte das noch
Jahrzehnte später ein anderes der zwölf Mitglieder, der Schweizer Historiker G o n z a g u e d e R e y n o l d. Er schrieb Anfang der
sechziger Jahre in seinen Memoiren, daß "un Allemand authentique", ein
"wirklicher Deutscher", damals - im Frühjahr 1922 - für die Völkerbunds-

länder völlig unannehmbar gewesen wäre.[2]

Selbst mit der Kandidatur Einsteins war es den Verantwortlichen anscheinend nicht ganz geheuer gewesen. P a u l   P a i n l e v é, Mathematiker, Politiker, ab und an französischer Kriegsminister und Regierungs-
kreisen immer nahestehend, hielt es für angebracht, etwaigem Protest der
französischen Öffentlichkeit vorzubeugen, - dem Vorwurf vor allem, der
Völkerbundsrat begehe mit Einsteins Ernennung einen Akt unangebrachter
Versöhnlerei. In französischen Tageszeitungen wies er im Frühjahr 1922
darauf hin, daß es sich bei Einstein eben nicht um einen Deutschen wie
jeden anderen handele: er sei Internationalist, Pazifist, und während
des Krieges *der* Abscheu der Alldeutschen gewesen. Auch der glühendste
französische Patriot sollte also an seiner Kandidatur nichts auszusetzen
haben.[3]

Painlevé irrte sich wohl, wenn er meinte, Kritiker würden etwas an der
Person Einsteins auszusetzen haben (und wenn, dann wäre das aus anderen
Gründen geschehen). In der etwas verblasenen Konzeption der Kommission
als einem Kollegium erlauchter Geister, sollten die zwölf Mitglieder
gleichzeitig die grossen Kulturkreise der Menschheit vertreten, das
heißt, die Gemeinschaft der Kulturvölker, wie man damals noch sagte.[4]
Kritiker stießen sich also vielmehr an der Möglichkeit, daß die Gegenwart eines Vertreters des "deutschen Kulturkreises" - und das sollte
Einstein ja sein - die deutsche Kultur und die deutsche Wissenschaft
wieder irgendwie hoffähig machen könnte, nachdem man sie doch gerade
recht wirkungsvoll in Acht und Bann geschlagen hatte; daß sie den deutschen Professoren wieder den Rücken stärken könnte, wo man doch gerade
dabei war, ihnen für ihre Arroganz eine Lektion zu erteilen, indem man
sie praktisch vom internationalen Wissenschaftsbetrieb ausschloss.

Diese Befürchtungen waren unbegründet. Einstein war nicht der Mann,
durch den die Mehrheit der deutschen Professoren die "deutsche Wissenschaft" oder die "deutsche Kultur", das heißt, sich selbst, in der Welt
angemessen vertreten sah. Es fand sich außerdem schnell heraus, daß die
Bedeutung der Mitgliedschaft Einsteins in der Kommission für geistige
Zusammenarbeit vom Völkerbundssekretariat je nach taktischem Dafürhal-
ten manipuliert wurde: während man bei beunruhigten Anfragen aus den
Entente-Ländern versicherte, Einstein sei ja gar kein Deutscher, sondern
habe einen Schweizer Paß, sei Zionist und überhaupt meistens in Amerika,
wurde Vorwürfen wie dem, daß die Kommission eben ganz wie der Völkerbund bloß ein Klüngel der Siegermächte sei, mit dem Hinweis begegnet,
daß doch die deutsche Wissenschaft besonders glanzvoll durch Albert
Einstein vertreten sei.[5]

In der großen Rede- und Schreibeschlacht, die seit 1914 zwischen den
Intellektuellen, den Hochschulprofessoren vor allem, der kriegführenden
Länder ausgebrochen war und weiter tobte, war die Völkerbundskommission
für internationale geistige Zusammenarbeit allerdings bloß ein Neben-
schauplatz. Hauptkonzentrationspunkt des "Boycotts" und sogenannten "Gegen-Boycotts" war eine andere Organisation, der Internationale Forschungsrat. Dieser Forschungsrat war 1919 in Brüssel unter Ausschluß der Mittelmächte gegründet worden (er wurde später, 1931, zum Internationalen
Rat wissenschaftlicher Vereinigungen - *ICSU* - reformiert).[6]

Dieser - auf den ersten Blick und vor allem aus unserer heutigen Perspektive ziemlich harmlos erscheinende - Dachverband war die treibende
Kraft gegen die Wiederaufnahme der Beziehungen mit deutschen Wissen-
schafts- und Hochschulkreisen und, dementsprechend, deren *bête noire*.
Die Verbitterung war heftig und langlebig.

Denn in der Mitte der zwanziger Jahre hatte die deutsche Außenpolitik
ihrer Nachkriegsisolierung schließlich überwunden, - nicht so die deutschen Akademien und wissenschaftlichen Verbände. Während im Oktober
1925 die Locarno-Verträge unterzeichnet wurden, Deutschland ein Jahr
später Mitglied des Völkerbundes wurde, weigerten sich die Sprecher dieser Akademien und Verbände beharrlich, dem Internationalen Forschungsrat
beizutreten, obwohl dessen Generalversammlung sie im Juni 1926 offiziell
und einstimmig dazu eingeladen hatte.

Warum sträubten sich die Akademien und repräsentativen Körperschaften
der deutschen Wissenschaft, unter ihnen vor allem der Hochschulverband?
Warum widersetzten sie sich so hartnäckig, mit soviel Pathos, dem Druck
des Auswärtigen Amts, dem es so verzweifelt daran lag, seine Thesen von
den friedfertigen Eliten der Republik endlich mit einer versöhnlichen
Geste aus dem Gelehrtenlager illustrieren zu können?

Bevor ich versuche, das Ausmaß von Isolierung und Kooperation im internationalen Wissenschaftsbetrieb der zwanziger Jahre abzuschätzen und
mir über seine Tragweite Gedanken zu machen, möchte ich kurz auf die
Perspektive dieser Untersuchung eingehen, das heißt: erklären, warum
mich das überhaupt interessiert, meine Problematik beschreiben. Universalität und Gemeinschaftlichkeit sind - siehe   R o b e r t   M e r t o n
- soziale Normen der Wissenschaft. Es wird die Frage zu stellen sein:
wie wirkt sich Isolierung, also das Fehlen von solcher Gemeinschaftlichkeit, auf die wissenschaftliche Betätigung, auf die wissenschaftliche
Produktion aus? Dann, nächste Frage: wenn dies Gemeinschaftlichkeit
also offenbar ihre Grenzen hat, wovon werden diese Grenzen bestimmt?
Gibt es Verhaltensnormen, die stärker sind als die der Wissenschaft?

Diese Frage wird im Folgenden natürlich mit Beziehung auf die Normen *politischen* Verhaltens gestellt. Wir sind daran gewöhnt, internationale wissenschaftliche Zusammenarbeit als politischen Faktor, sogar als mögliches Instrument der Außenpolitik zu sehen: als Wegbereiter politischer Zusammenarbeit, als konfliktreduzierendes oder -retardierendes, jedenfalls paralleles Kommunikations-System. Im Leitartikel-Stil gilt längst der Grundsatz: Wissenschaft einigt, Politik trennt. Feiner drückt sich das als Hypothese aus: wissenschaftliche Zusammenarbeit ist ein verständigungsfördernder Faktor, einmal wegen der sich ergebenden Sachzwänge, zum anderen auf Grund von berufsspezifischem Verhalten und berufsspezifischer Gesinnung, beides Produkte wissenschaftlicher Denkweise und wissenschaftlicher Praxis. Weltbürgertum wachse Wissenschaftlern natürlich zu, schreibt J o s e p h N e e d h a m, - "scientists are naturally more world-minded".[7]

Die zwanziger Jahre schienen sich da als ein Bündel von Fallstudien geradezu aufzudrängen.

Im Jahre 1919 konnte der Ausschluß der Akademien der Mittelmächte von den neuen wissenschaftlichen Nachkriegs-Organisationen eigentlich niemanden überraschen: er fügte sich nahtlos an die Drohungen und üblen Versprechungen, mit denen man sich über die Fronten hinweg vier Jahre lang bombardiert hatte. Es wäre aber zu einfach, wollte man dieses Debakel der internationalen Gelehrten-Republik nur aus den Vorfällen des Ersten Weltkrieges heraus interpretieren. Schon E r n s t T r o e l t s c h hatte, noch im Krieg, bemerkt, daß in den Vorkriegsjahren "ja auch nicht ... die Umarmungen so innig gemeint waren".[8] Das waren sie allerdings nicht. Die starke Zunahme von internationalen wissenschaftlichen Vereinigungen und Kongressen in den letzten Jahrzehnten vor 1914, die so gern als Illustration eines triumphierenden wissenschaftlichen Internationalismus' hingestellt wird, war gar nicht wissenschafts-spezifisch. Diese Zunahme war nur Teil eines Gesamtphänomens der Internationalisierung, das sich in der Zunahme derartiger Kongresse und Gesellschaften auf allen Gebieten ausprägte, von den Handlungsreisenden bis zu Vereinen zum Schutz junger Mädchen,[9] und das einen viel machtvolleren Ausdruck fand, zum Beispiel, in der internationalen Schiedsgerichtsbewegung.

Der wissenschaftliche Internationalismus der Vorkriegszeit war ein Internationalismus von Patrioten, das läßt sich auch durch Hinweise auf die Notwendigkeit der Spannung von Zusammenarbeit und Wettbewerb in der Forschung nicht wegdiskutieren. Wenn H e r m a n n D i e l s, Rektor der Berliner Universität, 1906 in einer Festrede erklärte, daß es in diesem Kampf - er meint den Austausch geistiger Güter - weder Sieger

noch Besiegte gäbe,[10] so hat er das sicher aufrichtig gemeint. Jedenfalls gab es Stolze und Besorgte, Triumphierende und Frustrierte.

Auch die internationale Assoziation der Akademien, 1899 gegründet, war
nicht in erster Linie ein Zeugnis und Werkzeug der Internationalität
der Wissenschaft gewesen. Das Hauptziel der Assoziation war nicht, politische Grenzen für die Forschung zu überbrücken, - dazu bedurfte es
keiner internationalen Organisation. Die alte Assoziation war wohl vor
allem zu Koordinierungszwecken geschaffen worden, aber auch, weil sich
die Gründer davon gewisse Finanzierungsaussichten und einen größeren
Einfluß auf Entscheidungen über nationale Beteiligung an internationalen Unternehmungen versprach. In der steigenden Zahl gelehrter Gesellschaften und deren internationalen Kongressen war den Akademien eine
Konkurrenz entstanden, die nicht nur ihren Alleinvertretungsanspruch
der Wissenschaft ihres Landes gefährdete, sondern auch mehr und mehr in
den Genuß öffentlicher Gelder kam. Hier sollte die Gründung der Internationalen Assoziation eine Autorität schaffen, die den Akademien auf
nationaler Ebene mindestens ein Mitspracherecht bei Entscheidungen über
internationale wissenschaftliche Unternehmungen sicherte, vor allem was
deren Repräsentationsanspruch und Finanzierung betraf.[11]

Trotz aller Rhetorik von der wissenschaftlichen Zusammenarbeit als dem
Unterpfand des Friedens, hatten weder diese Vereinigung noch die privaten Verbindungen einzelner Gelehrter bei Kriegsausbruch der vater-
ländischen Pflicht eine höhere Solidarität entgegenzusetzen gehabt. Die
Brücken in der Welt der Wissenschaft brachen so schnell zusammen wie die
der Diplomatie. Im Gegenteil, die Gelehrten aller Lager gerieten in ihrer
Mehrzahl sehr schnell, redend und schreibend, in den Sog der "geistigen
Mobilmachung". Es kann hier nicht einmal andeutungsweise auf die Ströme
von Gift und Galle eingegangen werden, die den "Geistigen", wie Einstein
sie dann später nannte, da aus der Feder flossen. Man schickte Orden
unter Protest zurück, strich sich gegenseitig aus Mitgliederlisten von
Akademien und gelehrten Gesellschaften, bewies die Existenz unüberwind-
licher Gegensätze hinab bis ins Paläolithikum, bezichtigte sich gegenseitig des Mangels an Genie, an Gewissen, aller möglichen Untaten, vom
Plagiat bis zur Kapitulation vor der politischen Macht und zum Eintreten
für Verbrechen: P a u l E h r l i c h habe das Festessen noch
nicht verdaut gehabt, hieß es da, das man ihm zu Ehren in Paris gegeben
habe, als er die deutschen Grausamkeiten in Belgien und Nordfrankreich
gutgeheißen habe..,[12] - eine direkte Anspielung auf den fatalen "Aufruf an die Kulturwelt" vom Oktober 1914, den Ehrlich unterzeichnet hatte,
und R ö n t g e n und P l a n c k, und viele andere.[13]

Natürlich hatte es Ausnahmen gegeben. Einstein war nicht die einzige.
Aber viele waren es nicht. Auf beiden Seiten war auch versucht worden,
irgendeinen Zusammenhang zu retten mit der Erinnerung an freundschaft-
liche, mindestens zivilisierte Beziehungen zu ausländischen Kollegen
in der Vorkriegszeit. Oft geschah das durch die Theorie von den "zwei
England", den "zwei Deutschland", der Existenz einer "Friedenspartei",
deren Anhänger aber schweigen müßten.[14] Oder man nahm einfach an, mit
den früher so wohl zu leidenden, vernünftigen Kollegen sei eine Veränderung vorgegangen, und fand dafür mitunter abenteuerliche Erklärungen
(W i l l i a m   R a m s a y , der englische Chemiker und Nobelpreisträger, schrieb an seinen Kollegen I r a   R e m s e n nach Balti-
more, er habe von zuverlässiger Seite gehört, die Deutschen hätten zu
85 Prozent die Syphilis, und - selbst wenn normale berufliche Tätigkeit aufrechterhalten werden könnte, schiene diese Krankheit häufig zu
seelischen Störungen zu führen..)[15]

Es war wohl allen, Beteiligten und Beobachtern klar, daß mit dem völ-
kerrechtlichen Ende des Krieges nicht automatisch die kollegialen Beziehungen unter den Gelehrten wieder da aufgenommen werden könnten, wo
man sie im Juli 1914 gelassen hatte. In ihrer Mehrheit zerbrachen sie
sich kaum den Kopf darüber, wie die internationalen Beziehungen in der
Wissenschaft sich später im Einzelnen gestalten sollten, - aus Enttäu-
schung oder weil sie ganz andere Sorgen hatten. Gerade in dieser Frage
setzte sich dann aber schnell die ganz präzise Konzeption einer Minderheit durch.

Im Oktober 1918 - der Krieg war noch nicht zuende - versammelten sich
Vertreter der interalliierten Akademien der Wissenschaften in London,
um die Gründung einer neuen, internationalen Wissenschaftsorganisation
vorzubereiten. Einen Monat später, also kurz nach dem Waffenstillstand,
trafen sie sich wieder, in Paris. Im Juli 1919 wurde dann in Brüssel,
in Gegenwart des belgischen Königs, der Internationale Forschungsrat
gegründet.[16]

Im Katalog der Aufgaben der neuen Organisation fanden sich die üblichen
Hinweise auf die Notwendigkeit, Zusammenarbeit zu erleichtern, For-
schungsprojekte zu koordinieren und vor allem jene Verbindung zu fördern
- das war neu -, die sich im Laufe des Krieges als so nützlich heraus-
gestellt hatte: die Zusammenarbeit zwischen Universitäten, Industrie und
Regierung, die sogenannte "kooperative Forschung" (ein Steckenpferd
übrigens des Vertreters der amerikanischen Akademie, G e o r g e
E l l e r y   H a l e).[17] Das Hauptziel der Organisation war politischer Natur und in einer Entschließung der Pariser Akademie schon Ende

September 1918 klar formuliert worden:

> *Die Akademie*, hieß es da, *ist der Meinung, daß persönliche Beziehungen zwischen Gelehrten der kriegführenden Lager unmöglich sind, solange die Mittelmächte nicht wieder in den Kreis der Kulturnationen zurückgekehrt sind, das heißt, solange sie nicht für die Verbrechen gezahlt und gebüßt haben, durch die sie sich außerhalb der Menschheit gestellt haben"*.[18]

Etwas weniger hochtrabend ausgedrückt, besagte das: Deutschland und Österreich, ihre wissenschaftlichen Institutionen und Einzelpersonen, sollten vom internationalen Wissenschaftsbetrieb auf möglichst lange Zeit ausgeschlossen werden.

Wie die alte Internationale Vereinigung (die für erloschen erklärt wurde), war auch der Internationale Forschungsrat eine Förderation von Akademien oder entsprechenden Forschungsorganisation oder, wenn es in einem Land nichts von beidem gab, zuständigen Behörden. Entscheidungsbefugnis war in einem Exekutiv-Komitee konzentriert, in dem - wie im Fünfer-Rat der Friedenskonferenz - die Großmächte vertreten waren: A r t h u r S c h u s t e r von der Royal Society, E m i l e P i c a r d von der Pariser Académie des Sciences, G e o r g e s L e c o i n t e von der Académie Royale de Belgique, V i t o V o l t e r r a von der Akademie der Lincei, und G e o r g e E l l e r y H a l e von der amerikanischen National Academy of Sciences.[19]

Mit den konkurrierenden Fach-Unionen, die der alten Vereinigung der Akademien noch soviel Kopfzerbrechen gemacht hatten, wurden diesmal nicht viele Umstände gemacht: sie waren in die Struktur des Rates mit einbezogen; seine Satzungen galten auch für Fach-Verbände als verbindlich. Der Ausschluß der Akademien der Mittelmächte war ebenfalls in den Satzungen verankert. Satzungsänderungen wurden für die nächsten zwölf Jahre praktisch unmöglich gemacht, indem man sie an Dreiviertel-Mehrheiten band und außerdem eine Stimmengewichtung einführte, die den interalliierten Akademien in jedem Fall eine Blockierungsmöglichkeit gab.

Die Organisatoren hatten sich beeilt und dafür gesorgt, daß die Satzungen fertig waren, bevor die neutralen Akademien billigerweise an der Formulierung hätten beteiligt werden müssen, das heißt, vor Kriegsende. Zwar waren die wichtigsten neutralen Akademien nach Brüssel zur Gründungsversammlung eingeladen worden, - eine Diskussion der Satzung ließen die Veranstalter jedoch nicht mehr zu. Bis Ende des Jahres 1920 waren diese neutralen Akademien dem Internationalen Forschungsrat beigetreten. Damit schien, zur großen Erleichterung der Gründer, eine große Gefahr gebannt, die Gefahr nämlich, die deutschen Akademien könnten die alte Internationale Vereinigung wieder aufleben lassen, die neutralen Akademien zum Beitritt bewegen und damit den Spieß umdrehen: nicht die deutschen, sondern die interalliierten Akademien wären dann in die Isolierung geraten.[20]

In den ersten Nachkriegsjahren setzte der Internationale Forschungsrat denn auch seine Ziele weitgehend durch: beinahe 80 Prozent aller internationalen wissenschaftlichen Kongresse fanden ohne deutsche oder österreichische Beteiligung statt. Internationale Institute, Zentralbüros, wurden aus Deutschland und Österreich abgezogen und anderswo wieder aufgemacht, ohne deutsche Beteiligung. Der 1918 gelegentlich geäußerte Vorschlag, eventuellen "Flaumachern" oder "Überläufern" unter Wissenschaftlern aus den alliierten Ländern durch Aufstellung einer Art "schwarzer Liste" zu begegnen, hatte sich nicht durchgesetzt. Aber es gelang den entscheidenden Gegnern deutscher Beteiligung immer wieder, die Einladungen deutscher Vertreter zu verhindern oder bereits eingeladene wieder ausladen zu lassen, wenn irgendjemand in dieser Hinsicht nachlässig gewesen war.[21]

Nun soll hier kein Mißverständnis entstehen: daß dieser Boycott in den ersten Jahren so wirksam werden konnte, war nicht nur eine Frage - und nicht einmal in erster Linie eine Frage - sorgfältig abgefaßter Satzungen. Daß die Ausschluß-Direktiven des Internationalen Forschungsrats sich erst einmal so erfolgreich durchsetzen konnten, lag vor allem daran, daß sie ganz dem politischen Klima dieser Jahre entsprachen. Deutschlands Verantwortung am Krieg wurde nicht ernsthaft in Frage gestellt. Die deutschen Professoren, als Stand, hatten es fertiggebracht, sich überall unbeliebt zu machen, nicht nur bei denen, die sie ohnehin nie gemocht hatten, sondern auch in neutralen, oft traditionell deutschfreundlichen Ländern. Unbeliebt genug jedenfalls, daß kaum jemand ihretwillen bereit war, einen Zusammenstoß vor allem mit französischen und belgischen Kollegen zu riskieren.

Mit anderen Worten heißt das: der Erfolg des "Boycotts" der deutschen Wissenschaft war nur ein Aspekt der deutschen Isolierung schlechthin. Aus eben diesem Grunde wurde es für den Internationalen Forschungsrat und die in ihm maßgebende Gruppe der Unversöhnlichen zunehmend schwieriger, den Respekt des Ausschluß-Paragraphen durchzusetzen, in dem Maße nämlich, als sich das allgemeine politische Klima beruhigte. Es war außerdem nie ein Geheimnis gewesen, daß die neutralen Akademien dem Rat 1919-1920 von vornherein mit der Absicht beigetreten waren, von innen heraus für eine Normalisierung der Beziehungen zu arbeiten. Emile Picard hatte sich nicht getäuscht, als er im Herbst 1918 warnte, die Neutralen seien die Hintertür, durch die die Deutschen wieder auf die internationale Szene dringen würden.[22] Die starre Unversöhnlichkeit der "alten Garde" und die ausgeklügelten Sperrklauseln machten es aber schwierig, die Aufhebung des Ausschluß-Paragraphen auf demokratischem Wege zu erreichen. Noch in der zweiten Generalversammlung im Juli 1925 wurde ein Antrag auf Aufhebung des Ausschluß-Paragraphen abgelehnt. Durch die 1918 beschlossene Stimmengewichtung war die erforderliche Dreiviertel-Mehrheit nicht zustandegekommen: Frankreich, Belgien, Polen, die Tschechoslowakei, Marokko und Ägypten hatten dagegengestimmt.[23]

Doch dieser Sieg der "alten Garde" war trügerisch: noch in der gleichen Sitzung erklärte H e n d r i k A. L o r e n t z , daß er schwarz sähe für die Zukunft der Organisation, wenn sie sich nicht in kürzester Zeit dazu entschlösse, volle Internationalität herzustellen. Diesmal, im Juli 1925, hatte auch die Delegation der Royal Society mit den Neutralen gestimmt.[24] Obendrein bekamen die Unversöhnlichen mehr und mehr politischen Druck zu spüren: die Regierungen der Locarno-Mächte bekamen es satt, sich mit den Querelen ihrer gelehrten Körperschaften zu befassen, deren ständig an die Öffentlichkeit gebrachte Entrüstung zudem Wasser auf die Mühlen einer nationalistischen Agitation war, die der Entspannungspolitik nicht dienlich war, besonders in Deutschland und Frankreich.

Schließlich, im Sommer 1926, mußte das Exekutiv-Komitee eine außerordentliche Versammlung einberufen, auf der - in einer kurzen Sitzung - einstimmig die Aufhebung des Ausschluß-Paragraphen beschlossen wurde. E r n e s t R u t h e r f o r d , Leiter der englischen Delegation, brachte dann den Antrag ein, die Akademien Deutschlands, Österreichs, Ungarns und Bulgariens zur Mitgliedschaft einzuladen. Die Einladung wurde ebenfalls einstimmig angenommen.[25]

Die Reaktion in Deutschland entsprach den Befürchtungen der Pessimisten: das Kartell der Akademien nahm die Einladung nicht an, lehnte sie aber auch nicht rundheraus ab. Ihre Vertreter stellten Bedingungen: Reform der Statuten des Forschungsrats (Dezentralisierung, ein Sitz für einen deutschen Vertreter im Exekutiv-Komitee, Gleichberechtigung der deutschen Sprache), aber auch eine Ehrenerklärung der bisherigen Mitglieder, daß die 1919 gegen die deutsche Wissenschaft vorgebrachten Anklagen und ihre Ächtung ungerechtfertigt waren.[26] Jahrelang wurde über diese Bedingungen verhandelt. Obwohl das Auswärtige Amt, im Zusammengehen übrigens mit dem Reichsinnenministerium und dem Preußischen Kultusministerium, sich immer wieder bemühte, die Vertreter der großen wissenschaftlichen Körperschaften zum Beitritt zu bewegen, blieben die hartnäckig bei ihrer kaum verschleierten Weigerung. Sie wählten die Isolierung - Warum?

Ich möchte dafür drei Hauptgründe anführen:

1.　　– Zuerst psychologische: die tatsächliche Verbitterung über die
　　　　erlittenen Beleidigungen;

2.　　– zweitens, die Möglichkeit, den Boycott in der Praxis zu unter-
　　　　laufen, das heißt, dem "Bannfluch" des Internationalen For-
　　　　schungsrats zum Trotz, internationale Zusammenarbeit wieder
　　　　aufzunehmen;

3.　　– drittens: innenpolitische Opportunität.

Da gegen die deutschen Akademien gerichtete Ausschlußpolitik wurde von
Erklärungen begleitet, die an Outriertheit denen der Kriegszeit nicht
nachstanden und auf deutscher Seite Reaktionen in der gleichen Tonart
hervorriefen. Was das Klima so nachhaltig vergiftete, war wohl vor allem
die leidenschaftliche Absicht zu kränken, der Vorsatz zu demütigen. Die
Frage des Ausschlußes und der möglichen Wiederaufnahme der deutschen
Gelehrten in die internationalen Wissenschaftsorganisationen wurde von
deren Vertretern mit Vorliebe als eine Frage von Schuld und Sühne hingestellt. Man gab vor, öffentliche Reue-Erklärungen oder andere Beweise
der Läuterung zu erwarten, – eine Forderung, die dann prompt von deutscher Seite retourniert wurde, als der Internationale Forschungsrat
1925 – 26 auf eine versöhnlichere Linie einschwenkte. Die Diskussion
über die Wiederaufnahme internationaler Wissenschaftsbeziehungen orien-
tierte sich, unübersehbar, an der Kriegsschulddebatte.

Hinzu kam, daß die "deutsche Wissenschaft" in diesem Zusammenhang und
unter Mobilisierung historischer Reminiszenzen[27] zum Einsatz verlorener
politischer Macht, zur letzten Bastion verlorener Größe hochstilisiert
wurde, deren Integrität also eifersüchtig gehütet werden mußte. Ein
"Gegen-Boycott" wurde propagiert. Genau wie der Erfolg der Ausschluß-
direktiven von ihrer einheitlichen Befolgung abhing, so ging es auch
beim "Gegen-Boycott", der "Strategie des leeren Sessels", um die Bewahrung einer lückenlosen gemeinsamen Front. In einem Merkblatt von 1923
verlangte denn auch zum Beispiel der Auslandsausschuß des Hochschulverbandes, daß – wenn es sich um Entente-Länder handelte – jeder Hochschullehrer auch *private* Annäherungsversuche sorgfältig zu prüfen habe und in
jedem Fall auf absoluter Gegenseitigkeit bestehen müsse; daß er immer
wieder auf das an seinem Volk begangene Verbrechen hinzuweisen habe und
nie vergessen sollte, daß die deutsche Ehre und Würde nicht bloßen
wissenschaftlichen Interessen, wie Informationen, geopfert werden dürfe.[28]

Man war mit den Mitteln, die gemeinsame Front zusammenzuhalten, nicht
eben kleinlich. Als G e r h a r d v o n S c h u l z e - G a e v e r n i t z ,
der Volkswirtschaftler, 1926 beim Internationalen Institut für geistige

Zusammenarbeit in Paris eingestellt wurde, desavouierte ihn die "deutsche Gelehrtenwelt" öffentlich in den Mitteilungen des Hochschulverbandes.[29] Aber nicht nur die Extremisten sorgten sich um die Bewahrung
einer gemeinsamen Haltung. Auch unermüdliche Vermittler wie   F r i t z
H a b e r   wollten es auf einen Bruch in der deutschen Professorenschaft nicht ankommen lassen. Als gegen 1929 wieder einmal offensichtlich wurde, daß sich das Kartell nicht zum Beitritt würde entschließen
können, gab er auf: die Einigkeit der deutschen Wissenschaft stehe ihm
höher als eine Beschleunigung der Eingliederung Deutschlands in die internationalen wissenschaftlichen Organisationen.[30]

War diese Haltung unsachlich? verstiegen? Wurden hier wissenschaftliche
Normen, vielleicht der wissenschaftliche Fortschritt, politischen,
nationalstaatlichen Anliegen geopfert? Hier erhebt sich nun die Frage,
inwieweit "Boycott" und "Gegenboycott" wissenschaftliche Tätigkeit tatsächlich beeinträchtigt haben. Hatte die Isolierung, die die maßgebende
deutschen Gelehrtenkreise dann nach 1926 aus freien Stücken und mit
einer Art düsteren Stolzes aufrechterhielten, Nachteile für die Wissenschaft in Deutschland?

Diese Frage muß auf verschiedenen Ebenen gestellt und beantwortet werden.

Daß Deutschland vor allem in den frühen zwanziger Jahren vom internationalen Wissenschaftsbetrieb empfindlich abgeschnitten war, läßt sich
nicht bestreiten. Wieviel von dieser Isolierung auf das Konto der Boycott-Direktiven des Internationalen Forschungsrats ging, ist schon sehr
viel schwieriger auszumachen. Politische Unruhe, Devisenmangel, Versor-
gungsschwierigkeiten, der finanzielle Zusammenbruch durch die Inflation,
- diese Fakoren müssen in Rechnung gestellt werden, wenn von der "Not
der deutschen Wissenschaft" die Rede ist. Der Rückgang der Zahl ausländischer Zeitschriften in deutschen Bibliotheken war dramatisch;[31]
er war zum Teil sicher auf das Erlöschen vor dem Krieg abgeschlossener
Austauschabkommen zurückzuführen. Aber daß der Abbruch solcher Austauschabkommen in jedem Fall auf die Boycott-Direktiven zurückzuführen war,
ist mehr als fraglich.

Es ist ebenfalls unbestreitbar, daß der Boycott während der ersten
Nachkriegsjahre deutsche Beteiligung an internationalen Vereinigungen,
Kongressen und anderen öffentlichen Veranstaltungen weitgehend verhindert
hat. Deutsche Beteiligung blieb auch in späteren Jahren eine heikle Angelegenheit, - das geht schon daraus hervor, daß man stets sehr viel
Aufhebens davon machte, wenn eine Veranstaltung mit deutscher Beteiligung reibungslos verlaufen war.

In vielen Fällen waren aber persönliche Beziehungen zwischen einzelnen

Forschern schon früh, jedenfalls diskret, wieder angeknüpft worden. Wäre der Boycott ganz und gar hermetisch gewesen, hätten die Wortführer von Boycott und Gegen-Boycott auch sicher nicht so ununterbrochen vor den Leichtgläubigen, den Flaumachern und den Überläufern gewarnt.

Es kam auch zur Aufnahme von Beziehungen, die über rein persönlich-freundschaftliche Gesten hinausgingen: Spenden kamen aus dem Ausland, aus den Vereinigten Staaten zum Beispiel: Bücher, Zeitschriften, Apparate, Lebensmittel, aber auch Geld, das heißt, Devisen. Seit 1923 erhielt die Notgemeinschaft der deutschen Wissenschaft Beträge von der General Electric, von der Rockefeller-Stiftung; Geld kam auch, für die chemische Forschung, aus Japan.[32] Derartige Hilfsaktionen wurden in der deutschen Öffentlichkeit um so mehr herausgestrichen, als sie aus ehemals feindlichen Ländern kamen. Es handelte sich hier - und das war, vom materiellen Wert der Spenden ganz abgesehen - um Gesten, die sich als Solidaritätsbeweise gegenüber der deutschen Wissenschaft und dementsprechend als eine Schlappe für die Boycott-Politik auslegen ließen.[33]

Die Autorität des Internationalen Forschungsrats wurde nicht nur durch derartige Solidaritätsbeweise gegenüber der deutschen Forschung in Frage gestellt. Er sah sich sehr bald Angriffen aus den Reihen der Wissenschaftler in den eigenen Mitgliedsländern ausgesetzt, die dem Rat seinen Hang zur Hyper-Organisation vorwarfen und ihm - und den in ihm vertretenen nationalen Institutionen - das Recht absprachen, die ganze wissenschaftliche Tätigkeit durch weitreichende Entscheidungen gängeln zu wollen. Daß es sich bei diesen Protesten weniger um Sympathie-Kundgebungen für die ostrazisierten Deutschen als den Ausdruck interner Spannungen in der verschiedenen Gemeinden der Wissenschaft handelte, dürfte den deutschen Beobachtern nicht entgangen sein, - trotzdem wurden derartige Proteste der Öffentlichkeit stets als moralische Unterstützung der deutschen Wissenschaft gegen die Boycott-Bewegung dargeboten.[34]

Die Boycott-Politik stieß auch auf Schwierigkeiten, die in der Natur der Sache liegen. So erwies es sich zum Beispiel als unmöglich, die deutsche Referaten-Literatur von heute auf morgen durch eine gleichwertige Produktion in den alliierten Ländern zu ersetzen oder von den an deutschen Verlegern orientierten Märkten zu verdrängen. Auch die aus Deutschland abgezogenen Zentralbüros ließen sich nicht so rigoros von deutscher Mitarbeit abschneiden, wie es beabsichtigt war. Schon 1922 hatte der internationale astronomische Nachrichtendienst - vor dem Kriege in Kiel - von der Sternwarte in Uccle in Belgien nach Kopenhagen verlegt werden müssen. Der Direktor der Kopenhagener Sternwarte stand den Direktiven des Forschungsrats mehr als gleichgültig gegenüber und beherbergte eine Zweigstelle des Kieler Observatoriums.[35]

Stark resümierend und ohne auf Details einzugehen, ließe sich also feststellen, daß es im Lauf der Zeit mehr und mehr deutsche Wissenschaftlern gelang, sozusagen am Herrschaftsanspruch des Internationalen Forschungsrats vorbei, den Anschluß an die internationale Zusammenarbeit
wiederherzustellen. Nach 1925 - 26 wurde der Ausschluß deutscher Mitgliedschaft oder deutscher Teilnehmer mehr und mehr zur Ausnahme. Auch
die Verbesserung der wirtschaftlichen Lage trug sicher ihren Teil zur
Normalisierung des internationalen Wissenschaftsverkehrs bei: es wurde
mehr veröffentlicht, mehr ausgetauscht, und man reiste wieder. Schließlich konnte der Generalsekretär der Kaiser-Wilhelm-Gesellschaft die
Einrichtung des Harnackhauses auch mit dessen Rolle als internationalem
Treffpunkt rechtfertigen, und überhaupt für die Gesellschaft die repräsentativen Aufgaben eines "Außenministeriums der Wissenschaft" be-
anspruchen.[36] Das war zwar ziemlich überheblich, aber - 1929 - schon
wieder glaubhaft.

Es stellt sich nun die nächste Frage: selbst wenn man von einer fühl-
baren Beeinträchtigung der internationalen Zusammenarbeit nur in der
ersten Hälfte der zwanziger Jahre sprechen kann, - ist diese Beeinträchtigung der deutschen wissenschaftlichen Leistung, ihrer Substanz,
abträglich gewesen? So pauschal läßt sich das natürlich nicht beantworten. Hier müßte fachspezifisch nachgefragt werden. Auf empirisch-quanti-
tative Methoden ist hier wenig Verlaß (Anteil an der Weltproduktion
wissenschaftlicher Artikel, Anteil der Zitate zu deutschen Artikeln,
Anteil an Auslandsstudenten, usw.). Es ist hier gar nicht nötig, auf
die Frage der allgemeinen Zuverlässigkeit dieser Methoden einzugehen.
Sie würden im vorliegenden Fall auf beträchtliche Hindernisse stoßen:
die Periode des Boycotts ist relativ kurz, - nur etwa sechs bis acht
Jahre. Es kommt hinzu, daß sich die Boycott-Maßnahmen als Faktor praktisch kaum isolieren lassen von den Auswirkungen anderer Belastungen,
wie wirtschaftlichen Schwierigkeiten, politischen Spannungen.

Es gibt kaum zeitgenössische Versuche, die Situation ganz sachlich zu
beurteilen. Solange der Boycott dauerte, schlugen die Wellen der Empörung
zu hoch. Später fand man die ganze Sache wohl so peinlich, daß man so
wenig wie möglich daran rühren mochte. Immerhin stellten Mitglieder
der Göttinger Akademie 1927 fest, daß die Maßnahmen des Internationalen
Forschungsrats für ihre jeweiligen Disziplinen nur eine geringe Tragweite gehabt hatten. Rat und Unionen hätten sich auf organisatorische
Maßnahmen beschränkt. Internationale Zusammenarbeit in der Wissenschaft
aber entstehe in der Hauptsache spontan, erklärte einer der befragten

Herren; vor allem die Fragen, die jetzt gerade in den meisten Disziplinen die größte Aufmerksamkeit erregten, bedürften zu ihrer Erforschung garkeiner wie immer gearteten internationalen Infrastruktur.[37]

Sonst ist mit zeitgenössischen Stimmen nicht viel anzufangen. Da wurden
einerseits die "Diktatur des Internationalen Forschungsrats" und die
"brutale Tyrannei der Entente" angeprangert, die die deutsche Wissenschaft zu erdrosseln versuche,[38] andererseits den Brüssler Organisationen jede Bedeutung abgesprochen. W i l h e l m   R ö n t g e n
ging 1920 so weit, die deutsche Wissenschaft zum Ausschluß von internationalen Kongressen geradezu zu beglückwünschen: die deutschen Gelehrten wären dabei sowieso immer ausgebeutet worden von Ausländern,
die sich ihre Ideen angeeignet hätten; und das Bemühen der Veranstalter,
bei den Vortragsprogrammen ein gewisses "internationales Gleichgewicht"
zu bewahren, hätte bloß dazu geführt, die tatsächliche deutsche Über-
legenheit zu verschleiern.[39] Albert Einstein - obwohl ganz am anderen
Ende des politischen Spektrums - fand auch, daß internationale Kongresse
für die wissenschaftliche Tätigkeit nur untergeordnete Bedeutung hätten.
Für ihn lag die wichtigste Konsequenz des Boycotts auf psychologischem
Gebiet: er verpasse den deutschen Professoren eine Lektion, die sie
nötig hätten.[40]

P a u l   F o r m a n   hat schon vor ein paar Jahren darauf hingewiesen, daß die politisch inspirierte, leidenschaftliche Unversöhnlichkeit, diese freiwillige Isolierung, an der die offiziellen Vertreter
der deutschen Wissenschaft festhielten, überhaupt nur möglich waren, weil
inoffiziell und sozusagen auf ebener Erde, durchaus zusammengearbeitet
wurde. Mit anderen Worten: weil *inoffiziell* die Norm der Gemeinschaftlichkeit aufrechterhalten wurde, konnte man sich erlauben, sie *offiziell*
lautstark zu verwerfen.[41] Wenn der Boycott also wirklich nicht hermetisch war, wenn er unterlaufen werden konnte, wenn die internationalen
Betätigungen, auf die er sich besonders auswirkte (wie Gesellschaften
und Kongresse) für wissenschaftliche Arbeit gar nicht einhellig als
wichtig angesehen wurden, dann muß man zu dem Schluß kommen, daß Boycott und Gegen-Boycott im wesentlichen seelischen Bedürfnissen entsprachen.

Dieses Urteil sollte aber wenigstens vorläufig eingeschränkt werden.
Einmal, weil fachspezifische Untersuchungen fehlen; zum anderen, weil
internationale Zusammenarbeit, der Austausch von Wissen und Erfahrung,
über den Informations-Charakter hinaus auch unübersehbaren Symbolwert
haben, und sei es (um hier dem Vorwurf zu entgehen, das sei eben auch
"psychologisch") in der Mechanik beruflicher Erfolgsbestätigung:

Berufungen an ausländische Akademien, Wahl zu Ämtern in internationalen
Organisationen, Einladungen zu Kongressen, usw. Wir wissen noch viel zu
wenig über den Zusammenhang zwischen dem System beruflicher Erfolgsbestätigung, institutioneller Entwicklung und der Entwicklung der Wissenschaften als Disziplinen, um dieses Problem als irrelevant zu betrachten.

Trotzdem bleibt befremdlich, daß die Sprecher der deutschen Gelehrtenwelt auch nach dem Widerruf des Boycotts weiter so penetrant das er-
littene Unrecht beschworen und unentwegt auf Genugtuung bestanden. Mit
dem Wunsch allein, den Internationalen Forschungsrat dadurch zu zerstören, läßt sich das nicht erklären. Der Grund dafür liegt meiner
Meinung nach auf einem Gebiet, das mit der Wissenschaft an sich gar
nichts mehr zu tun hat, aber dennoch - und vielleich eben deshalb - das
Verhältnis zwischen Wissenschaft und Politik besonders gut erhellt:
das Anprangern des Boycotts und die leidenschaftlichen Appelle zur
Wahrung der deutschen Würde erwiesen sich als wertvolle Instrumente anti-
republikanischer Agitation.

In einer Gesellschaft, in der der Professorenstand noch immer großen
Respekt einflößte, fanden die bewegten Schilderungen der erlittenen
Schmach erheblichen Widerhall. Die Tageszeitungen öffneten ihre Spal-
ten, oft das Titelblatt, der Berichterstattung über den Boycott, obwohl
diese Angelegenheit im Grunde nur eine verschwindend kleine Minderheit
betraf. Aber diese Probleme der internationalen Wissenschaftsorganisation gaben anti-republikanischen Kreisen die Möglichkeit, die Regierung
- wenigstens bei den sogenannten "Gebildeten" - zu diskreditieren. Bedurfte es denn anderer Beweise für die Würdelosigkeit dieser Regierung
als den Druck, den sie auf ihre Gelehrten ausübte, um sie in die "Boy-
cott-Organisationen" zu pressen, ungeachtet der infamen Beleidigungen,
die diese den deutschen Männern der Wissenschaft zugefügt hatten?

Die Wortführer der deutschen Wissenschaft widersetzten sich dem Druck
des Auswärtigen Amts nicht, weil sie die Ziele der deutschen Außenpolitik nicht verstanden,[42] sondern weil sie diese Ziele - Ziele der "Er-
füllungspolitik" - mißbilligten. Es ist hier nicht der Ort, die Diskussion auszudehnen auf das Verhältnis der deutschen Universitäten zum
Weimarer Staat. Das ist bekannt. Wichtig ist nur festzustellen, daß
dieses Verhältnis auch hier, in der scheinbar weit abseits liegenden
Frage internationaler Kooperation in der Wissenschaft, ganz wesentlich
bestimmend wirkte.

Auf diesem Hintergrund müssen nicht nur die zahlreichen scharf nationalistischen und die deutsche Regierung gelegentlich kompromittierenden
Kundgebungen deutscher Gelehrter an die Adresse des Auslands verstanden

werden, sondern zum Beispiel auch die etwas überraschende Begeisterung
für enge Beziehungen mit den sowjetrussischen Kollegen. Vor allem die
Berliner Akademie und die Notgemeinschaft hatten relativ schnell Beziehungen mit der russischen Akademie wieder in Gang gebracht - Literatur wurde geschickt, Delegationen wurden entsandt.[43] Die Deutsche Chemische Gesellschaft feierte den Abschluß der Rapallo-Verträge im Frühjahr 1922 in Gegenwart sowjetrussischer Chemiker. Auch die Leningrader
Akademie war nicht zur Mitgliedschaft im Internationalen Forschungsrat
aufgefordert worden, und bis zu einem gewissen Grade wurden diese Beziehungen zu den russischen Kollegen durchaus in der Perspektive einer
Solidarität und Gegen-Boycott-Strategie gesehen.[44] Der Berliner Patho-
loge L u b a r s c h   ging zum Beispiel so weit vorzuschlagen, die
Redaktionen deutscher Fachzeitschriften sollten bei der Beurteilung von
Aufsätzen russischer Kollegen etwas großzügigere Maßstäbe anlegen, um
dadurch der Anziehungskraft entgegenzuwirken, die die Wissenschaft der
westlichen Demokratien womöglich auf sie ausüben könnte.[45]

Auch diese Beziehungen zur sowjetrussischen Gelehrtenschaft müssen in
einen größeren Zusammenhang gestellt werden. Wir wissen, daß schon früh
Kontakte zwischen der Weimarer Republik und Sowjetrußland hergestellt
wurden - militärische, industrielle, wirtschaftliche.[46] Die Rapallo-
Politik schuf die seltene Situation der Übereinstimmung von offizieller
deutscher Außenpolitik und Gelehrtenmeinung. Dennoch war diese Übereinstimmung nur scheinbar: was für die deutsche Regierung ein diploma-
tischer Schachzug war, wurde von den an bevorzugten Beziehungen mit Rußland interessierten Gelehrten in einer Koalitionsperspektive gesehen,
deren anti-westliche Akzente unverkennbar waren. Natürlich konnte man
sich auf Traditionen berufen. Aber die Promoteure des deutsch-sowjeti-
schen wissenschaftlichen Rapprochements gaben den verschiedenen Veranstaltungen in den folgenden Jahren bestimmt auch deshalb soviel Aufmerksamkeit, weil sie sich dadurch um so nachdrücklicher distanzierten
von der politischen Linie des Auswärtigen Amts, dem es vor allem an einer
Wiederaufnahme der Beziehungen mit den internationalen Nachkriegsorgani-
sationen lag.

Gegen die *West-Orientierung* der Regierung (die ja nicht nur die internationale Zusammenarbeit betraf sondern auch ein Votum für das parla-
mentarische System bedeutete) setzten diese Gelehrten eine *Ost-Orientie-
rung*, die sie in ihrer Verblendung wohl weniger an der Gegenwart als
an gewissen Traditionen preußisch-deutscher Außenpolitik banden, - ein
Tauroggen für die deutsche Wissenschaft.

Ich hatte zu Anfang dieses Textes meine Problematik beschrieben und gefragt: Hat die Unterbrechung des internationalen Wissenschaftsverkehrs, die Unterbrechung der "Gemeinschaftlichkeit", wissenschaftliche Tätigkeit beeinträchtigt? Meine Antwort darauf ist hinhaltend: ja und nein.

Die nächste Frage war: setzt sich, im Konfliktfall, die Norm der Gemeinschaftlichkeit der Wissenschaft durch gegen andere Verhaltensnormen, zum Beispiel politische? In der Regel also ganz sicherlich nicht.

Es wird aus dem gerade gesagten ganz offensichtlich, daß es hier garnicht so sehr um die Frage geht, ob die Universalität der Wissenschaft politischen Konflikten eine höhere Solidarität entgegensetzen kann. Hier, in Deutschland in der zweiten Hälfte der zwanziger Jahre ist die Norm der Gemeinschaftlichkeit der Wissenschaft nicht *Gegenstand* einer Debatte, ist wissenschaftliche Zusammenarbeit nicht *Gegenstand eines Normenkonflikts*, hier wurde die Norm wissenschaftlicher Internationalität als Instrument innenpolitischer Agitation manipuliert.

Hat diese Untersuchung zu weit weggeführt von dem, was wir unter *wissenschaftlicher Arbeit* verstehen? War sie zu sehr an die Geschichte von *Organisationen* gebunden? Mir ist durchaus bewußt, daß wissenschaftliche Organisationen nicht stellvertretend für die gesamte wissenschaftliche Zusammenarbeit stehen können, die weit darüber hinausgeht. Aber wenn nicht das Ganze, so sind sie doch Teil - sichtbar, nachprüfbar, und immer noch die besten Anhaltspunkte, um vorsichtige Kommentare zu einem so umfassenden Thema zu geben, wie es uns hier gestellt war.

Anmerkungen

1. Einstein an Krüss, 16. September 1928 (Handakten Krüss, Staatsbibliothek, Völkerbund Coop. Int. Allgem. III).

2. *Mes mémoires III*, Genf, 1963, S. 409.

3. *Le Matin*, 23. März 1922; *Le Petit Parisien*, 1. April 1922.

4. Über die Internationale Kommission für geistige Zusammenarbeit, siehe B. Schroeder-Gudehus, *Les scientifiques et la paix*, Montréal, 1978, S. 175 und *passim* mit weiteren Literaturangaben.

5. *Ibid.*, S. 176, 191 - 195.

6. Gleichzeitig war auf dem Gebiet der Geisteswissenschaften die Internationale Akademische Union gegründet worden. Beiden Organisationen trat Deutschland vor dem Zweiten Weltkrieg nicht mehr bei.

7. Joseph Needham, *The Place of Science and International Scientific Cooperation in Post-War World Organisation*, 1945, S. 3.

8.   *Südd. Monatshefte*, Oktober 1915, S. 131 - 132.

9.   Siehe z.B. die von der Union des Associations Internationales
     veröffentlichten Zusammenstellungen: *Les Congrès internationaux
     de 1681 à 1899*. Brüssel, 1960, 76 S.; *Les 1978 organisations
     internationales fondées depuis le Congrès de Vienne*, Brüssel,
     1957, 204 S.

10. Hermann Diels, *Internationale Aufgaben der Universität*, Berlin,
     1906.

11. Siehe u.a. die Denkschrift Friedrich von Hartels, zitiert bei
     W. His, "Zur Vorgeschichte des Kartells und der Internationalen
     Assoziation der Akademien", *Berichte über die Verh. der Kgl.
     Sächsischen Gesellschaft der Wissenschaften zu Leipzig*, Math.-
     phys. Klasse, <u>54</u>, Sonderheft, 1902, 33 S.

12. E. Gaucher, "La thérapeutique commerciale des Allemands", in
     G. Petit, M. Leudet, *Les Allemands et la Science*, Paris, 1916,
     S. 165.

13. Abgedruckt in Hermann Kellermann, *Der Krieg der Geister*, Weimar,
     Dresden, 1915, S. 64 - 69.

14. Siehe B. Schroeder-Gudehus, *op. cit.*, S. 88.

15. *Ibid.*, S. 88 - 89 (Archiv Ira Remsen, Johns Hopkins University,
     Baltimore).

16. International Research Council. Constitutive Assembly held at
     Brussels, July 18th to July 28th, 1919. *Reports of Proceedings*,
     London, 1919, 286 S. - Gründungsmitglieder waren die Akademien
     (oder entsprechende Körperschaften) folgender Länder: Belgien,
     Brasilien, Vereinigte Staaten von Amerika, England, Australien,
     Canada, Neuseeland, Südafrika, Frankreich, Griechenland, Italien,
     Japan, Polen, Portugal, Rumänien, Serbien.

17. Siehe z.B. G. E. Hale, "The possibilities of cooperation in re-
     search", in R. M. Yerkes, Hrsg., *The New World of Science*,
     Freeport, 1920, S. 393 - 404.

18. République Française, Journal officiel, Chambre des députés,
     Documents, No. 284 du 18 octobre 1918, S. 9095.

19. Siehe, auch für das folgende, B. Schroeder-Gudehus, *op. cit.*,
     S. 101 - 160.

20. Conférences des Académies des sciences inter-alliées (deuxième
     session) tenue à Paris en novembre 1918, compte rendu, Bulletin
     de la classe des sciences de l'Académie royale de Belgique,
     1919, S. 63 - 81, und Sitzungsprotokoll (Archiv Académie des
     Sciences, Paris, S. 113).

21. - wie es anläßlich des Internationalen Geographie-Kongresses in
     Kairo (1925) der Fall war (B. Schroeder-Gudehus, op. cit., S.
     153 - 159).

22. Conférence des Académies des sciences inter-alliées tenue à Paris,
     26 - 29 novembre 1918. Sitzungsprotokoll (Archiv Académie des
     Sciences) der Sitzung vom 27. November, S. 38.

23. International Research Council, Third Assembly, Brussels 1925, *Reports of Proceedings*, London, 1925, S. 12.

24. *Ibid.*, S. 6 - 7, 10.

25. International Research Council, Assemblée générale extraordinaire, le 29 juin 1926, *Procès verbal de la séance*, 1926, S. 1 - 6.

26. Denkschriften 1926 - 27, zitiert in B. Schroeder-Gudehus, *op.cit.*, S. 277 - 278.

27. - wie zum Beispiel die Gründung der Berliner Universität und der Friedrich Wilhelm III. zugeschriebene Ausspruch, der Staat müsse nach Verlust der materiellen auf die geistigen Kräfte zurückgreifen.

28. Auslandsmerkblatt des Deutschen Hochschulverbandes, *Mitteilungen des VDH*, III, 3. Sonderheft, Dezember 1923, S. 3 - 4.

29. *Mitteilungen des VDH*, V (9 - 10), Oktober 1925, S. 164 - 165.

30. Notiz der Kulturabteilung des Auswärtigen Amts vom 26. Juni 1929 (AA KultAbt. VIw42/2/6372).

31. So konnte 1921 die Staatsbibliothek nur 150 der 2300 Abonnements erneuern, die sie vor dem Kriege hatte aufgeben können.

32. Siehe dazu vor allem die Jahresberichte der Notgemeinschaft der Deutschen Wissenschaft.

33. Die *Mitteilungen* des Hochschulverbandes berichteten laufend über derartige Spenden.

34. Offene Briefe englischer Gelehrter, wie z.B. die von G. H. Hardy, *(Manchester Guardian*, 3. Juni 1924), Lowes Dickinson *(Manchester Guardian*, 23. Februar 1924), Gilber Murray *The Times*, 5. März 1924) wurden durch die in der Organisation des "Gegen-Boycotts" aktive Reichszentrale für wissenschaftliche Berichterstattung (Karl Kerkhof) an Tageszeitungen und Zeitschriften versandt und dort auch veröffentlicht.

35. Siehe B. Schroeder-Gudehus, *op. cit.*, S. 145 - 146.

36. F. Glum an F. Haber, 3. Juni 1929 (KWG Archiv, I A 18).

37. "Einige Äußerungen über die wissenschaftliche Betätigung du Conseil international de recherches bzw. der diesem angegliederten Unionen, veranlaßt von der mathematisch-physikalischen Klasse der Gesellschaft der Wissenschaften zu Göttingen", zitiert bei P. Forman, "Scientific internationalism and the German physicists in the Weimar Republic", *Isis*, 64, Juni 1973, S. 178.

38. Siehe z.B. Georg Karo, "Der Krieg der Wissenschaft gegen Deutschland", *Südd. Monatshefte*, Mai 1919, S. 167; Bericht des Auslandsausschusses des Hochschulverbandes, anläßlich des 4. Deutschen Hochschultages, *Mitteilungen des VDH*, V (2), Februar 1925, S. 34.

39. *Unabhängige Nationalkorrespondenz*, v. 16. Mai 1919 (zitiert bei H. Wehberg, *Wider den Aufruf der 93!* Berlin 1920, S. 32 - 34.

40. A. Einstein an H. A. Lorentz im Sommer 1919, zitiert bei P. Forman, "Scientific Internationalism...", loc. cit., S. 177 - 178.

41. *loc. cit.*, S. 179 - 80.

42. Seit dem Jahr 1923 lud das Auswärtige Amt (oft im Zusammengehen mit dem Reichsinnenministerium und dem Preußischen Kultusministerium) Vertreter der wissenschaftlichen Körperschaften (Universitäten, Kaiser-Wilhelm-Gesellschaft, Notgemeinschaft der Deutschen Wissenschaft, Hochschulverband usw.) zu Informationssitzungen ein, auf denen versucht wurde, ihnen die Ziele der deutschen Außenpolitik zu erklären und ihre Unterstützung dafür zu gewinnen (siehe B. Schroeder-Gudehus, *op. cit., passim*).

43. Vergl. Wolfgang Schlicker, *Die Berliner Akademie der Wissenschaften in der Zeit des Imperialismus.*, Teil II, Berlin, 1975, S. 131 - 139, 215 - 227 und *passim*.

44. *Ibid.*, siehe auch U. Kretzschmar, "Die russische Naturforscherwoche in Deutschland, 19. - 25. Juni 1927", *Jahrbuch für Geschichte der UdSSR und der volksdemokratischen Länder Europas*, 9, 1966, S. 102.

45. O. Lubarsch, *Ein bewegtes Gelehrtenleben*, Berlin, 1931, S. 369.

46. Vergl. u.a. A. Pogge v. Strandmann, "Großindustrie und Rapallopolitik", *Historische Zeitschrift*, 222 (2), 1976, S. 265-341.

EINSTEIN UND DEUTSCHLAND

Armin Hermann, Universität Stuttgart

Einsteins Urteil über Deutschland und die Deutschen hat in den fünfziger
Jahren eine politische Rolle gespielt in der weltweiten Diskussion über
die Beteiligung der Bundesrepublik an der sogenanten "Europäischen Ver-
teidigungsgemeinschaft" und später der NATO. Im In- und Ausland ist von
Gegnern der Politik Konrad Adenauers Einstein als Kronzeuge gegen den
preußisch-deutschen Militarismus zitiert worden, ausgesprochen oder un-
ausgesprochen nach dem Motto: "Plus ça change, plus ça reste la même
chose."

Bei dem Festakt in der Kongreßhalle Berlin am 1. März 1979 hat sich Bundespräsident Walter Scheel mit Einsteins Urteil über Deutschland auseinandergesetzt;[1] ganz offenbar sieht er die Vergangenheitsbewältigung als
ein eminent politisches Problem für das heutige Deutschland an.

Auch wenn für eine spätere Generation die deutsche Geschichte in der ersten Hälfte des 20. Jahrhunderts nicht mehr von politischer Relevanz sein
mag, das Thema "Einstein und Deutschland" wird seine Bedeutung behalten:
Ist es nicht Deutschlands wegen, dann Einsteins wegen. Durch Banesh Hoff-
manns Einstein-Biographie[2] ist uns eine Einstein-Karikatur der Washing-
ton Post bekannt geworden. Es handelt sich um einen Blick in das Universum, von irgendeinem imaginären Punkt aus, und man sieht im Hintergrund
das planetarische Staubkorn, das wir Erde nennen. Von diesem unbedeutenden
Himmelskörper halten die Menschen der Zukunft (oder irgendwelche neuen
Spezies) nichts anderes hervorhebenswert als: "Albert Einstein lived
here."

In der Tat: Schon jetzt kann man sagen, daß Einstein einer der ganz wenigen Gestalten unseres Jahrhunderts ist, der die Zeiten überdauern wird.
Was er geleistet hat, was er erlebt hat, wie er geurteil hat über Theorien, über Ideologien, über Menschen, über Völker, das wird noch in Hun-
derten von Jahren registriert werden, und es wird wohl in einem Ausmaß,
von dem wir uns heute noch keinen Begriff machen, das Urteil bestimmen.

Wir kommen damit zum Thema: Einstein und Deutschland. Bekanntlich ist
Einstein in Ulm geboren und in München aufgewachsen. Über diese Jahre
sind Einsteins eigene Erinnerungen die einzige Quelle. Er hat sich nur
sehr sporadisch und kurz geäußert; wir können aber schließen, daß die in
München verbrachten Jahre seine Einstellung zu Deutschland wesentlich geprägt haben:

> *Die Lehrerschaft der Volksschule war liberal und machte keine*
> *konfessionellen Unterschiede. Unter den Gymnasiallehrern waren*
> *einige Antisemiten, hauptsächlich einer, der den Reserveoffizier*
> *herauskehrte. Unter den Kindern war besonders in der Volksschule*
> *der Antisemitismus lebendig ... Tätliche Angriffe und Beschimp-*
> *fungen auf dem Schulweg waren häufig, aber meist nicht gar so*
> *bösartig gemeint. Sie genügten aber, um ein lebhaftiges Gefühl*
> *des Fremdseins schon im Kinde zu festigen.*[3]

1894 zog die Familie nach Mailand. Der 16jährige Einstein blieb in München, um am Luitpold-Gymnasium noch sein Abitur abzulegen. Der scheue
und verträumte junge Mann fand sich aber weder mit den Klassenkameraden
zurecht noch mit den Lehrern:

> *Als ich in der 7. Klasse des Luitpold-Gymnasiums war, [d.h. mit*
> *16 Jahren,] ließ mich der Klassenlehrer kommen und äußerte den*
> *Wunsch, ich möchte die Schule verlassen. Auf meine Bemerkung, daß*
> *ich mir doch nichts hätte zuschulden kommen lassen, antwortete er*
> *nur: 'Ihre bloße Anwesenheit verdirbt mir den Respekt in der*
> *Klasse.'*[4]

Einstein verließ die Schule und verließ München. Die Bedeutung, die dieser Schritt für ihn hatte, kommt in seinem Antrag zum Ausdruck, aus der
deutschen Staatsangehörigkeit entlassen zu werden. Er ging in die Schweiz,
machte sein Abitur in Aarau, studierte an der späteren Eidgenössischen
Technischen Hochschule und erwarb das Bürgerrecht von Zürich und damit
die Schweizer Staatsangehörigkeit. Einen Arbeitsplatz fand Einstein 1902
am "Eidgenössischen Amt für geistiges Eigentum". So wirkte er, als er
1905 seine ersten großen Arbeiten veröffentlichte, am Patentamt in Bern
als technischer Vorprüfer.

In der scientific community der Physiker war Einstein ganz unbekannt. Man
muß staunen, daß so ungewöhnliche Arbeiten wie die Spezielle Relativitätstheorie und die Lichtquantenhypothese überhaupt veröffentlicht wurden. In Deutschland ist Einstein viel Übles geschehen. Es darf aber auch
nicht vergessen werden, daß es vor allem deutsche Gelehrte waren, die ihn
zur Anerkennung gebracht haben.

Als erster war Paul Drude mit den Abhandlungen Einsteins befaßt. Er war
damals der Herausgeber der führenden deutschen Fachzeitschrift "Annalen
der Physik". Kollegen rühmten sein "gesteigertes Verantwortlichkeitsge-
fühl". Als abschreckendes Beispiel stand ihm Hans Christian Poggendorff
vor Augen, der 1841 als zuständiger Annalen-Redakteur den ersten Aufsatz
von Julius Robert Mayer mit den Grundgedanken zum Energieprinzip als
"unphysikalische Spekulation" nicht veröffentlicht hatte.

Mit auf dem Titelblatt der Zeitschrift stand als "Mitwirkender" bei der
Redaktion Max Planck. Er erhielt aber nur gelegentlich Manuskripte zur
Begutachtung. Ob Planck die Arbeiten Einsteins *vor* dem Druck gesehen hat,

ist deshalb ungewiß. Sicher ist, daß er sich vor allen anderen Kollegen gründlich mit Einsteins Arbeiten auseinandersetzte.

Plancks Interesse zog rasch weitere Kreise. So war die Spezielle Relativitätstheorie bereits 1908 im Kreise der führenden deutschen theoretischen Physiker und Mathematiker anerkannt. Daraufhin wurden auch die Quantenarbeiten Einsteins ernster genommen. Schon 1909 erhielt Einstein die Aufforderung, vermutlich auf Initiative Plancks, vor dem Forum der "Deutschen Naturforscher und Ärzte" einen großen Überblicksvortrag zu geben.

Von 1910 an beschäftigte sich Walther Nernst mit der Vorbereitung einer internationalen "Quantenkonferenz". Der durch die neuen Ergebnisse geschaffene "Zustand der Theorie ist ein lückenhafter", schrieb Planck, und die Not gebiete, "sich zusammenzutun und gemeinsam auf Abhilfe zu sinnen."[5] Ähnlich wie heute Staatsmänner in einer politischen und wirtschaftlichen Krise versammelten sich damals eine Woche lang die führenden Physiker, um im kleinen Kreis die nötigen Reformen zu diskutieren.

Etwa seit der gleichen Zeit liefen Bemühungen, Einstein nach Berlin zu berufen. In der Staatsbibliothek Preußischer Kulturbesitz findet sich eine Briefkarte, "Nernst an Collegen" mit dem Datum vom 31. Juli 1910:

> *Wegen Einstein habe ich mich erkundigt, aber noch keine Nachricht erhalten. Übrigens ist die betreffende Angelegenheit wohl unabhängig davon, ob E. jetzt etwas besser gestellt ist.*[6]

Im Frühsommer 1913 fuhren Planck und Nernst nach Zürich, um die entscheidende Unterredung zu führen. Sie brachten einen Vorschlag mit, der dem jungen Einstein die höchsten Ehren verhieß. Von heute aus betrachtet stellt dieser Vorschlag der Weitsicht der Preußischen Unterrichtsverwaltung und der Berliner Physiker ein glänzendes Zeugnis aus.

Mit 35 Jahren erhielt Einstein als ordentliches hauptamtliches Mitglied der Preußischen Akademie der Wissenschaften und zusätzlich als Direktor des "Kaiser-Wilhelm-Instituts für physikalische Forschung" die beste Stellung, die damals ein Gelehrter im Deutschen Reiche haben konnte.

Erst in den dreißiger Jahren trat dieses "KWI für Physik" in die Wirklichkeit mit einem eigenen Gebäude. Peter Debye und Werner Heisenberg waren die ersten Institutschefs. Das Institut hatte aber schon vom Zeitpunkt der juristischen Gründung 1917 Bedeutung. Der umfangreiche Briefwechsel des Instituts mit Physikern aus ganz Deutschland, der heute in der neuen Berliner Einrichtung "Archiv und Bibliothek der Max-Planck-Gesellschaft" verwahrt wird, zeigt: Einstein und sein späterer Stellvertreter Max von Laue haben Anträge von Kollegen entgegengenommen, geprüft und ggf. aus ihrem Institutsetat finanziell gefördert. Dazu gehörten u.a.

die wichtigen Strukturuntersuchungen von Peter Debye und die noch wichtigeren Quantenarbeiten von Max Born.

Man geht in der Annahme sicher nicht fehl, daß dieses Modell Friedrich Schmidt-Ott 1920 mit zur Gründung der Notgemeinschaft veranlaßt hat, der heutigen Deutschen Forschungsgemeinschaft. Hier wird im großen Stile das gleiche getan, was damals, im kleinen und auf die Physik beschränkt, Einstein und Laue geleistet haben.

An der Universität, der Technischen Hochschule, der Physikalisch-Technischen-Reichsanstalt, den Kaiser-Wilhelm-Instituten und den Forschungsinstituten der Industrie wirkte damals eine große Zahl von Physikern, wie es sie sonst an keinem anderen Ort der Welt gab. Immer wieder führten die Sitzungen der Deutschen Physikalischen Gesellschaft und das Physikalische Kolloquium die Gelehrten zusammen. Die Zusammenarbeit war kollegial und freundschaftlich. "Alle Beteiligten haben sich so gut bei der Sache verhalten", schrieb Einstein nicht nur einmal: "Insbesondere kann ich Planck nicht genug rühmen."

Einstein wirkte in der Deutschen Physikalischen Gesellschaft; jahrelang saß er im Vorstand, zeitweise amtierte er als 1. Vorsitzender. In dieser Eigenschaft bereitete er die Feier zum 60. Geburtstag von Planck 1918 vor. An Sommerfeld schrieb er:

> *Ich freue mich schon heute auf den Abend, weil ich Planck sehr*
> *lieb habe und er sich sicher freuen wird, wenn er sieht, wie*
> *gern wir ihn alle haben und wie alle seine Lebensarbeit hoch-*
> *halten.*[7]

Einmal hatte der Berliner Vorstand einen Beschluß gefaßt, der die Gründung einer neuen Zeitschrift betraf, der "Zeitschrift für Physik". Auf Kritik aus Würzburg und München schrieb Einstein in seiner humorvollen Art:

> *Von der Ferne sieht alles schief und suspekt aus, besonders wenn*
> *es von den verflixten Berlinern kommt. Und doch sind wir (beinah)*
> *alle sanft wie Lämmer und verschüchtert durch unser böses Renommee.*
> *Säße ich woanders, so würde ich natürlich auch gegen die gewalt-*
> *tätigen Berliner losziehen.*[8]

Über den Ersten Weltkrieg war Einstein tief unglücklich - schon zu einer Zeit, als die Kollegen sich noch im Rausch des Nationalgefühls befanden. Für Einstein war der Krieg nichts anderes als die "Intrigen armseliger Menschen".[9]

> *Die internationale Katastrophe lastet sehr auf mir internationalem*
> *Menschen,*

sagte er zu Paul Ehrenfest:

Man begreift schwer beim Erleben dieser 'großen Zeit', daß man
dieser verrückten, verkommenen Spezies angehört, die sich Wil-
lensfreiheit zuschreibt. Wenn es doch irgendwo eine Insel für
die Wohlwollenden und Besonnenen gäbe, da wollte ich auch glü-
hender Patriot sein. [10]

Nach dem Zusammenbruch des Kaiserreiches verfolgte er die politische Entwicklung mit großem Optimismus - wieder im Gegensatz zur Mehrheit der
Kollegen. Da Arnold Sommerfeld die Auffassung Einsteins nicht verstehen
konnte, schrieb er Ende 1918 verwundert an den verehrten Freund:

Ich höre von Kossel, daß Sie an die neue Zeit glauben und an ihr
mitarbeiten wollen!

Einstein erwiderte:

Es ist wahr, daß ich von dieser Zeit mir was erhoffe, trotz der
vielen häßlichen Dinge, die sie im einzelnen bringt. Ich sehe die
politische und wirtschaftliche Organisation unseres Planeten vor-
schreiten ... Auch die mir so widerwärtige Militärwirtschaft wird
so ziemlich verschwinden. Wenn nun die Übergangszeit gerade für uns
ziemlich drückend wird, so ist es nach meiner Meinung - offen sei's
gesagt - nicht ganz unverdient. Ich bin aber der festen Überzeugung,
daß kulturliebende Deutsche auf ihr Vaterland bald wieder so stolz
sein dürfen wie je - mit mehr Grund als vor 1914. [11]

Nach dem Ersten Weltkrieg spielte bekanntlich die Frage der Kriegsschuld
eine große Rolle. Auch Sommerfeld ging 1921 in den "Münchner Neuesten
Nachrichten" auf das vieldiskutierte Thema ein und geißelte die alliierte
Kriegspropaganda gegen Deutschland. Als er Einstein einen Aufsatz mit der
Bitte übersandte, Einstein möge seinen Einfluß geltend machen, daß auch
in einer englischen Zeitschrift ein Abdruck zustande käme, antwortete
Einstein:

Nun zum Zeitungsartikel. Ich bedaure es offen gesagt, daß Sie ihn
geschrieben haben. Dies ist überhaupt nur infolge der durch den
Krieg geschaffenen Isolierung möglich. Kein gebildeter Engländer
glaubt an die Kriegsmärchen oder legt Wert auf solche kleinliche
Dinge. Ich habe bei meinem Aufenthalt in England die Erfahrung ge-
macht, daß die dortigen Gelehrten meist vorurteilsfreier und objek-
tiver sind als unsere deutschen ... Wenn Sie dort gewesen wären,
würden Sie sicher auch fühlen, daß es nicht am Platze ist, den
Leuten dort mit solchen Lappalien zu kommen. Wie das große Publi-
kum denkt, weiß ich nicht. Aber bei uns ist auch ungeheuer gelogen
worden ohne Dementi, und es wäre gewiß wenig fruchtbar, wenn die
ganze während des Krieges angesammelte schmutzige Wäsche nun mit
vereinten Kräften ans Tageslicht gezogen würde. Jedenfalls kann
ich meine Hand dazu nicht bieten und bitte Sie im Interesse des
wiederherzustellenden guten internationalen Einvernehmens, diese
unfruchtbare Sache liegen zu lassen. In Amerika und England habe
ich überall ehrlichen Verständigungswillen, Hochachtung für die
geistigen Arbeiter Deutschlands ... gefunden. Also weg mit dem
alten Groll. Man kann es, ohne sich das Geringste zu vergeben! [12]

Wie Sommerfeld konstatiert hatte, glaubte also Einstein nicht nur an die
neue Zeit, sondern wollte "an ihr mitarbeiten". Der in Kreisen des Bür-

gertums weit verbreitete Antisemitismus, den er auch persönlich zu spüren
bekam, und die herrschenden antidemokratischen und monarchisch-konservativen Auffassungen stimmten Einstein aber zunehmend skeptischer. Aus
Rostock berichtete er einmal:

> *Ich war einige Tage in Rostock bei Gelegenheit der Jubiläums-*
> *feier der Universität, hörte dort bei diesem Anlaß arge poli-*
> *tische Hetzreden und sah recht Ergötzliches in Kleinstaat-*
> *Politik ... Als Festsaal stand nur das Theater zur Verfügung,*
> *wodurch der Feier etwas Komödienhaftes gegeben wurde. Reizend*
> *war da zu sehen, wie in zwei Proszeniumslogen untereinander*
> *die Männer der alten und der neuen Regierung saßen. Natürlich*
> *wurde die neue von den akademischen Größen mit Nadelstichen*
> *aller erdenklichen Art traktiert, dem Ex-Großherzog eine nicht*
> *endenwollende Ovation dargebracht. Gegen die angestammte*
> *Knechts-Seele hilft keine Revolution!*[13]

Nach dem Ende des Ersten Weltkrieges brach eine ungeheure Publicity über
Einstein herein. Jeder wollte wissen, was diese geheimnisvolle "Relativität" eigentlich bedeutet.

> *Mir selbst war es stets unverständlich,*

hat Einstein dazu gesagt,

> *warum die Relativitätstheorie mit ihren dem praktischen Leben*
> *so entfernten Begriffen und Problemstellungen in den breitesten*
> *Schichten der Bevölkerung eine so lebhafte, ja leidenschaftliche*
> *Resonanz gefunden hat ... Was konnte hier [bei der Relativitäts-*
> *theorie] die große und nachhaltige psychische Wirkung veranlassen?*
> *Ich habe bisher keine wirklich überzeugende Antwort auf diese Fra-*
> *ge zu hören bekommen.*[14]

Soweit Einstein selbst. Ich will das Phänomen zu erklären versuchen. Einstein hatte vor dem Ersten Weltkrieg zunächst bei den Physikern, dann
auch bei den Philosophen für genügend Aufsehen gesorgt. Die von ihm eingeleitete wissenschaftliche Revolution erregte die Experten.

Dann kam der Erste Weltkrieg und die militärische Niederlage von 1918.
Die Menschen in Deutschland waren im tiefsten erschüttert. Denn fast alle
empfanden sie national. Am meisten verbitterte den deutschen Bürger, daß
Deutschlands Ansehen von den Alliierten wirklich oder vermeintlich in den
Schmutz gezogen wurde. Als deshalb britische Gelehrte die Allgemeine Relativitätstheorie bestätigten, wurde das als nationale Genugtuung empfunden.

Dies änderte sich auf der extrem rechten Seite des politischen Spektrums
als bekannt wurde, daß Einstein Jude war und nicht national, sondern international dachte. Für den Umschwung auf der rechten Seite aber gibt es
noch einen weiteren Grund: In den ersten Jahren der Weimarer Republik
entfalteten sich politische und künstlerische Aktivitäten, die Konservative als schlimme Verfallserscheinungen betrachteten.

In Berlin bildete sich ein Verein gegen die Relativitätstheorie, der sich polemisch "Arbeitsgemeinschaft deutscher Naturforscher zur Erhaltung reiner Wissenschaft" nannte. Ihr Anführer bezeichnete die Relativitätstheorie als eine "Massensuggestion", Produkt einer geistig verwirrten Zeit, wie sie anderes Abstoßende schon die Menge hervorgebracht habe. So steigerte sich der Demagoge bis zu dem Satz: Die Relativitätstheorie ist wissenschaftlicher Dadaismus.

Damit war die Verbindung hergestellt zwischen "entarteter Wissenschaft" und dem, was später einmal, während des Dritten Reiches, "entartete Kunst" heißen sollte. So wurde also das sogenannte "gesunde Volksempfinden" mobilisiert. Diese Taktik wurde später vom Dritten Reich zur Meisterschaft entwickelt.

All diese Aktivitäten provozierten Gegenkundgebungen, und so kam es, daß Einstein und seine Relativitätstheorie für Jahre ein Thema blieb, das die Öffentlichkeit erregte.

Am 27. August 1920 meldete das Berliner Tageblatt, daß die "Arbeitsgemeinschaft deutscher Naturforscher" offenbar bereits ihr Hauptziel erreicht habe:

> *Albert Einstein, angewidert von den alldeutschen Anrempelungen und den pseudowissenschaftlichen Methoden seiner Gegner, will der Reichshauptstadt [und Deutschland] den Rücken kehren. So also steht es im Jahre 1920 um die geistige Kultur Berlins! Ein deutscher Gelehrter von Weltruf, ... dessen Werk über die Relativitätstheorie als eines der ersten deutschen Bücher nach dem Kriege in englischer Sprache erscheint: ein solcher Mann wird aus der Stadt, die sich für das Zentrum deutscher Geistesbildung hält, herausgeekelt. Eine Schande!*[15]

Max von Laue, Heinrich Rubens und Walther Nernst sandten eine sehr deutliche Verlautbarung an die Berliner Zeitungen, in der sie sich mit Einstein solidarisierten. Ebenso gab der preußische Kultusminister Konrad Haenisch eine energische öffentliche Erklärung ab.

Max von Laue berichtete auch nach München an Arnold Sommerfeld, der damals als Vorsitzender der Deutschen Physikalischen Gesellschaft amtierte.

> *Wenn Einstein Deutschland tatsächlich verlassen würde,*

so schrieb Laue,

> *erlebten wir zu allem sonstigen Unglück also auch noch, daß national sein wollende Kreise einen Mann vertreiben, auf den Deutschland so stolz sein konnte, wie nur auf ganz wenige. Man kommt sich manchmal vor, als lebte man in einem Tollhaus.*[16]

Sofort wandte sich Sommerfeld in einem Brief an Einstein:

> *Mit wahrer Wut habe ich, als Mensch und als Vorsitzender der Physikalischen Gesellschaft, die Berliner Hetze gegen Sie*

verfolgt.[17]

Für die Physiker brach eine Welt zusammen.

Ich war so überzeugt gewesen,

so schilderte es Werner Heisenberg in seiner Autobiographie,

> *daß wenigstens die Wissenschaft vom Streit der politischen Meinungen vollständig ferngehalten werden könnte. Nun sah ich, daß selbst das wissenschaftliche Leben durch böse politische Leidenschaften infiziert und entstellt werden kann.* [18]

Dieses politische Schlüsselerlebnis hatte Heisenberg 1922 bei der Versammlung der Deutschen Naturforscher und Ärzte in Leipzig. Bei der Tagung verteilten die Assistenten Philipp Lenards einen roten Handzettel, der die Unterschrift ihres Meisters trug. Die Gründung der "Arbeitsgemeinschaft" hatten Lenard und andere veranlaßt, nun auch öffentlich aufzutreten.

Die Nobelpreisträger Philipp Lenard und Johannes Stark wurden die Exponenten der später sogenannten "Deutschen Physik". Wenn auch diese kleine Gruppe in der Wissenschaft ohne Einfluß blieb, so hat doch die ständige Agitation Einstein das Leben in Berlin verbittert. Mehrfach hat Einstein deshalb erwogen, Deutschland zu verlassen. Wie tief verwundet er tatsächlich war, haben seine Freunde erst verstanden, als es schon zu spät war.

Ich möchte jetzt doch daran erinnern,

schrieb Einstein 1933 an Max Planck, nachdem er seine Stellung in Berlin niedergelegt hatte,

> *daß ich Deutschlands Ansehen in all diesen Jahren nur genützt habe und daß ich mich niemals daran gekehrt habe, daß in der Rechtspresse systematisch gegen mich gehetzt wurde, ohne daß es jemand für der Mühe wert gehalten hat, für mich einzutreten.* [19]

Das war wohl nicht ganz gerecht, aber sehr verständlich in der Bitternis der Erlebnisse von 1933. Einsteins stärkste inner Bindung wurde nun die zum jüdischen Volk.

Bei der Härte des jüdischen Schicksals,

schrieb Einstein 1938 an Wolfgang Pauli,

ist meine Bereitschaft zu helfen eine unbedingte. [20]

Seine größte Sorge war nun die Machthybris der Nationalsozialisten. Einstein war ein überzeugter Pazifist, aber er wußte, daß dem aggressiven Dritten Reich gegenüber nichts anderes am Platze war als Festigkeit und militärische Stärke. Von der Furcht getrieben, Hitler könnte die Atomenergie zu einer politischen Erpressung größten Ausmaßes nutzen, übermittelte er kurz vor Ausbruch des Zeiten Weltkrieges dem amerikanischen Präsidenten eine Warnung. Einsteins berühmter Brief vom 2. August 1939 gab mit den Anstoß zum Bau der amerikanischen Atombombe.

Als er später erfuhr, daß es in Deutschland gar keine ernsthaften Versuche gegeben hat, zu einer Atombombe zu kommen, hat Einstein zutiefst bedauert, daß er diesen Brief unterzeichnet hat. Nach Kriegsende sah er es als seine Aufgabe, die Menschen über die Gefahren des Wettrüstens aufzuklären, und leidenschaftlich plädierte er für eine Verständigung zwischen den Großmächten. Auch in anderen Fällen setzte er seinen Namen ein, wenn es galt, eine in seinen Augen gefährliche Entwicklung zu verhindern. So hat er scharf auf die Pläne zur Wiederbewaffnung der Bundesrepublik ragiert.

Die Erfahrungen als Schüler in München, die Hetzkampagne gegen ihn in den zwanziger Jahren und die Judenverfolgung im Dritten Reich prägten sein Deutschland-Bild.

> *Du weißt,*

schrieb er an Max Born,

> *daß ich nie besonders günstig über die Deutschen dachte (in moralischer und politischer Beziehung). Ich muß aber gestehen, daß sie mich doch einigermaßen überrascht haben durch den Grad ihrer Brutalität und - Feigheit.*[21]

Von seinem Pauschalurteil ausgenommen hat Einstein nur einige wenige alte Kollegen. Von Max von Laue pflegte er zu sagen, daß er "nicht nur ein Kopf, sondern auch ein Kerl sei." Seine Bewunderung für Max Planck, der so ganz anders veranlagt war als er, änderte sich nicht. Weiterhin herzlich gesinnt blieb er auch den Freunden Otto Hahn und Arnold Sommerfeld, die für ihn zu "den paar Einzelnen" zählten[22], "die in dem Bereiche der Möglichkeit standhaft geblieben" waren.

Offiziell aber wollte er nichts mehr mit Deutschen zu tun haben, und er lehnte es ab, seine Mitgliedschaft in den deutschen Akademien zu erneuern.

> *Von Max von Laue, Rudolf Ladenburg und anderen Kollegen werden Sie gehört haben,*

schrieb Otto Hahn als Präsident der neuen Max-Planck-Gesellschaft am 18. Dezember 1948 an den alten, nun so weit entfernten Kollegen nach Princeton,

> *daß wir hier in Göttingen im Februar 1948 die Max-Planck-Gesellschaft zur Förderung der Wissenschaften ... gegründet haben. Die Max-Planck-Gesellschaft soll an die Tradition der Kaiser-Wilhelm-Gesellschaft anknüpfen ... Auf meine Bitte sind James Franck, Otto Meyerhof, Rudolf Ladenburg, Richard Goldschmidt und andere ... nunmehr als Auswärtige Wissenschaftliche Mit-*

> *glieder der neuen Max-Planck-Gesellschaft beigetreten. Ich*
> *möchte Sie fragen, ob auch Sie sich zu demselben Schritt*
> *entschließen können. Dem Senat unserer Gesellschaft und mir*
> *selbst wäre dies natürlich eine große Freude und zugleich*
> *auch Ehre ...*[23]

Otto Hahn ahnte wohl nicht, daß Einstein, der heitere und humorvolle Freund von früher, zwischen sich und Deutschland einen endgültigen Trennungsstrich gezogen hatte:

> *Nachdem die Deutschen meine jüdischen Brüder in Europa hinge-*[24]
> *mordet haben, will ich nichts mehr mit Deutschen zu tun haben.*

Die Antwort Einsteins an Otto Hahn ist ein Dokument. Für die Deutschen ein betrübliches.

> *Ich empfinde es schmerzlich,*

schrieb Einstein,

> *daß ich gerade Ihnen, das heißt einem der wenigen, die aufrecht*
> *geblieben sind und ihr Bestes taten während dieser bösen Jahre,*
> *eine Absage senden muß. Aber es geht nicht anders. Die Verbrechen*
> *der Deutschen sind wirklich das Abscheulichste, was die Geschichte*
> *der sogenannten zivilisierten Nationen aufzuweisen hat. Die Hal-*
> *tung der deutschen Intellektuellen - als Klasse betrachtet - war*
> *nicht besser als die des Pöbels. Nicht einmal Reue und ein ehrli-*
> *cher Wille zeigt sich, das Wenige wieder gutzumachen, was nach*
> *dem riesenhaften Morden noch gutzumachen wäre. Unter diesen Um-*
> *ständen fühle ich eine unwiderstehliche Aversion dagegen, an*
> *irgendeiner Sache beteiligt zu sein, die ein Stück des deutschen*
> *öffentlichen Lebens verkörpert, einfach aus Reinlichkeitsbedürf-*
> *nis. Sie werden es schon verstehen und wissen, daß dies nichts*
> *zu tun hat mit den Beziehungen zwischen uns beiden, die für mich*
> *stets erfreulich gewesen sind. Ich sende Ihnen meine herzlich-*
> *sten Grüße und Wünsche für fruchtbare und frohe Arbeit.*[25]

Es kam Einstein darauf an, klar und unmißverständlich seine Absage zu formulieren. Trotzdem ist der Brief nicht ohne Wärme. Einstein war ein ehrlicher Mensch, der die herzliche Freundschaft von einst nicht vergessen hatte.

In der Sache war die Antwort niederschmetternd. Nach seinen Erfahrungen in der Weimarer Zeit war und blieb Einstein überzeugt, daß die Entwicklung in der Bundesrepublik nicht zu einer wahren Demokratie führen könne. Auch als sich im deutschen Volk die Anzeichen für einen Bewußtseinswandel häuften, änderte Einstein seine Meinung nicht.

Wie ist das zu verstehen? Wohl aus den großen Hoffnungen, die er sich nach dem Ende des Kaiserreiches gemacht hat, und die so bitter enttäuscht wurden.

Auf einer Postkarte, die er am 11. Nobember 1918, am Tag des Waffenstillstandes, an seine Mutter geschrieben hat, kommt die Freude über die Revolution zum Ausdruck:

Sorge Dich nicht. Bisher ging alles glatt, ja imposant. Die
jetzige Leitung scheint ihrer Aufgabe wirklich gewachsen zu
sein. Ich bin glücklich über die Entwicklung der Sache. Jetzt
wird es mir erst recht wohl hier. Die Pleite hat Wunder getan. [26]

Als die Studenten den Rektor der Universität für abgesetzt erklärten,
holten die Professoren Einstein zu Hilfe. Mit Max Born und Max Wertheimer
fuhr er zum Reichstag ("mit einer Tram", wie er später erwähnte). Dort
tagten die revolutionären Studentenkomitees. Einstein warnte vor einem
sowjetischen Räte-System und plädierte entschieden für eine Demokratie
westlichen Zuschnitts:

Rückhaltlose Anerkennung gebührt unseren jetzigen sozial-
demokratischen Führern. Im stolzen Bewußtsein der werbenden
Kraft der von ihnen vertretenden Gedanken haben sie sich be-
reits für die Einberufung der gesetzgebenden Versammlung ent-
schlossen. Damit haben sie gezeigt, daß sie das demokratische
Ideal hochhalten. Möge es ihnen gelingen, uns aus den ernsten
Schwierigkeiten herauszuführen, in die wir durch die Sünden
und Halbheiten ihrer Vorgänger hineingeraten sind. [27]

Wie bereits erwähnt: Einstein glaubte an die neue Zeit und wollte an ihr
mitarbeiten. Aber sein Optimismus verflog bald. Nach 1933 kreiste der
Briefwechsel Einsteins mit seinem Kollegen und Freund Max Born immer wieder um die Frage: War das Schicksal des deutschen Volkes, von der "Haß-
und Gewaltseuche" des Nationalsozialismus ergriffen zu werden, etwas Un-
vermeidliches, Unausweichliches gewesen? Eine solche Auffassung lag Einstein später nahe. Auch in der Wissenschaft wollte er auf strenger Deter-
miniertheit beharren, obwohl sich, mit angebahnt durch seine früheren
Auffassungen und insbesondere durch seine Quantenarbeit von 1917, eine
andere Interpretation durchzusetzen begann. Einstein meinte später, daß
das triebhafte Verhalten der Menschen in politischen Dingen geeignet sei,
den Glauben an den Determinismus in der Physik wieder recht lebendig zu
machen.

Vielleicht ist es erlaubt, den Vergleich zwischen der Wissenschaft und
der Politik noch weiterzuführen. Das Zusammenleben der Menschen im Staat
wird bestimmt von den "politischen Kräften". Ebenso spricht man in der
Physik von "Kräften" und zählt vier Arten: die Schwerkraft, die elektromagnetische Kraft, die radioaktive und die starke Kernkraft. Wenn man die
Eigenschaften all dieser Kräfte kennt und ihren Zusammenhang, so meint
man heute, hat man das Grundproblem der Physik gelöst.

Auf dem Gebiet der Physik besaß Einstein ursprünglich einen geradezu un-
glaublichen Sinn für die Wirklichkeit. Als er jedoch in den vierziger und
fünfziger Jahren nach einer "Einheitlichen Feldtheorie" suchte, hat er
sich auf die elektromagnetischen Kräfte und die Schwerkraft beschränkt,
die später entdeckten starken und die radioaktiven Kernkräfte hat er

nicht mehr in seine Betrachtungen einbezogen, obwohl diese doch das Bild entscheidend veränderten.

Ebenso in der Politik. Viel früher als andere Beobachter hatte Einstein ein sicheres Urteil über den Nationalsozialismus und über die Gefahren, die der jungen Weimarer Republik drohten. Als aber nach dem Ende des Zweiten Weltkrieges eine ganz neue Entwicklung einsetzte, hat er den starken demokratischen Kräften in Deutschland keine Rolle mehr in seinem Urteil zugebilligt.

Als Theodor Heuss 1951 den Orden "Pour le mérite" für Wissenschaften und Künste erneuerte, gehörte Einstein zu den vier alten Mitgliedern, die aus den zwanziger Jahren noch am Leben waren. Auf die Anfrage des deutschen Bundespräsidenten, ob er bereit sei, wieder beizutreten, erwiderte Einstein:

> *Nach dem Massenmord, den die Deutschen an dem jüdischen Volk begangen haben, ist es ... evident, daß ein selbstbewußter Jude nicht mehr mit irgendeiner deutschen offiziellen Veranstaltung oder Institution verbunden sein will.*[28]

Wie die anderen Konsequenzen seiner Geschichte muß Deutschland auch dieses Urteil, diese Verurteilung tragen. Bundespräsident Walter Scheel hat am 1. März 1979 in Berlin bei der großen Feier das Verhältnis Einsteins zu Deutschland behandelt. Er hat dabei nichts beschönigt: Auch im Falle Einsteins wird jetzt endlich die deutsche Vergangenheit mit allen Höhen und Tiefen nicht verdrängt, sondern wirklich bewältigt. Diese Bewältigung erlaubt es, auch die wenigen Punkte zu korrigieren, in denen Einsteins Urteil ungerecht war.

Einstein hat bis zu seinem Tode 1955 nicht an eine Entwicklung zur Demokratie in Deutschland glauben wollen.

> *Es ist unmöglich,*

sagte er,

> *aus den Kerlen dort ehrliche Demokraten zu machen.*

Scheel kommentierte:

> *Wir haben den Gegenbeweis angetreten.*

Das ist gesagt im Hinblick darauf, wie wir alle den 100. Geburtstag dieses großen Gelehrten und großen Menschen am besten feiern: Nicht indem wir alle seine Aussagen in Wissenschaft und Politik als für alle Zeiten unveränderliche Wahrheiten nehmen, sondern indem wir in seinem Sinne weiterstreben: Zu einer besseren Wissenschaft, zu einer besseren Gesellschaft.

Anmerkungen

1. Presse- und Informationsamt der Bundesregierung. Bulletin Nr. 27, S. 237-241. 6. März 1979

2. Banesh Hoffmann: "Albert Einstein. Creator and Rebel". London 1972. Hier S. 263. Deutsch: Dietikon-Zürich 1976

3. Banesh Hoffmann: "Einstein und der Zionismus". In: Peter C. Aichelburg und Roman U. Sexl: "Albert Einstein. Sein Einfluß auf Physik, Philosophie und Politik." Braunschweig/Wiesbaden 1979. Hier S. 179

4. Banesh Hoffmann: "Schöpfer und Rebell". Dietikon-Zürich 1976. Hier S. 23

5. Brief von Max Planck an Walther Nernst, 11. Juni 1910. Man vergl. Armin Hermann: "Frühgeschichte der Quantentheorie." Mosbach/Baden 1969. Hier S. 153

6. Man vergl. Armin Hermann: "Einstein in Berlin." In: Jahrbuch Preußischer Kulturbesitz. Jg. VIII/1970, S. 90-114. Hier S. 93

7. Brief von Albert Einstein an Arnold Sommerfeld, Ende Februar/Anfang März 1918. Veröffentlicht in "Albert Einstein/Arnold Sommerfeld. Briefwechsel. Sechzig Briefe aus dem goldenen Zeitalter der Physik." Basel/Stuttgart 1968. Hier S. 48

8. Ebd. S. 61

9. Ebd. S. 36

10. Albert Einstein: "Über den Frieden". Bern 1975. Hier S. 20

11. Brief von Einstein an Sommerfeld, 6. Dezember 1918. In: Einstein/Sommerfeld - Briefwechsel. S. 55

12. Brief von Einstein an Sommerfeld, 13. Juli 1921. In: Einstein/Sommerfeld - Briefwechsel. S. 83f.

13. Brief von Einstein an Max Born, 9. Dezember 1919. Veröffentlicht in "Albert Einstein/Hedwig und Max Born. Briefwechsel 1916-1955." München 1969, S. 38f.

14. Philipp Frank: "Einstein. Sein Leben und seine Zeit." Braunschweig 1979. Hier Vorwort von Albert Einstein

15. Berliner Tageblatt, 27. August 1920/Morgenausgabe

16. Brief von Max von Laue an Arnold Sommerfeld, 27. August 1920. Bibliothek des Deutschen Museums. Handschriftensammlung. Nachlaß Sommerfeld

17. Brief von Sommerfeld an Einstein, 3. September 1920. In: Einstein/Sommerfeld - Briefwechsel. S. 65

18. Werner Heisenberg: "Der Teil und das Ganze. Gespräche im Umkreis der Atomphysik." München 1969. Hier S. 67

19. Brief von Einstein an Max Planck, 6. April 1933. Man vergl. "Max Planck in Selbstzeugnissen und Bilddokumenten". Rowohlts Bildmonographien Nr. 198. Reinbek 1973. Hier S. 79

20. Brief von Einstein an Wolfgang Pauli, ohne Datum, wahrscheinlich September 1938

21. Brief von Einstein an Max Born, 30. Mai 1933. In: Einstein/Born-Briefwechsel. S. 160

22. Brief von Einstein an Arnold Sommerfeld, 14. Dezember 1946. In: Einstein/Sommerfeld - Briefwechsel. S. 121

23. Man vergleiche Armin Hermann: "Die neue Physik. Der Weg in das Atomzeitalter." München 1979. Hier S. 117

24. Anm. 22

25. Anm. 23, S. 118

26. Albert Einstein: "Über den Frieden". Bern 1975. Hier S. 43

27. Ebd. S. 45

28. Ebd. S. 575

Wolfgang Pauli

Wissenschaftlicher Briefwechsel mit Bohr,
Einstein, Heisenberg u. a. – Band I: 1919-1929
Scientific Correspondence with Bohr, Einstein,
Heisenberg a. o. – Volume I: 1919-1929

Herausgeber/Editors: A. Hermann, K. v. Meyenn,
V. F. Weisskopf

1979. 1 Faksimile, 34 Abbildungen, 6 Tabellen.
XLVII, 577 Seiten. Briefe in Deutsch, Dänisch
und Englisch. (Sources in the History of
Mathematics and Physical Sciences, Volume 2)
ISBN 3-540-08962-4

Inhaltsübersicht: Das Jahr 1919: Auseinander-
setzung mit der Allgemeinen Relativitätstheorie. –
Das Jahr 1920: "Relativitätsartikel" und erste Arbeiten zur Atomphysik. – Das Jahr 1921: Dissertation über das Wasserstoffmolekülion. – Das Jahr
1922: Göttingen – Hamburg – Kopenhagen. –
Das Jahr 1923: Anomaler Zeemaneffekt. – Das
Jahr 1924: Weg zum Ausschließungsprinzip. –
Das Jahr 1925: "Quantenartikel" und Göttinger
Matrizenmechanik. – Das Jahr 1926: Rotierendes
Elektron und Verallgemeinerungen der Quantenmechanik. – Das Jahr 1927: Kopenhagener Inter-
pretation und Quantenelektrodynamik. – Das Jahr
1928: Berufung nach Zürich – Schwierigkeiten in
der Quantenelektrodynamik. – Das Jahr 1929:
Systematischer Aufbau der Quantenfeldtheorie. –
Anhang.

J. A. Wheeler

Einsteins Vision

Wie steht es heute mit Einsteins Vision, alles als
Geometrie aufzufassen?

1968. 10 Abbildungen, 1 Porträt. VII, 108 Seiten
ISBN 3-540-04389-6

"...Das Buch zeichnet sich durch einen lebendigen,
anschaulichen Stil in der Darstellung aus, der allen
denen den Zugang zu den Dingen erleichtert, die
mit der schwierigen Materie schon etwas vertraut
sind. Die aus persönlicher Bekanntschaft in Princeton erwachsene Verehrung für Albert Einstein, die
Begeisterung für sein Lebenswerk, dessen Weiter-
führung und Vertiefung kommen in dem Buch so
stark zum Ausdruck, daß die Lektüre selbst dann
Genuß und Gewinn ist, wenn es nicht bis in alle
Einzelheiten verstanden wird."
Die Sterne

Texts and Monographs in Physics

Editors: W. Beiglböck, M. Goldhaber, E. H. Lieb,
W. Thirring

W. Rindler
Essential Relativity
Special, General, and Cosmological
Second Edition. 1977. 44 figures. XV, 284 pages
ISBN 3-540-07970-X

A. Böhm
Quantum Mechanics
1979. 105 figures, 7 tables. XVII, 522 pages
ISBN 3-540-08862-8

O. Bratteli, D. W. Robinson
Operator Algebras and Quantum Statistical Mechanics
Volume 1
C* and W*-algebras. Symmetry Groups.
Decomposition of States.
1979. XII, 500 pages
ISBN 3-540-09187-4

H. M. Pilkuhn
Relativistic Particle Physics
1979. 85 figures. XII, 427 pages
ISBN 3-540-09348-6

M. D. Scadron
Advanced Quantum Theory and Its Applications Through Feynman Diagrams
1979. 78 figures, 1 table. XIV, 386 pages
ISBN 3-540-09045-2

Springer-Verlag
Berlin
Heidelberg
New York

Selected Issues from

Lecture Notes in Mathematics